Mutualism

Mutualism

EDITED BY

Judith L. Bronstein
Department of Ecology and Evolutionary Biology,
University of Arizona, USA

OXFORD
UNIVERSITY PRESS

OXFORD
UNIVERSITY PRESS

Great Clarendon Street, Oxford, OX2 6DP,
United Kingdom

Oxford University Press is a department of the University of Oxford.
It furthers the University's objective of excellence in research, scholarship,
and education by publishing worldwide. Oxford is a registered trade mark of
Oxford University Press in the UK and in certain other countries

First Edition published in 2015

Published in the United States of America by Oxford University Press
198 Madison Avenue, New York, NY 10016, United States of America

British Library Cataloguing in Publication Data
Data available

Library of Congress Control Number: 2015937786

ISBN 978–0–19–967565–4 (hbk.)
ISBN 978–0–19–967566–1 (pbk.)

Contents

Contributors

Erol Akçay Department of Biology, University of Pennsylvania, Philadelphia, PA, USA

Bruce Anderson Department of Botany and Zoology, University of Stellenbosch, South Africa

Jordi Bascompte Institute of Evolutionary Biology and Environmental Studies, University of Zurich, Switzerland

Carina A. Baskett Department of Plant Biology, Michigan State University, East Lansing, MI, USA

Brendan Bohannan Center for Ecology and Evolutionary Biology, University of Oregon, Eugene, OR, USA

Judith L. Bronstein Department of Ecology and Evolutionary Biology, University of Arizona, Tucson, AZ, USA

Emilio M. Bruna Department of Wildlife Ecology and Conservation, and Center for Latin American Studies, University of Florida, Gainesville, FL, USA

Redouan Bshary Laboratoire d'éco-éthologie, Université De Neuchâtel, Neuchâtel, Switzerland

Daniel P. Cariveau Department of Ecology, Evolution, and Natural Resources, Rutgers University, New Brunswick, NJ, USA

Daniel F. Doak Department of Environmental Studies, University of Colorado-Boulder, Boulder, CO, USA

Angela E. Douglas Department of Entomology, Cornell University, Ithaca, NY, USA

Megan E. Frederickson Department of Ecology and Evolutionary Biology, University of Toronto, Canada

Maren L. Friesen Department of Plant Biology, Michigan State University, East Lansing, MI, USA

Tamar L. Goulet Department of Biology, University of Mississippi, University, MS, USA

Roger Guimerà Department d'Enginyeria Química, Universitat Rovira i Virgili, Tarragona, Spain

Tina Harrison Department of Ecology, Evolution, and Natural Resources, Rutgers University, New Brunswick, NJ, USA

Jason D. Hoeksema Department of Biology, University of Mississippi, University, MS, USA

J. Nathaniel Holland Department of Biology and Biochemistry, University of Houston, Houston, TX, USA

Robert D. Holt Department of Biology, University of Florida, Gainesville, FL, USA

Jiri Hulcr School of Forest Resources and Conservation, University of Florida, Gainesville, FL, USA

Gregory Hurst Department of Evolution, Ecology and Behaviour, University of Liverpool, UK

Brian D. Inouye Department of Biological Science, The Florida State University, Tallahassee, FL, USA

Anthony R. Ives Department of Zoology, University of Wisconsin, Madison, WI, USA

John Jaenike Department of Biology, University of Rochester, Rochester, NY, USA

Nancy Collins Johnson School of Earth Sciences and Environmental Sustainability, Northern Arizona University, Flagstaff, AZ, USA

Atsushi Kawakita Center for Ecological Research, Kyoto University, Otsu, Shiga, Japan

E. Toby Kiers Department of Ecological Science, Vrije Universiteit, Amsterdam, Netherlands

David Kleijn Alterra, Wageningen University and Research Centre, Wageningen, The Netherlands

Conrad C. Labandeira Department of Paleobiology, National Museum of Natural History, Washington, DC, USA

Anna-Liisa Laine Center-of-Excellence in Metapopulation Biology, Department of Biosciences, University of Helsinki, Finland

Louis J. Lamit School of Forest Resources and Environmental Science, Michigan Technological University, Houghton, MI, USA

Jonathan R. Leake Department of Animal and Plant Sciences, The University of Sheffield, Sheffield, UK

Charlotte T. Lee Biology Department, Duke University, Durham, NC, USA

Silvia B. Lomáscolo National Scientific and Technical Research Council, Buenos Aires F.D., Argentina

Molly MacLeod Department of Ecology, Evolution, and Natural Resources, Rutgers University, New Brunswick, NJ, USA

Kevin McCann Department of Integrative Biology, University of Guelph, Guelph, Ontario, Canada

John McCutcheon Division of Biological Sciences, University of Montana, Missoula, MT, USA

Tom E.X. Miller Department of Ecology and Evolutionary Biology, Rice University, Houston, TX, USA

Ulrich G. Mueller Section of Integrative Biology, University of Texas at Austin, Austin, TX, USA

Michele K. Nishiguchi Department of Biology, New Mexico State University, Las Cruces, NM, USA

Scott L. Nuismer Department of Biological Sciences, University of Idaho, Moscow, ID, USA

Jens M. Olesen Department of Bioscience, Aarhus University, Denmark

Todd M. Palmer Department of Biology, University of Florida, Gainesville, FL, USA

John M. Pandolfi Australian Research Council Centre of Excellence for Coral Reef Studies, School of Biological Sciences, University of Queensland, St Lucia, Australia

Stephanie Porter Department of Biology, University of California, Riverside, CA, USA

Elizabeth G. Pringle Michigan Society of Fellows, University of Michigan, Ann Arbor, MI, USA

John Rawls Department of Molecular Genetics and Microbiology, Duke University, Durham, NC, USA

Joel L. Sachs Department of Biology, University of California, Riverside, CA, USA

Eugenia M. Sampayo Australian Research Council Centre of Excellence for Coral Reef Studies, School of Biological Sciences, University of Queensland, St Lucia, Australia

Robert L. Sanford, Jr. School of Earth Sciences and Environmental Sustainability, Northern Arizona University, Flagstaff, AZ, USA

Douglas W. Schemske Department of Plant Biology, Michigan State University, East Lansing, MI, USA

Kari A. Segraves Department of Biology, Syracuse University, Syracuse, NY, USA

Ellen Simms Department of Integrative Biology, University of California, Berkeley, CA, USA

Diana L. Six Department of Ecosystem and Conservation Sciences, University of Montana, Missoula, MT, USA

Michael Stat The Oceans Institute and Centre for Microscopy, Characterisation and Analysis, The University of Western Australia, Crawley, Australia

Adrian Stier Department of Zoology, University of British Columbia, Vancouver, BC, Canada

Erin L. Treiber Department of Plant Biology, University of Minnesota, Saint Paul, MN, USA

Jason Tylianakis School of Biological Sciences, University of Canterbury, Christchurch, New Zealand

George D. Weiblen Bell Museum and Department of Plant Biology, University of Minnesota, Saint Paul, MN, USA

Noah Whiteman Department of Ecology and Evolutionary Biology, University of Arizona, Tucson, USA

Thomas G. Whitham Department of Biological Sciences and Merriam-Powell Center for Environmental Research, Northern Arizona University, Flagstaff, AZ, USA

Peter Wilf Department of Geosciences, Pennsylvania State University, University Park, PA, USA

Rachael Winfree Department of Ecology, Evolution, and Natural Resources, Rutgers University, New Brunswick, NJ, USA

J. Timothy Wootton Department of Ecology and Evolutionary Biology, University of Chicago, Chicago, IL, USA

SECTION 1

Introduction to Section 1

Judith L. Bronstein

Mutualisms, interactions between two species that benefit both of them, have long captured the public imagination. Humans are undeniably attracted by the idea of cooperation in nature. For thousands of years, we have been seeking explanations for its occurrence in other organisms, often imposing our own motivations and mores in an effort to explain what we see. Spectacular natural history stories abound; some of them are even true.

However, the importance of mutualisms lies much deeper than simply providing material for philosophical treatises and endless natural history documentaries. Their influence transcends levels of biological organization from cells to populations, communities, and ecosystems. Mutualistic symbioses were key to the origin of eukaryotic cells, and perhaps to the invasion of the land. Mutualisms occur in every aquatic and terrestrial habitat; indeed, ecologists now believe that almost every species on Earth is involved directly or indirectly in one or more of these interactions. Mutualisms are crucial to the reproduction and survival of many plants and animals, and to nutrient cycles in ecosystems. Moreover, the ecosystem services mutualists provide (e.g., seed dispersal, pollination, and carbon, nitrogen, and phosphorus cycles resulting from plant–microbe interactions) are leading mutualisms to be increasingly considered conservation priorities, at the same time as acute risks to their ecological and evolutionary persistence are being identified.

In light of our deeply ingrained fascination with their natural history and their more recently recognized ecological and evolutionary importance, it is surprising how little conceptually oriented literature on mutualism yet exists. Perhaps their natural history appeal has been part of the problem: mutualisms have long been treated in a piecemeal fashion, as a loose set of fascinating, individually unique, relationships. The literature on certain forms of mutualism (for example, plant/pollinator and plant/mycorrhizal interactions) is rather massive. Yet, despite the fact that researchers in these fields often address similar questions, efforts to compare, contrast, and synthesize features of different mutualisms are still rare. Conversely, most models of mutualism evolution and population dynamics have such general features that they are difficult to tie to the biology of any particular interaction. Clearly, to further our understanding of mutualism, there is a pressing need to identify the ecological and evolutionary features that unite and that divide mutualisms, and to place them in clear relation to other pairwise, interspecific interactions. This volume attempts to fill that gap.

The first section lays the groundwork by reviewing the general state of our knowledge on mutualism. Chapter 1, by Judith Bronstein, offers an intellectual history of the field. Two main topics are covered. First, this chapter considers how we define and categorize the universe of interspecific interactions in general, and of beneficial interactions in particular. If we hope to reach a common understanding of mutualism, then we need to start by agreeing what set of interactions we are talking about; Bronstein lays out a series of confusions that have characterized the literature, then clarifies the language that is adopted by authors in this volume. She then reviews the study of mutualism, from its distant origins to its exciting present. In Chapter 2,

Angela Douglas focuses on what makes symbiotic mutualisms, i.e., those involving intimate contact between partners, similar and different from non-symbiotic ones. Ranging widely from the population to the ecosystem scale and across both ecological and evolutionary phenomena, Douglas introduces many of the themes that are developed in later chapters of this volume. Highlighted in this chapter are three particularly intriguing and less well-known microbial symbioses, each illustrating a principle that carries over to other mutualisms: *Wolbachia*–arthropod associations, which clearly range from mutualistic to antagonistic (Gregory Hurst, Box 2.1); the human microbiome, a unit that functions not only at the level of pairwise interactions but as communities and ecosystems as well (John Rawls, Box 2.2); and bacterial-derived organelles, whose symbiotic origins challenge and deepen our understanding of the evolution of multicellular life on earth (John McCutcheon, Box 2.3).

CHAPTER 1

The study of mutualism

Judith L. Bronstein

1.1 Introduction

Mutualism is everywhere. In tropical rainforests, the vast majority of flowering plants depend on mutualistic pollinators and seed dispersers. Deserts are dominated by nitrogen-fixing legumes, lichens dominate tundra habitats, and most northern hardwood forest and grassland plant species depend on mycorrhizal fungi to survive and persist. In the ocean, both coral and deep-sea vent communities are rich with mutualisms; coral itself is the product of a mutualistic symbiosis. Organisms from every kingdom have mutualists, and certain kingdoms may in fact have originated as a consequence of symbiotic innovations. Other key events in the history of life on earth have been linked to mutualism as well, including the origin of the eukaryotic cell, invasion of the land, and the radiation of the angiosperms.

Based on these observations, there should be no doubt that mutualism has played a central role in both the ecology and the evolution of life on Earth. However, until quite recently there has been no unanimity among biologists on this point. For example, Daniel Janzen (1985) contended that "mutualisms are the most omnipresent of any organism-to-organism interaction. All terrestrial higher plants, vertebrates, and arthropods are involved in one diffuse mutualism and many are involved in several." Yet, George Williams (1966) believed that "really good examples of mutualism are relatively rare," and Robert May (1976) considered them to be relatively inconspicuous features of most ecosystems. Striking differences of opinion have coexisted regarding the importance of mutualism as an evolutionary force as well. Lynn Margulis (1991) expressed the view that most organisms evolved as (mutualistic) "symbiotic complexes involving uncounted numbers of live entities integrated in diverse ways in an unstudied fashion." She contended that the composite nature of individuals invalidates entire fields of study, including cladistics and much of neo-Darwinian evolutionary theory. Others responded that evolutionary innovation is far more likely to come about from processes acting within species, rather than via novel symbioses (e.g., Law 1991).

Attitudes like these about the importance of mutualism were largely based on biological intuition rather than on hard data. For instance, here are two arguments based on thought experiments: what would the world look like without mutualism?

> In short, a world without the four common mutualisms [pollination, seed dispersal, protection, and nutritional mutualisms] would be less species-rich, less ecologically and life-form diverse, richer in extreme specialists and incompetent generalists, and have population patterns based more on monospecific stands (lower within-habitat diversity). Competition would be more intense among the survivors, and the survivors would expend more resources on it than they do now. Seed and flower predators would play an even larger role than at present in moulding the reproductive biology of plants. (Janzen 1985)

> Without symbiosis, the nature of life on Earth would be unrecognizable from that which is found today. This is not just because symbiosis was crucial to the evolution of eukaryotes from their prokaryote ancestors. It is also that most modern terrestrial ecosystems are critically dependent on symbiosis: 90 percent of land plants in nature are mycorrhizal, and virtually all mammalian and insect herbivores would starve without their cellulose-digesting symbionts. (Smith 1991)

Mutualism. Edited by Judith L. Bronstein.

Some will find these scenarios convincing enough, or consistent enough with their own impressions, to accept that mutualism deserves serious scientific attention. Many others find the natural history of mutualism to be fascinating enough to merit study irrespective of its ubiquity. Others still would prefer to see some quantitative evidence for its significance. Some data, although not very much, are available. At least two retrospective analyses have tallied up positive interactions discovered incidentally during community-wide surveys of competition. Stone and Roberts (1991) concluded that 19%–50% of all interactions in a community are mutualistic. Based upon a partially overlapping list of surveys, Dodds (1997) argued more conservatively for a figure of 1%–15%. A few researchers have used either literature or field surveys to count up particular kinds of mutualisms in particular habitats. Thus, for instance, 45% of all British insect species may be involved in at least one mutualism (Price 1984); probably the majority of tropical arboreal ants derive significant portions of their diets from nutritional mutualisms with plants and other insects (Davidson 1997); mycorrhizal fungi and nitrogen-fixing bacteria are responsible for 5%–20% (grassland and savannah) to 80% (temperate and boreal forests) of all nitrogen, and up to 75% of phosphorus, acquired by plants annually (van der Heijden et al. 2008); and about 88% of flowering plant species worldwide are animal-pollinated (Ollerton et al. 2011).

Accepting that mutualism is a very common interaction still leaves the most interesting questions unaddressed, however. Is it biologically important, and if so, how? Are there certain communities or certain clades in which mutualism is particularly prevalent or particularly important, and if so, what might explain those patterns? How do mutualisms evolve, and under what conditions can they persist? Will mutualisms be able to persist in a rapidly changing world, and how can we intervene to make sure that this happens? This volume explores the answers that are emerging from the young but vigorous field of mutualism studies. It also points to conclusions not yet reached, and questions not yet asked.

The rest of this chapter focuses on how mutualism has been and currently is being defined and studied. First, I deal with semantic issues. How can the universe of interactions between species be divided up, and then how do we define and group the mutually beneficial ones? I will summarize the terminology that authors use throughout this volume. I then review the history of mutualism studies. While we have long had a surprisingly deep knowledge of many individual mutualisms, a synthetic and conceptual perspective on mutualism as a whole was much slower to develop. Mutualism as a discipline has now taken off. This will be evident in the subsequent chapters of this book.

1.2 The semantics of mutualism

1.2.1 Classifying interactions

Classification schemes are often proposed early in the study of biological phenomena, and can subsequently influence how we think about and subsequently investigate them. They can lead certain phenomena to be disproportionately studied, or to be unjustifiably ignored; they can call attention to similarities and differences, but can also lead them to be overlooked. Classifications of species interactions are no exception.

The most common way to categorize interspecific interactions is to use an "interaction grid," a summary of the net effects of different interactions on each of the two participants (Figure 1.1a). Within this grid, net effect is categorized as plus (beneficial to the participant), minus (harmful to it), or zero (neutral); hence, mutualism is categorized as a plus/plus interaction. The interaction grid was first proposed by Edward Haskell (1947) as an attempt to develop a classification system for human social behaviors. It was introduced into ecology by Eugene Odum in his popular 1953 textbook and rapidly became a fixture of the literature. It tended to be presented a little differently each time. For instance, not all authors included the zero/zero interaction. The names for certain interactions were also inconsistent: for example, the minus/minus interaction was variously termed struggle (Haskell 1947), synnecrosis (Haskell 1949), antibiosis (Dindal 1975), and, most commonly, competition (Odum 1953). Furthermore, some divided up the grid more finely than others. Dindal (1975), for example, subdivided the interaction grid into over twenty distinct kinds of interactions.

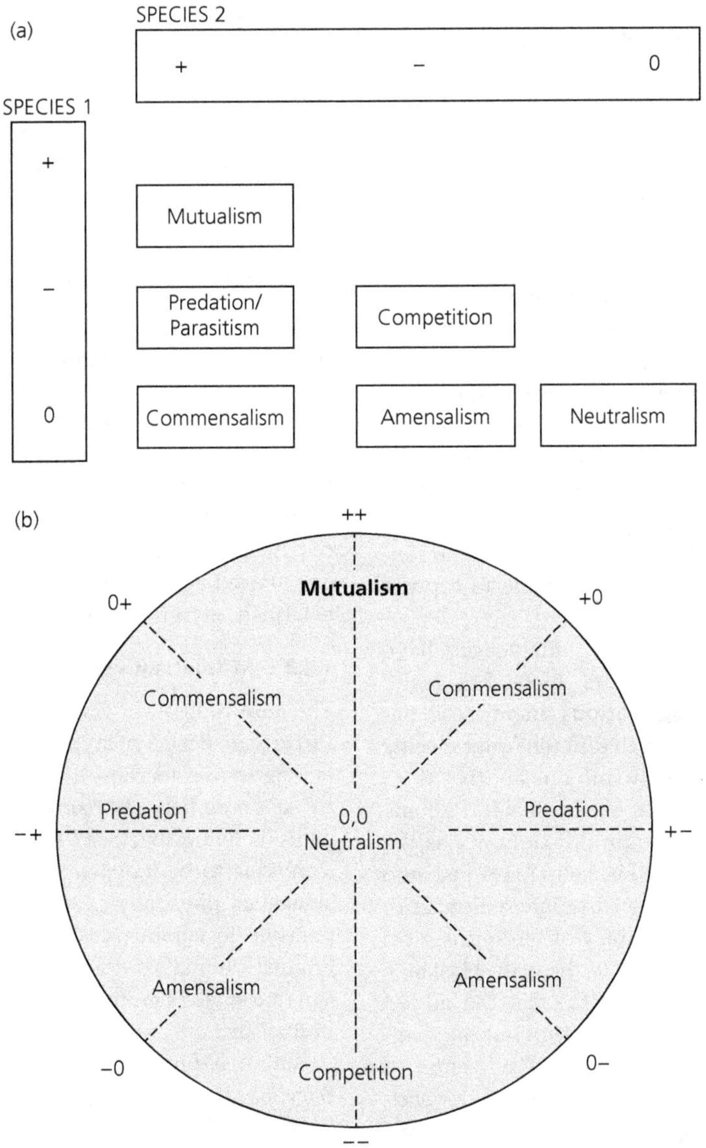

Figure 1.1 (a) The interaction grid. This is the most popular method of displaying the universe of possible interspecific interactions. Each interaction is placed according to its net effect (positive, negative, or neutral) on each of two participants. (b) The coaction compass. First proposed by Haskell (1949), it illustrates how different interactions grade into one another. Moving away from the center along a radius increases the magnitude but not the direction of the net effect of the interaction for one or both species; moving around the circumference shifts the direction of the net effect for one or both species.

By the mid-1980s, the interaction grid had come under heavy attack. First, there is considerable ambiguity about what exactly a plus, minus, or zero means. A plus, for example, could signify that the population of one species increases in the presence of the other, or else that fitness of individuals of one species is higher in the presence of the other species than in its absence. Templeton and Gilbert (1985) and Abrams (1987) discussed common situations in which these concepts of benefit will be contradictory. Second, the grid confuses mechanisms of interactions with their effects. For instance, even though there are minus/minus interactions that are not competition and plus/minus interactions that are neither predation nor parasitism, these are the names that have come to be associated with those spots on the grid. Arthur and Mitchell (1989) suggested renaming plus/minus interactions "contramensalism" to remove the implication of mechanism, whereas Abrams (1987) advocated shifting to a classification scheme explicitly based on mechanism. Finally, the grid's typological approach leads one almost inevitably to see different forms of interaction as both static and fundamentally different from each other. This is clearly incorrect. Interactions can be beneficial at some places, times, and time scales, and detrimental at others (Chapter 10). Furthermore, as covered in Chapters 3 through 6, interactions can shift between mutualistic and antagonistic over evolutionary time. The interaction grid is unable to capture any of this variation.

Other schemes for classifying interactions have been proposed in recent years. None has really caught on. These include various attempts to depict a continuum along which different interspecific interactions grade smoothly into one another (e.g., Dindal 1975, Price 1984, Bronstein 1994a). Perhaps the most powerful continuum illustration was the earliest one, presented by Haskell (1949) just two years after he first proposed the interaction grid. Beyond escaping the inaccuracy of the typological view of interactions inherent to the grid, Haskell's "coaction compass" (Figure 1.1b) has the added benefit of displaying not only the sign but the magnitude of an interaction's net effect. The center of the compass is denoted as the (0,0) point. Depending on the radius along which one moves toward the circumference, the positive or negative effect grows progressively stronger for one or both participants. The coaction compass was given a quantitative foundation by Leary (1985), who used it to illustrate how a given interaction could shift in outcome and intensity over time. This approach addresses one of Abrams' (1987) most serious criticisms of the interaction grid: it does not specify the time scale over which the outcome is measured. Using Leary's method, one can map a given interaction at successive times to track if and how its outcome shifts. For instance, in a classic paper, Mattson and Addy (1975) used it to map temporal shifts in the intensity and outcome of interactions between two forest trees and their insect herbivores. Despite the compass' utility, however, biologists who study interactions remain essentially unaware of its existence. Ecology textbooks that illustrate interaction categories still use some version of a grid.

1.2.2 Defining mutualism

One might conclude from this discussion that mutualism could straightforwardly be defined as the interspecific interaction occupying the plus/plus zone on the grid, continuum, or compass. In fact, definitions of mutualism have been inconsistent for well over a hundred years. This is not strictly a semantic annoyance. If we have any hope of identifying the ecological and evolutionary phenomena shared across mutualisms, it is crucial first to establish what interactions we are talking about.

1.2.2.1 Mutualism versus symbiosis

The term mutualism was introduced into biology by Pierre van Beneden in 1873 to mean "mutual aid" among species (Section 1.3.3). Symbiosis was defined by Anton de Bary six years later to mean "unlike organisms living together." De Bary explicitly defined symbiosis to be independent of outcome; that is, it subsumed physically close interactions ranging from parasitic to commensal to mutualistic. Using these original definitions, then, some but not all mutualisms are symbiotic, and some but not all symbioses are mutualistic.

Within a few years of van Beneden's and de Bary's work, however, the tendency to assume a simple equivalency between symbiosis and mutualism had already begun (Lewin 1982). Goff (1982) suggests that the origin of this confusion was that de Bary's original discussion of symbiosis focused largely on examples of mutualistic symbioses such as lichens. In any case, the confusion was great enough by 1937 that a committee of the American Society of Parasitologists was given the charge of recommending proper usage of the two terms. In its final report (Hertig et al. 1937), the committee reiterated de Bary's original definition of symbiosis, but then recommended that since symbiosis was so commonly used to mean mutualism, researchers simply be sure to define what they meant by the terms whenever they used them. Indiscriminate use of the two terms has persisted, however (Douglas 2010).

The lack of standard definitions has led to inconsistency regarding which mutualisms are considered to be symbiotic. In this volume, symbiotic mutualisms are defined as those in which there is prolonged physical intimacy between partner species (Sections 1.2.3.2 and 1.2.4). Other authors, however, have used the term symbiosis to describe a wide variety of other mutualisms as well. In discussing obligate ant–plant protection mutualisms, Davidson and McKey (1993) argue that when an interaction is relatively close physically and its fitness consequences to the participants have not entirely been worked out, it is safer to call it a symbiosis than a mutualism. It is certainly true that naming an interaction before its net effects have been measured can lead one to make erroneous assumptions about it. However, since the vast majority of interactions have been quantified from the perspective of one of the two partners only and within a limited range of conditions, locations, and moments in time, such a stringent criterion would prevent us from categorizing almost any interaction on earth.

1.2.2.2 Mutualism versus protocooperation

Many researchers have been reluctant to call an interaction a mutualism unless they have evidence that the species have shared a long coevolutionary history, or have at least some level of obligate dependence. For more generalized, facultative, and/or ephemeral mutualisms they have felt more comfortable with other terms, including symbiosis, cooperation, and facilitation. This has added another layer of confusion, since each of those terms has a different and widely accepted definition of its own.

This restrictive definition of mutualism may have originated with Odum's ecology textbooks. In the edition that first popularized the interaction grid, Odum (1953) defined mutualism as an interaction "in which growth and survival of both populations is benefited and neither can survive under natural conditions without the other." In contrast, when both populations benefit but the association is not obligate, Odum termed the interaction "protocooperation" (in later editions, simply "cooperation"), co-opting a terminology that had previously been used primarily for intraspecific interactions. This distinction based on obligacy was maintained in ecology textbooks at least through the late 1970s. More recent texts have shifted to the broader definition of mutualism used by the authors of this volume, which focuses on reciprocal net benefits and is independent of obligacy or shared evolutionary history.

1.2.2.3 Mutualism versus facilitation

Facilitation is an interaction in which the presence of one species alters the environment in a way that enhances growth, survival, or reproduction of a second, neighboring species (Callaway 2007). The study of facilitation is a young field that emerged out of a different historical background from research on mutualism. From its inception, it has focused on interactions in which certain plant individuals experience higher performance in the presence of a neighboring species. Facilitation was also recognized early on as a potentially important mechanism of plant succession. Most facilitation research continues to focus on plant–plant interactions, although it has also been identified among other groups of organisms, including those occupying different trophic levels. Importantly, the idea of facilitation emerged out of an interest in the role of competition and the forces that can counter it in structuring ecological communities. In contrast, competition played a limited role in the origin of mutualism as a field of study.

In spite of a thriving literature on facilitation, the definition of the concept itself has been remarkably fluid, particularly with regard to its reciprocal effects. This has left the relationship between mutualism and facilitation rather murky. As I discuss elsewhere (Bronstein 2009a), some consider facilitation to be an interaction that benefits one species and does not harm the other; that is, it can be neutral or beneficial to the facilitating species, but not detrimental. To others, facilitation can have any possible outcome (beneficial, neutral, or even harmful) to the facilitator; the interaction is facilitative as long as the effect on the other species is positive. By any definition, some but not all cases of facilitation will be mutualistic; see Pugnaire et al. (1996) for a good example.

1.2.2.4 Mutualism versus cooperation

Of course, reciprocally beneficial interactions occur not only between but also within species. At this point we have a better conceptual and theoretical understanding of intraspecific phenomena, although there are far more data pertaining to interspecific ones. Some of the same principles apply to these two classes of interactions. For instance, cheating can be expected to carry an advantage in both kinds of beneficial interactions, although the processes by which cheaters are weeded out or tolerated may differ (Chapter 5). There is at least one central difference, however. Kin selection is often a driving force underlying the evolution of intraspecific cooperation (Dugatkin 1997). In contrast, interspecific benefits are generally assumed to have evolved in the absence of this evolutionary mechanism (but see Chapter 4).

Intraspecific and interspecific benefits have attracted attention from rather different research communities, and have rarely been juxtaposed in a way that highlights their similarities and differences. The search for such patterns has unfortunately become confounded by inconsistent use of terminologies to describe the two classes of phenomena (West et al. 2007). Most commonly, mutualism is used to denote interspecific benefit, whereas cooperation is used somewhat more vaguely to denote benefits in a within-species context. However, as we have seen, cooperation (or protocooperation) has also been used to refer to the subset of mutualisms that are not obligate. Other researchers have referred to *all* forms of mutualism as cooperation, possibly following the broad usage of this term in Axelrod and Hamilton's (1981) seminal paper (see the introduction to Section 2). Cooperation is also used more casually to denote the attribute of mutualism that contrasts with "conflict" (e.g., Herre et al. 1999).

Conversely, various subsets of interactions *within* species have sometimes been referred to as mutualisms. In particular, intraspecific interactions in which cooperation is an incidental result of individually selfish behaviors have been dubbed "byproduct mutualisms" (Dugatkin 1997). This is particularly confusing since the same phenomenon describes some (interspecific) mutualisms as well (Chapters 2 and 6).

1.2.3 Categorizing mutualisms

Several vast bodies of literature focus not on the unified phenomenon of mutualism, but on one or another particular form of it. Several have received book-length treatments of their own (e.g., Rico-Gray and Oliveira 2007 on ant–plant protection; Stadler and Dixon 2008 on ant–insect protection; Douglas 2010 on symbioses; Willmer 2011 on pollination). The current volume, in contrast, focuses on concepts that cut across different mutualisms. However, many of the best-understood and most frequently cited examples are discussed herein, often in great detail. Table 1.1 highlights those that are treated in the greatest depth, and where in this volume the information can be found.

It is easy to identify ecological and evolutionary phenomena that characterize subsets of mutualisms, if not all of them. In this section I highlight three common approaches to subdividing mutualisms into classes with distinct properties.

Table 1.1 Iconic mutualisms and their coverage in this book. Mutualisms are grouped in the traditional categories discussed in Section 1.2.3.1. Many of these mutualisms (referred to here by shorthand names) are discussed repeatedly in this volume; I have highlighted the location(s) where each is considered in the greatest depth.

Transportation mutualisms	
Pollination	Box 14.1
Fig–fig wasp	Chapter 3, 7
Yucca–yucca moth	Box 7.3
Seed dispersal	Box 10.2
Protection mutualisms	
Ant–plant protection	Box 6.1, Box 9.2
Bioluminescence	Box 5.3
Cleaning	Box 6.2
Nutritional mutualisms	
Legume–rhizobium	Box 4.2, Box 5.2
Mycorrhizal	Box 12.1
Fungus-growing ants	Box 5.2
Bark beetle–fungus	Box 13.1
Defensive endosymbioses	Box 10.1
Coral–algae symbioses	Box 9.1, Box 10.3, Box 13.2
Gut microbiome	Box 2.2, Box 14.2
Organelles	Box 2.3

1.2.3.1 The benefits of mutualism

The usual way to group mutualism is according to the nature of the benefits that partners exchange. In the most common scheme, almost all benefits fall into one of just three categories. *Transportation* includes movement of partners themselves, but also movement of partners' gametes, as when animals transfer pollen between flowers on different individuals. *Protection* involves the provision of defense against the biotic and abiotic environment. It includes aggression toward or consumption of the partner's enemies, as in ant protection mutualisms, as well as conferring resistance to enemies, as in the case of certain endophytic fungi that render the plants they inhabit distasteful or even fatal to herbivores. Finally, *nutrition* is shorthand for the provision of one or more limiting nutrients to the partner. Algae provide photosynthate to their fungal associates in lichens, *Rhizobium* bacteria produce fixed nitrogen for host plants, mycorrhizal fungi provide host plants with phosphate, and so on. Mutualisms can then be grouped further according to which of these three benefits are exchanged by the two partners. For example, an exchange of nutrition for protection characterizes protection mutualisms between ants and plants, ants and aphids, as well as cleaners and their hosts.

This approach to categorizing mutualisms may have first been proposed by Janzen (1985) in his magisterial overview of the natural history of mutualism. A weakness is that it uses so few categories that many important distinctions are lost. For instance, unlike nutritional exchanges, which involve different limiting nutrients (Chapter 12), the exchange of protection for protection (e.g., in Müllerian mimicry) is a symmetrical interaction: each individual is better protected if it associates with a well-protected individual. Moreover, some kinds of benefits are difficult to place within this scheme: joint attraction of shared mutualists (e.g., Feldman et al. 2004) is not included, for example, nor are mutualisms in which one species augments its partner's competitive abilities (Clay et al. 1993). Some mutualisms confer multiple kinds of benefits to a given partner, and thus are hard to fit into this scheme. For example, mycorrhizae facilitate nutrient uptake, but may also protect plants against colonization by root pathogens (Chapter 2).

There are other ways to group mutualisms according to their benefits. For example, Addicott (1984) drew a distinction between mutualisms in which the benefits are direct (i.e., involving direct contact between two partners) versus indirect (accrued via the effects of one mutualist on a third species). In this scheme, mutualisms involving the direct benefits of nutrition and transportation appear more similar to each other than either is to mutualisms involving protection, since most protective mutualisms are beneficial only in the context of the two species' interaction with a third party, the natural enemy of one of the partners (Chapters 9 and 10). In another approach, the benefits of mutualism are grouped into commodities of exchange in a "biological market" (Chapter 4) and divided into rewards (e.g., nectar) and services (e.g., transport of pollen). Mutualisms can then be grouped according to whether they involve the exchange of rewards for rewards, rewards for services, or services for services. Going further, Connor (1995) grouped mutualisms according to the mechanisms by which organisms obtain these commodities. He argued that an individual could obtain commodities in one of three ways: by investing in a partner at a cost to oneself, by preying upon or parasitizing a partner, or by receiving incidental effects from a partner's self-serving behaviors. There are six possible combinations, and hence six possible origins of mutualisms.

In Chapter 8, J. Nathaniel Holland offers another highly informative scheme. It treats mutualists as either consumers or resources, with the exchange of commodities either bidirectional or unidirectional (Table 8.1). As he discusses, this approach has the advantage of helping to fold mutualisms into the broader study of consumer-resource dynamics.

1.2.3.2 Symbiotic versus non-symbiotic mutualisms

Symbionts confer a wide range of benefits to their partners, but they can be broadly categorized as either nutrition or protection. The larger proportion (and certainly the best-known) of nutritional mutualisms are symbiotic. Most ecological and evolutionary generalizations about mutualism apply to both symbiotic and non-symbiotic interactions. As Angela Douglas discusses in Chapter 2, however, some unique phenomena arise within symbioses as a direct consequence of persistent contact, and hence of repeated interactions involving the same individuals. Among these is the genomic deterioration of symbionts, linked to the obligately vertical transmission (parent to offspring) of small symbiont populations. Broadly speaking, symbiosis research has been characterized by a closer focus on mechanism than one sees in the broader mutualism research program. In part for these reasons, the study of symbiosis has had a research agenda not closely aligned with the study of mutualism as a whole. This is beginning to change (Chapters 2 and 12).

1.2.3.3 Obligacy and specificity

At one time it was common to categorize mutualisms based on specificity and degree of mutual dependence. Thus, they were termed specialized (each species has a single partner species) or generalized (each species is associated with multiple partner species), and either obligate (each species cannot survive and/or reproduce without its partner) or facultative (each species performs better in the partner's presence but experiences some success without it). While strong patterns of obligacy and specificity do exist and are still of great interest today (see in particular Chapters 3 and 11), this approach to categorizing mutualisms is used much less now than in the past. This is in part a consequence of an increased recognition that obligacy and specificity (a) are not dichotomies but continua (Ollerton 2006), and (b) can be strikingly asymmetric for different members of an association. These observations have stimulated extensive new theoretical development (Chapter 11); they have also altered our understanding of how and when mutualists coevolve, how mutualistic communities assemble, and which mutualisms are most threatened by anthropogenic change. Importantly, however, ecological and evolutionary models of mutualism still largely focus on pairwise obligate interactions (Chapters 4 and 8). Furthermore, although far more data exist on generalized mutualisms as a whole, studies of a handful of highly specialized ones, such as the fig and yucca pollination mutualisms and obligate ant–plant defensive mutualisms, have contributed disproportionately to our understanding.

1.2.4 Suggested terminological standards

Sections 1.2.2 and 1.2.3 introduced readers to some of the terminological confusions regarding mutualism. As it is easy to get lost in a morass of conflicting definitions, it needs to be stressed why this is more than a purely semantic issue. Not all generalizations that have been published about "mutualism" are relevant to the group of relationships that forms the focus of this volume. Conversely, a great deal of information about the interactions we cover can be found in older literature that seems actively to avoid use of the term mutualism.

Here are the definitions adopted by the authors of this volume. It generally follows the original definitions of de Bary and van Beneden, and is consistent with recommendations of most recent authors who have grappled with issues of terminology.

Mutualism refers to all mutually beneficial, interspecific interactions, regardless of their specificity, intimacy, or evolutionary history. Appropriate modifiers can group sets of mutualisms, according to characteristics such as these:

- Degree of dependency: mutualism ranges from *obligate* to *facultative* for each of the two partner species (Section 1.2.3.3).
- Degree of specificity: mutualism can range from *specialized* (at the extreme, *species-specific*) to *generalized*, for each of the two partner species (Section 1.2.3.3).
- Degree of physical integration: mutualism is *symbiotic* when the two partners are in intimate physical contact, often physiologically integrated, for most or all of their life cycles. All other mutualisms are *non-symbiotic or non-persistent*. Thus, some but not all mutualisms are symbiotic, and some but not all symbioses are mutualistic (Section 1.2.3.2). The great majority of symbioses involve large, long-lived hosts associated with populations of small organisms (symbionts).
- Shared evolutionary history: in *coevolved* mutualisms, there has been reciprocal evolutionary change in the context of the interaction. *Non-coevolved* mutualisms include those in which only one partner has undergone evolutionary change, and those in which neither partner has (Chapter 7).

- Nature of the exchanged benefits: mutualisms can be further subdivided based on the commodities that the partners exchange (Section 1.2.3.1).

Facilitation is a positive effect of one species on another, without reference to the reciprocal effect; it is usually the result of a modification of the environment by one of the two species. If each species facilitates the other, their interaction is a mutualism (Section 1.2.2.3).

Cooperation is a fairly loose term used to describe mutually beneficial interactions within, not between, species. When clearly defined, it can also be employed to group mutually beneficial interactions at the within- and between-species levels (Section 1.2.2.4).

1.3 A history of mutualism studies

Mutualism, unlike symbiosis (Sapp 1994), has never been given a full historical treatment. Here I briefly summarize the evolution of the mutualism concept (see also Boucher 1985a, Bronstein 1998, 2009b, and Douglas 2010) and attempt to explain how the current mutualism research program has come into existence.

1.3.1 Mutualism before 1850

The oldest and perhaps most culturally universal ecological theory is the "balance of nature" (Egerton 1973, Boucher 1985a, Sapp 1994), the idea that natural forces prevent species from either becoming too abundant or going extinct. This concept can be found in even the most ancient scientific writings, and mutualisms were commonly cited as evidence for it. For example, Herodotus discussed how plovers removed leeches from crocodiles' mouths ("The crocodile enjoys this, and never, in consequence, hurts the bird."). Aristotle, Cicero, Pliny, and others repeated this story and added others, drawing moral lessons that showed the importance of "friendships" in maintaining nature's balance. Mutualisms, then, were well known long before they had a name.

The idea that the harmony of society mirrored an underlying harmony of nature persisted throughout the Middle Ages, at least in the West. Just as each person was believed to have a preordained role assigned by the Creator, so nature was seen as well: plants provided food for animals, which in turn decayed in order to fertilize the soil for plants (Boucher 1985a).

This view of mutualism persisted through the eighteenth century. Mutualisms were cited as important forms of evidence for Divine Providence and perfect adaptation in natural theology. Linnaeus in particular wrote eloquently of how nature resembled "a well regulated state in which every individual has his proper employment and subsistence, and a proper gradation of offices and officers is appointed to correct and restrain every detrimental excess." In this context, he discussed how animals were created to serve plants by both feeding upon them (to "preserve a due proportion among vegetables") and dispersing their seeds.

1.3.2 Mutualism and Darwinism

The well-ordered universe postulated by natural theology began to unravel during the upheavals of the Industrial Revolution of the mid-nineteenth century. Ideas of competition and struggle as forces of progress came to pervade political and economic analyses of the time, spurred most notably by the work of Thomas Malthus. At the same time, de Candolle and others, possibly stimulated by an acquaintance with Malthus' work, were documenting the importance of competition as a force in nature (Egerton 1977). These lines of thought came together to form the underlying thesis of Charles Darwin's life work.

Darwin's influence has often been interpreted as deflecting scientific attention away from beneficial phenomena in nature. Boucher (1985a), for instance, states that "while Darwin devoted a good deal of time to studying such mutualisms as insect pollination of flowers and decomposition of leaf litter by earthworms, and mentioned the dispersal of mistletoe seeds by birds in the beginning of the Origin, there is no doubt that his fundamental contribution was the demonstration that adaptation and speciation can be explained by competition." Sapp (1994) similarly dismisses Darwin's contributions to our understanding of mutualism, arguing that his "fundamental ecological and evolutionary thinking and that of those who followed him was that adaptation and speciation could be explained by conflict and competition."

These evaluations surely downplay Darwin's considerations of mutualism. Accounting for mutualisms was in fact critically important to him, because their existence posed a serious challenge to his contention that organisms did not perform actions strictly to benefit others. One of the better-known passages from the *Origin* occurs in the context of a discussion of pollination and seed dispersal: "Natural selection cannot possibly produce any modification in a species exclusively for the good of another species; though throughout nature one species incessantly takes advantage of, and profits by, the structures of others . . . If it could be proved that any part of the structure of any one species had been formed for the exclusive good of another species, it would annihilate my theory, for such could not have been produced through natural selection." Darwin therefore discussed at great length the evolution of fruits and flowers, showing how traits that benefit animals function first and foremost to increase plants' own reproductive success. He pointed out cases in which the interests of mutualists could come into conflict, discussing how cheating behaviors such as nectar-robbing might become favored. It was also in the context of mutualism that he explicitly pointed to the possibility of coevolution: "Thus I can understand how a flower and a bee might slowly become, either simultaneously or one after the other, modified and adapted to each other in the most perfect manner, by the continued preservation of all the individuals which presented slight deviations of structure mutually favourable to each other." These themes dominate Darwin's only book devoted entirely to interspecific interactions, his work on orchid pollination.

In showing that mutualisms could emerge in nature strictly by selfish actions, limited by costs and driven by conflicts of interest between partners, Darwin laid the groundwork for our current view of these interactions. However, with few exceptions, his remarkably modern view of mutualism doesn't seem to have been influential in his own time.

1.3.3 Mutualism in the nineteenth century: political and biological contexts

Boucher (1985a) provides a political perspective on the conceptualization of mutualism in nineteenth-century Europe. He describes how resistance developed to the competition and progress-through-struggle themes of natural and social sciences during this period, particularly in socialist and Marxist circles. Pierre-Joseph Proudhon, a founder of socialist and anarchist movements during the middle of the century, publicized the idea of *mutuellisme*, a socialist ideal involving just and equal exchange made possible by eliminating the unfair advantage gained by capitalists through their inheritance of property. Thus, the term mutualism was in common use before it entered biology (Sapp 1994).

Pierre van Beneden, a Belgian zoologist, defined mutualism in a biological context in a communication to the Royal Academy of Belgium in 1873. In *Animal Parasites and Messmates*, a popular work published in 1876, van Beneden stated, "There is mutual aid in many species, with services being repaid with good behaviour or in kind, and *mutualism* can well take its place beside *commensalism*." (Boucher [1985a] argues that he must have been familiar with Proudhon's work.) Examples of mutualisms that he chose to discuss included insects that live on mammals, crustaceans that live on fish, and—harking back to Herodotus—birds that pick crocodiles' teeth clean.

The idea of symbiosis was conceived in the late 1870s independently by Albert Bernhard Frank in Leipzig and Anton de Bary in Strasbourg. (De Bary is widely credited as the originator of the concept, but Sapp [1994] points out that Frank published his definition of *symbiotismus* a year earlier.) The concept arose in the course of debates on whether the alga and fungus that formed lichens were parasitic, commensal, or mutualistic; "symbiosis" was coined by both authors as a term that would acknowledge the closeness of interactions like these without implying their outcome. De Bary (1879) thought that there was good evidence that lichens in fact could span the full range from parasitic to mutualistic. He and many others argued that the same was likely to be true of other symbioses as well. Hence, it was recognized very early in the scientific study of interspecific interactions that a continuum view of net effects was more appropriate than the typological view that we often casually adopt today (see Section 1.2.1).

De Bary and Frank discussed many other examples of symbioses that they considered mutualistic, particularly among algae and higher plants. Many prominent researchers of their time, however, strongly discounted the idea that mutualism was a common outcome of symbiosis. For example, Eugen Warming, author of the first plant ecology textbook (1895), interpreted all intimate associations to be parasitisms and doubted that true reciprocity of benefit even existed, at least among plants. In a fascinating article, Roscoe Pound (1893) derided the tendency to assume that most symbioses were mutualistic, a view he considered unscientific and sentimental. He pointed out that rigorous criteria were not being used to test the direction of net effects within interactions. This point is equally valid today, although at the time it led him to ridicule the idea that mycorrhizae benefited their host plants.

Pound (1893) clearly followed Darwin's lead in recognizing that mutualisms can result from conflicting selection pressures on mutually exploitative species: "It is not necessary, as Frank seems to think, in order to establish mutualism to show that the organisms do no injury to each other. Mutualism of the kind we meet with in the vegetable kingdom involves sacrifice on the part of the host. The parasite is not there gratuitously. It is there to steal from its host the living it is hereditarily and constitutionally indisposed to make for itself. If the host gains any advantage from the relation, it can only do so by sacrificing – by giving the parasite the benefit of its labor that it may subsist . . . Practically, in some instances, symbiosis seems to result in mutual advantage. In all cases it results advantageously to one of the parties, and we can never be sure that the other would not have been nearly as well off, if left to itself." As a footnote, not long after publishing this article in *The American Naturalist* (at age 23), Roscoe Pound (1870–1964) quit botany and obtained his law degree. He ultimately became Dean of the Harvard Law School and one of the leading legal scholars of the twentieth century.

1.3.4 The waning (or flowering?) of mutualism studies

In the late nineteenth and early twentieth centuries the idea of mutualism was picked up enthusiastically by popular writers and, interestingly, by natural and social scientists who rejected the role of natural selection in evolution. (Pierre van Beneden was among these.) Particular attention has been paid to mutualism and symbiosis as important political, social, and scientific concepts in Russia during this period (Boucher 1985a, Todes 1989, Khakina et al. 1992, Sapp 1994). In striking contrast, prominent ecologists like Fredric Clements and Arthur Tansley, whose work shaped the study of ecology as a whole, gave only passing attention to mutualism. (Tobey [1981] argues that intellectual origin of Clements' "superorganism" concept lay in Social Darwinism and had little to do with any notion of cooperation.) Hence, this period has been seen as one in which interest in the scientific study of mutualism nearly ceased.

In another way, however, this was one of the most important periods in the history of mutualism research, because it was during this time that most of our best-known mutualisms were first discovered and studied. From the 1840s through the 1880s, lichens, mycorrhizae, and plant/*Rhizobium* mutualisms were examined in depth. By the 1880s, many researchers had come to accept that the symbiotic algae within marine animals were mutualistic. Some were even arguing that the eukaryotic cell itself was the result of a mutualistic symbiosis (Sapp 1994). Huge compendia on pollination biology were published during this period (Chapter 11). It was also an era of exploration, in which the biological richness of tropical habitats first came to be appreciated. With it came the discovery of new and ever more intriguing mutualisms. Thomas Belt, Odoardo Beccari, and others documented diverse interactions between plants and the protective ants inhabiting them in the Old and New World tropics, although until the early 1970s most ecologists dismissed the likelihood that these interactions were in fact mutually beneficial (Bronstein 1998).

By the turn of the century, hundreds of articles had been published on a wide diversity of mutualisms. Boucher et al. (1982) argue that by this point there was "a general recognition of the fundamental similarity of interactions ranging from mycorrhizae to cleaner fish."

1.3.5 Mutualism coalesces as a research field

Despite the growing natural history literature, mutualism *as a concept* was not prominent during the first three-quarters of the twentieth century. Evolutionary biologists of the time were concerned with the genetic basis of individual traits, and did not turn their attention to any form of interspecific interaction until mid-century. Once they did, they primarily focused on antagonisms such as herbivory and host–pathogen interactions, whose genetics were already under investigation. Population ecologists of the era were motivated by applied problems, particularly animal population fluctuations, and, by extension, predator–prey dynamics (Crowcroft 1991). As Kingsland (1985) has documented, applied interest in how antagonisms could generate population fluctuations also stimulated the development of mathematical and theoretical population biology, by researchers such as Alfred Lotka, Vito Volterra, A.J. Nicholson, V.A. Bailey, Raymond Pearl, G.F. Gause, G. Evelyn Hutchinson, and Robert MacArthur. Mutualism was not missing entirely from ecology during this period, however. Warder C. Allee and his colleagues were intensely interested in cooperation (Mitman 1992), although primarily within rather than between species. They argued for its importance in a series of influential textbooks.

Several hypotheses were posed to explain the relative lack of attention paid to mutualism during the first half of the twentieth century. Boucher (1985a) and others argue for a true decline in mutualism studies during this period, correlated with a growing (and, in their opinion, excessive) focus on the individual as the target of natural selection. My own analysis suggests that mutualism studies did not in fact wane during this period. Rather, interest in antagonisms exploded, linked to their significance as applied problems that biologists were attempting (and receiving funding) to solve. Natural history studies of mutualism continued unabated (e.g., Robertson 1895, Wheeler 1907, Bequaert 1922, Ridley 1930, Way 1963, Grant and Grant 1965). Economically important mutualisms were studied in particular depth.

The first really striking conceptual advances began in the late 1960s. Daniel Janzen tested naturalists' early conjectures about the mutualistic nature of certain ant/plant interactions using rigorous quantitative experiments (Janzen 1966, 1967). His work on these and other plant–animal interactions infused the study of mutualism with an appreciation for experimentation and an evolutionary viewpoint that it retains today. During the 1970s, many researchers grappled with how mutualisms might influence population growth (e.g., Vandermeer and Boucher 1978); the first significant evolutionary model of mutualism was also proposed (Roughgarden 1975). New arguments were made that interactions appearing to be purely antagonistic might actually be mutualisms (e.g., Owen and Wiegert 1976), while conversely, some putative mutualisms possibly weren't (Losey 1972). Mutualism began to be appreciated as a force that could structure ecological communities (Heithaus 1974, Gilbert 1975). Great interest developed in the significance of mutualism in major events in the history of life on earth, including the origin of the eukaryotic cell (Margulis 1970), the rise of the angiosperms (Mulcahy 1979), and the invasion of the land by plants (Pirozynski and Malloch 1975). Boucher et al. published the first general review of mutualism biology in 1982.

As the study of mutualism coalesced, a number of researchers tried to puzzle out why ecologists seemed to have remained so persistently uninterested in it. Among the many hypotheses suggested were these:

- Culture: Keddy (1989) suggested that "scientists can only draw models from the possibilities of which they are aware, and perhaps ecology has been hampered by restricted access to individuals (and ideas) offering co-operative models for society and nature."
- Political connotations: mutualism came to be seen as a political and social ideal touted enthusiastically by socialists and anarchists, leading many western scientists to shy away from it (Boucher 1985a, Margulis 1990).
- Gender: males are more aggressive than females, it was argued, and seek out aggression in nature;

if females had dominated the upper echelons of science, nurturing cooperative interactions might have received the emphasis they deserve (Margulis 1990).

- Taxonomic bias: most ecological research has been conducted on birds and mammals, which may be affected by competition and predation compared to more dominant but less studied taxa (Keddy 1989).
- Geographic bias: mutualisms in the temperate zone are relatively inconspicuous, and it is only recently that ecological research has been conducted in the tropics (May 1982).
- Theoretical problems: early mathematical models of mutualism produced nonsensical results and so were set aside in favor of competition and predation (May 1982).
- Excitement: watching something kill something else may simply be more exciting than watching cooperation in action (Keddy 1989).

Rather than analyzing the merits of these hypotheses, most of which seem rather dubious to me, it may be more relevant to ask: On what evidence did so many thoughtful biologists conclude that science had ignored mutualism? The only real evidence for neglect was the undeniably paltry coverage of mutualism in textbooks of ecology of that time (Keddy 1989), a situation that has not appreciably changed in the intervening years (Bronstein, unpubl. data). However, there is no law saying that textbooks must proportionally reflect either interest in or understanding of the phenomena they cover. If we wish to know how much or how little mutualism is studied, or how much or how little we now understand about it, it seems more relevant to examine the primary research literature. My own analysis (Bronstein 1994b) revealed that about a quarter of all articles on interspecific interactions appearing in ten major ecology and evolution journals during 1986–1990 focused on mutualism. This survey highlighted the existence of a rather massive body of literature, focused heavily on a few mutualisms; in particular, studies of pollination made up over half the surveyed articles. However, the literature of the period was largely descriptive. Two-thirds of all mutualism articles, for example, focused primarily on identifying partners and describing how they interacted. This was a rapidly expanding mutualism literature, but not a conceptually oriented one.

Douglas Boucher edited the first (and until now, only) volume on the biology of mutualism as a whole in 1985 (Boucher 1985b). Examination of its coverage provides insight into a field striving to identify key concepts such as the costs of mutualism, cheating, the differences between symbiotic and non-symbiotic mutualisms, and the role of coevolution. Many chapter authors—with the exception, interestingly, of Daniel Janzen—argued that it was past time for mutualism to coalesce as a field.

1.3.6 Conceptual synthesis and new themes

Four central concepts began to emerge by the early 1990s. They were stimulated not only by the Boucher volume, many of whose chapters are still quite valuable today, but by multifaceted, question-oriented research initiated by ecologists on three obligate mutualisms: the fig pollination mutualism (Chapters 3 and 7), the yucca pollination mutualism (Box 7.2), and tropical ant/plant protection mutualisms (Box 6.1 and Chapter 9). At the same time, the first sustained attention was being paid to the evolution of mutualism, primarily by biologists interested in understanding cooperation more generally (see Chapter 4). These and other concepts have begun now to coalesce to form the basis for our current understanding of mutualism.

1.3.6.1 Mutualisms not only confer benefits but inflict costs

The fundamental interest of mutualism, both from an ecological and an evolutionary perspective, undoubtedly lies in the benefits they involve (Section 1.2.3.1). However, we now recognize that partners in most putative mutualisms (although not all; see Chapter 2) incur costs as well. The interaction is beneficial to both partners when its net effect—the benefits minus the costs—is positive (Chapter 8). Costs of mutualism are a miscellany of investments, negative consequences, and missed opportunities that reduce the potential benefits of mutualism to their actual, observed effects. They can be easy to identify, but are often difficult or impossible to measure. For example, the net effect of biotic pollination to a certain plant could in theory, but probably not in practice, be calculated as some measure of the potential fecundity advantage it derives from pollen delivery and removal by animals, minus the costs it incurs in producing attractants and nectar rewards for those visitors. To the visitors, the benefit of this interaction would be the energetic value of the nectar it can obtain from a visit, minus the costs invested in finding and consuming it.

The idea of costs suffuses current research on mutualism. Costs help us to understand traits that might appear puzzling if we considered mutualism to involve benefits only. These include traits mutualists use to defend themselves against antagonistic aspects of their partners' behavior, not only adaptations to attract and to reward those partners (Chapter 6). Costs help explain when and where cheaters, which gain the benefits of mutualism without reciprocating, will arise (Chapters 4 and 6). Population dynamics of mutualism result not only in functional responses of benefits, but functional responses of costs as well (Chapter 8). Finally, the fact that costs are often contingent upon the local environment helps explain why mutualisms vary so much across space and time (Chapters 10 and 13).

1.3.6.2 Mutualisms vary greatly in time and space

As individual mutualisms came to be studied in multiple locations and across time, their remarkable variability became abundantly clear (e.g., Herrera 1988, Jordano 1994). This stimulated development of conceptual frameworks designed to interpret the implications of this variation. On the ecological side came the recognition that both benefits and costs exhibit conditionality or context dependency (Thompson 1988, Bronstein 1994a). Outcomes can vary from more to less mutualistic if benefits exceed costs everywhere, or can range to commensal or antagonistic if costs equal or exceed benefits. Subsequent research has focused on identifying which mutualisms are more and less likely to be context dependent, as well as what the broader implications of this phenomenon may be (Chapter 10). On the evolutionary side, documentation of variable outcomes in mutualism contributed to the development of coevolutionary theory (Chapter 7), notably the Geographic Mosaic of Coevolution (Thompson 2005). Modern theories of mutualism evolution recognize and often focus on spatial and temporal variation in selection regimes (Chapter 5).

1.3.6.3 Mutualisms cannot be understood in isolation from their ecological contexts

Although a large body of work on generalized pollination systems was already well situated within community ecology by the 1990s, most hypothesis-driven empirical investigations, as well as the vast majority of theoretical studies, focused on obligate, highly specialized mutualisms. Growing recognition that extreme specificity is usually the exception rather than the rule, that mutualists vary in the costs and benefits they confer, and that the outcomes of mutualism often vary due to the actions of other species, began forcing mutualism researchers to think more broadly. This consideration is evident in research at scales ranging from the human microbiome (Boxes 2.2 and 14.2) to the African savanna (Chapter 9). One profound realization of recent years is that while *mutualists* may be specialized or generalized, the same cannot be said for *mutualisms*. Rather, partners differ in their relative degrees of dependence on any one partner (Chapter 11). Cross-connections among mutualisms via shared partners weave communities together, with potentially profound consequences for coevolutionary processes (Chapters 7 and 11), community and ecosystem function (Chapters 9 and 12), and the conservation of biodiversity (Chapters 11, 13, and 14).

1.3.6.4 Mutualisms are under threat

Janzen (1974) pointed out the potential fragility of mutualisms to anthropogenic change long before it was on anyone else's radar. The modern field of conservation ecology began to emerge around the same time Boucher's mutualism volume was published. The potential risks of losing mutualisms were not ignored; an early and influential paper by John Terborgh (1986), for example, posited the risks of disrupting "keystone mutualisms" in tropical rainforests. Unfortunately, the literature can no longer speak in hypotheticals. A large body of research now exists documenting the relative sensitivity of different kinds of mutualisms to different types of anthropogenic threats. Pollination systems have been given particularly close scrutiny, in part because of their importance to agriculture and thus the human food supply (Box 14.1); so has the mutualistic symbiosis between corals and *Symbiodinium* dinoflagellates, which forms the backbone of coral reef ecosystems (Boxes 9.1, 10.3, and 13.2). As discussed in Chapters 13 and 14, new research is turning these case studies into predictive frameworks and associated recommendations for conservationists and land managers aiming to conserve and to restore mutualisms. This literature takes as its foundation the three concepts just discussed: the existence of costs as well as benefits of mutualism; spatiotemporal variation; and the importance of community context. In contrast to much of the conservation literature, it is also building explicit links to evolutionary biology. For example, there is growing recognition that rapid evolution can potentially rescue mutualisms (Chapter 13).

1.4 This volume, and beyond

The goal of this volume is to provide an up-to-date introduction to the full scope of mutualism studies today. It is both ecological and evolutionary in orientation, and covers scales ranging from the cell to the ecosystem. It is organized conceptually rather than by type of mutualism, reflecting the current state of this field. Section 1 introduces themes that run through the volume as a whole. The chapters in Section 2 are evolutionary in focus; they cover macroevolutionary, microevolutionary, and coevolutionary dynamics, with a special focus on the "cheater problem" and the challenges it poses for mutualisms. Section 3 covers population, community, and ecosystem ecology, as well as context dependency, a phenomenon of relevance to all these levels. Finally, Section 4 covers conservation and restoration of mutualism. Each chapter, written by authorities on this topic, seeks to introduce readers to its orienting concepts as well to exciting recent discoveries. However, each chapter has two other goals: to lay out unanswered questions and challenges for the future, and to introduce readers to new tools—conceptual, theoretical, and empirical—that can be used to do so. Each chapter is accompanied by highlighted, boxed material, mostly written by other expert contributors, that address particular concepts, tools, or case examples in greater depth.

Of course, coverage of topics as broad as those in this volume will by necessity be somewhat personal and idiosyncratic. However, we believe these chapters provide an entrance into the literature that will allow readers to explore more deeply individual topics that catch their interest.

The material presented here will soon fall out of date. In fact, the authors of this volume hope that it does: this would be a strong signal of a healthy, thriving, and rapidly developing field of study. We hope that the next book on mutualism does not take thirty years to appear.

References

Abrams, P.A. (1987). On classifying interactions between populations. *Oecologia*, **73**, 272–281.

Addicott, J.F. (1984). Mutualistic interactions in population and community processes. In P.W. Price, C.N. Slobodchikoff, and B.S. Gaud (eds.), *A New Ecology: Novel Approaches to Interactive Systems*, pp. 437–455. New York: John Wiley and Sons.

Arthur, W., and Mitchell, P. (1989). A revised scheme for the classification of population interactions. *Oikos*, **56**, 141–143.

Axelrod, R., and Hamilton, W.D. (1981). The evolution of cooperation. *Science*, **211**, 1390–1396.

Bequaert, J. (1922). Ants in their diverse relations to the plant world. *Bulletin of the American Museum of Natural History*, **45**, 333–583.

Boucher, D.H. (1985a). The idea of mutualism, past and future. In D.H. Boucher (ed.), *The Biology of Mutualism*, pp. 1–28. New York: Oxford University Press.

Boucher, D.H. (ed.). (1985b). *The Biology of Mutualism*. New York: Oxford University Press.

Boucher, D.H., James, S., and Keeler, K.H. (1982). The ecology of mutualism. *Annual Review of Ecology and Systematics*, **13**, 315–347.

Bronstein, J.L. (1994a). Conditional outcomes in mutualistic interactions. *Trends in Ecology and Evolution*, **9**, 214–217.

Bronstein, J.L. (1994b). Our current understanding of mutualism. *Quarterly Review of Biology*, **69**, 31–51.

Bronstein, J.L. (1998). The contribution of ant-plant protection studies to our understanding of mutualism. *Biotropica*, **30**, 150–161.

Bronstein, J.L. (2009a). The evolution of facilitation and mutualism. *Journal of Ecology*, **97**, 1160–1170.

Bronstein, J.L. (2009b). Mutualism and Symbiosis. *Princeton Guide to Ecology*. Princeton: Princeton University Press.

Callaway, R.M. (2007). *Positive Interactions and Interdependence in Plant Communities*. Dordrecht: Springer.

Clay, K., Marks, S., and Cheplick, G.P. (1993). Effects of insect herbivory and fungal endophyte infection on competitive interactions among grasses. *Ecology*, **74**, 1767–1777.

Connor, R.C. (1995). The benefits of mutualism: a conceptual framework. *Biological Reviews*, **70**, 427–457.

Crowcroft, P. (1991). *Elton's Ecologists. A History of the Bureau of Animal Population*. Chicago: University of Chicago Press.

Davidson, D.W. (1997). The role of resource imbalances in the evolutionary ecology of tropical arboreal ants. *Biological Journal of the Linnean Society*, **61**, 153–181.

Davidson, D.W., and McKey, D. (1993). The evolutionary ecology of symbiotic ant-plant relationships. *Journal of Hymenoptera Research*, **2**, 13–83.

de Bary, A. (1879). *Die Erscheinung der Symbiose*. Strassburg: Karl J. Trübner.

Dindal, D.L. (1975). Symbiosis: nomenclature and proposed classification. *The Biologist*, **57**, 129–142.

Dodds, W.K. (1997). Interspecific interactions: constructing a general neutral model for interaction type. *Oikos*, **78**, 377–383.

Douglas, A.E. (2010). *The Symbiotic Habit*. Princeton: Princeton University Press.

Dugatkin, L.A. (1997). *Cooperation Among Animals. An Evolutionary Perspective*. New York: Oxford University Press.

Egerton, F.N. (1973). Changing concepts of the balance of nature. *Quarterly Review of Biology*, **48**, 322–350.

Egerton, F.N. (1977). A bibliographical guide to the history of general ecology and population ecology. *History of Science*, **15**, 189–215.

Feldman, T.S., Morris, W.F., and Wilson, W.G. (2004). When can two plant species facilitate each other's pollination? *Oikos*, **105**, 197–207.

Gilbert, L.E. (1975). Ecological consequences of a coevolved mutualism between butterflies and plants. In L.E. Gilbert and P.H. Raven (eds.), *Coevolution of Animals and Plants*, pp. 210–240. Austin: University of Texas Press.

Goff, L. (1982). Symbiosis and parasitism: another viewpoint. *Bioscience*, **32**, 255–256.

Grant, V., and Grant, K. (1965). *Flower Pollination in the Phlox Family*. New York: Columbia University Press.

Haskell, E.F. (1947). A natural classification of societies. *Transactions of the New York Academy of Sciences*, **29**, 186–196.

Haskell, E.F. (1949). A clarification of social science. *Main Currents in Modern Thought*, **7**, 45–51.

Heithaus, E.R. (1974). The role of plant-pollinator interactions in determining community structure. *Annals of the Missouri Botanical Gardens*, **61**, 675–691.

Herre, E.A., Knowlton, N., Mueller, U.G., and Rehner, S.A. (1999). The evolution of mutualisms: exploring the paths between conflict and cooperation. *Trends in Ecology and Evolution*, **14**, 49–53.

Herrera, C.M. (1988). Variation in mutualisms: the spatio-temporal mosaic of a pollinator assemblage. *Biological Journal of the Linnean Society*, **35**, 95–125.

Hertig, M., Taliaferro, W.H., and Schwartz, B. (1937). Report of the committee on terminology. *Journal of Parasitology*, **23**, 325–329.

Janzen, D.H. (1966). Co-evolution of mutualism between ants and acacias in Central America. *Evolution*, **20**, 249–275.

Janzen, D.H. (1967). Interaction of the bull's horn acacia (*Acacia cornigera L.*) with an ant inhabitant (*Pseudomyrmex ferruginea F. Smith*) in eastern Mexico. *University of Kansas Science Bulletin*, **47**, 315–558.

Janzen, D.H. (1974). The deflowering of Central America. *Natural History*, **83**, 48–53.

Janzen, D.H. (1985). The natural history of mutualisms. In D.H. Boucher (ed.), *The Biology of Mutualism*, pp. 40–99. New York: Oxford University Press.

Jordano, P. (1994). Spatial and temporal variation in the avian-frugivore assemblage of *Prunus mahaleb*: patterns and consequences. *Oikos*, **71**, 479–491.

Keddy, P.A. (1989). *Competition*. New York: Chapman and Hall.

Khakina, L.N., Margulis, L., and McMenamin, M. (eds.). (1992). *Concepts of Symbiogenesis. History of Symbiosis as an Evolutionary Mechanism*. New Haven: Yale University Press.

Kingsland, S.E. (1985). *Modeling Nature. Episodes in the History of Population Ecology*. Chicago: University of Chicago Press.

Law, R. (1991). The symbiotic phenotype: origins and evolution. In L. Margulis and R. Fester (eds.), *Symbiosis as a Source of Evolutionary Innovation*, pp. 57–71. Cambridge, MA: MIT Press.

Leary, R.A. (1985). *Interaction Theory in Forest Ecology and Management*. Dordrecht: Martinus Nijhoff/Dr. W. Junk Publishers.

Lewin, R.A. (1982). Symbiosis and parasitism—definitions and evaluations. *Bioscience*, **32**, 254–260.

Losey, G.S. (1972). The ecological importance of cleaning symbiosis. *Copeia*, **1972**, 820–833.

Margulis, L. (1970). *Origin of Eukaryotic Cells*. New Haven: Yale University Press.

Margulis, L. (1990). Words as battle cries—symbiogenesis and the new field of endocytobiology. *Bioscience*, **40**, 673–677.

Margulis, L. (1991). Symbiogenesis and symbionticism. In L. Margulis and R. Fester (eds.), *Symbiosis as a Source of Evolutionary Innovation*, pp. 1–14. Cambridge, MA: MIT Press.

Mattson, W.J., and Addy, N.D. (1975). Phytophagous insects as regulators of forest primary production. *Science*, **190**, 515–522.

May, R.M. (1976). Models for two interacting populations. In R.M. May (ed.), *Theoretical Ecology: Principles and Applications*, pp. 49–70. Philadelphia: Saunders.

May, R.M. (1982). Mutualistic interactions among species. *Nature*, **296**, 803–804.

Mitman, G. (1992). *The State of Nature. Ecology, Community and American Social Thought*. Chicago: University of Chicago Press.

Mulcahy, D.L. (1979). The rise of the angiosperms: a genecological factor. *Science*, **206**, 20–23.

Odum, E.P. (1953). *Fundamentals of Ecology*. Philadelphia: Saunders.

Ollerton, J. (2006). "Biological barter": patterns of specialization compared across different mutualisms. In N.M. Waser and J. Ollerton (eds.), *Plant-Pollinator Interactions: From Specialization to Generalization*, pp. 411–435. Chicago: University of Chicago Press.

Ollerton, J., Winfree, R., and Tarrant, S. (2011). How many flowering plants are pollinated by animals? *Oikos*, **120**, 321–326.

Owen, D.F., and Wiegert, R.G. (1976). Do consumers maximize plant fitness? *Oikos*, **27**, 488–492.

Pirozynski, K.A., and Malloch, D.W. (1975). The origin of land plants: a matter of mycotrophism. *Biosystems*, **6**, 153–164.

Pound, R. (1893). Symbiosis and mutualism. *American Naturalist*, **27**, 509–520.

Price, P.W. (1984). *Insect Ecology* (second edition). New York: John Wiley and Sons.

Pugnaire, F.I., Haase, P., and Puigdefábregas, J. (1996). Facilitation between higher plant species in a semiarid environment. *Ecology*, **77**, 1420–1426.

Rico-Gray, V., and Oliveira, P.S. (2007). *The Ecology and Evolution of Ant-Plant Interactions*. Chicago: University of Chicago Press.

Ridley, H.N. (1930). *Dispersal of Plants Throughout the World*. Kent: L. Reeve and Company.

Robertson, C. (1895). The philosophy of flower seasons, and the phenological relations of the entomophilous flora and the anthophilous insect fauna. *American Naturalist*, **29**, 97–117.

Roughgarden, J. (1975). Evolution of marine symbiosis—a simple cost-benefit model. *Ecology*, **56**, 1201–1208.

Sapp, J. (1994). *Evolution by Association: A History of Symbiosis*. New York: Oxford University Press.

Smith, D.C. (1991). Foreword. In L. Margulis and R. Fester (eds.), *Symbiosis as a Source of Evolutionary Innovation*, pp. ix–x. Cambridge, MA: The MIT Press.

Stadler, B., and Dixon, A.F.G. (2008). *Mutualism: Ants and Their Insect Partners*. Cambridge: Cambridge University Press.

Stone, L., and Roberts, A. (1991). Conditions for a species to gain advantage from the presence of competitors. *Ecology*, **72**, 1964–1972.

Templeton, A.R., and Gilbert, L.E. (1985). Population genetics and the coevolution of mutualism. In D.H. Boucher (ed.), *The Biology of Mutualism*, pp. 128–144. New York: Oxford University Press.

Terborgh, J. (1986). Keystone plant resources in the tropical forest. In M.E. Soulé (ed.), *Conservation Biology: The Science of Scarcity and Diversity*, pp. 330–344. Sunderland, MA: Sinauer.

Thompson, J.N. (1988). Variation in interspecific interactions. *Annual Review of Ecology and Systematics*, **19**, 65–87.

Thompson, J.N. (2005). *The Geographic Mosaic of Coevolution*. Chicago: University of Chicago Press.

Tobey, R.C. (1981). *Saving the Prairies. The Life Cycle of the Founding School of American Plant Ecology 1885–1955*. Berkeley: University of California Press.

Todes, D. (1989). *Darwin Without Malthus. The Struggle for Existence in Russian Evolutionary Thought*. New York: Oxford University Press.

van Beneden, P.J. (1876). *Animals and Their Messmates*. London: Henry S. King.

van der Heijden, M. G. A., R. D. Bardgett, and N. M. van Straalen. 2008. The unseen majority: soil microbes as drivers of plant diversity and productivity in terrestrial ecosystems. Ecology Letters **11**, 296–310.

Vandermeer, J.H., & Boucher, D.H. (1978). Varieties of mutualistic interaction in population models. Journal of Theoretical Biology, **74**, 549–558.

Way, M.J. (1963). Mutualism between ants and honeydew-producing homoptera. Annual Review of Entomology, **8**, 307–344.

West, S.A., Griffin, A.S., & Gardner, A. (2007). Social semantics: altruism, cooperation, mutualism, strong reciprocity and group selection. Journal of Evolutionary Biology, **20**, 415–432.

Wheeler, W.W. (1907). The fungus-growing ants of North America. Bulletin of the American Museum of Natural History, **33**, 669–807.

Williams, G.C. (1966). *Adaptation and Natural Selection*. Princeton: Princeton University Press.

Willmer, P. (2011). *Pollination and Floral Ecology*. Princeton: Princeton University Press.

CHAPTER 2

The special case of symbioses: mutualisms with persistent contact

Angela E. Douglas

2.1 Introduction

This chapter concerns symbiotic mutualisms, i.e., persistent associations between organisms of different species from which all participants derive benefit. These associations lie at the intersection between the overlapping categories of mutualisms and symbioses. While mutualisms, by definition, involve reciprocal benefits, the term symbiosis, as originally defined, refers to "any persistent association between species," including mutualistic and parasitic relationships that involve direct, sustained interspecific contact. (Symbioses have also been interpreted as exclusively mutualistic, but this perspective can be unhelpful because the pattern of benefit/harm is often unquantified or varies with environmental circumstance; see Douglas (2010) and Chapter 1 for a detailed discussion of this issue.) Symbiotic mutualisms can contribute a rich and diverse array of examples that illuminate and advance our understanding of mutualisms in general. The great majority of symbiotic mutualisms involve organisms of different size. Often a single large host is associated with a population of small organisms (symbionts) with short generation time. As a result, a single host may interact with multiple generations of its partners.

From the perspective of mutualisms in general, mutualistic symbioses are special; because the participants in symbioses are in persistent contact, an organism inescapably experiences the consequences of its interactions with its partners. If a participant fails to provide a benefit, it has to contend with the resultant reduction in physiological condition and services rendered by its partners, with likely negative consequences for its own fitness. The intuitive expectation that persistent contact tends to ameliorate antagonistic traits and promote mutualistic traits in both ecological and evolutionary time has been articulated formally in multiple theoretical scenarios and demonstrated experimentally in various systems (Bull et al. 1991, Sachs and Wilcox 2006).

The parallels and contrasts between symbiotic and non-persistent mutualisms are addressed in this chapter, within the framework of three topics. The various types of interactions in symbiotic mutualisms, including reciprocity, byproduct interactions, and addiction, are considered first, and these illustrate the commonality between symbiotic and non-persistent mutualisms. The second topic addresses the defining trait of symbioses, persistent contact between the partners, including their resilience to environmental crisis. Finally, partner fidelity within symbioses is considered. Although many aspects of partner fidelity and its variation have parallels with non-persistent mutualisms, certain consequences are unique to symbiotic mutualisms. In particular, genomic deterioration and the evolutionary transition to organelles, a consequence of obligate vertical transmission over very many host generations in certain associations, have no parallel among non-symbiotic mutualisms. As the examples in this chapter illustrate, symbiotic mutualisms share many features with non-persistent mutualisms, but also display some specific traits that set them apart, from the perspectives of both their ecology and evolution.

Mutualism. Edited by Judith L. Bronstein.

2.2 Interactions between symbiotic partners

The great majority of symbiotic mutualisms are founded on the acquisition of nutrients or protection from natural enemies. Examples include the various plant root symbioses with microorganisms (e.g., rhizobia, mycorrhizal fungi, and the actinobacterium *Frankia*); the gut symbioses with cellulose-degrading microorganisms in many animals, including ruminant mammals and termite insects; the photosynthetic symbioses in various aquatic animals (e.g., corals) and fungi (the lichens); and the myrmecophytic plants (see Chapter 9) which house protective or nutrient-provisioning ant colonies. In this section, three key processes are reviewed: reciprocity, byproduct mutualism, and addiction. All three can occur in both symbiotic and non-persistent mutualisms, illustrating the overlap between these two types of association. Research on symbiotic mutualisms has made important contributions to our understanding of these processes.

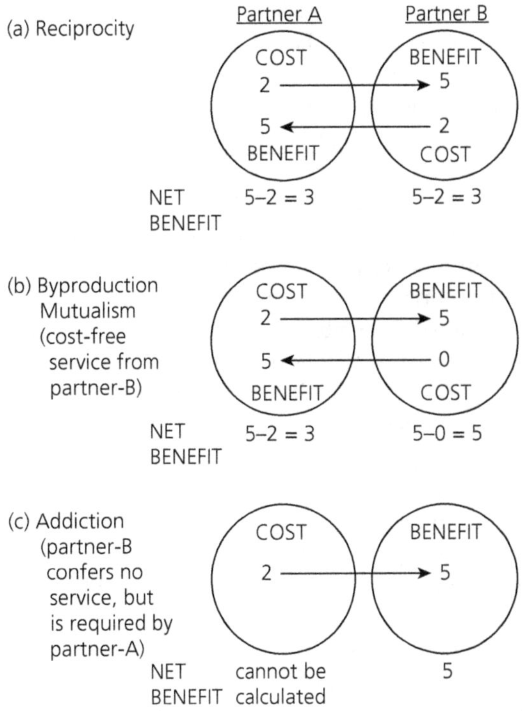

Figure 2.1 The multiple processes that can occur in symbiotic mutualisms, here shown as two organisms (partner A and partner B) with costs and benefits quantified in arbitrary units (0–5). (a) Reciprocity, in which both organisms benefit more from the service provided by their partner (benefit is 3 units) than the service they provide to the partner (cost is 2 units). (b) Byproduct mutualism, in which the service provided by partner B is cost-free. (c) Addiction, in which partner A is dependent on partner B, without deriving any benefit.

2.2.1 The services underpinning reciprocity

Reciprocity is central to our understanding of mutualisms, including symbiotic mutualisms. Specifically, each organism provides its partners with a service that is more valuable to the recipient than the donor, such that the net benefit of the interaction is greater than the cost for all the participating organisms (Figure 2.1). In both symbiotic and non-symbiotic mutualisms, reciprocated services include one or more of the following: enhanced access to nutrients, protection against antagonists, and transportation.

Symbiotic organisms can gain privileged access to food. For example, ants resident on myrmecophytic plants consume food items, such as lipid-rich bodies and extrafloral nectar produced by the host plant. In other symbioses, especially involving microorganisms, nutrients are exchanged directly between the cells of the partners. Various microorganisms possess metabolic capabilities absent from their host, e.g., nitrogen fixation by rhizobia in leguminous plants, photosynthesis by algae or chemosynthesis by bacteria in marine invertebrates, and essential amino acid synthesis by bacteria in some insects. These microorganisms overproduce specific products of their metabolic capabilities (ammonia, glucose, amino acids, etc.) and release them to their host (Day et al. 2001, Burriesci et al. 2012, Macdonald et al. 2012).

Symbioses with protective function are underlain by a great diversity of processes, including prior residency by non-pathogens (competition for adhesion sites), activation of the host immune system functions that are more deleterious to the invader than the resident (e.g., Diaz-Albiter et al. 2012), toxin production (e.g., alkaloids in endophytic fungi of grasses, polyketide or proteinaceous toxins in bacterial symbionts of various invertebrate animals), and defensive behavioral traits (as in ant partners of plants and insects) (Piel 2002, Pierce et al. 2002, Saikkonen et al. 2006, Ferrari and Vavre 2011). Some symbionts can protect their hosts against abiotic stress, including elevated temperature (Montllor et al. 2002, Marquez et al. 2007), hypoxia (Finlay et al. 1996), and toxic metals in the environment (Read 1991, Sharples et al. 2000).

The most widely studied transportation mutualisms provide for pollination and seed dispersal, the great majority of which involve contact too brief to be classified as symbioses. Exceptionally, however, some insect pollinators develop to adulthood and mate within their plant host in brood-site pollination symbioses (figs/fig wasp, yucca/yucca moth mutualisms; Sakai 2002). In addition, the transport of various microorganisms is enhanced by the many mobile animals that regularly shed viable microbial symbionts (gut microbiota in animal feces, light organ symbionts of squid; Malloch and Blackwell 1992, Sekirov et al. 2010, Wier et al. 2010).

Although many mutualistic symbioses can be defined by the reciprocal exchange of nutritional, protective, and transportation services, some associations include other functions. For example, the luminescent bacterial symbionts may not only protect their fish or squid host from predators, for example through counterillumination (Sonke et al. 2004), but can also contribute to intraspecific communication in some hosts, e.g., flashlight fish (Anomalopidae) and anglerfish (Morin et al. 1975, Herring 2007).

An important developing line of research involves the harnessing of a thorough mechanistic understanding of mutualistic symbioses to understand the selection pressures that dictate the abundance and distribution of these associations. One system for which this approach is being spearheaded is the association between plant roots and arbuscular-mycorrhizal fungi. The fungal partner obtains its carbon needs from plant-derived sugars, obviating the need for saprophytic exploitation of chemically-complex substrates, such as cellulose and lignin, in the soil. In return, the fungus provides the plant with inorganic phosphate which, because of its low diffusion coefficient, rapidly becomes depleted in the immediate vicinity of roots. The very fine, branching fungal hyphae are more effective than the relatively short, non-branching root hairs at circumventing these zones of phosphate depletion. In short, the symbiosis alleviates the genetic, physiological, and anatomical constraints on the capacity of the fungus to gain carbon, and the plant to gain phosphate. In mycorrhizal symbioses, as in other systems, reciprocity is expected to require the coupling of services rendered to benefits received by each partner. Specifically, the experiments described above demonstrate that both the fungus and plant are able to discriminate between partners that provide nutrients at high and low rates. As Figure 2.2 illustrates, phosphate supply from the fungus to plant roots is increased by high sugar, and

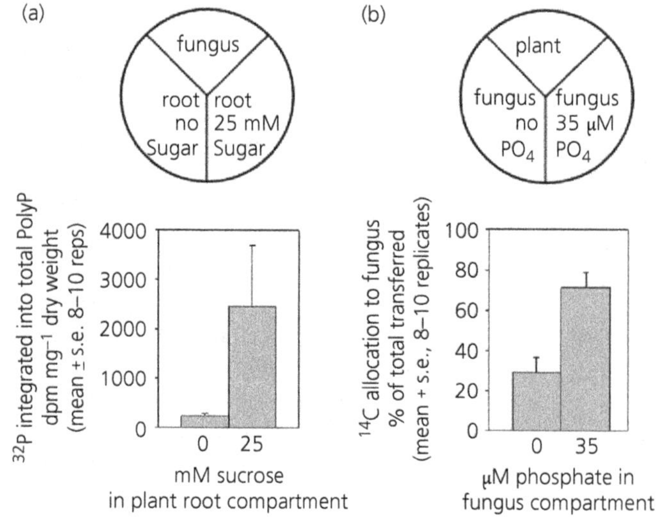

Figure 2.2 Reciprocity in the association between the arbuscular mycorrhizal fungus *Glomus intraradices* and triple-split root system of the plant *Medicago truncatulus*: each partner (fungus and plant root) allocates more resource to the partner that provides greater reciprocated resource. (a) The fungus provides more phosphate to the plant root with more carbon (sugar). (b) The plant root provides more carbon to the fungus with more phosphate. [Modified from Figures 2A and 2C of (Kiers et al. 2011).]

sugar supply to the fungus is promoted by high phosphate (Kiers et al. 2011).

Importantly, nutrient exchange is not the only service provided in the mycorrhizal symbioses. In many associations, the fungus confers protection against fungal pathogens of plant roots. For example, mycorrhizal fungi play a crucial role protecting the annual grass species *Vulpia ciliata* from fungal pathogens, and apparently make no contribution to the plant phosphate nutrition (Newsham et al. 1995). As this example illustrates, symbiotic mutualisms can be functionally diverse and variable in the reciprocated services provided. An important corollary of these considerations is that symbiotic mutualisms are underpinned by multiple interactions, often occurring simultaneously. Some of these interactions are not necessarily founded on reciprocity. In some systems, organisms provide services to their partners without cost (Figure 2.1b), or depend on the association, even when the benefit is slight or not discernible (Figure 2.1c).

2.2.2 Byproduct mutualisms

"One organism may enjoy benefits that are *incidental* to or a *by-product* of the self-serving traits of another organism" (Connor 1995). Such cost-free services are commonly described as byproduct mutualism, and are frequently regarded as important interactions promoting both the evolutionary origin and persistence of mutualisms displaying reciprocity (Connor 1995).

A classic example is provided by various crabs that inhabit empty mollusk shells colonized by sea anemones. As the crab feeds, often messily, stray food particles are consumed by the sea anemone, without any effect on the crab (Manuel, 1988). No-cost interactions may also be widespread in associations involving microorganisms that can utilize the waste products of metabolism in their partners. Symbiotic algae in corals can consume the waste ammonia of their host, utilizing it as a source of nitrogen for growth (Pernice et al. 2012); the resultant reduction in the potentially toxic ammonia load in the coral tissues is advantageous to the host. Similarly, hydrogen transfer from various anaerobic ciliates to their intracellular methanogenic bacteria both provides a substrate for ATP-yielding methanogenesis in the bacteria and promotes sustained fermentation metabolism in the ciliate host (Fenchel and Finlay 1995).

Let us be clear about our terminology. A cost-free service is one that incurs no *extra* cost for being made available to the partner. The organism is anticipated to incur some cost in displaying the trait. The crab expends energy to capture and process its prey, the algal symbiont consumes resources to synthesize the enzymes required for ammonia assimilation, and so on. In some instances, it can be difficult to determine whether a trait expressed in a symbiosis is entirely self-serving or involves extra cost. The difficulty can be illustrated by the protective symbioses between rove beetles of the genus *Pederus* and their bacterial endosymbionts *Pseudomonas*. The bacteria synthesize a distasteful and toxic polyketide, known as pederin, which is released into the insect hemolymph (Piel 2002). On mechanical damage to the beetle, caused for example by a predator, the pederin is released from the hemolymph causing damage (e.g., acutely irritating dermatitis to human skin) and deterrence. Because the continued survival of the beetle host is selectively advantageous for the bacterium, pederin synthesis by *Pseudomonas* can be considered as a trait that protects the resource supply for the bacteria. The critical unanswered question is whether the cost of defense for *Pseudomonas* is greater in symbiosis with the insect than for cells in the free-living condition. Furthermore, the overlap in selective interest between host and symbiont can select for changes in the protective functions of symbiotic organisms, to act specifically against certain antagonists of the partner while minimizing the toxicity to the partner. A likely example is the protection of the pea aphid *Acyrthosiphon pisum* against the specialist parasitoid *Aphidius ervi* by phage-borne protein toxins in the bacterial symbiont *Hamiltonella defensa* (Oliver et al. 2009).

2.2.3 Addiction in symbioses

An extra dimension to our understanding of the interactions in symbioses comes from a consideration of dependence. Enhanced fitness in the presence of a symbiotic partner is commonly interpreted as evidence for benefit, and this is the starting point for study of services provided by the partner. However, there is increasing evidence that dependence does not have to be founded on benefit. Some organisms can require an association through addiction, i.e., dependence on the partner in the absence of any discernible benefit (Aanen and Hoekstra 2007).

Symbioses have played a crucial role in the development of our understanding of addiction in biological associations. Of particular interest is the relationship between the parasitic wasp *Asobara tabida* and the bacterium *Wolbachia*. *Wolbachia* is renowned as a reproductive parasite, but is increasingly appreciated to have a rich and variable relationship with its animal hosts, with evidence for nutritional and protective functions. Greg Hurst describes these various interactions in Box 2.1. Even among the arthropod associations with *Wolbachia*, the interaction with *A. tabida* is distinctive. When *Wolbachia* is eliminated by antibiotic treatment, *A. tabida* is reproductively sterile with egg production halted by massive apoptosis (programmed cell death) of the nurse cells in the ovary. There are indications that *Wolbachia* inhibits apoptosis, perhaps by perturbing iron metabolism (Kremer et al. 2009). It has been argued that the host constitutively up-regulates apoptotic signaling in response to *Wolbachia*-mediated interference (Pannebakker et al. 2007), such that apoptosis is excessive in the absence of the inhibitory signal from *Wolbachia*. In this way, *A. tabida* has become dependent on *Wolbachia* without deriving any discernible benefit.

Box 2.1 The changing face of *Wolbachia*

Gregory Hurst

Symbionts have commonly been classified as parasitic or beneficial. However, it is now clear that there is no sharp distinction between nice and nasty, but rather a continuum from deleterious to beneficial. This is nicely typified by *Wolbachia*, one of the best-studied heritable symbionts. Classically, *Wolbachia* was considered a reproductive parasite. Now, it is recognized that some members of the clade are beneficial, or combine beneficial and parasitic properties, and may evolve rapidly from costly to advantageous symbiont in short periods of time.

The traditional view of *Wolbachia* has been summarized in Werren (1999). *Wolbachia* was maternally inherited, present in upward of 20% of arthropod species, and was a reproductive parasite: that is to say, it spread through the host population by manipulating host reproduction. Commonly, *Wolbachia* caused cross incompatibility, with *Wolbachia*-infected males being incompatible with uninfected females, or females carrying a different strain. Elsewhere, *Wolbachia* was responsible for cases of feminization in isopod crustaceans, parthenogenesis induction in many haplodiploid insects, and male-killing in flies, butterflies, and beetles. These studies established *Wolbachia* as the quintessential reproductive parasite—a maternally inherited microbe that invaded populations through enhancing either the production or survival of infected female hosts.

This caricature of *Wolbachia* as "just" a reproductive parasite has changed recently.

First, it was discovered that *Wolbachia* was widely present in filarial nematodes, and the filarial nematodes required infection for fertility. Indeed, the congruent host and symbiont phylogenies indicate they are long term symbiotic partners (Bandi et al. 1998). The requirement for filariae is indicated by sensitivity of worm fertility to administration of antibiotics, raising the prospect of *Wolbachia* being used as the target of anti-filarial treatment strategies (Slatko et al. 2010). *Wolbachia* have recently also been observed to be obligate symbionts of bedbugs, in this case supplying B vitamins (Hosokawa et al. 2010). Again, antibiotic treatment sterilizes the host.

Second, it is quite common to find *Wolbachia* strains in arthropods, such as *Drosophila melanogaster*, that do not affect host sex/sex ratio (they were present in both sexes), and produce little or no cross incompatibility (Hoffmann et al. 1998). Symbionts are unlikely to persist by genetic drift. The presence of these symbiont infections at intermediate prevalence represents a "smoking gun" that symbioses between *Wolbachia* and their arthropod hosts go beyond reproductive parasitism.

Recent functional studies have demonstrated diverse physiological roles for *Wolbachia*. Studies of *Wolbachia* in *D. melanogaster* have shown that the symbiont provides tolerance to dsRNA virus infection, increasing host life span when confronted with this natural enemy (Hedgges et al. 2008).

Box 2.1 *Continued*

This trait has since been observed in a variety of *Wolbachia* strains, and most importantly is portable—that is to say, it is maintained when the *Wolbachia* is placed in a novel host. This has therefore been developed as a mechanism of blocking the transmission of arboviruses such as dengue (Walker et al. 2011). Further work on the symbiosis in *D. melanogaster* was motivated by the presence of a heme pathway in the wMel genome, indicating that the infection was beneficial to the host under conditions of both low and high iron availability (though not under intermediate levels; Brownlie et al. 2009). Thus, *Wolbachia* acts as a subtle nutritional symbiont.

It is clear that the symbiosis between *Wolbachia* and its host is quite evolutionarily labile. Alteration in the direction of the symbiosis from parasitic to mutualistic has been observed in *D. simulans* (Weeks et al. 2007). *Wolbachia* strain wRi invaded southern California *D. simulans* in the early 1980s, with infection reaching the north of the state by 1990. In 1988, the fitness impact of *Wolbachia* was measured, and infected females were observed to produce 10% fewer progeny than uninfected comparators. The same measurement was conducted in 2002, but this time *Wolbachia*-infected females were either as or more fecund than uninfected comparators. This alteration was associated with a change in *Wolbachia* rather than host. The simplest interpretation of this data is that *Wolbachia* fitness was dependent on that of the female host, and strains that reduced impact and contributed to host physiology were favored.

In retrospect, these patterns are unsurprising. Fitness of a maternally inherited symbiont like *Wolbachia* is the joint product of multiple interactions, some of which are deleterious to the host (e.g., reproductive manipulation) and others advantageous (e.g., protection from pathogens).

References

Bandi, C., Anderson, T.J.C., Genchi, C., and Blaxter, M.L. (1998). Phylogeny of *Wolbachia* in filarial nematodes. *Proceedings of the Royal Society of London B*, **265**, 2407–2413.

Brownlie, J.C., Cass, B.N., Riegler, M. et al. (2009). Evidence for metabolic provisioning by a common invertebrate endosymbiont, *Wolbachia pipientis*, during periods of nutritional stress. *PLoS Pathogens* **5**, e1000368.

Hedges, L.M., Brownlie, J.C., O'Neill, S.L., and Johnson, K.N. (2008). *Wolbachia* and virus protection in insects. *Science*, **322**, 702–702.

Hoffmann, A.A., Hercus, M., and Dagher, H. (1998). Population dynamics of the *Wolbachia* infection causing cytoplasmic incompatibility in *Drosophila melanogaster*. *Genetics*, **148**, 221–231.

Hosokawa, T., Koga, R, Kikuchi, Y., Meng, X.Y., and Fukatsu, T. (2010). *Wolbachia* as a bacteriocyte-associated nutritional mutualist. *Proceedings of the National Academy of Sciences of the United States of America*, **107**, 769–774.

Slatko, B.E., Taylor, M.J., and Foster, J.M. (2010). The *Wolbachia* endosymbiont as an anti-filarial nematode target. *Symbiosis*, **51**, 55–65.

Walker, T., Johnson, P.H., Moreira, L.A. et al. (2011). The wMel *Wolbachia* strain blocks dengue and invades caged *Aedes aegypti* populations. *Nature*, **476**, 450–453.

Weeks, A.R., Turelli, M., Harcombe, W.R., Reynolds, K.T., and Hoffmann, A.A. (2007). From parasite to mutualist: rapid evolution of *Wolbachia* in natural populations of *Drosophila*. *PLoS Biology*, **5**, 997–1005.

Werren, J.H. (1999). Biology of *Wolbachia*. *Annual Review of Entomology*, **42**, 587–609.

Dependence without benefit (addiction) may be widespread in associations traditionally described as mutualistic, including interactions between animals and their resident gut microbiota. For example, acetic acid, a metabolite produced by the gut microbiota in *Drosophila*, has been implicated in the regulation of insulin/Tor signaling network (the master regulator of cell growth and energy homeostasis) in fruit flies (Shin et al. 2011); and butyrate, a fermentation product of the microbiota in the mammalian (including human) colon, regulates the cell growth and differentiation of colonic epithelial cells, influencing their propensity for cancerous growth (Wong et al. 2006). Further research is needed to establish the incidence of addictive interactions in symbioses that are usually interpreted as mutualistic, and particularly their role in promoting the persistence of associations under conditions where the selective advantage of reciprocal interactions is low.

In the next section, we turn to the defining feature of symbiotic mutualisms: the persistent contact between the partners, permitting repeated interactions between the same individuals. The persistence of symbioses is linked to their common trait of resilience to perturbation.

2.3 The persistence of symbioses

2.3.1 Open and closed symbioses

The partners in symbiotic mutualisms are in persistent contact and so, by definition, their interaction is sustained over time. The identity of the organisms in a symbiosis is determined by the probability of making contact, as dictated by their abundance and distribution in the shared habitat, and by their biological compatibility. In many symbioses, among-partner communication facilitating contact and discrimination among different potential partners is mediated chemically, and the molecules mediating signal exchange have been elucidated in several systems (Bright and Bulgheresi 2010).

Symbioses can conveniently be defined as either "closed" or "open." In closed systems, the persistence of the symbiosis and the biomass of the interacting partners are shaped by the ecological processes of birth and death. Open systems combine birth and death with immigration and emigration (gain and loss from the association), such that the participants in an association can change over time. In this way, symbioses differ from non-persistent associations, which are shaped, at least proximally, by immigration and emigration exclusively.

Bacterial symbioses in cockroaches illustrate the distinction between open and closed systems. All cockroaches bear a dense and diverse community of bacteria in the hindgut, and an intracellular bacterium *Blattabacterium*, restricted to specialized cells called bacteriocytes, associated with the fat body. The relationship with *Blattabacterium* is closed, in that its persistence depends on balanced proliferation of bacterial cells and the bacteriocytes within one individual insect, and the vertical transmission of bacterial cells to offspring hosts via the egg; no microorganisms other than *Blattabacterium* can gain access to the bacteriocytes. By contrast, the association with the gut bacteria is open, in that the birth and death of bacteria and gut cells lining the lumen is accompanied by the regular import of novel bacteria with food, and shedding of bacterial cells in the feces.

Some symbioses are open for a very restricted period of time, during which the association is formed, after which they become closed. An example is the symbiosis between the bobtail squid *Euprymna scolopes* and luminescent bacteria *Vibrio fisheri*, described in Box 5.3. Bacterial cells in the water column are channeled into the light organ by densely cilated epithelial flaps. Once the light organ is colonized, the flaps degenerate, preventing subsequent colonization (Koropatnick et al. 2004).

It is increasingly appreciated that the identity of the partners in open symbioses can be explained using the basic principles of population ecology and community ecology. As John Rawls describes in Box 2.2, these approaches are particularly being applied to investigate the dynamics of the microbiota in animal guts, linked to the Human Microbiome Project, MetaHIT, and other large consortial initiatives on the resident microbiota in humans (Human Microbiome Project 2012a).

2.3.2 The resilience of symbioses

There is considerable current interest in the response of open symbioses to abrupt changes in environmental conditions that cause a substantial reduction in microbiota population sizes. Are associations resilient, returning to the predisturbance condition? Are some partners eliminated during the disturbance? And is there recruitment of new partners with different traits, including taxa resistant to subsequent disturbance events? Over a decade or more, these questions have been asked of the symbiosis between corals and their symbiotic algae *Symbiodinium*, revealing that there is no single pattern in response to environmental stress (see Boxes 10.3 and 13.2). The context for the research on resilience is coral bleaching, which is caused by mass expulsion of algal cells in response to elevated sea temperature and other environmental stressors. The responses are very variable. Among those corals in which the algal population recovers, both taxonomic composition and bleaching resistance of the recovered algal cells has sometimes been shown to be altered and sometimes to remain unchanged. Analogous questions are now starting to be asked of the gut microbiota in humans, particularly in relation to the response to antibiotic treatments that dramatically deplete the microbial populations (Box 2.2).

The resilience of symbioses to abrupt environmental change is very variable, and the processes shaping this variation are currently poorly understood. Ecological approaches have great potential to help

Box 2.2 The human microbiome

John Rawls

How should we define the human body? Is it defined only by our own human cells and genome? Our growing understanding of the microbes that inhabit the surfaces of our body suggests that a broader definition may be warranted. Humans and other vertebrates are devoid of microbes at birth, and all exposed body surfaces subsequently become colonized by complex communities of microbes including bacteria, archaea, eukaryotes, and viruses. By the time we reach adulthood, the numbers of microbial cells that inhabit the human body outnumber our own human cells by at least an order of magnitude. Moreover, the collective genomes of these microbial cells encode genes that outnumber our human genes by two orders of magnitude (Human Microbiome Consortium, 2012, Qin et al. 2010). Far from being inert passengers, these microbial communities and their associated genomes (collectively called the human microbiome) interact with many aspects of our biology including exclusion of potential pathogens, modulation of our immune system, synthesis of essential vitamins and amino acids, digestion of dietary nutrients, even modifying our behavior (Sekirov et al. 2010, Hooper et al. 2012, Cryan and Dinan 2012). This has led to the conceptualization of the human body as an ecosystem, comprising both human cells and microbial cells.

Our understanding of the human microbiome has been revolutionized in recent years through the advent of high-throughput sequencing of microbial DNA which permits the analysis of microbial diversity without the need for cultivation. Microbial diversity within a specific human body site varies according to location, ranging from the highest in the mouth and colon to the lowest in the vagina. There appear to be no bacterial taxa found in all individuals and body sites, although most body sites are dominated by a single or few signature bacterial taxa. Differentiation among host individuals is relatively high but varies according to body site, ranging from the highest in regions of the skin to the lowest in the mouth (Human Microbiome Consortium 2012). The bacterial communities are clustered according to anatomy, suggesting a rich landscape of microbial habitats across the human body. While the diversity of bacterial taxa among individuals is relatively high, the metabolic pathways encoded within their respective genomes appear to be relatively constant (Human Microbiome Consortium 2012), indicating that taxonomic diversity may not always correlate with functional diversity. Within-host variation over time is generally lower than inter-individual variation, suggesting that each individual possesses a unique microbiome that is relatively stable. The causes of high inter-individual variation remain unclear but might be attributable to physiologic and environmental variation among anatomical locations and among hosts, as well as sampling bias. The microbiota is not more similar between identical twins than fraternal twins, suggesting that microbiome composition is influenced more by shared environmental exposures associated with kinship, rather than genotype (Yatsunenko et al. 2012). Diet has been implicated as a major environmental factor shaping microbiome composition in humans and other mammals (Flint et al. 2012). The causes and consequences of variation in the human microbiome remain active areas of scientific inquiry, and are expected to have direct relevance for human health.

Although your microbiome may be unique, it is not impervious to change. In fact, many perturbations are known to impact the microbiome. The microbiome can be remarkably resilient to temporary perturbations. For example, exposure to antibiotics results in rapid alterations to the gut microbiome which are largely restored within a few days after exposure (Costello et al. 2012). However, chronic perturbations such as changes in diet can result in rapid and enduring changes to microbiome composition, suggesting that each individual's microbiome can achieve multiple alternative states of equilibrium (Flint et al. 2012). Interestingly, the microbiomes of people with inflammatory bowel diseases (Turnbaugh et al. 2006), obesity (Flint et al. 2012), autism (Cryan and Dinan 2012), and several other diseases differ in distinctive ways from the microbiomes of healthy people. These results raise the question of causation: are the differences in microbiome the cause or the consequence of the disease? Research using germ-free animal hosts can help to answer this question. For example, rodent models of human inflammatory bowel disease are asymptomatic when raised under germ-free conditions (Manichanh et al. 2012), and germ-free mice administered a gut microbiome from obese donors gain more weight than mice receiving a gut microbiome from lean donors (Turnbaugh et al. 2006). The human microbiome offers a unique opportunity to explore the fundamental ecological principles underlying the assembly, maintenance, variation, and resilience of communities of symbionts, as well as the specific contributions of these microbial communities to our health.

References

Costello, E.K., Stagaman, K., Dethlefsen, L., Bohannan, B.J., and Relman, D.A. (2012). The application of ecological theory toward an understanding of the human microbiome. *Science*, **336**, 1255–1262.

continued

Box 2.2 *Continued*

Cryan, J.F. and Dinan, T.G. (2012). Mind-altering microorganisms: the impact of the gut microbiota on brain and behaviour. *Nature Reviews Neuroscience*, **13**, 701–712.

Flint, H.J., Scott, K.P., Louis, P., and Duncan, S.H. (2012). The role of the gut microbiota in nutrition and health. *Nature Reviews in Gastroenterology and Hepatology*, **9**, 577–589.

Hooper, L.V., Littman, D.R., and Macpherson, A.J. (2012). Interactions between the microbiota and the immune system. *Science*, **336**, 1268–1273.

Manichanh, C., Borruel, N., Casellas, F., and Guarner, F. (2012). The gut microbiota in IBD. *Nature Reviews in Gastroenterology and Hepatology*, **9**, 599–608.

Qin, J., Li, R., Raes, J., Arumugam, M et al. (2010). A human gut microbial gene catalogue established by metagenomic sequencing. *Nature*, **464**, 59–65.

Sekirov, I., Russell, S.L., Antunes, L. C., and Finlay, B.B. (2010). Gut microbiota in health and disease. *Physiological Reviews* **90**, 859–904.

The Human Microbiome Consortium. (2012). Structure, function and diversity of the healthy human microbiome. *Nature*, **486**, 207–214.

Turnbaugh, P.J., Ley, R.E., Mahowald, M.A., Magrini, V., Mardis, E.R., and Gordon, J.I. (2006). An obesity-associated gut microbiome with increased capacity for energy harvest. *Nature*, **444**, 1027–1131.

Yatsunenko, T., Rey, F.E., Manary, M.J. et al. (2012). Human gut microbiome viewed across age and geography. *Nature* **486**, 222–227.

resolve these uncertainties. This is illustrated by a population-based modeling analysis of the response of gut microbiota to antibiotic treatment (Bucci et al. 2012). This two-group model investigated the antibiotic-induced switch from a microbial community dominated by antibiotic-susceptible bacteria to antibiotic-tolerant bacteria. The tolerant bacteria could continue to dominate the community for long periods after the antibiotic treatment had ended because coexistence between the two groups in the model was unstable. Understanding the resilience of open symbioses to environmental crisis represents an important issue for future research, not only from the academic perspective of understanding the ecology of these symbioses, but also in urgent practical ways: for example, to optimize antibiotic treatment protocols to protect the resident microbiota that positively contribute to human health (see also Box 14.2); and to explain and predict the impacts of climate changes on coral bleaching and coral health.

Not all environmental crises are of external origin. In some associations, environmental crises are a normal part of the development or even the daily life of the host, and the resilience of the association (and the variation in that resilience) is an evolved trait. Notably, developmental disturbance is integral to the biology of gut symbioses in insects. In all insects, the foregut and hindgut epithelium is bounded by a protective chitinous layer that is lost at every larval molt. The extrusion of the chitin skeleton from the gut is a major disturbance for the substantial microbiota in the hindgut of many insects. In termites, the resilience of the resident hindgut microbiota to the habitat disruption at molts is exceptionally low, requiring recolonization from the microbiota of other colony members after every molt. There is virtually no information on the resilience of most other hindgut symbioses in insects, even though variation in resilience to these developmental crises could have major, poorly understood impacts on the fitness of insect developmental stages subsequent to a molt.

A second source of developmental disturbance affects the many insects that undergo complete metamorphosis, with complete dissolution of the gut of the larval insect. In some insects (e.g., ants), the adult gut is apparently not colonized by members of the larval gut microbiota (J. Russell, pers. comm.). In *Drosophila melanogaster*, however, bacterial cells associated with the larval gut are retained, probably associated with the serosal membranes in the puparium, and subsequently infect the adult gut, thus mediating transmission across metamorphosis (Bakula 1969, Ridley et al. 2012). The ecological implications of the variation among insects in adult dependence on an exogenous source of gut microbiota have not been studied.

One symbiosis that is exquisitely resilient to perturbation is the association between the squid *Euprymna scolopes* and luminescent bacteria *Vibrio fischeri*, located in tubules of the squid light organ (see Box 5.3). The light organ is connected to the surrounding seawater via canals, but the association is apparently closed to incoming bacteria subsequent to the initial infection in the juvenile animal. The adult expels 95% of the bacterial cells from the light organ every day at dawn, and the residual bacteria repopulate the light organ during the day, to reach sufficient density for the induction of luminescence in the following night (Wier et al. 2010). The population processes underpinning the daily resilience of this symbiosis are tightly orchestrated by the daily switches in supply of different carbon compounds from the host (Wier et al. 2010).

Daily cycles in proliferation and expulsion of algal symbionts in other marine invertebrates have also been reported (Moran et al. 1993, Martinson et al. 2011, Shade and Handelsman 2012), but the incidence of such diurnal rhythms, and their significance for the persistence of symbioses, have not been investigated widely in other symbioses.

2.4 Partner fidelity and its consequences

Partner fidelity (discussed in more detail in Chapters 4 and 6) refers to the number or duration of interactions between the same two (or more) individuals of the different species participating in the associations. As in other mutualisms, there is considerable variation in the degree of partner fidelity among symbioses. Over the lifetime of an individual organism, partner fidelity tends to be higher in closed associations than open associations. Across generations, partner fidelity is promoted by both processes that promote contact between the partners and recognition events that discriminate among candidate partners. Partner fidelity is particularly promoted by vertical transmission, i.e., where one partner (usually a microbial symbiont) is transferred from parent (usually mother) to offspring. Very commonly, vertically transmitted symbionts are incorporated into the egg, distributed to the target organ or cell during embryogenesis, and the resultant symbiosis is closed to invasion by alternative organisms. Vertical transmission is the norm for many symbioses involving intracellular microorganisms, especially in invertebrate animals (Bright and Bulgheresi 2010), and the incidence and cellular patterns of transmission have been studied in particular detail in insects (Douglas 1989, Degnan et al. 2004).

An instructive approach to investigate partner fidelity is to compare the topology of the phylogeny of the interacting partners (see also Chapter 3 and Box 7.2). Poor phylogenetic congruence is indicative of multiple partner switches over evolutionary time. Although particularly likely in open symbioses with low specificity, partners with phylogenies that are topologically incongruent may routinely display vertical transmission because phylogenetic congruence can be disrupted by infrequent horizontal transmission events (i.e., transfer between unrelated partners). This is illustrated by *Wolbachia* bacteria, whose phylogeny does not match that of its insect hosts, despite a high incidence of vertical transmission (O'Neill et al. 1992, Vavre et al. 1999). Similarly, congruent phylogeny is not necessarily indicative of vertical transmission, and can, in principle, be sustained by efficient horizontal transmission and high specificity. One further issue that requires careful consideration is the time scale of evolutionary diversification in the lineages of the partner organisms. As discussed in Chapter 3, the phylogeny of two organisms can be congruent either because of cospeciation (i.e., a consequence of high partner fidelity) or because the evolutionary diversification of one partner tracks preexisting traits of the other partner that vary with phylogeny (i.e., the cause of high partner fidelity). In summary, phylogenetic analysis of partners in symbioses can play a valuable role in investigating the degree of partner fidelity of symbioses in ecological time, but the phylogenetic approach does require careful interpretation, ideally assisted by information on the biology of the interacting organisms.

One area of active current interest relates to the notion of a core microbiota in open symbioses, such as the gut microbiota of animals. There has been an expectation that a subset of the microbiota is common to all individuals of one host taxon (usually a species), and that this subset coevolves with the host lineage, generating a congruent phylogeny between hosts and microbial community (Hamady and Knight 2009, Shade and Handelsman 2012). Evidence for a core gut microbiota has been obtained in multiple systems (Ochman et al. 2010, Martinson et al. 2011, Roeselers et al. 2011, Brucker and Bordenstein 2012), but a core microbiota appears to be absent from marine sponges, *Drosophila*, humans,

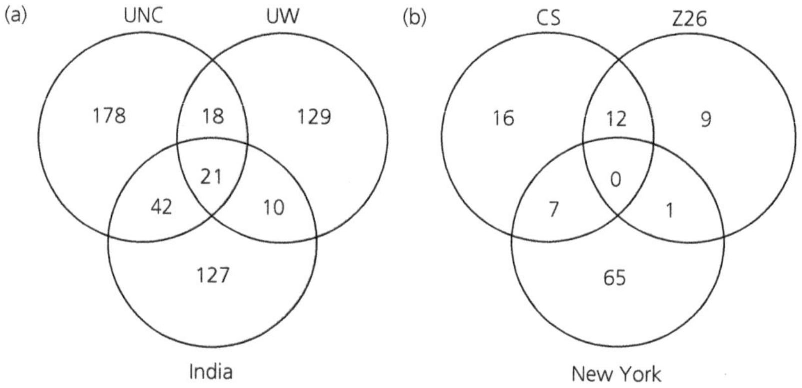

Figure 2.3 A core microbiota can be identified in some, but not all, symbioses, by the criterion of number of bacterial operational taxonomic units (OTUs; 16S rRNA gene sequences at 97% sequence identity cutoff) shared among conspecific animal hosts. (a) Core gut microbiota present in all samples of zebra fish *Danio rerio* collected from the field (India) and maintained in laboratory culture at University of North Carolina (UNC) and University of Washington (UW): the core comprises 21/525 (4%) OTUs shared across the three sets of samples (data from Roeselers et al. 2011). (b) Absence of core gut microbiota of *Drosophila melanogaster*, comprising two strains maintained under uniform conditions in laboratory culture (CS, Canton S; Z26, Zimbabwe 26) and a sample from a wild population collected from apples (New York, USA) in September 2010: none of 110 OTUs were shared among the three datasets (data from Wong et al. 2013).

and other mammals (Nikoh et al. 2010, Human Microbiome Project 2012b, Wong et al. 2013). The reasons for this variation (some examples of which are shown in Figure 2.3) are uncertain. Partner fidelity, especially in open systems, can be constrained by sampling bias (each host contacts a small proportion of the total compatible microbiota) and abrupt changes in the microbiota induced by developmental or environmental factors, such that any one host can interact with multiple alternative taxa.

An informative contrast to the open symbioses with low partner fidelity is provided by the obligately vertically transmitted symbioses, in which each individual host has derived its complement of symbionts from its parent with absolute fidelity, in some systems for tens to hundreds of millions of years. This condition occurs, for example, in the symbioses between various insects and their intracellular microorganisms (Moran et al. 1993, Chen et al. 1999, Degnan et al. 2004, Urban and Cryan 2012). Restriction to the host habitat leads to relaxed selection on genes with function only in the free-living habitat, resulting in gene loss (Hongoh and Ishikawa 2000; see also Chapter 5). This effect is compounded by the accumulation of mildly deleterious mutations, as a result of the relatively small numbers of symbionts transmitted via the egg, and bottlenecking of the symbiont population (Moran 1996, Funk et al. 2001). As a consequence, obligately vertically transmitted symbionts can have much reduced genomes, many with genome size less than 1 Mb and gene content at 20% or less of free-living bacteria. As John McCutcheon describes in Box 2.3, the genome sizes of some symbionts are comparable to those of mitochondria and plastids, raising the possibility that they have made the evolutionary transition from bacterium to organelle.

The function of one obligately vertically transmitted bacterium, *Buchnera aphidicola*, has been studied very intensively from a genomic perspective. Despite its very small genome size, ranging from 0.42 Mb to 0.64 Mb (Shigenobu et al. 2000, Perez-Brocal et al. 2006), *Buchnera* is genetically independent of its aphid host, in the sense that the host nucleus does not bear genes of *Buchnera* origin (Nikoh et al. 2010) and host-derived proteins are not targeted to the *Buchnera* cells (Poliakov et al. 2011). Nevertheless, this bacterium has an absolute metabolic dependence on the insect host because it has complex nutritional requirements and has shared metabolic pathways with the host, involving intermediate metabolites transferred across the host–symbiont interface (Hansen and Moran 2011, Poliakov et al. 2011, Macdonald et al. 2012). It is

Box 2.3 Genomic decay and bacterial-derived organelles

John McCutcheon

A critical event—if not the defining event—in the origin of eukaryotes was the establishment of a stable symbiosis between the proto-eukaryotic cell and a bacterium from the class α-proteobacteria. This symbiosis eventually led to the formation of the mitochondrion, the first of two primary cellular organelles derived from endosymbiotic bacteria (Martin and Kowallik 1999, Margulis 1970). The second endosymbiosis resulting in the establishment of an organelle involved a cyanobacterium, which led to the formation of the plastid, and was the event that defined the origin of all photosynthetic eukaryotes. These two fundamental evolutionary transitions, both involving bacterial endosymbioses, have clear and important implications for life as we know it today. A modern eukaryotic cell is not simply a eukaryotic cell; it is the product of a symbiosis between two or three lineages of organisms, each with its own extant genome.

The coevolution of mitochondria and plastids with their hosts has been occurring for at least a billion years, and these intimate and stable relationships have had profound impacts on their genomes. Most organellar genomes have become very small, and have lost the majority of genes present in their bacterial ancestors. Many genes were lost due to redundancy with the host cell; if a metabolite can be readily obtained from the host, a symbiont no longer needs to retain the machinery to make it. But it is also clear that many organelle-encoded genes were not simply lost but were transferred to the host genome (Timmis et al. 2004). These transferred genes are now under control of the host nucleus, and their gene products are transferred back into mitochondria and plastids via complex transport systems.

Have organelles originated only twice? This of course depends on one's definition of organelle. There is a critical lack of consensus on this issue (Keeling and Archibald 2008), compounded by recent data from two different fields (Nowack et al. 2008, Nakayama and Ishida 2009, McCutcheon and Moran 2012). The first example is from the amoeba *Paulinella chromataphora*, which has independently acquired a photosynthetic cyanobacterium that provides its host with photosynthate, obviating the need to feed on environmental bacteria for nutrition. Similar to mitochondria and chloroplasts, the *Paulinella* symbiont is dependent on its host for survival, and has transferred critical genes to the host nucleus (Nakayam and Ishida 2009). The second example is from the nutritional endosymbionts of sap-feeding insects, several of which have genomes that substantially overlap organelles and viruses in terms of both genome size and number of encoded genes (McCutcheon and Moran 2012). In at least two sap-feeding insects, massive endosymbiotic gene loss is apparently enabled by horizontal transfer of bacterial genes to the host insect genome (Husnik et al. 2013, Sloan et al. 2014). While this is similar to what is observed in mitochondria and plastid evolution, an important difference is that most, but not all (Sloan et al. 2014), of these genes do not come from the resident nutritional endosymbiotic bacteria. Remarkably, most of these horizontally transferred genes have been acquired from various historical bacterial infections. This evolutionary mix-and-match pattern parallels what is seen in mitochondria in certain protists, where the enzymes used to synthesize heme are scattered among several cellular compartments and are pieced together from diverse phylogenetic origins (Kořený et al. 2013).

References

Husnik, F., Nikoh, N., Koga, R et al. (2013). Horizontal gene transfer from diverse bacteria to an insect genome enables a tripartite nested mealybug symbiosis. *Cell*, **153**, 1567–1578.

Keeling, P.J. and Archibald, J.M. (2008). Organelle evolution: what's in a name? *Current Biology*, **18**, R345–R347.

Kořený, L., Oborník, M., and Lukeš, J. (2013). Make it, take it, or leave it: heme metabolism of parasites. *PLoS Pathogens*, **9**, e1003088.

Margulis, L. (1970). Origin of Eukaryotic Cells. New Haven: Yale University Press.

Martin, W. and Kowallik, K.V. (1999). Annotated English translation of Mereschkowsky's 1905 paper "Uber Natur und Ursprung der Chromatophoren im Pflanzenreiche." *European Journal of Phycology*, **34**, 287–295.

McCutcheon, J.P. and Moran, N.A. (2012). Extreme genome reduction in symbiotic bacteria. *Nature Reviews Microbiology*, **10**, 13–26.

Nakayama, T. and Ishida, K.I. (2009). Another acquisition of a primary photosynthetic organelle is underway in *Paulinella chromatophora. Current Biology*, **19**, R284–R285.

Nowack, E.C.M., Melkonian, M., and Glöckner, G. (2008). Chromatophore genome sequence of *Paulinella* sheds light on acquisition of photosynthesis by eukaryotes. *Current Biology*, **18**, 410–418.

Sloan, D.B., Nakabachi, A., Richards, S. et al. (2014). Parallel histories of horizontal gene transfer facilitated extreme reduction of endosymbiont genomes in sap-feeding insects. *Molecular Biology and Evolution*, **31**, 857–871.

Timmis, J.N., Ayliffe, M.A., Huang, C.Y., and Martin, W. (2004). Endosymbiotic gene transfer: organelle genomes forge eukaryotic chromosomes. *Nature Reviews Genetics*, **5**, 123–135.

possible that the host genomes of symbionts with yet smaller genomes bear genes derived from their bacterial symbionts (see Box 2.3).

2.5 Concluding comments

Symbiotic mutualisms display striking parallels and contrasts with non-symbiotic mutualisms. Comparable services (nutrition, protection, and transportation) dominate both symbiotic and non-symbiotic mutualisms, often through tightly reciprocated interactions; and both types of mutualism vary widely in their partner fidelity, in both the physiological time of a single individual and evolutionary time among related species. For example, persistent associations in some open symbioses may involve frequently changing casts of partners that sustain the interactions with their host, while some ancient, closed associations have evolved metabolic or other dependencies involving exquisite sharing of elements in multi-step processes. These persistent interactions over millions of years have resulted in evolutionary trajectories without parallel in non-persistent mutualisms, most notably the evolutionary transition of bacteria to organelles.

These complex and varied symbioses can be considered as ecosystems nested within and interacting with the wider ecosystem in which the association is embedded. We can only reach a full understanding of the conditions and resources in the outer ecosystem that shape the abundance and distribution of a symbiosis by integrating these factors with the dynamics of the inner ecosystem. This is a major challenge for the future of symbiosis research.

References

Aanen, D.K., and Hoekstra, R.F. (2007). The evolution of obligate mutualism: if you can't beat 'em, join 'em. *Trends in Ecology and Evolution*, **22**, 506–509.

Bakula, M. (1969). Persistence of a microbial flora during postembryogenesis of *Drosophila melanogaster*. *Journal of Invertebrate Pathology*, **14**, 365–374.

Bright, M., and Bulgheresi, S. (2010). A complex journey: transmission of microbial symbionts. *Nature Reviews Microbiology*, **8**, 218–230.

Brucker, R.M., and Bordenstein, S.R. (2012). The roles of host evolutionary relationships (genus: *Nasonia*) and development in structuring microbial communities. *Evolution*, **66**, 349–362.

Bucci, V., Bradde, S., Biroli, G., and Xavier, J.B. (2012). Social interaction, noise and antibiotic-mediated switches in the intestinal microbiota. *PLoS Computational Biology*, **8**, e1002497.

Bull, J. J., Molineux, I.J., and Rice, W.R. (1991). Selection of benevolence in a host-parasite system. *Evolution*, **45**, 875–882.

Burriesci, M. S., Raab, T.K., and Pringle, J.R. (2012). Evidence that glucose is the major transferred metabolite in dinoflagellate-cnidarian symbiosis. *Journal of Experimental Biology*, **215**, 3467–3487.

Chen, X., Li, S., and Aksoy, S. (1999). Concordant evolution of a symbiont with its host insect species: molecular phylogeny of genus *Glossina* and its bacteriome-associated endosymbiont, *Wigglesworthia glossinidia*. *Journal of Molecular Evolution*, **48**, 49–58.

Connor, R.C. (1995). The benefits of mutualism: a conceptual framework. *Biological Reviews*, **70**, 427–457.

Day, D.A., Poole, P.S., Tyerman, S.D., and Rosendahl, L. (2001). Ammonia and amino acid transport across symbiotic membranes in nitrogen-fixing legume nodules. *Cellular and Molecular Life Sciences*, **58**, 61–71.

Degnan, P. H., Lazarus, A.B., Brock, C.D., and Wernegreen, J.J. (2004). Host-symbiont stability and fast evolutionary rates in an ant-bacterium association: cospeciation of *Camponotus* species and their endosymbionts, *Candidatus Blochmannia*. *Systematic Biology*, **53**, 95–110.

Diaz-Albiter, H., Sant'Anna, M.R., Genta, F.A., and Dillon, R.J. (2012). Reactive oxygen species-mediated immunity against *Leishmania mexicana* and *Serratia marcescens* in the sand phlebotomine fly *Lutzomyia longipalpis*. *Journal of Biological Chemistry*, **287**, 23995–24003.

Douglas, A.E. (1989). Mycetocyte symbiosis in insects. *Biological Reviews*, **64**, 409–434.

Douglas, A.E. (2010). *The Symbiotic Habit*. Princeton, NJ: Princeton University Press.

Fenchel, T., and Finlay, B.J. (1995). *Ecology and Evolution in Anoxic Worlds*. Oxford: Oxford University Press.

Ferrari, J., and Vavre, F. (2011). Bacterial symbionts in insects or the story of communities affecting communities. *Philosophical Transactions of the Royal Society of London* B, **366**, 1389–1400.

Finlay, B.J., Maberly, S.C., and Esteban, G.F. (1996). Spectacular abundance of ciliates in anoxic pond water: contribution of symbiont photosynthesis to host respiratory oxygen requirements. *FEMS Microbiology Ecology*, **20**, 229–235.

Funk, D.J., Wernegreen, J.J., and Moran, N.A. (2001). Intraspecific variation in symbiont genomes: bottlenecks and the aphid-buchnera association. *Genetics*, **157**, 477–489.

Hamady, M., and Knight, R. (2009). Microbial community profiling for human microbiome projects: tools, techniques, and challenges. *Genome Research*, **19**, 1141–1152.

Hansen, A.K., and Moran, N.A. (2011). Aphid genome expression reveals host-symbiont cooperation in the production of amino acids. *Proceedings of the National Academy of Sciences USA*, **108**, 2849–2854.

Herring, P.J. (2007). Sex with the lights on? A review of bioluminescent sexual dimorphism in the sea. *Journal of the Marine Biological Association UK*, **87**, 829–842.

Hongoh, Y., and Ishikawa, H. (2000). Evolutionary studies on uricases of fungal endosymbionts of aphids and planthoppers. *Journal of Molecular Evolution*, **51**, 265–277.

Human Microbiome Consortium. (2012a). A framework for human microbiome research. *Nature*, **486**, 215–221.

Human Microbiome Consortium. (2012b). Structure, function and diversity of the healthy human microbiome. *Nature*, **486**, 207–214.

Kiers, E.T., Duhamel, M., Beesetty, Y et al. (2011). Reciprocal rewards stabilize cooperation in the mycorrhizal symbiosis. *Science*, **333**, 880–882.

Koropatnick, T.A., Engle, J.T., Apicella, M.A., Stabb, E.V., Goldman, W.E., and McFall-Ngai, M.J. (2004). Microbial factor-mediated development in a host-bacterial mutualism. *Science*, **306**, 1186–1188.

Kremer, N., Voronin, D., Charif, D., Mavingui, P., Mollereau, B., and Vavre, F. (2009). *Wolbachia* interferes with ferritin expression and iron metabolism in insects. *PLoS Pathogens*, **5**, e1000630.

Macdonald, S.J., Lin, G.G., Russell, C.W., Thomas, G.H., and Douglas, A.E. (2012). The central role of the host cell in symbiotic nitrogen metabolism. *Proceedings of the Royal Society of London* B, **279**, 2965–2973.

Malloch, D., and Blackwell, M. (1992). Dispersal of fungal diaspores. In G.C. Carroll and D.T. Wicklow (eds.), *The Fungal Community: Its Organization and Role in the Ecosystem*, pp. 147–171. New York: Marcel Dekker Inc.

Manuel, R. (1988). *British Anthozoa*. London: Linnean Society of London.

Marquez, L.M., Redman, R.S., Rodriguez, R.J., and Roossinck, M.J. (2007). A virus in a fungus in a plant: three-way symbiosis required for thermal tolerance. *Science*, **315**, 513–515.

Martinson, V.G., Danforth, B.N., Minckley, R.L., Rueppell, O., Tingek, S., and Moran, N.A. (2011). A simple and distinctive microbiota associated with honey bees and bumble bees. *Molecular Ecology*, **20**, 619–628.

Montllor, C.B., Maxmen, A., and Purcell, A.H. (2002). Facultative bacterial endosymbionts benefit pea aphids *Acyrthosiphon pisum* under heat stress. *Ecological Entomology*, **27**, 189–195.

Moran, N.A. (1996). Accelerated evolution and Muller's rachet in endosymbiotic bacteria. *Proceedings of the National Academy of Sciences USA*, **93**, 2873–2878.

Moran, N., Munson, M.A., Baumann, P., and Ishikawa, H. (1993). A molecular clock in endosymbiotic bacteria is calibrated using the insect hosts. *Proceedings of the Royal Society of London* B, **253**, 167–171.

Morin, J.G., Harrington, A., Nealson, K., Krieger, N., Baldwin, T.O., and Hastings, J.W. (1975). Light for all reasons: verstility in behavioral repertoire of flashlight fish. *Science*, **190**, 74–76.

Newsham, K.K., Fitter, A.H., and Watkinson, A.R. (1995). Arbuscular mycorrhiza protect an annual grass from root pathogenic fungi in the field. *Journal of Ecology*, **83**, 431–438.

Nikoh, N., McCutcheon, J.P., Kudo, T., Miyagishima, S.Y., Moran, N.A., and Nakabachi, A. (2010). Bacterial genes in the aphid genome: absence of functional gene transfer from *Buchnera* to its host. *PLoS Genetics*, **6**, e1000827.

Ochman, H., Worobey, M., Kuo, C.H. et al. (2010). Evolutionary relationships of wild hominids recapitulated by gut microbial communities. *PLoS Biology*, **8**, e1000546.

Oliver, K.M., Degnan, J.P., Hunter, M.S., and Moran, N.A. (2009). Bacteriophages encode factors required for protection in a symbiotic mutualism. *Science*, **325**, 992–994.

O'Neill, S.L., Giordano, R., Colbert, A.M., Karr, T.L., and Robertson, H.M. (1992). 16S rRNA phylogenetic analysis of the bacterial endosymbionts associated with cytoplasmic incompatibility in insects. *Proceedings of the National Academy of Sciences USA*, **89**, 2699–2702.

Pannebakker, B.A., Loppin, B., Elemans, C.P., Humblot, L., and Vavre, F. (2007). Parasitic inhibition of cell death facilitates symbiosis. *Proceedings of the National Academy of Sciences USA*, **104**, 213–215.

Perez-Brocal, V., Gil, R., Ramos, S. et al. (2006). A small microbial genome: the end of a long symbiotic relationship? *Science*, **314**, 312–313.

Pernice, M., Meibom, A., Van Den Heuvel, A. et al. (2012). A single-cell view of ammonium assimilation in coral-dinoflagellate symbiosis. *The ISME Journal*, **6**, 1314–1324.

Piel, J. (2002). A polyketide synthase-peptide synthetase gene cluster from an uncultured bacterial symbiont of *Paederus* beetles. *Proceedings of the National Academy of Sciences USA*, **99**, 14002–14007.

Pierce, N.E., Braby, M.F., Heath, A. et al. (2002). The ecology and evolution of ant association in the *Lycaenidae* (Lepidoptera). *Annual Reviews of Entomology*, **47**, 733–771.

Poliakov, A., Russell, C.W., Ponnala, L. et al. (2011). Large-scale label-free quantitative proteomics of the pea aphid-*Buchnera* symbiosis. *Molecular and Cellular Proteomics*, **10**, M110 007039.

Read, D.J. (1991). Mycorrhizas in ecosystems. *Experientia*, **47**, 376–390.

Ridley, E.V., Wong, A.C., Westmiller, S., and Douglas, A.E. (2012). Impact of the resident microbiota on the nutritional phenotype of *Drosophila melanogaster*. *PLoS One*, **7**, e36765.

Roeselers, G., Mittge, E.K.,Stephens, W.Z. et al. (2011). Evidence for a core gut microbiota in the zebrafish. *The ISME Journal*, **5**, 1595–1608.

Sachs, J.L., and Wilcox, T.P. (2006). A shift to parasitism in the jellyfish symbiont *Symbiodinium microadriaticum*. *Proceedings of the Royal Society of London* B, **273**, 425–429.

Saikkonen, K., Lehtonen, P., Helander, M., Koricheva, J., and Faeth, S.H. (2006). Model systems in ecology: dissecting the endophyte-grass literature. *Trends in Plant Science*, **11**, 428–433.

Sakai, S. (2002). A review of brood-site pollination mutualism: plants providing breeding sites for their pollinators. *Journal of Plant Research*, **115**, 161–168.

Sekirov, I., Russell, S.L., Antunes, L.C., and Finlay, B.B. (2010). Gut microbiota in health and disease. *Physiological Reviews*, **90**, 859–904.

Shade, A., and Handelsman, J. (2012). Beyond the Venn diagram: the hunt for a core microbiome. *Environmental Microbiology*, **14**, 4–12.

Sharples, J.M., Meharg, A.A., Chambers, S.M., and Cairney, J.W. (2000). Symbiotic solution to arsenic contamination. *Nature*, **404**, 951–952.

Shigenobu, S., Watanabe, H., Hattori, M., Sakaki, Y., and Ishikawa, H. (2000). Genome sequence of the endocellular bacterial symbiont of aphids *Buchnera sp*. APS. *Nature*, **407**, 81–86.

Shin, S. C., Kim, S.H., and You, H. (2011). *Drosophila* microbiome modulates host developmental and metabolic homeostasis via insulin signaling. *Science*, **334**, 670–674.

Sonke, J., Widder, E.A., and Mobley, C.D. (2004). Propagation and perception of bioluminescence: factors affecting counterillumination as a cryptic strategy. *Biological Bulletin*, **207**, 1–16.

Urban, J.M., and Cryan, J.R. (2012). Two ancient bacterial endosymbionts have coevolved with the planthoppers (Insecta: Hemiptera: Fulgoroidea). *BMC Evolutionary Biology*, **12**, 87.

Vavre, F., Fleury, F., Lepetit, D. Fouillet, P., and Bouletreau, M. (1999). Phylogenetic evidence for horizontal transmission of *Wolbachia* in host-parasitoid associations. *Molecular Biology and Evolution*, **16**, 1711–1723.

Wier, A.M., Nyholm, S.V., Mandel, M.J. et al. (2010). Transcriptional patterns in both host and bacterium underlie a daily rhythm of anatomical and metabolic change in a beneficial symbiosis. *Proceedings of the National Academy of Sciences USA*, **107**, 2259–2264.

Wong, C.N., Chaston, J.M., and Douglas, A.E. (2013). The inconstant gut microbiota of *Drosophila* species revealed by 16S rRNA gene analysis. *The ISME Journal*, **7**, 1922–1932.

Wong, J.M., de Souza, R., Kendall, C.W., Emam, A., and Jenkins, D.J. (2006). Colonic health: fermentation and short chain fatty acids. *Journal of Clinical Gastroenterology*, **40**, 235–243.

SECTION 2

Introduction to Section 2

Judith L. Bronstein

Charles Darwin clearly articulated the challenge of understanding how beneficial interactions, both within and between species, can evolve (Chapter 1). One might therefore have expected to encounter mutualism as an early target of study by evolutionary biologists. This did not happen. With the exception of some prescient early writings (Chapter 1), minimal attention was paid to the genetics, evolution, or diversification of mutualism until the second half of the twentieth century. To the extent that these topics were addressed at all, it generally involved adopting theory or extrapolating empirical conclusions from studies of other interactions, particularly within-species cooperation and among-species antagonism.

There was no single moment at which the study of mutualism evolution took off, but we can perhaps identify two central events. The first was the birth of the concept of coevolution, described in general outline by Darwin, developed in the late 1950s by plant pathologists, then elaborated further in the 1960s for plant–herbivore interactions. Daniel Janzen then seized upon the concept in the late 1960s in his studies of highly specialized mutualisms between ants and acacias. Janzen demonstrated that coevolution could be framed as a set of field-testable hypotheses. His insights forged a bridge between studies of the ecology and evolution of mutualism (particularly evident in Chapter 3) that remains a hallmark of this field to the present day. A rich diversity of coevolved mutualisms has been highlighted in many chapters and much of the boxed material in this volume; indeed, they include many of our most seductive examples of cooperation in nature. Mutualistic coevolution is treated in depth in Chapter 7 by Bruce Anderson. He outlines the diversity of processes by which coevolution occurs, as well as sets of conditions and types of mutualisms in which it would not be expected. He also covers the patterns coevolution can produce. These include the well-known geographic mosaic developed in the influential work of John Thompson, as well as trait mismatches, an intriguing phenomenon given a theoretical footing by Scott Nuismer in Box 7.1. Phylogenetic analysis holds great promise for studying coevolution, although as both Anderson and Kari Segraves in Box 7.2 caution, it can be deceptively easy to interpret cophylogenies as saying more about coevolution than they actually do.

The second key event in the history of this field was the development of game theory and its extension into mutualism. The formative paper was Axelrod and Hamilton's 1981 classic, "The evolution of cooperation." As discussed in Chapter 4 by Erol Akçay and Chapter 6 by Joel Sachs, their simple and intuitive model, which suggested that any type of cooperation should have an exceedingly difficult time persisting in the face of a "temptation to cheat," led to an outpouring of research aimed at explaining how it can be so common in nature regardless. Akçay argues that understanding how cheating is coped with is the start of essentially all evolutionary theory in this field. He provides a unique and thorough review of the questions that have been asked about how mutualism arises and the conditions under which it can persist, as well as the theoretical tools available to investigate these issues. Sachs explores cheating in mutualism in great depth from a theoretical angle, but also an empirical one: attention to the "cheater problem" led to an outpouring of work devoted to determining who cheaters are, from where they arise, and the conditions under which they persist. Two of the best-understood and most influential model systems involve (a) plants, their highly specialized ant defenders, and ants that provide no protection (discussed in Box 6.1 by Megan Frederickson), and (b) fish that exploit the mutualism between cleaners and their hosts (Box 6.2 by Redouan Bshary). The former system has inspired ecological theory on cheating as well: for example, different aspects of mutualist–cheater coexistence in ant–plant systems are modeled by Daniel Doak in Box 9.2 and by Charlotte Lee, Brian Inouye, and Tom Miller in Box 9.3.

Where does mutualism come from, and what happens next? Two chapters in this section take a microevolutionary (Chapter 5) and macroevolutionary (Chapter 3) perspective on these challenging questions. In Chapter 5, Carina Baskett and Douglas Schemske focus on how traits involved in mutualisms can be expected to arise and to evolve differently from those relevant in interactions with the abiotic environment. They stress how little we know at present about the genetic basis of mutualism and of mutualistic traits. A few model systems, though, are providing rich and exciting insights into mutualism evolution: the fungus-growing ant mutualism (Box 5.1 by Ulrich Mueller); legumes and their nitrogen-fixing bacteria (Box 5.2 by Maren Friesen; see also Box 4.2 by Stephanie Porter and Ellen Simms); and squid and their bioluminescent bacteria (Box 5.3 by Michele Nishiguchi). Chapter 3 by George Weiblen and Erin Treiber looks at the evolution and diversification of mutualism and mutualistic traits in deep time. The first review to our knowledge on this topic, this chapter offers strong evidence that mutualism has many origins that have often, but not always, resulted in adaptive diversification.

A wide diversity of tools and sources of information, many of them quite new, are being applied to understand the genetics, evolution, and diversification of mutualism. In addition to new theoretical approaches such as adaptive dynamics (highlighted by Erol Akçay in Box 4.1) are empirical data emerging from the fossil record (Box 3.1 by Peter Wilf and Conrad Labandeira), phylogenetic analysis (Box 3.2 by Jiri Hulcr and Box 7.2 by Kari Segraves), and genomics (Box 7.3 by Noah Whiteman). This fusion of new concepts developed specifically for mutualism and new approaches with which to explore them suggests that the day may not be far off when we can recognize a more unified theory of mutualism evolution.

CHAPTER 3

Evolutionary origins and diversification of mutualism

George D. Weiblen and Erin L. Treiber

3.1 Webs in the tree of life

In this chapter, we describe how investigating the past may contribute to our understanding of the evolutionary and ecological processes that shape mutualism. Species interactions are considered to have the potential to generate new ecological opportunities that might have fostered the diversification of lineages in the past (Farrell et al. 1992), and mutually beneficial partnerships are occasionally cited to account for the rise of particular lineages to ecological dominance. Ancient mutualisms, for example the associations of land plants with ectomycorrhizal fungi (Remy et al. 1994) or animal pollinators, seem like plausible candidates for diversification in partnership. Although insect pollination mutualisms have long been implicated in flowering plant diversification (Darwin 1862), insect lineages comprising the contemporary pollinators of flowering plants were already highly diverse prior to the origin of angiosperms (Labandeira and Sepkoski 1993). This example illustrates how a phylogenetic perspective is needed to understand whether or how past mutualisms might have shaped contemporary patterns of species diversity. In this chapter, we aim to interpret the kinds of evidence that suggest when mutualisms appeared and how beneficial partnerships might have tipped the balance of speciation and extinction to promote diversification.

Figure 3.1 illustrates an extreme case of diversification in partnership (Cruaud et al. 2012). Molecular phylogenies and ecological associations for approximately two hundred closely related plant species and their specialized, obligate pollinators are depicted without taxonomic names so that we might focus our attention instead on general problems of interpretation. A glance at this web in the tree of life suggests the parallel diversification of plant and pollinator clades, as well as instances in which ancient pollinators might have diversified after switching from one host plant lineage to another. For example, a lineage of seven plant species at the upper left is pollinated exclusively by a particular insect lineage, and congruent branching patterns in the two phylogenies suggests diversification in partnership (see arrows in Figure 3.1). However, the plant lineage at the upper left descends from the base of the tree and appears to be more ancient than the associated pollinator lineage. Regardless of whether more recent diversification catalyzed by host shifts or ancient codiversification (parallel cladogenesis) might have dominated the history of this mutualism, the multiplicity of lineages engaged in partnership is impressive. In fact, most evidence for extensive patterns of codiversification involves exceptionally specialized and obligate mutualisms. Might this reflect a coincidence, a bias in sampling, or a biological reality?

This chapter examines this question by comparing evidence for origins and diversification across the spectrum of species interactions, from mutualism to antagonism. We regard mutualisms as reciprocal exploitations that nonetheless provide benefits

Mutualism. Edited by Judith L. Bronstein.

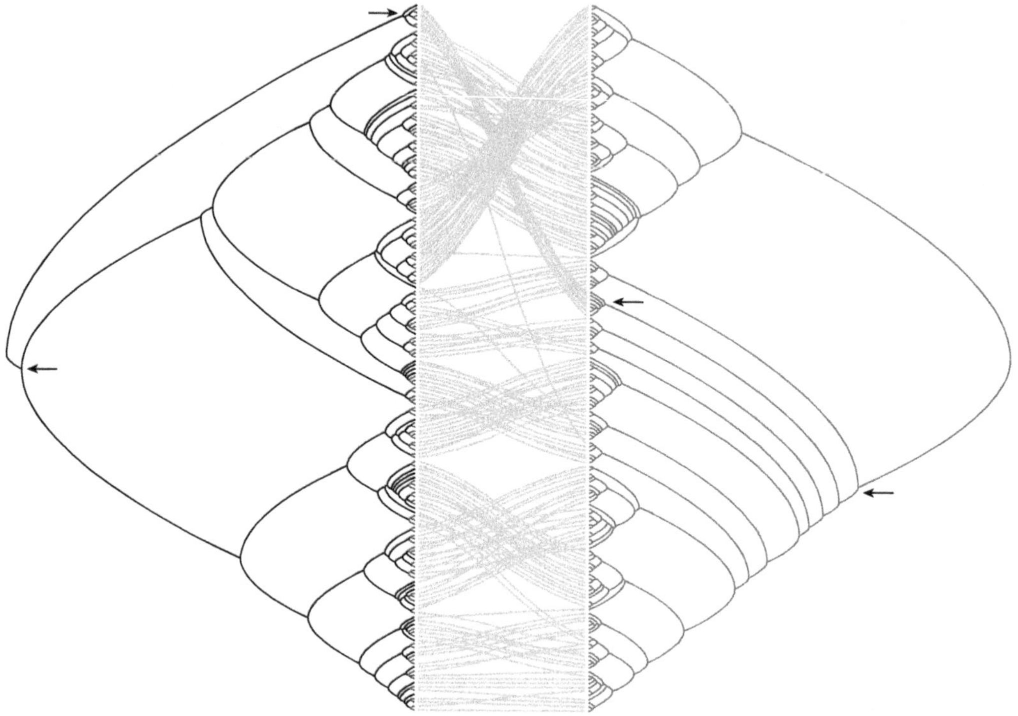

Figure 3.1 Cophylogeny of figs and fig wasps redrawn from the supplementary information in Cruaud et al. (2012). A molecular phylogeny estimate for 186 *Ficus* species (Moraceae) based on five genes (at left) is paired with an independent phylogeny estimate for fig-pollinating wasps (Agaonidae) based on six genes (at right) with observed species-specific associations (middle). Arrows point to a pair of congruent figs and pollinator clades that appear to have originated at different times.

to each partner, and we distinguish mutualism from cooperation, where the former is reserved for positive interactions among different species while the latter refers to positive interactions among individuals of the same species (Boucher 1992; see Chapter 1). The exact nature of associations among species may vary at the level of populations, over time, and according to ecological context, so that classifying particular species interactions only as mutualistic, antagonistic, commensal, etc. is not always possible (Thompson 1988). However, analyzing such variation in light of genetic change within species populations, phylogenetic patterns among lineages, and fossil evidence may shed light on evolutionary conditions that supported the origin and diversification of mutualism. The objective of this chapter is to review approaches, limitations, and opportunities for identifying such conditions from a macroevolutionary perspective.

3.2 Investigating origin and diversification

3.2.1 The fossil record

Paleontology provides a compelling, but largely incomplete, record of mutualism that serves to illustrate the kinds of insights that can be gained about species interactions from the history of life. Wilf and Labandeira (Box 3.1) describe how fossil evidence is compared with contemporary ecology to shed light on the early history and persistence of mutualisms over time. For example, fossils point to ancient origins of mutualism at least 1.5 bya in the case of mitochondria, 450 mya for mycorrhizal symbiosis, and 250 mya for pollination of seed plants by insects. The range of mutualisms for which we have a paleontological record remains rather narrow, and simultaneous preservation of partners is so rare that interactions are typically inferred from other sources

Box 3.1 The fossil record of mutualisms

Peter Wilf and
Conrad C. Labandeira

Interspecies interactions are well represented in the fossil record, especially those involving plants and insects. Most of the known associations were antagonistic, as the voluminous record of insect herbivory over the past ~410 million years notably shows (Labandeira 2002). Several interactions show remarkable levels of early behavioral specialization by insects. These include Middle Jurassic leaf mimicry (Wang et al. 2012) and rolled-leaf hispine beetle feeding on gingers, an extant association, from the late Cretaceous (Wilf et al. 2000).

Well-supported evidence for mutually beneficial associations is rare but extremely informative. Endosymbiosis in macroorganisms is ancient. Many species in the oldest diverse animal faunas, from the Ediacaran Period (631–541 Ma), were large, flattened, and frondose, implicating photosynthetic endosymbiosis that was perhaps analogous to the mid-Mesozoic origin of zooxanthellae symbioses in corals. Arbuscular mycorrhizae are found through the entire record of land plants when preservation is sufficient, starting with the ~410 Ma Rhynie Chert (Remy et al. 1994). This deposit also contains anatomically preserved lichens that represent cyanobacterial-algal symbioses (Taylor et al. 1997).

Pollination is one of the most important mutualisms and probably accounts for the evolution of a large share of today's plant and insect diversity. Pollination probably evolved through the consumption of spores, pollen, and exudates from plant reproductive structures such as pollen drops (Labandeira 2002). The large prepollen of many late Paleozoic seed plants exceeded the size range typically transported by wind and indicates a likely role for entomophily (Taylor and Millay 1979). Insect guts containing prepollen and pollen date to the Early Permian (Krassilov and Rasnitsyn 1997) and are well documented from the mid-Mesozoic (Labandeira 2002). Pollination of cycads and bennettites ("cycadeoids") by beetles was presumably widespread during the Jurassic, based on beetle pollination of many living cycads and extensive borings found in fossil bennettite cones (Crepet 1974). Mid-Mesozoic true flies, scorpionflies (Box 3.1, Figure 1), and lacewings bore prominent siphonate proboscides that in all likelihood functioned to access fluid rewards. Mesozoic gymnosperms offered these rewards through angiosperm-like reproductive structures with recessed ovules, which

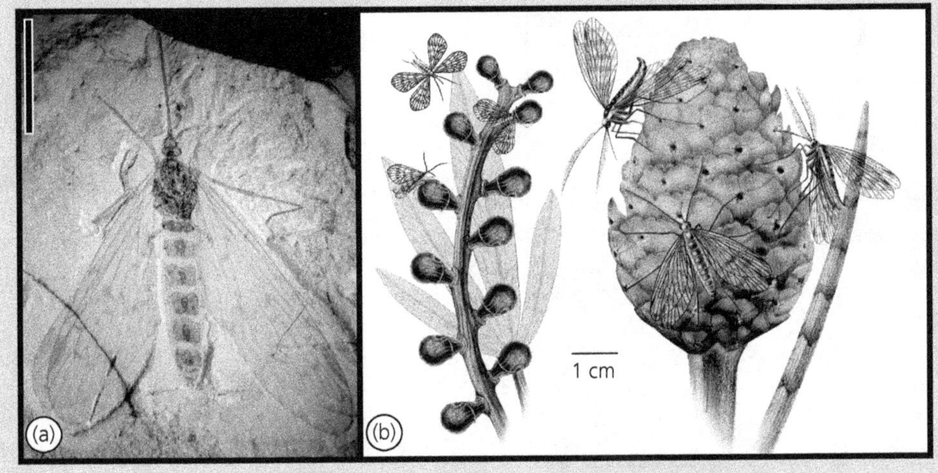

Figure 1 Selected paleontological reconstructions of mutualism (Labandeira 2010). (a) The long-proboscate mesopsychid scorpionfly *Lichnomesopsyche gloriae* from the Middle Jurassic of Inner Mongolia, China, a probable pollinator of some gymnospermous plants. (b) Reconstructions of two mid-Mesozoic pollinator associations between mesopsychid scorpionflies and their inferred host plants. At left, *L. gloriae* feeds on the pollen drops of *Caytonia sewardi* via an integumental tube under the lower lip of the cupule. At right is another mesopsychid, *Vitimopsyche kozlovi*, from the mid Early Cretaceous of Liaoning, China, feeding on catchment funnel nectaries from the ovulate cone of a cheirolepidiaceous conifer and achieving pollination in the process. Original art by Mary Parrish (See Plate 1 for color version).

continued

Box 3.1 *Continued*

were often connected to the outer female organ surfaces by tubular or channeled structures. These lineages, representing three unrelated orders of insects, constituted a long-proboscid, mid-Mesozoic guild of likely pollinating insects prior to the significant ecological presence of angiosperms (Labandeira 2010).

From Early Cretaceous amber, female thrips are known with specialized collecting structures (ring setae) that amassed large loads of gymnosperm pollen, strongly indicating pollination (Peñalver et al. 2006), as well as an ancestral bee species that harbored another distinctive specialization (branched hairs) for accessing pollen (Danforth and Poinar 2011). By the early Late Cretaceous, there is significant evidence for specialized pollination of angiosperms by diverse insects, inferred primarily from fossil flowers that show nectaries, resin ducts, bilateral symmetry, and pollen often presented as multiple attached grains and bearing sticky surfaces (Crepet 1996). Other anthophilic Cretaceous insects included basal moths, several lineages of small, pollinating beetles, and the earliest occurrence of a modern bee, a stingless honey bee worker from one of the deposits that yielded much of the floral evidence for pollination (Michener and Grimaldi 1988). During the Paleogene, several groups of pollinating flies are known, such as long-proboscate, hovering tanglevein and flower flies, and many other specialized groups including diverse bees and butterflies. One of the most celebrated pollination mutualisms as emphasized in this chapter, figs and fig wasps, has a fossil record starting from 34 Ma, consisting of spectacularly preserved female wasps with attached fig pollen (Box 3.1, Figure 2; Compton et al. 2010, Peñalver et al. 2006).

Figure 2 Fossil evidence of mutualism (Compton et al. 2010). (a) Holotype of the 34-million-year-old agaonid wasp, *"Ponera" minuta*, from the late Eocene of the Isle of Wight, United Kingdom (scale bar, 500 μm). (b) The pollen pocket, located on the anterior thorax, is enlarged under SEM, showing lodged *Ficus* pollen (scale bar, 20 μm). Used with permission from Royal Society Publishing and The Natural History Museum, London.

Seed dispersal is perhaps the most significant mutualism involving vertebrates, in all likelihood accounting for significant diversifications in frugivores and plants. Some Paleozoic wetland plants had large, robust seeds, up to ca. 11 cm long. These seeds had large nucelli that probably were nutritious, although no likely tetrapod dispersers are known. However, fish were well diversified and had reached large body sizes by this time; speculatively, fish may have played a dispersal role, especially in flooded forests (see also Tiffney 1986). There is a similar lack of direct evidence for dispersers of Mesozoic seed plants, but it seems quite likely that many groups, especially those with large seeds such as cycads, ginkgophytes, and some conifers, provided forage to the diverse large-bodied reptiles of this interval. Large angiosperm fruits and seeds do not become significant in the record until after the end-Cretaceous extinction (66 Ma), along with modern birds and characteristically frugivorous, arboreal mammals (Wing and Tiffney 1987). Rare Eocene conifers also exhibited large, fleshy, specialized cone receptacles identical to those of living podocarps with bird-dispersed seeds (Wilf 2012). Eocene mammalian frugivory is spectacularly displayed by body fossils with gut contents and coprolites from the Messel oil shales of Germany (Schaal and Ziegler 1992). More recently, grasslands represent a mutualism with vertebrate herbivores at a global scale, to the extent that herbivores participate in maintaining biome structure and composition through grazing, disturbance, and consequent selective pressures that favor particular plant lineages and adaptations. Widespread grasslands and characteristic vertebrate adaptations were present by the late Oligocene to Miocene on many continents, but the existence of older grassland biomes is contentious. It is of considerable interest that in Patagonia, the presence of high-crowned teeth (hypsodonty) in diverse mammal

Box 3.1 *Continued*

clades by 26 Ma and earlier has long been taken as evidence for precocious South American grasslands, but recent pollen data show that grasses were not abundant there before about 2 Ma (Palazzesi and Barreda 2012).

The fossil record of interactions is, of course, incomplete, and inferences about the nature of past associations, whether mutualistic or antagonistic, require multiple lines of evidence. It is even more challenging to gain insight from the record as to whether or how past associations might have influenced the diversification of lineages in partnership. Although many extant associations are preserved, extinct associations become more prevalent with geologic age, especially before the Cenozoic. Beyond establishing minimum ages, the value of particular fossil occurrences for broader evolutionary studies, such as their widespread use in anchoring phylogenetic analyses of cladogenesis through time, depends on many factors and is dramatically improved with well-understood geologic context and absolute dating of the fossils, temporally and spatially dense sampling, and rigorous use of morphological characters to support identifications and behavioral interpretations.

References

Compton, S.G., Ball, A.D., Collinson, M.E., Hayes, P., Rasnitsyn, A.P., and Ross, A.J. (2010). Ancient fig wasps indicate at least 34 Myr of stasis in their mutualism with fig trees. *Biology Letters*, **6**, 838–842.

Crepet, W.L. (1974). Investigations of North American cycadeoids: the reproductive biology of Cycadeoidea. *Palaeontographica Abteilung*, **148B**, 144–169.

Crepet, W.L. (1996). Timing in the evolution of derived floral characters: Upper Cretaceous (Turonian) taxa with tricolpate and tricolpate-derived pollen. *Review of Palaeobotany and Palynology*, **90**, 339–359.

Danforth, B.N., and Poinar, G.O. (2011). Morphology, classification, and antiquity of *Melittosphex burmensis* (Apoidea: Melittosphecidae) and implications for early bee evolution. *Journal of Paleontology*, **85**, 882–891.

Krassilov, V.A., and Rasnitsyn, A.P. (1997). Pollen in the guts of Permian insects: first evidence of pollinivory and its evolutionary significance. *Lethaia*, **29**, 369–372.

Labandeira, C.C. (2002). The history of associations between plants and animals. In C.M. Herrera and O. Pellmyr (eds.), *Plant-Animal Interactions: An Evolutionary Approach*, pp. 26–74. Oxford, UK: Blackwell Science.

Labandeira, C.C. (2010). The pollination of mid Mesozoic seed plants and the early history of long-proboscid insects. *Annals of the Missouri Botanical Garden*, **97**, 469–513.

Michener, C.D., and Grimaldi, D.A. (1988). The oldest fossil bee: apoid history, evolutionary stasis, and antiquity of social behavior. *Proceedings of the National Academy of Sciences USA*, **85**, 6424–6426.

Palazzesi, L., and Barreda, V. (2012). Fossil pollen records reveal a late rise of open-habitat ecosystems in Patagonia. *Nature Communications*, **3**, 1294.

Peñalver, E., Engel, M.S., and Grimaldi, D.A. (2006). Fig wasps in Dominican amber (Hymenoptera: Agaonidae). *American Museum Novitates*, **3541**, 1–16.

Remy, W., Taylor, T.N., Hass, H., and Kerp, H. (1994). Four hundred-million-year-old vesicular arbuscular mycorrhizae. *Proceedings of the National Academy of Sciences USA*, **91**, 11841–11843.

Schaal, S., and Ziegler, W. (1992). *Messel: An Insight into the History of Life and of the Earth*. Oxford: Clarendon Press.

Taylor, T.N., Hass, H., and Kerp, H. (1997). A cyanolichen from the lower Devonian Rhynie Chert. *American Journal of Botany*, **84**, 992–1004.

Taylor, T.N., and Millay, M.A. (1979). Pollination biology and reproduction in early seed plants. *Review of Palaeobotany and Palynology*, **27**, 329–355.

Tiffney, B.H. (1986). Evolution of seed dispersal syndromes according to the fossil record. In D.R. Murray (ed.), *Seed Dispersal*, pp. 273–305. Sydney: Academic Press.

Wang, Y., Labandeira, C.C., Shih, C., Ding, Q., Wang, C., Zhao, Y., and Ren, D. (2012). Jurassic mimicry between a hangingfly and a ginkgo from China. *Proceedings of the National Academy of Sciences USA*, **109**, 20514–20519.

Wilf, P. (2012). Rainforest conifers of Eocene Patagonia: attached cones and foliage of the extant Southeast-Asian and Australasian genus *Dacrycarpus* (Podocarpaceae). *American Journal of Botany*, **99**, 562–584.

Wilf, P., Labandeira, C.C., Kress, W.J., Staines, C.L., Windsor, D.M., Allen, A.L., and Johnson, K.R. (2000). Timing the radiations of leaf beetles: hispines on gingers from latest Cretaceous to recent. *Science*, **289**, 291–294.

Wing, S.L., and Tiffney, B.H. (1987). The reciprocal interaction of angiosperm evolution and tetrapod herbivory. *Review of Palaeobotany and Palynology*, **50**, 179–210.

of evidence. When morphological, anatomical, or chemical features resembling those of a contemporary mutualism are observed in the fossil record, it is postulated that ancestors may have interacted in a similar manner. For example, evidence that insects pollinated early seed plants includes the simultaneous occurrence of elongate insect proboscides and tubular channels in gymnosperm cones during the Mesozoic (Figure 1 in Box 3.1). Although the coexistence of potential partners can sometimes be firmly established by paleontological records, fossils are for the most part silent about past ecology such as whether ancient mutualisms were facultative or obligate. At the same time, fossils are indispensable for calibrating molecular phylogenies based on DNA sequence divergence to estimate dates of origin (Arbogast et al. 2002) and to infer the evolution of morphological traits (Donoghue et al. 1989).

3.2.2 Phylogeny estimation

The earliest phylogenetic studies of mutualism did not consider the evolutionary history of both partners simultaneously, as in Figure 3.1, but instead examined the phylogeny of one partner in relation to the mutualistic traits or associations of the other (Remy et al. 1994, Pellmyr and Thompson 1992, Hibbett et al. 2000). We begin by considering the challenges of estimating phylogeny and inferring ancestral conditions in the first place, before examining what further insight might be gained by comparing the phylogenies of interacting lineages. The tendency for descendants to carry on as their ancestors once did (i.e., phylogenetic conservatism) offers hope that we might be able to identify the origin of a mutualism in the tree of life by mapping the distribution of traits and species associations on a phylogeny. In the case of land plants, for example, the similar anatomy of fossilized roots and contemporary plants exhibiting ectomycorrhizal associations (Remy et al. 1994) suggests that this mutualism dates from the Ordovician period. Not so long ago, simple parsimony was the primary vehicle for inferring origins of mutualism (Pellmyr et al. 1996). More recent advances in phylogenetic theory and computation have fueled an explosion of new statistical and analytical methods that expand the potential to reconstruct histories of mutualism. For example, maximum likelihood estimation can be used to evaluate the probability of ancestral mutualism given a phylogeny and models of evolutionary transition among alternative conditions such as parasitism or commensalism. Hibbett et al. (2000) applied such a model to suggest that ectomycorrhizal symbioses involving basidiomycete fungi have been unstable and dynamic since the Ordovician, considering that the ectomycorrhizal condition appears to have arisen repeatedly from ancestors that derived their nutrition from dead, decaying wood (saprophytes). One advantage of this approach over parsimony has to do with how uncertainty about ancestral states is handled. A maximum likelihood estimate explicitly models the probabilities of any and all possible ancestral scenarios, however improbable some scenarios might appear to be. Baum and Smith (2013) provide an accessible introduction to these methods while their practical application and theoretical development are treated by Lemey et al. (2009) and Felsenstein (2004) respectively.

Bayesian approaches have become especially popular, wherein an evolutionary model, DNA sequences, dates from the fossil record, and other prior information are used simultaneously to estimate phylogeny, divergence times, and model parameters. The posterior probability distributions that result from Bayesian inference can be used to statistically test alternative hypotheses about the evolution of interactions. For example, a recent analysis of diverse fungal life histories found it significantly more probable that fungi feeding exclusively on dead wood (saprophytes) evolved from pathogens that infect and destroy living plant tissues or from the commensal associates of foliage (endophytes) than vice versa (Arnold et al. 2009). Uncertainty surrounding such inferences is often sufficiently large that alternative scenarios cannot be rejected. Much of the explanatory zeal that characterized the early literature on coevolution of mutualism (Chapter 7) is likely to be tempered as more sophisticated modeling grapples with the complexity and uncertainty surrounding past events.

When interactions among species are highly specialized and pairwise, it may be possible to

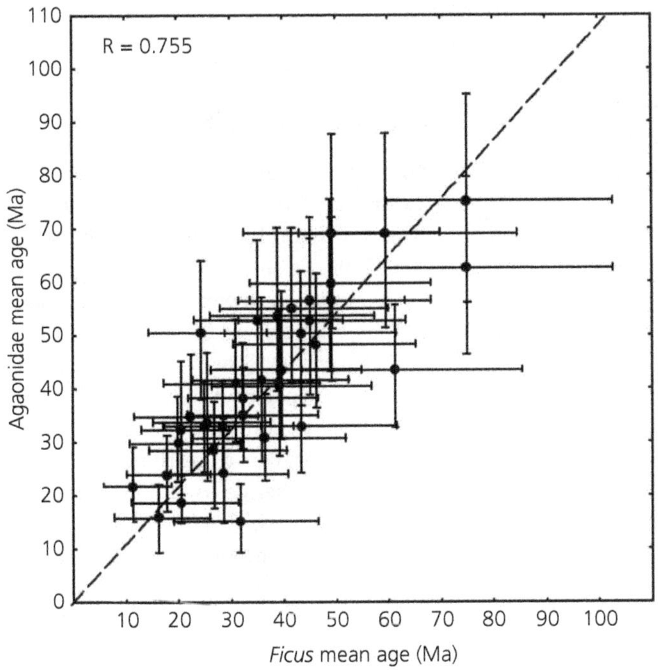

Figure 3.2 Correlated divergence times of figs and pollinating fig wasps (Cruaud et al. 2012). Error bars indicate 95% confidence intervals for the minimum ages of plant lineages that are uniquely associated with particular pollinator lineages. Used with permission from Oxford University Press.

compare independent fossil-calibrated estimates of molecular divergence times for lineages in partnership (Figure 3.2). Such "double-dating" evaluates the alternative hypotheses of codiversification and asynchronous speciation. For example, Silvieus et al. (2008) compared divergence times among fig trees, obligate pollinators, and non-pollinating seed-eating parasites, which can be nearly as host-specific as the pollinators of certain lineages of figs. Age estimates for clades of fig species and their wasp pollinators supported codiversification, while the timing of speciation in parasites appeared to be asynchronous with respect to host plant speciation. In this case, unique insight on diversification was obtained through the comparison of contrasting life histories in a similar ecological context (e.g., pollinators and parasites associated with the same host). However, this approach is not applicable to a wide array of mutualisms that are either facultative or generalized on at least one side of the interaction. It is possible that partnerships often arise, persist, or dissolve with little or no impact on the diversification of either lineage. A number of other problems with "double-dating" are noteworthy. First, as illustrated by the large error bars in Figure 3.2, uncertainty associated with age estimates reduces our power to reject the null hypothesis of asynchronous divergence. Second, relatively weak permutation tests are employed instead of linear models because age estimates for hierarchically nested clades are autocorrelated. Lastly, accurate and independent fossil evidence or geological dates for partner lineages are often unavailable (see Box 3.1).

Phylogenetic approaches to identifying shifts in diversification rate are also popular (Paradis 2011). Maximum likelihood models of speciation and extinction (Stadler 2010) have been employed to suggest that mutualism promotes diversification across a broad spectrum of interactions (Litsios et al. 2012, Gomez and Verdu 2012, Stireman et al. 2010). Examples of symbiotic dinoflagellates, mycorrhizal fungi, gut endosymbionts, cleaner fish, pollinators, and seed dispersers will serve to illustrate this central point in Section 3.4. We will also aim to describe how models that estimate diversification rates and patterns of character evolution simultaneously (Maddison et al. 2007) provide further opportunity to test the hypothesis of mutualism as a driver of diversity. For example, are gains or losses of mutualistic traits associated with shifts in rates of speciation or extinction? In rare circumstances, it may even be possible to identify the genetic changes responsible for such shifts (Chapter 5). Wolfe et al. (2014) associated the origin and expansion of ectomycorrhizal symbiosis in agaric mushrooms with the loss of the particular metabolic pathway that is required to derive nutrition from decaying wood.

Beyond phylogeny and the fossil record, the ecology of contemporary interactions and actual observations of evolution in species populations must also inform our interpretation of the past (Boucher et al. 1982). A synthetic understanding of the causes and consequences of mutualism will require a view of history that is consistent with microevolutionary processes (Chapter 5) and experimental evidence. In the remainder of this chapter, we aim to explore the extent of macroevolutionary patterns with this broader goal in mind.

3.3 Origins of mutualism

We have outlined how the fossil record and phylogeny can establish minimum ages for lineages in partnership and trigger speculation on the origin of mutualism. For example, the origin of land plants has been hypothesized to involve a fungal symbiosis that enabled plants to contend with the terrestrial challenge of efficiently acquiring mineral nutrients while avoiding desiccation (Pirozynski and Malloch 1975). The symbiotic functions implied by this hypothesis are based on the similar appearance of fungal hyphae in fossilized early land plants when compared to those found in contemporary ectomycorrhizae. That some reference to modern ecology is nearly always invoked in such historical interpretations raises the question of whether we could ever recognize the origin of a mutualism with no modern analog. Nonetheless, phylogenetic evidence that potential partners were at least likely to have been contemporaneous is valuable and, in rare cases of fossil copreservation, it may even be possible to establish the antiquity of particular partnerships. In this section, we examine in greater depth the phylogenetic evidence for opportunism and antagonism as alternate paths to the origin of mutualism.

3.3.1 Chance and opportunity

When partner phenotypes appear to be coupled, as in the example of elongate insect proboscides matching the tubular channels of gymnosperm cones (Figure 3.2), we are tempted to regard the exchange of benefits among mutualists as necessarily the result of adaptation where an interaction has precipitated some kind of evolutionary genetic change in one or more partners. (Chapter 5 provides examples of this phenomenon and how to recognize it.) However, the origin of a mutualism need not coincide with genetic change in either partner, and mutually beneficial associations can originate from unexpected encounters among species. The expansion of a species' geographic range, for example, could immediately engage partners in the exchange of benefits with no prior history of interaction. Janzen (1980) described the likely scenario in which a vertebrate frugivore newly colonizes the habitat of a tree species already possessing fruit traits associated with vertebrate seed dispersal. Mutualism is observed among partners with little or no history of association in the case of exotic invasive plant species dispersed by native birds (Gleditsch and Carlos 2011, Deckers et al. 2007, Foster and Robinson 2007), where interactions happen to confer mutual benefit by chance. Such opportunities for mutualism are likely to be influenced by particular traits regardless of how the organisms came to possess them. In other words, mutualism may arise from coincidental combinations of preexisting adaptations and not necessarily by the operation of selective processes optimizing interactions around mutual benefit.

It may be especially challenging to identify evolutionary origins of mutualism among generalists. Fleshy-fruited plants that enclose their propagules in colorful, nutritious pulp often simultaneously attract and reward a broad range of vertebrates, including bats, birds, and primates, any of which may be able to deposit viable seeds in favorable environments (Herrera and Pellmyr 2002). Vertebrate foraging and perching behavior, functional morphologies that swallow intact seeds, and gut metabolism favoring rapid passage further appear to facilitate seed dispersal. Whether such traits evolved in the context of interactions among particular species or clades is likely to remain obscure because the very adaptations that appear to foster opportunities for mutualism among generalists are often resilient to changes in partner affiliation (see also Box 10.2 on the context dependency of seed dispersal interactions).

3.3.2 From antagonism to mutualism and back

Mutualisms may originate from antagonistic interactions such as parasitism and predation (Thompson 1982, Thompson 1994) and the collapse of mutualism also has the potential to give way to parasitic or predatory interactions (Kiers et al. 2010). In this section, we consider the kinds of evidence and theory that support such evolutionary transitions. We begin with examples of bacteria in which diverse life histories and a multiplicity of associations with eukaryotic hosts provide an opportunity to observe general patterns. Molecular phylogeny suggests that mutualistic bacteria have evolved more often from parasitic than from free-living ancestors (Sachs et al. 2014). There appear to have been numerous independent origins of mutualism in bacteria with little or no ecological genomic predisposition for particular hosts, and horizontal transfer of genes may have often accompanied the evolution of mutualistic associations, especially involving nitrogen fixation (Sachs et al. 2011). Contrary to theoretical models predicting that mutualists could be vulnerable to extinction and reversal to parasitism (Chapter 4), empirical data suggest that once bacterial lineages evolve to be mutualists, they rarely return to parasitic or free-living states (Sachs et al. 2011).

The phylogenetic hypothesis that mutualists are descended from antagonistic ancestors is best evaluated in simple cases, where uncertainty about ancestral states is low. Parrots, for example, comprise a large clade of seed predators with beaks capable of cracking most seeds. Vulturine parrots endemic to the island of New Guinea, however, have unusually soft beaks and pass intact seeds through the gut while maintaining a diet of fruit pulp instead of seeds (Mack and Wright 1998). The derived phylogenetic position of vulturines suggests that changes in behavior, diet, and morphology were associated with the loss of a seed predatory habit and the origin of a seed dispersal mutualism. This particular transition from predator to disperser almost certainly involved prior associations with the same or similar plant lineages, whereas other transitions from antagonism to mutualism appear to be accompanied by radical host shifts. For example, the cleaner wrasse belongs to a family of reef fish in which the habit of cleaning other fish species appears to have evolved only once from a coral-feeding habit (Cowman et al. 2009). In none of the cases of cleaner wrasse, vulturine parrots, or symbiotic bacteria does phylogeny suggest evolutionary reversals to antagonism.

Obligate pollination mutualisms involving seed predators are among the most striking cases in which phylogeny and ecology point to antagonistic origins of mutualism. Examples include wasps (Agaonidae: Chalicidoidea) that pollinate the flowers of figs (*Ficus*), and various seed-eating moths that pollinate *Yucca* plants, *Senita* cacti, and *Glochidion* trees (Moe et al. 2012). The insects lay eggs in a fraction of the flowers they pollinate where their offspring prey on developing seeds. They appear to have retained the highly specialized, seed-predatory life histories of their ancestors while having also acquired adaptations for pollen delivery that serve the reproductive interests of both partners (Pellmyr and Thompson 1992). The extreme species specificity of obligate mutualisms, in which neither partner persists without the other, provides an opportunity to infer ancient host associations with a degree of confidence that is rarely the case in generalized or facultative mutualisms. Identifying the origins of mutualism is therefore limited not only by the extent of systematic knowledge and the power of phylogenetic methods but also by our incapacity to recognize likely participants in ancient associations and the precise nature of their interactions.

Theory on the evolution of virulence predicts the evolution of reduced antagonism (and perhaps shifts to mutualism) in cases of parasitism where host fidelity is high and the availability of alternate hosts is low (Chapter 4). A tendency toward mutualism is thought to arise when partners are maternally inherited, or are otherwise vertically transmitted to offspring. Alternatively, parasites are predicted to be more virulent when horizontal modes of infecting new hosts are prevalent. Ecological or genetic change altering the predominant mode of parasite transmission could trigger evolutionary transitions to or from mutualism. An experimental test of this theory manipulated the transmission mode of a jellyfish photosymbiont (Sachs and Wilcox 2006). Horizontally transmitted algae reduced the growth rate of the host and were more prolific compared to vertically transmitted algae in accord with the prediction of increased virulence. It remains unclear whether altering the mode of transmission alone would be sufficient to trigger the breakdown of mutualism and an evolutionary shift to parasitism.

Interactions among fig wasps and their nematode parasites illustrate how the evolution of less virulent genotypes may be favored in scenarios of vertical transmission when potential hosts are few (Herre et al. 1999). Nematodes are transported among figs by female wasps, and opportunities for colonizing new hosts depend on the number of females that pollinate and lay eggs in a given fig. Nematode reproductive success is closely tied to that of a particular wasp in figs pollinated by a single female where nematodes have no choice other than to infect her offspring. In agreement with theory on the evolution of virulence, Herre (1993) demonstrated that nematodes of singly pollinated fig species were less virulent than those of multiply pollinated fig species. However, a transition to mutualism in which vertically transmitted nematodes increase the fitness of their host wasps has yet to be observed. The fact that horizontal transmission of mutualistic partners is common in nature is also a challenge for virulence theory (Sachs and Wilcox 2006).

Microbes may provide the most compelling evidence for the rapid evolution of mutualism from antagonism. For example, *Wolbachia* (see Box 2.1) are maternally inherited bacteria known to confer cytoplasmic incompatibility between infected and uninfected invertebrates such that reduced fecundity is observed among infected females who mate with uninfected males. As a *Wolbachia* infection spread among populations of the fruit fly *Drosophila simulans*, increases in fecundity among infected individuals were observed over time (Weeks et al. 2007). The evolution of *Wolbachia* from a parasitic, host fecundity-decreasing strain to a fecundity-increasing strain supports the prediction from virulence theory that transitions to mutualism may be favored under conditions of vertical transmission in which opportunities for parasites to colonize new hosts are curtailed.

Few mutualisms lend themselves to such elegant tests of virulence theory, and evidence of origins, whether antagonistic or otherwise, is unavailable for some of the most obvious partnerships on earth. Endosymbiosis among prokaryotes, for example, is thought to have produced eukaryotic life when mitochondria and plastids evolved from interactions among free-living cyanobacteria and proteobacteria, respectively (Kutschera and Niklas 2005). We do not know whether ancestral prokaryotic associations might have shifted from parasitism to mutualism prior to or concurrent with the evolutionary events that caused mitochondria to develop an integral metabolic role in the eukaryotic cell. Perhaps experimental evolution, molecular genetics, and cell biology will provide mechanistic insights in the future.

Context dependency of species interactions is a further source of frustration in attempting to embed the origin of mutualism in antagonistic interactions. Costs and benefits for partners are likely to vary among ecological conditions as influenced by third parties and the abiotic environment (Chapter 10). For example, toxic alkaloids produced by fungal plant pathogens (Clavicipitaceae) are thought to defend infected grasses against herbivores (Faeth 2002, Rodriguez et al. 2009), but the nature of this interaction could range from mutualism to antagonism depending on the extent of herbivory. Associations of ectomycorrhizal fungi with host plants also vary from parasitic to mutualistic according to the ecological context (Hoeksema et al. 2010), and experimental evidence from the nitrogen-fixing legume–*Rhizobium* symbiosis suggests that non-mutualistic rhizobia may be selected when nitrogen is not limiting. Progress in ecological genetics has highlighted the importance of context in shaping the dynamics of interactions, and how even some very ancient partnerships continue to traverse a continuum between antagonism and mutualism (Heath and Stinchcombe 2013). It might be worthwhile to consider the additional possibility that some mutualisms might trace their origins to commensal situations in which one partner benefits without affecting the other, or amensalism, in which a partner is harmed without benefiting the other.

3.4 Diversification of mutualism

The possibility that mutually beneficial interactions might promote species diversification has attracted the interest of evolutionary biologists since Darwin (Dodd et al. 1999, Vamosi and Vamosi 2010). Empirical examples of the phenomenon are for the most part limited to cases in which mutualism either catalyzes a new ecological opportunity or enforces the evolution of reproductive isolation (Hembry et al. 2014). Increased diversification is generally attributed to either an elevated rate of speciation or a decreased extinction rate. Whether and how species associations affect extinction rates are likely to depend on the nature, frequency, specificity, and persistence of partnerships through evolutionary time. For example, defensive mutualisms could contribute to the persistence of species populations by reducing rates of predation whereas mutualists that are wholly dependent on their partners for survival might share an increased risk of extinction. Mutualism could also elevate rates of speciation either indirectly by expanding the range of ecological opportunities available to lineages in partnership or directly by influencing the reproductive isolation and divergence of partnered populations.

The alternative possibility, that mutualism could hinder species diversification, is also worthy of consideration. Interdependence among species is commonly expected to increase the risk of extinction despite relatively little evidence to support the idea (Dunn et al. 2009). Perhaps coextinction is pervasive but so rapid that it leaves no trace, or mutualisms might be more resilient than is assumed by models of coexistence. Efforts to develop theoretical models of divergence in mutualisms along these lines have been relatively modest compared to the extensive theory on mutualism origin and maintenance (see Chapter 4). An alternative approach involves empirical comparisons of relative diversity. Considering that lineages diverging from a common ancestor have had equal time to diversify, evidence for a shift in the diversification rate can be gleaned by comparing the relative size of a mutualistic lineage to a non-mutualistic sister group. Insight may also be gained by comparing the relative diversity of lineages engaged in particular mutualisms. A review of patterns of diversity among relatives and partners aims to examine whether mutualism has special or predictable consequences for diversification.

3.4.1 Comparing relatives

The comparison of closely related clades either engaged in mutualism or not may identify differences in relative diversity that could be attributed to mutualism. Mutualisms involving symbiotic fungi, dinoflagellates, gut endosymbionts, cleaners, pollinators, and seed dispersers suggest conditions and circumstances for speculation. Our consideration of these examples represents but the first step toward understanding potential influences of beneficial interactions on divergence and speciation. A recent study of ambrosia gall midges, for example, examined how the acquisition of fungal symbionts might catalyze lineage diversification. These insects induce galls in their host plants by laying eggs and by inoculating their hosts with pathogenic fungal spores where midge larvae feed on fungal hyphae inside of the galls. Joy (2013) compared lineages of ambrosia gall midges that form a mutualism with fungi to non-mutualistic lineages and the number of host plant taxa in the diet of each. These comparisons suggest not only that the mutualism between gall midges and fungi catalyzed diversification, but that the mutualistic lineage also had a sevenfold increase in the range of host plants used, consistent with the idea of mutualism facilitating niche expansion.

The fig pollination mutualism provides an example of how a shift in diversification may result from the rapid evolution of reproductive isolation (Moe et al. 2012). *Ficus* is ten times more numerous in species than its closest relatives in the tribe Castilleae of the mulberry family (Moraceae) that participate in a similarly specialized insect brood-site pollination mutualism. The accelerated rate of speciation in *Ficus* compared to Castilleae can be attributed to both the nature and accessibility of floral rewards. The life history of thrips pollinators of Castilleae involves pollen-feeding throughout life and multiple visits to inflorescences per generation. In comparison, the fig ovules that serve as brood sites for fig wasps are enclosed by inflorescences in a manner that limits each pollinator to a single visit per lifetime, which carries a severe fitness consequence for choosing an unfavorable fig (Moe and Weiblen 2012). Fig wasps might, therefore, be highly inclined to pollinate figs resembling their birth fig, with the further consequence of restricted gene flow among figs exhibiting divergent chemosensory attractants (Moe et al. 2012). In this scenario, accelerated speciation in figs need not necessarily be the result of mutualism facilitating niche expansion; rather, diversification could be a byproduct of extreme specialization in which the reproductive interests of partners are inextricably aligned.

Mutualism with sea anemones has been implicated in the diversification of clownfish (Litsios et al. 2012). Clownfish secrete protective mucus that permits their close association with anemones despite the presence of poisonous nematocysts that aid anemones in preying upon other small fish. Molecular phylogeny supported the monophyly of clownfish, with inferred rates of speciation significantly higher than in the recently identified sister group of damselfish. However, there are comparable cases in which mutualistic clades are no more species rich than their non-mutualistic relatives. For example, specialized cleaner fish (*Labroides*) are no more diverse than their antagonistic relatives (Labrichthyines) after ~10 My of mutualism (Cowman et al. 2009). That only five *Labroides* species are known, compared to ten species of coral-feeding Labrichthyines (Parenti and Randall 2000), indicates that the acquisition of a mutualistic habit need not necessarily result in diversification.

Molecular genetic analysis has revealed extensive species diversity among *Symbiodinium* dinoflagellates (Sampayo et al. 2009), the photosynthetic symbionts ("zooxanthellae") of marine invertebrates including anemones, corals, jellyfish, sponges, flatworms, and mollusks, in which products of photosynthesis and inorganic nutrients are exchanged (see Box 10.3 for examples). Although fossil-calibrated molecular dating suggests that *Symbiodinium* is ancient, at least of early Eocene origin (Pochon et al. 2006), it remains unclear whether *Symbiodinium* is more or less diverse than other free-living or endoparasitic dinoflagellates. Relationships among major dinoflagellate clades are poorly resolved, and taxonomic knowledge of most groups is inadequate to either estimate or compare species diversity. Conversely, we might compare the diversity of marine invertebrate lineages with "zooxanthellae" to closely related lineages without such symbionts. For example, are anthozoans (corals and sea anemones) hosting *Symbiodinium* more diverse than their non-zooxanthellate relatives? Thousands of mutualistic soft coral species appear to be closely related to a few hundred species of plankton-feeding sea pens (Pennatulacea). However, once again, appropriate sister group comparisons are hindered by poor phylogenetic resolution and clade support (McFadden et al. 2006). Efforts to compare relatives are likely to be aided by phylogenomic approaches that hold promise for a more highly resolved tree of life.

3.4.2 Comparing partners

In Box 3.2, Jiri Hulcr describes methods for comparing the relative diversity of partners. He emphasizes examples in which phylogenetic patterns are suggestive of codiversification. Although such patterns can be impressive, the accumulation or loss of species in partnered lineages is also likely be affected by intrinsic and extrinsic factors such that the diversification of partners need not be coupled. For example, comparable numbers of fig species and fig wasp species suggests parallel diversification (Figure 3.2), but the pattern is less striking in other brood-site pollination mutualisms such as yuccas and yucca moths (Althoff et al. 2012, Smith et al. 2008). Many mutualisms, in fact, exhibit highly asymmetrical patterns of diversity among partner lineages. In the lichen symbiosis, numerous lineages of ascomycete fungi engage a relatively small set of algae and cyanobacteria as photobionts (Arnold et al. 2009). If we assume that species concepts are biologically meaningful, then it follows that rates of diversification are substantially higher for the lichenized fungi than for their algal photobionts. Mutualism has been implicated in the diversification of ascomycetes by enabling these organisms to extend the range of environments suitable for growth and survival. In other partnerships, the diversification of one lineage or the other may have occurred long prior to the origin of the mutualism. For example, the oxpecker bird cleans ticks and ectoparasites from dozens of ungulate species (Nunney 1993) that evidently radiated in Africa long before the oxpecker evolved its highly specialized diet.

Asymmetrical diversity among mutualistic partners (see Chapter 11) more often involves a large number of associates with a small number of hosts. For example, in the case of fungal symbionts and vascular plants, approximately thirty species of alder trees (*Alnus*) are facultatively associated with hundreds of ectomycorrhizal fungi (Walker et al. 2014). Sampling ectomycorrhizal communities across the geographic distribution of a given alder species suggests a degree of phylogeographic concordance between hosts and symbionts that is better explained by comigration than by codiversification (Kennedy et al. 2011). The whistling thorn (*Acacia drepanolobium*), a long-lived East African tree hosting four different ant symbionts, illustrates how unequal numbers of partners could be selected and maintained in a mutualism. Each ant symbiont provides a different quality of service during host plant development and whistling thorns achieve the greatest overall benefit when all four are present (Palmer et al. 2010). Codiversification, therefore, appears to be but one of several possibilities.

Fungus-farming ants (see Box 5.1) were once thought to have codiversified with their symbiotic fungi (Currie et al. 2003). However, more recent phylogenetic evidence suggests that multiple lineages of fungi were domesticated independently (Dentinger et al. 2009). The discovery of an unexpected branch in the fungal tree of life that is also engaged in the mutualism with fungus-farming ants reminds us to exercise caution in interpreting patterns of codiversification when sampling is incomplete. In another example, Hongoh et al. (2005) sampled termite gut symbionts from multiple host species and found that congeneric termites harbored similar bacteria regardless of the individual, colony, location, or host species. The authors suggest that the gut microbiome may have diversified in parallel with termite lineages, but, as in early studies of fungus gardening ants, sampling was rather limited.

Returning to the extreme case of figs and their pollinating wasps, Weiblen and Bush (2002) reported the first phylogenetic comparison of obligate mutualists and antagonists interacting with the same set of hosts. Although the extent of species specificity in fig pollination has been questioned (Machado et al. 2005), most subsequent comparisons of pollinating and non-pollinating wasps have shown that speciation in figs appears to be more closely coupled with pollinator diversification than with speciation in the non-pollinating parasites (Marussich and Machado 2007, Silvieus et al. 2008; but see also Jousselin et al. 2008). This pattern might be best explained by the impact of selection acting on pollinator host recognition (see Section 3.2.1) for the evolution of plant reproductive isolation (Moe and Weiblen 2012). However, in such highly specialized mutualisms, where partners are often codistributed, it is difficult to disentangle the relative influence of geographic isolation and selection on speciation (Hembry et al. 2014). It seems likely that these factors interact to produce phylogenetic patterns.

Mechanistic explanations for patterns of species diversity can be confusing if the distinction between trait and lineage diversification is not made very explicit (see Chapter 7). For example, cophylogeny has little to do with coevolution (meaning correlated trait evolution in the strict sense), although the two have often been conflated (see examples in Box 7.1). However tempting it is to interpret congruent phylogenetic patterns among obligate mutualists as evidence of coevolution, such patterns need not be explained by reciprocal selection. This is why we have limited the scope of our chapter thus far to questions of when and where mutualisms arose, how often they may have been lost, and whether rates of partner speciation or extinction correlate through time. We conclude our treatment by considering how to evaluate hypotheses that particular traits have influenced the origin and diversification of mutualism.

Box 3.2 Phylogenetics in symbiology

Jiri Hulcr

Detecting a symbiotic relationship between two organisms or two groups of organisms can be carried out by experiments, such as the removal or manipulation of one partner to measure effects on the other partner. Where experiments are not feasible, or where complex symbiotic systems are studied, the inference of pattern may provide some insight. One such approach is, of course, phylogenetics. Many phylogenetic methods are available to measure the extent and even statistical significance of co-occurrence between putative symbionts across populations, between species, or in evolutionary time.

The phylogenetic approach with the longest tradition in symbiology includes tests of congruence between phylogenies of two groups of putatively symbiotic organisms (species, or populations within species). This approach is based on the assumption that a stable symbiosis may lead to shared evolutionary paths such that phylogenies of the groups are predicted to be congruent. A number of methods exist to test for congruence of evolutionary histories; for example, congruence between two cladograms using programs such as TreeMap (Charleston 2011) and Jane (Conow et al. 2010) or between data matrices underlying phylogeny estimates as with the incongruence length difference test (Farris et al. 1994), likelihood ratio tests (Huelsenbeck and Bull 1996), and the Shimodaira–Hasegawa test (Huelsenbeck et al. 1997), or the relative timing of cladogenesis. Examples of the latter are the studies of Rønsted et al. (2005) and Cruaud et al. (2012), which compared phylogenies of figs and fig wasps, the specialized fig pollinators. These analyses supported the prediction of parallel phylogenies of symbionts both in terms of the pattern and timing of cladogenesis.

Increasingly, many symbiotic systems are shown to be not bipartite but multipartite. For example, communities of microbial symbionts of animal intestines often include thousands of operational taxonomic units (OTUs), a product of diffuse community coevolution rather than stepwise cocladogenesis. From the perspective of phylogenetic analysis, symbiotic communities differ from one-on-one symbioses in two main respects. First, it is not possible, or at least not informative, to test for cophylogeny between the macrosymbiont and each of the thousands of suspected microsymbionts. Second, symbionts differ widely in the specificity of their association. In such complex communities, the test statistic in no longer the congruence of two phylogenies, but whether the distribution of the symbiotic communities on the macrosymbiont phylogeny (or population network) is non-random. Symbiont

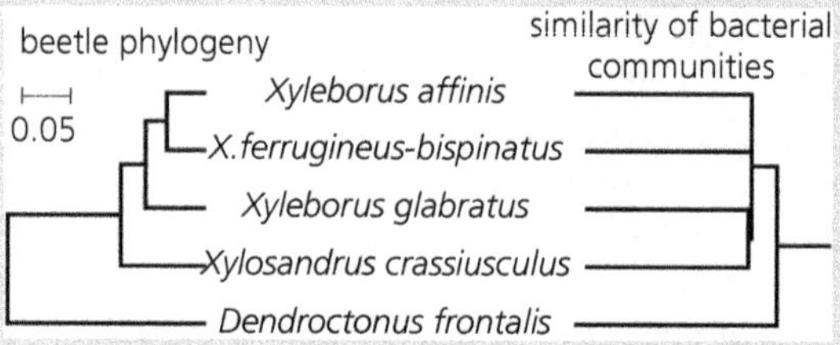

Figure 1 Microsymbiotic communities are often highly similar among closely related macrosymbionts. For example, bacteria are shared among *Xyloborus* and *Xylosandrus* ambrosia beetles to a greater extent than with a more distant relative (Hulcr et al. 2012). Such analyses of complex symbiotic systems are increasingly possible due to the integration of phylogenetics, statistics, and high-throughput DNA sequencing. The scale refers to patristic phylogenetic distances between beetle species, derived from concatenated sequences of 28S ribosomal and cytochrome oxidase I mitochondrial DNA. Reproduced under the Creative Commons Attribution license.

community similarity can be measured by indices that are entirely non-phylogenetic (e.g., the Bray–Curtis index), or they can take into account the phylogenetic relatedness of the microsymbionts involved (e.g., UNIFRAC; Lozupone and Knight 2005). Correlation between community similarities and phylogenetic distances may be tested by methods that correct for non-independence among phylogenetic distances (for example the Mantel test, or permutational multivariate analysis of variance; Oksanen et al. 2011).

High-throughput DNA sequence data and multiplexing of samples allows for the integration of phylogenies, community similarity measures, and environmental factors in statistically robust sampling designs enabling analyses of complex symbiotic systems (Box 3.2, Figure 1). Hulcr et al. (2012) tested for signatures of symbiosis between ambrosia beetles and diverse communities of bacteria without prior taxonomic knowledge of these microbes. By comparing the effect of the host beetle phylogeny to that of beetle provenance using the similarity among the respective bacterial communities, it was shown that the dominant bacterial taxa are vertically transmitted, rather than acquired from the environment. However, no bacterial taxa appeared to be obligate symbionts.

Species-level and population-level phylogenetic methods serve to detect patterns of association and have limited capacity to identify mechanisms of symbiosis. However, phylogenetics may be used to illuminate mechanisms of a symbiosis when focused on genes that encode or regulate symbiotic functions. For example, Moulin et al. (2004)

Box 3.2 *Continued*

examined the phylogeny of *Bradyrhizobium*, nitrogen-fixing bacteria that live in symbiosis with plants. According to housekeeping and ribosomal genes with no particular symbiotic function, the symbiotic habit evolved several times independently among these bacteria. On the other hand, genes involved in the signaling pathways of symbiosis appear to be monophyletic and only loosely congruent with the organismal phylogeny. This suggests that mutualistic function may not have evolved independently but rather might have arisen by lateral transfer of genes facilitating the spread of symbiotic capacity among mutualistic *Bradyrhizobium*.

Despite the power of phylogenetics as an inferential tool in symbiology, there remain essential questions that analyses of pattern are unable to answer. For example, measuring the fitness costs and benefits of associations to understand whether associates are mutualists, commensals, or parasites generally will require experimental study of biological processes.

References

Charleston, M.A. (2011). TreeMap 3.0. <http://www.cs.usyd.edu.au/~mcharles/software/treemap/treemap3.html>.

Conow, C., Fielder, D., Ovadia, Y., and Libeskind-Hadas, R. (2010). Jane: a new tool for the cophylogeny reconstruction problem. *Algorithms for Molecular Biology*, **5**, 16.

Cruaud, A., Rønsted, N., Chantarasuwan, B., et al. (2012). An extreme case of plant-insect co-diversification: figs and fig-pollinating wasps. *Systematic Biology*, **61**, 1029–1047.

Farris, J.S., Kallersjo, M., Kluge, A.G., and Bult, C. (1994). Testing significance of incongruence. *Cladistics*, **10**, 315–319.

Huelsenbeck, J.P., and Bull, J.J. (1996). A likelihood ratio test to detect conflicting phylogenetic signal. *Systematic Biology*, **45**, 92–98.

Huelsenbeck, J.P., Rannala, B., and Yang, Z. (1997). Statistical tests of host-parasite cospeciation. *Evolution*, **51**, 410–419.

Hulcr, J., Rountree, N., Diamond, S.E., Stelinski, L., Fierer, N., and Dunn, R.R. (2012). Mycangia of ambrosia beetles host communities of bacteria. *Microbial Ecology*, **64**, 784–793.

Lozupone, C., and Knight, R. (2005). UNIFRAC: a new phylogenetic method for comparing microbial communities. *Applied and Environmental Microbiology*, **71**, 8228–8235.

Moulin, L., Béna, G., Boivin-Masson, C., and Stępkowski, T. (2004). Phylogenetic analyses of symbiotic nodulation genes support vertical and lateral gene co-transfer within the Bradyrhizobium genus. *Molecular Phylogenetics and Evolution*, **30**, 720–732.

Oksanen, J., Blanchet, F. G., Kindt, R., et al. (2011). vegan: Community Ecology R Package. <http://cran.r-project.org/package=vegan>.

Rønsted, N., Weiblen, G.D., Cook, J.M., Salamin, N., Machado, C.A., and Savolainen, V. (2005). 60 million years of co-divergence in the fig-wasp symbiosis. *Proceeding of the Royal Society of London Series B*, **272**, 2593–2599.

3.5 Innovation and adaptation

It is popular to invoke adaptive trait evolution when interpreting repeated origins of mutualism. New mutualistic phenotypes have the potential to increase fitness, affect reproductive isolation, or enable partners to occupy new niches. For example, lipid-rich appendages attached to seeds (elaiosomes) are regarded as adaptations for seed dispersal by ants (myrmecochory) that have arisen independently in many different plant lineages (Lengyel et al. 2009). Lengyel et al. (2009) compared diversification rates among elaisome-bearing plant lineages to non-myrmecochorous sister groups and attributed the greater diversification of ant-dispersed lineages to either a reduced extinction rate or to ant-mediated barriers to gene flow affecting plant speciation. A similar example involves elaborate presentations of floral rewards to animal pollinators that incidentally might have accelerated the evolution of reproductive isolation among plant populations. Hodges and Arnold (1995) proposed that the evolution of "nectar spurs" (specialized tubular structures often varying among relatives in coloration, form, and the localization of nectar) may have triggered the diversification of floral form and the rapid evolution of reproductive isolation by limiting floral attractiveness to particular kinds of visitors such as hummingbirds, bees, or butterflies. Such hypotheses are both appealing and difficult to test.

At the very least, we would like to know whether the gain or loss of mutualistic traits is significantly correlated with increasing or decreasing rates of diversification. Maddison et al. (2007) introduced a binary-state speciation and extinction model (BiSSE) to specifically address such questions. Unlike methods that indirectly compare diversification rates among sister clades of mutualists and non-mutualists, BiSSE directly examines whether changes in character states along branches of a phylogenetic tree are correlated with shifts in rates of diversification among clades. A further improvement to this model includes the potential for trait evolution to coincide with speciation as well as to occur along branches of the tree (Goldberg and Igic 2012). Although statistical power is low when taxon samples are fewer than several hundred or when the frequency of alternative character states is uneven (Davis et al. 2013), these models have the potential to identify mutualistic traits associated with diversification. They have supported key innovation hypotheses involving mutualism between gall midges and fungi (Stireman et al. 2010), primates and plants (Gomez and Verdu 2012), and sea anemones and clownfish (Litsios et al. 2012).

If adaptive traits are associated with the origin of mutualism, then perhaps the coevolution of mutualistic traits could play a role in maintaining partnerships over time. Theory predicts the collapse of mutualism in which partners might either defect or assume antagonistic relations under conditions in which the cost of participating in the mutualism exceeds the benefit (Chapters 4, 6 and 13). It has been popular to measure correlations among interacting traits and to attribute the matching of mutualistic traits to "coevolutionary complementarity" through which partnerships are maintained by reciprocal selection (Chapter 7). Phylogenetic methods cannot easily disentangle correlation and causation, but we might yet gain some insight by comparing the strength of trait correlations within lineages (e.g., allometric relationships) to mutualistic trait correlations among partnered lineages (i.e., phenotype matching). When the latter correlations are stronger than simple allometry, reciprocal selection is a potential explanation that may be tested by experiment. For example, lengths of fig wasp ovipositors that interact with fig flowers in the exchange of pollination services for seed provisioning to offspring are more highly correlated with flower size than with wasp body size (Weiblen 2004). Patterns of phenotypic variation in the obligate moth pollinators of the Joshua tree (*Yucca brevifolia*) also match geographic variation in floral forms despite substantial gene flow between varieties (Yoder et al. 2013). Intriguingly, a theoretical model of trait coevolution suggested that mutualism might, in fact, favor the conservation of phenotypes, whereas antagonism favors trait diversification (Yoder and Nuismer 2010). Empirical evidence of fitness differences influencing oviposition preference according to the larval performance of pollinating seed predators (Moe and Weiblen 2012) is consistent with this model in suggesting the conservation of matching phenotypes in such mutualisms (Godsoe et al. 2010).

Insect herbivores and their microbial associates hold promise as systems in which to evaluate the possibility of mutualism as a catalyst for adaptive diversification (Janson et al. 2008). Dietary specialization is common among phytophagous insects such that opportunities for diversification may arise through the colonization and exploitation of novel host plants. Microbial mutualists that confer benefits to their insect hosts by assimilating toxic or recalcitrant compounds in the diet or by protecting hosts from natural enemies could affect the balance of speciation and extinction. Microbes might either promote or prevent diversification by influencing the palatability of host plants to herbivores, or by modifying fitness trade-offs associated with the colonization of new host plants, or by altering tritrophic interactions (Janson et al. 2008). Insect life histories seem especially amenable to experimental studies that manipulate host plants in conjunction with the gut microbiome to measure fitness and gauge the extent of adaptive evolution.

3.6 Conclusion

Mutualism is often regarded as an engine for ecological opportunity, a source of new adaptive landscapes, and a stage for coevolutionary dynamics. Patterns emerging from investigations of the past suggest processes affecting the origin and diversification of mutualism. Paleontology provides essential but incomplete information. Phylogenetic models and comparative methods have potential to bridge gaps in knowledge insofar as inferences are robust to uncertainty. Mutualism can arise through coincidental combinations of preexisting adaptations, from antagonistic interactions such as parasitism or predation, or perhaps even from amensalism. Comparisons of relative species diversity between mutualistic and non-mutualistic sister groups indicate that mutualism is sometimes associated with diversification, but lineages in partnership need not diversify in parallel. The repeated evolution of mutualism is often interpreted in terms of adaptive traits. Although such hypotheses are difficult to test, a promising approach may be to apply phylogenetic models that simultaneously estimate character change within lineages and shifts in rates of speciation and extinction among lineages. Another intriguing avenue for comparative study in the context of mutualism is to examine the strength of trait correlations within lineages to that of interacting traits among lineages in partnership.

Future directions and opportunities for understanding the evolutionary history of mutualism will doubtless be advanced by the availability of genomic data. Key questions remain unanswered. Do mutualisms yield unique patterns of diversification over time or have other forms of species interaction such as parasitism diversified similarly? Perhaps it is not the beneficial nature of mutualism, but rather the extent of species specialization or reproductive isolating mechanisms that account for patterns of diversity. Ecological context, population density, generation time, and the geographic distribution of genetic variation in relevant traits are also worthy of consideration. When mutualisms are not maintained, should we expect extinction or antagonism? There appears to be relatively little evidence to support the idea that species interdependence increases the risk of extinction. Habitat loss and climate change are likely to impact mutualism in the future, but it remains unclear whether mutualists are more or less at risk than parasites or other non-mutualists in changing environments (see Chapter 12). The potential for decline of nitrogen-fixing mutualisms in response to anthropogenic nitrogen deposition is already a serious concern (Kiers et al. 2010). If we are to measure risks of extinction, we need to understand the interplay of specialization, partnership lability, and trophic cascades on the survival of species. Opportunities are emerging to investigate these questions and to better understand responses of mutualisms to global change, including the potential for extinction, ecosystem collapse, or unprecedented kinds of interaction.

Acknowledgments

We wish to thank J. Bronstein, K. Bushley, B. Anderson, C. Baskett, D. Schemske, and J. Yoder as well as the other contributors to this volume for their patience and guidance during the prolonged gestation of this chapter. The researcher exchange program at the Hawkesbury Institute for the Environment (University of Western Sydney) supported G.W.

References

Althoff, D.M., Segraves, K.A., Smith, C.I., Leebens-Mack, J. and Pellmyr, O. (2012). Geographic isolation trumps coevolution as a driver of yucca and yucca moth diversification. *Molecular Phylogenetics and Evolution*, **62**, 898–906.

Arbogast, B.S., Edwards, S.V., Wakeley, J., Beerli, P. and Slowinski, J.B. (2002). Estimating divergence times from molecular data on phylogenetic and population genetic timescales. *Annual Review of Ecology and Systematics*, **33**, 707–740.

Arnold, A.E., Miadlikowska, J., Higgins, K.L., Sarvate, S.D., Gugger, P.A.W., Hofstetter, V., Kauff, F. and Lutzoni, F. (2009). A phylogenetic estimation of trophic transition networks for ascomycetous fungi: Are lichens cradles of symbiotrophic fungal diversification? *Systematic Biology*, **58**, 283–297.

Baum, D.A., and Smith, S.D. (2013). *Tree Thinking: An Introduction to Phylogenetic Biology*. Greenwood Village, CO: Roberts & Co.

Boucher, D.H. (1992). Mutualism and cooperation. In Fox Keller, E. and Lloyd, A.M. (eds.) *Keywords in Evolutionary Biology*. Cambridge, MA: Harvard University Press.

Boucher, D.H., James, S., and Keeler, K.H. (1982). The ecology of mutualism. *Annual Review of Ecology and Systematics*, **13**, 315–347.

Cowman, P.F., Bellwood, D.R., and Van Herwerden, L. 2009. Dating the evolutionary origins of wrasse lineages (Labridae) and the rise of trophic novelty on coral reefs. *Molecular Phylogenetics and Evolution*, **52**, 621–631.

Cruaud, A., Rønsted, N., Chantarasuwan, B., et al. (2012). An extreme case of plant-insect co-diversification: figs and fig-pollinating wasps. *Systematic Biology*, **61**, 1029–1047.

Currie, C.R., Wong, B., Stuart, A.E., Schultz, T.R., Rehner, S.A., Mueller, U.G., Sung, G.-H., Spatafora, J.W., and Straus, N. A. (2003). Ancient tripartite coevolution in the attine ant-microbe symbiosis. *Science*, **299**, 386–388.

Darwin, C. (1862). *On the various contrivances by which British and foreign orchids are fertilised by insects, and on the good effects of intercrossing*. London: John Murray.

Davis, M.P., Midford, P., and Maddison, W.P. (2013). Exploring power and parameter estimation of the BiSSE method for analyzing species diversification. *BMC Evolutionary Biology*, **13**, 1–11.

Deckers, B., Verheyen, K., Vanhellemont, M., Maddens, E., Muys, B., and Hermy, M. (2007). Impact of avian frugivores on dispersal and recruitment of the invasive *Prunus serotina* in an agricultural landscape. *Biological Invasions*, **10**, 717–727.

Dentinger, B.T., Lodge, D.J., Munkacsi, A.B., Desjardin, D.E., and Mclaughlin, D.J. (2009). Phylogenetic placement of an unusual coral mushroom challenges the classic hypothesis of strict coevolution in the *Apterostigma pilosum* group ant-fungus mutualism. *Evolution*, **63**, 2172–2178.

Dodd, M.E., Silverton, J., and Chase, M.W. (1999). Phylogenetic analysis of trait evolution and species diversity variation among angiosperm families. *Evolution*, **53**, 732–744.

Donoghue, M.J., Doyle, J.A., Gauthier, J., Klugg, A.G., and Rowe, T. (1989). The importance of fossils in phylogeny reconstruction. *Annual Review of Ecology and Systematics*, **20**, 431–460.

Dunn, R.R., Harris, N.C., Colwell, R.K., Koh, L.P., and Sodhi, N.S. (2009). The sixth mass coextinction: are most endangered species parasites and mutualists? *Proceeding of the Royal Society B-Biological Sciences*, **276**, 3037–3045

Faeth, S.H. (2002). Are endophytic fungi defensive plant mutualists? *Oikos*, **98**, 25–36.

Farrell, B.D., Mitter, C., and Futuyma, D.J. (1992). Diversification at the insect-plant interface. *BioScience*, **42**, 34–42.

Felsenstein, J. (2004). *Inferring Phylogenies*. Sunderland, MA: Sinauer Associates.

Foster, J.T., and Robinson, S.K. (2007). Introduced birds and the fate of Hawaiian rainforests. *Conservation Biology*, **21**, 1248–1257.

Gleditsch, J.M., and Carlos, T.A. (2011). Fruit quantity of invasive shrubs predicts the abundance of common native avian frugivores in central Pennsylvania. *Diversity and Distributions*, **17**, 244–253.

Godsoe, W., Yoder, J.B., Smith, C.I., Drummond, C.S., and Pellmyr, O. (2010). Absence of population-level phenotype matching in an obligate pollination mutualism. *Journal of Evolutionary Biology*, **23**, 2739–2746.

Goldberg, E.E., and Igic, B. (2012). Tempo and mode in plant breeding system evolution. *Evolution*, **66**, 3701–3709.

Gomez, J.M., and Verdu, M. (2012). Mutualism with plants drives primate diversification. *Systematic Biology*, **4**, 567–577.

Heath, K.D., and Stinchcombe, J.R. (2013). Explaining mutualism variation: a new evolutionary paradox? *Evolution*, **68**, 309–317.

Hembry, D.H., Yoder, J.B., and Goodman, K.R. (2014). Coevolution and the diversification of life. *American Naturalist*, **184**.

Herre, E.A. (1993). Population structure and the evolution of virulence in nematode parasites of fig wasps. *Science*, **259**, 1442–1445.

Herre, E.A., Knowlton, N., Mueller, U.G., and Rehner, S.A. (1999). The evolution of mutualisms: exploring the paths between conflict and cooperation. *Trends in Ecology and Evolution*, **14**, 49–53.

Herrera, C.M., and Pellmyr, O. (eds.) (2002). *Plant-Animal Interactions: An Evolutionary Approach*. Oxford: Blackwell Science Ltd.

Hibbett, D.S., Gilbert, L.B., and Donoghue, M.J. (2000). Evolutionary instability of ectomycorrhizal symbioses in basidiomycetes. *Nature*, **407**, 506–508.

Hodges, S.A., and Arnold, M.L. (1995). Spurring plant diversification: are floral nectar spurs a key innovation? *Proceedings of the Royal Society of London Series B-Biological Sciences*, **262**, 343–348.

Hoeksema, J. D., Chaudhary, V. B., Gehring, C. A., et al. (2010). A meta-analysis of context-dependency in plant response to inoculation with mycorrhizal fungi. *Ecology Letters*, **13**, 394–407

Hongoh, Y., Deevong, P., Inoue, T., et al. (2005). Intra- and interspecific comparisons of bacterial diversity and community structure support coevolution of gut microbiota and termite host. *Applied Environmental Microbiology*, **71**, 6590–6599.

Janson, E.M., Stireman, J.O., Singer, M.S., and Abbot, P. (2008). Phytophagous insect-microbe mutualisms and adaptive evolutionary diversification. *Evolution*, **62**, 997–1012.

Janzen, D.H. (1980). When is it coevolution? *Evolution*, **34**, 611–612.

Jousselin, E., Van Noort, S., Berry, V., Rasplus, J.Y., Ronsted, N., Erasmus, J.C., and Greeff, J.M. (2008). One fig to bind them all: host conservatism in a fig wasp community unraveled by cospeciation analyses among pollinating and nonpollinating fig wasps. *Evolution*, **62**, 1777–1797.

Joy, J.B. (2013). Symbiosis catalyses niche expansion and diversification. *Proceedings of the Royal Society Series B*, **280**, 20122820.

Kennedy, P.G., Garibay-Orijel, R., Higgins, L.M., and Angeles-Arguiz, R. (2011). Ectomycorrhizal fungi in Mexican Alnus forests support the host co-migration hypothesis and continental-scale patterns in phylogeography. *Mycorrhiza*, **21**, 559–568.

Kiers, E.T., Palmer, T.M., Ives, A.R., Bruno, J.F., and Bronstein, J.L. (2010). Mutualisms in a changing world: an evolutionary perspective. *Ecology Letters*, **13**, 1459–1474.

Kutschera, U., and Niklas, K. (2005). Endosymbiosis, cell evolution, and speciation. *Theory in Biosciences*, **124**, 1–24.

Labandeira, C.C., and Sepkoski, J.J. (1993). Insect diversity in the fossil record. *Science*, **261**, 310–315.

Lemey, P., Salemi, M., and Vandamme, A.M. (eds.) (2009). *The Phylogenetic Handbook: A Practical Approach to Phylogenetic Analysis and Hypothesis Testing*. Cambridge: Cambridge University Press.

Lengyel, S., Gove, A. D., Latimer, A.M., Majer, J.D., and Dunn, R.R. (2009). Ants sow the seeds of global diversification in flowering plants. *PLoS One*, **4**, e5480.

Litsios, G., Sims, C., Wuest, R., Pearman, P., Zimmermann, N., and Salamin, N. (2012). Mutualism with sea anemones triggered the adaptive radiation of clownfishes. *BMC Evolutionary Biology*, **12**, 1–15.

Machado, C.A., Robbins, N., Gilbert, T.P., and Herre, E.A. (2005). Critical review of host specificity and its coevolutionary implications in the fig/fig-wasp mutualism. *Proceedings of the National Academy of Sciences of the United States of America*, **102**, 6558–6565.

Mack, A.L., and Wright, D.D. (1998). The vulturine parrot, *Psittrichas fulgidus*, a threatened New Guinea endemic: notes on its biology and conservation. *Bird Conservation International*, **8**, 185–194.

Maddison, W.P., Midford, P.E., and Otto, S.P. (2007). Estimating a binary character's effect on speciation and extinction. *Systematic Biology*, **56**, 701–710.

Marussich, W.A., and Machado, C.A. (2007). Host-specificity and coevolution among pollinating and nonpollinating New World fig wasps. *Molecular Ecology*, **16**, 1925–1946.

Mcfadden, C.S., France, S.C., Sánchez, J.A., and Alderslade, P. (2006). A molecular phylogenetic analysis of the Octocorallia (Cnidaria: Anthozoa) based on mitochondrial protein-coding sequences. *Molecular Phylogenetics & Evolution*, **41**, 513–527.

Moe, A.M., Clement, W.L., and Weiblen, G.D. (2012). Rapid evolution of pollinator-mediated plant reproductive isolation. In Singh, R.S., Xu, J., and Kulathinal, R.J. (eds.) *Evolution in the Fast Lane: Rapidly Evolving Genes and Genetic Systems*. Oxford: Oxford University Press.

Moe, A.M., and Weiblen, G.D. (2012). Pollinator-mediated reproductive isolation among dioecious fig species (*Ficus*, Moraceae) *Evolution*, **66**, 3710–3721.

Nunney, L. (1993). The influence of mating system and overlapping generations on effective population size. *Evolution*, **47**, 1329–1341.

Palmer, T.M., Doak, D.F., Stanton, M.L., Bronstein, J.L., Kiers, E.T., Young, T.P., Goheen, J.R., and Pringle, R.M. (2010). Synergy of multiple partners, including freeloaders, increases host fitness in a multispecies mutualism. *Proceedings of the National Academy of Sciences USA*, **107**, 17234–17239.

Paradis, E. (2011). Shift in diversification in sister-clade comparisons: A more powerful test. *Evolution*, **66**, 288–295.

Parenti, P., and Randall, J.E. (2000). An annotated checklist of the species of the labroid fish families Labridae and Scaridae. *Ichthyological Bulletin*, **68**, 1–97.

Pellmyr, O., and Thompson, J.N. (1992). Multiple occurrences of mutualism in the yucca moth lineage. *Proceedings of the National Academy of Sciences USA*, **89**, 2927–2929.

Pellmyr, O., Thompson, J.N., Brown, J.M., and Harrison, R.G. (1996). Evolution of pollination and mutualism in the yucca moth lineage. *American Naturalist*, **148**, 827–847.

Pirozynski, K.A., and Malloch, D.W. (1975). The origin of land plants: a matter of mycotrophism. *BioSystems*, **6**, 153–164.

Pochon, X., Montoya-Burgos, J.I., Stadelmann, B., and Pawlowski, J. (2006). Molecular phylogeny, evolutionary rates, and divergence timing of the symbiotic dinoflagellate genus *Symbiodinium*. *Molecular Phylogenetics and Evolution*, **38**, 20–30.

Remy, W., Taylor, T.N., Hass, H., and Kerp, H. (1994). Four hundred-million-year-old vesicular arbuscular mycorrhizae. *Proceedings of the National Academy of Sciences USA*, **91**, 11841–11843.

Rodriguez, R.J., White, J.F., Arnold, A.E., and Redman, R.S. (2009). Fungal endophytes: diversity and functional roles. *New Phytologist*, **182**, 314–330.

Sachs, J.L., Skophammer, R.G., Bansal, N., and Stajich, J.E. (2014). Evolutionary origins and diversification of proteobacterial mutualists. *Proceedings of the Royal Society B-Biological Sciences*, **281**, 20132146.

Sachs, J.L., Skophammer, R.G., and Regus, J.U. (2011). Evolutionary transitions in bacterial symbiosis. *Proceedings of the National Academy of Sciences USA*, **108**, 10800–10807.

Sachs, J.L., and Wilcox, T.P. (2006). A shift to parasitism in the jellyfish symbiont *Symbiodinium microadriaticum*. *Proceedings of the Royal Society B-Biological Sciences*, **273**, 425–429.

Sampayo, E.M., Dove, S., and Lajeunesse, T.C. (2009). Cohesive molecular genetic data delineate species diversity in the dinoflagellate genus *Symbiodinium*. *Molecular Ecology*, **18**, 500–519.

Silvieus, S.I., Clement, W.L., and Weiblen, G.D. (2008). Cophylogeny of figs, pollinators, gallers and parasitoids. In Tilmon, K.J. (ed.) *Specialization, Speciation, and Radiation: The Evolutionary Biology of Herbivorous Insects*. Berkeley, CA: University of California Press.

Smith, C.I., Pellmyr, O., Althoff, D.M., Balcazar-Lara, M., Leebens-Mack, J., and Segraves, K.A. (2008). Pattern and timing of diversification in *Yucca* (Agavaceae): Specialized pollination does not escalate rates of diversification. *Proceedings of the Royal Society B-Biological Sciences*, **275**, 249–258.

Stadler, T. (2010). Mammalian phylogeny reveals recent diversification rate shifts. *Proceedings of the National Academy of Sciences of the United States of America*, **108**, 6188–6192.

Stireman, J.O., Devlin, H., Carr, T.G., and Abbot, P. (2010). Evolutionary diversification of the gall midge genus *Asteromyia* (Cecidomyiidae) in a multitrophic ecological context. *Molecular Phylogenetics and Evolution*, **54**, 194–210.

Thompson, J.N. (1982). *Interaction and Coevolution*. New York: John Wiley & Sons.

Thompson, J.N. (1988). Variation in interspecific interactions. *Annual Review of Ecology and Systematics*, **19**, 65–87.

Thompson, J.N. (1994). *The Coevolutionary Process*. Chicago: University of Chicago Press.

Vamosi, J.C., and Vamosi, S.M. (2010). Key innovations within a geographical context in flowering plants: towards resolving Darwin's abominable mystery. *Ecology Letters*, **13**, 1270–1279.

Walker, J.K., Cohen, H., Higgins, L.M., and Kennedy, P.G. (2014). Testing the link between community structure and function for ectomycorrhizal fungi involved in a global tripartite symbiosis. *New Phytologist*, **202**, 287–296.

Weeks, A.R., Turelli, M., Harcombe, W.R., Reynolds, K.T., and Hoffmann, A.A. (2007). Rapid evolution of *Wolbachia* in natural populations of *Drosophila*. *PLoS Biology*, **5**, e114.

Weiblen, G.D. (2004). Correlated evolution in fig pollination. *Systematic Biology*, **53**, 128–139.

Weiblen, G.D., and Bush, G.L. (2002). Speciation in fig pollinators and parasites. *Molecular Ecology*, **11**, 1573–1578.

Wolfe, B.E., Tulloss, R.E., and Pringle, A. (2014). The irreversible loss of a decomposition pathway marks the single origin of an ectomycorrhizal symbiosis. *PLoS One*, **7**, e39597.

Yoder, J.B., and Nuismer, S.L. (2010). When does coevolution promote diversification? *The American Naturalist*, **176**, 802–817.

Yoder, J.B., Smith, C.I., Rowley, D.J., Flatz, R., Godsoe, W., Drummond, C., and Pellmyr, O. (2013). Effects of gene flow on phenotype matching between two varieties of Joshua tree (*Yucca brevifolia*; Agavaceae) and their pollinators. *Journal of Evolutionary Biology*, **26**, 1220–1233.

CHAPTER 4

Evolutionary models of mutualism

Erol Akçay

4.1 Introduction

Mutualisms are among the most important ecological interactions, yet folklore in the discipline holds that they were long neglected in favor of competition and prey–predator interactions. Fortunately, this situation has changed dramatically in the last three decades and the evolutionary theory of mutualisms has been attracting increasing interest. As a result, we now have a large enough collection of evolutionary models of mutualisms to warrant its own survey. Most of the present chapter focuses on the question that has been dominant in mutualism theory: how cooperation between species can evolve and be maintained through natural selection.

In many mutualisms, helping an individual of the partner species (e.g., providing protection, housing, or nutrients) is costly, i.e., all else being equal, it decreases the fitness of the individual providing it. Such costly helping behaviors (or phenotypes) are termed *investments*. Mutualisms based on investments are liable to be taken over by what are generally called cheaters (see Chapter 6): individuals that enjoy the benefit from the mutualism, but do not provide any benefits in return. An example would be a plant-ant that feeds on the extrafloral nectaries of its host tree, but does not provide any protection in return (see Box 6.1 and Chapter 9). In the absence of a means to control cheating, trees hosting such ants would get selected to reduce the amount of extrafloral nectar they produce, until cooperation disappears (or remains at vestigial levels). Many evolutionary models of mutualisms are concerned with how mutualisms can persist in the face of cheaters, or, equivalently, how mutualistic interactions resolve the inherent conflicts of interests between species. Accordingly, most of this chapter will deal with the modeling approaches used in this effort.

The twin goals of this chapter are to provide an overview of the rich theory of mutualism evolution as well as an introduction to the mathematical methods used in it. These methods are diverse, from adaptive dynamics to quantitative genetics, and with a whole host of diverse approaches inspired by economics. Many of the modeling approaches in mutualism theory parallel those that deal with cooperation within species, which is hardly surprising given the conceptual closeness of the questions and the fact that many theorists contribute to both areas. Yet, there are also some differences in emphasis. Although selection in structured populations (i.e., kin selection or multilevel selection) is the dominant explanation for within-species cooperation, its action in mutualisms is much less studied (see Section 4.3). Conversely, mutualism theory puts a lot more emphasis on flexible investment phenotypes (e.g., sanctions, partner choice; see Section 4.5) than within-species theory. On balance, there is scope for much better integration of within- and between-species cooperation theories and modeling approaches.

The chapter is organized around the biological concepts, discussing the relevant modeling methods in due order. The first four sections, making up the bulk of the chapter, are devoted to the different biological mechanisms that can resolve the cheater problem. Section 4.6 discusses the evolution of diversity, and Section 4.7 some of the exciting future directions for mutualism theory.

Mutualism. Edited by Judith L. Bronstein.

4.2 Vertical transmission of symbioses

In some mutualisms, such as those between the leafcutter ants and their fungal symbionts, the hosts acquire their symbionts (see Box 5.1) from their parents. Such "vertical" transmission of symbionts (as opposed to horizontal, or environmentally acquired) is expected to favor mutualisms (Fine 1975, Ewald, 1987): helping the host produce more progeny will also produce more symbiont progeny if symbionts are transferred to the offspring. Vertical transmission may have led to the evolution of two of the most important symbioses in the history of life: chloroplasts and mitochondria. On a less grand scale, maternally transmitted *Wolbachia* spreading through populations of *Drosophila simulans* in California has rapidly evolved from imposing a fecundity cost to infected females to conferring a fecundity benefit (Weeks et al., 2007). Thus, perfect (or high-fidelity) vertical transmission can in theory and in the real world resolve the conflict of interest between the host and symbiont and make the cheater problem disappear. But how do separate organisms evolve vertical transmission in the first place?

This question has been studied using the theory of virulence evolution in disease-causing organisms. Virulence is defined as an index of the harm a symbiont causes to the host, so a mutualist is simply a symbiont with negative virulence. Many virulence evolution models have to deal with the coupled ecological and evolutionary dynamics of host and symbiont populations. Generally, this is an intractable task, but in the last two decades, a framework called adaptive dynamics (AD) was developed that can yield analytical insight under some simplifying assumptions (Dieckmann and Law, 1996). The basics of AD are described in Box 1.

Box 4.1 Adaptive dynamics

Erol Akçay

Adaptive dynamics achieves tractability on the problem of eco-evolutionary dynamics by assuming that the ecological dynamics of populations are very fast compared to the rate mutations occur. The typical AD model starts with a monomorphic population (or multiple monomorphic populations for coevolutionary models) at its ecological equilibrium, introduces a rare mutant to this population, and calculates its "invasion fitness" (its rate of increase when rare) at this ecological equilibrium. In the case of epidemiological models, the ecological dynamics usually keep track of different classes of individuals separately (e.g., infected, susceptible, etc.). If the invasion fitness, denoted by $f(x_m, x_r)$ where x_r is the resident genotype's trait value and x_m the mutant's, is higher than the resident's increase rate, the mutant invades and takes over the population, and the procedure starts again. Assuming mutants differ very little from the residents, one can approximate this process by simple dynamics that follow the gradient of the invasion fitness function, i.e., $\frac{dx_r}{dt} \propto \frac{\partial f(x_m, x_r)}{\partial x_m}$, evaluated at $x_m = x_r$. If, for some trait value x_r, $\frac{\partial f(x_m, x_r)}{\partial x_m} = 0$ (at $x_m = x_r$), that value is said to be an evolutionarily singular point. If, further, $\frac{\partial^2 f(x_m, x_r)}{\partial x_m^2} < 0$, the trait value is evolutionarily stable: mutants to either side of the singular point will have lower fitness than the resident. Conversely, if $\frac{\partial^2 f(x_m, x_r)}{\partial x_m^2} > 0$, the resident trait value is a fitness minimum and mutants with both higher and lower trait values can invade the resident population. With sufficient mutational variability, the trait value will branch into two and start diverging (Metz et al., 1996). Finally, whether or not a population will evolve toward an evolutionarily singular point is determined by the cross derivative $\frac{\partial^2 f(x_m, x_r)}{\partial x_m \partial x_r} < 0$, which means that for values below the singular point, mutants with greater than the resident trait value have the advantage and invade, and above the singular point the converse is true. Such singular points are called convergent stable. A convergent stable singular point is called an evolutionarily stable strategy (ESS) if it is a fitness maximum, and an evolutionary branching point (EBP) if it is a fitness minimum. EBPs serve as a model for investigating the conditions under which selection can lead to

Box 4.1 *Continued*

diversification and adaptive speciation (Doebeli and Dieckmann, 2000), though their relevance to speciation dynamics in nature is debated (Doebeli et al. 2005, Gavrilets, 2005).

References

Doebeli, M., and Dieckmann, U. (2000). Evolutionary branching and sympatric speciation caused by different types of ecological interactions. *The American Naturalist*, **156**, S77–S101.

Doebeli, M., Dieckmann, U., Metz, J.A., and Tautz, D. (2005). What we have also learned: adaptive speciation is theoretically plausible. *Evolution*, **59**, 691–695.

Gavrilets, S. (2005). "Adaptive speciation" – it is not that easy: reply to Doebeli et al. *Evolution*, **59**, 696–699.

Metz, J.A.J., Geritz, S.A.H., Meszna, G., Jacobs, F.J., and Van Heerwaarden, J. (1996). Adaptive dynamics, a geometrical study of the consequences of nearly faithful reproduction. *Stochastic and Spatial Structures of Dynamical Systems*, **45**, 183–231.

Yamamura (1993) applies the AD methodology to a model where a symbiont can evolve to harm or help its host, with harm increasing the symbiont's own offspring production. The model confirms that high vertical transmission rates favor helping traits. However, how the transmission rate evolves depends on who controls it. The symbiont is always selected for higher vertical transmission rates, because there is no trade-off between vertical and horizontal transmission in the model. The host, on the other hand, is selected to lower the transmission rate when it is low, and increase it when it is already above a threshold. This model suggests that vertical transmission can evolve at first by a parasite overcoming host defenses and increasing its vertical transmission rate, becoming less virulent as this happens. When the threshold transmission rate is crossed both the host and the symbiont will be selected to ensure that vertical transmission happens with high fidelity, which should lead to maximum mutualistic benefit. The concordance of interest between the host and symbiont at the later stage of this process suggests that once a vertically transmitted mutualism evolves, it would be very hard to break up. This prediction is in agreement with recent data on the evolutionary transitions in symbioses that show there are more transitions from parasitism to mutualism than the other way around (Sachs et al., 2011).

More recent work by Ferdy and Godelle (2005) considers the trade-off between vertical and horizontal transmission. Such a trade-off is expected if traits that favor higher horizontal transmission (infection rates) reduce the fitness of the host, and thus vertical transmission rates, and is supported in many empirical studies (e.g., Stewart et al. 2005). Ferdy and Godelle allow the shape of the trade-off to be convex (i.e., decreasing marginal fecundity cost of horizontal transmission to the host) or concave (increasing marginal costs). With convex costs, they find that evolutionary branching into divergent strategies of high horizontal transmission-high virulence and low horizontal transmission-low virulence (or helping) can occur. This is a general phenomenon in AD models, where the shapes of costs and benefits and the fitness components they affect (e.g., fecundity of survivorship) determine the evolutionary outcome. In particular, evolutionary branching frequently occurs when costs are decelerating (are convex) or benefits are accelerating (concave) over some range of traits, as we will also see in the discussion on evolutionary diversification of mutualists in Section 4.6.

4.3 Population structure and genetic associations

One of the most important advances in evolutionary theory after the modern synthesis was the emergence of theories of selection under population structure, i.e., when individuals do not interact or disperse in a completely mixed manner. Hamilton (1964) famously showed that if individuals interact with other individuals that carry similar genotypes, phenotypes that help others even at a personal cost to oneself can evolve, provided that the coefficient of relatedness r (appropriately defined) is greater than the ratio of the personal cost c to the benefit b supplied, or $rb > c$. For mutualisms, population structure operates in two ways: first, population structure within a species can promote helping the other species if the return benefits are likely to go to related individuals. Second, population structure and limited dispersal can create association between mutualist genotypes, which also promotes cooperation.

There are many approaches to modeling evolution in structured populations (and to deriving versions of Hamilton's rule). The most rigorous models explicitly specify the population structure (most commonly a population structured in discrete demes), population regulation (most commonly fixed deme size), and the life cycle of the organisms (most commonly a Wright–Fisher or Moran process), and write down recursion equations for the probabilities of individuals being identical by descent (IBD) that can then be solved for determining relatedness. (Lehmann and Rousset (2010) provide a good overview of this approach.) Typically, such models assume that selection is weak, in which case relatedness can be approximated by the IBD probabilities under neutrality, and the demographical effects (such as local competition with related individuals) can be separated from the fertility effects of the traits (see Akçay and Van Cleve (2012) for an exposition of how this can be achieved). Practically no model of mutualisms has been built using this approach. Instead, models of spatial population structure in mutualism tend to use either a phenomenological approach (as in Bever and Simms (2000)), who use IBD probabilities as an exogenous parameter), or use individual-based simulations on a lattice (Doebeli and Knowlton 1998, Yamamura et al., 2004). Another powerful approach that has been very influential in social evolution theory is the Price equation, which will be discussed here in some detail.

4.3.1 The Price equation

The Price equation (Price 1970, 1972) expresses the total change in the mean value of a trait in a population at two time points, say times t_1 and t_2, as:

$$\Delta z = \mathrm{cov}(w, z) + \mathrm{E}(w\bar{\delta}). \tag{1}$$

Here, the first term is a covariance between an individual's trait at t_1, z, and its fitness, w, defined as the number of the individuals at t_2 that can be assigned to that particular individual at t_1. This term is usually interpreted as giving the effects of selection on the mean trait value. The second term is the expectation of fitness w and the difference between the mean trait of the descendants (alive at t_2) of a focal individual and that of the focal individual itself (measured at t_1), given by $\bar{\delta}$. This second term gives the effects of other processes, such as transmission bias, phenotypic change, and selection at different levels on the change in mean trait value. The Price equation is an exact identity, meaning that it holds anytime when one can identify an ancestral and a descendant population, measure traits of all individuals in them, and assign individuals from the descendant population to individuals of the ancestral population. In other words, the Price equation allows an elegant summary of the change in a population, if one already knows every single thing necessary to calculate the change. Therefore, in its exact form, the Price equation is not actually a model of evolution or a prediction of the change in the mean trait. Nonetheless, one can sacrifice the exactness of the Price equation and use its partitioning to organize and interpret the different processes influencing fitness when we do make models for those processes, such as one based on quantitative genetics (Frank, 1997).

One of the cases where the Price equation becomes most fruitful is when individual fitness is a function of the phenotypes of other individuals, and individuals interact non-randomly with each other. Suppose, for example, a population with pairwise interactions, in which fitness of a focal individual is given by an additive function of an individual's own phenotype and that of a partner, i.e.,

$$w_i = \beta_{w,z_i} z_i + \beta_{w,z_j} z_j + e, \tag{2}$$

where the βs give the regression coefficients of the fitness on z_i and z_j, respectively, and e is an error term independent of either phenotype. The last assumption makes this statistical model of fitness less general than the exact Price equation, so like many models we have sacrificed some generality in order to gain simplicity. Let us further assume that there is no transmission bias, i.e., the second term in equation (1) vanishes. Substituting (2) into (1), we obtain:

$$\begin{aligned}\Delta z &= \mathrm{cov}(w,z) = \mathrm{cov}(\beta_{w,z_i} z_i + \beta_{w,z_j} z_j + e, z_i) \\ &= \beta_{w,z_i}\mathrm{var}(z) + \beta_{w,z_j}\mathrm{cov}(z_j, z_i).\end{aligned} \tag{3}$$

Supposing that the phenotypes correspond directly to the genotypes of the individuals, or $z_i = g_i$, we can replace the phenotypic variance and covariance by the genotypic ones, and divide by the variance term to obtain:

$$\frac{\Delta z}{\mathrm{var}(g)} = \beta_{w,z_i} + \beta_{w,z_j} R, \tag{4}$$

where we have defined $R = \mathrm{cov}(g_j, g_i)/\mathrm{var}(g)$, or the regression of the partner's genotype on the focal individual's: one of the canonical definitions of relatedness (Frank, 1998). If the phenotype z is a costly helping behavior, β_{w,z_i} and β_{w,z_j} correspond to the cost $-c$ and benefit b, respectively, so the condition for the increase of such a behavior can be seen to be

$$Rb - c > 0,$$

which of course is the celebrated Hamilton rule (Hamilton, 1964). Another useful special case is when the phenotypes z are continuous, fitness is a smooth function of them, expressed as $w(z_i, z_j)$, and the amount of genetic variation around the population mean is small (which would happen with small-effect mutants and low mutation rate). In that case, we can use the following approximations: $\beta_{w,z_i} \sim \frac{\partial w}{\partial z_i}$ and $\beta_{w,z_j} \sim \frac{\partial w}{\partial z_j}$ (Taylor and Frank, 1996). This is one of the easiest ways to model phenotypic evolution in structured populations, provided the assumptions are palatable for the particular system.

In one of the very few applications of the Price equation to mutualisms, Frank (1994) uses it to derive an extension of Hamilton's rule, where a cooperative investment of an individual of, say, Species 1 benefits not only conspecific neighbors, but also individuals of a mutualist species (Species 2). In that case, the cooperative investment by Species 1 will be selected for if the following condition holds:

$$b_{1,1}R_{1,1} + b_{2,1}R_{2,1} - c_1 > 0, \tag{5}$$

where $b_{1,1}$ and $b_{2,1}$ are the benefit from a Species 1 individual's investment to conspecifics, and the benefit from Species 2's investment to a Species 1 individual, respectively. The coefficients $R_{1,1}$ and $R_{2,1}$ are again defined as regression coefficients between Species 1 individuals' investment genotypes and those of their conspecific or heterospecific interaction partners, respectively. Inequality (5) says that cooperation between species can evolve either if the mutualistic investment has "spillover" benefits to conspecifics, and relatedness is high enough [the first term in (5)], or if the investment from the partner species is beneficial enough for the focal species, and the cooperative genotypes from each species are non-randomly associated with each other [the second term in (5)]. Thus, for the evolution of mutualism in structured populations, both within- and between-species genetic associations are potentially important. Furthermore, mutualistic interactions are expected to generate between-species genetic associations, as Frank (1994) showed: when more mutualistic genotypes of both species co-occur in a patch, they will increase in abundance while populations in less mutualistic patches will decline. Such local buildup of highly mutualist populations is inherent in the population dynamics of mutualisms, but its effects have not been as widely studied, apart from a few individual-based simulation studies (e.g., Yamamura et al., 2004).

As in intraspecific models of social evolution, there is a tension between explicitly quantifying genetic association as a function of demographic quantities such as dispersal and fertility on one hand, and being able to derive analytically tractable models focusing on mechanisms that operate at the individual level. Where one chooses to be along this trade-off is a matter of what the model is primarily intended to do. As it stands, there is much work left to be done using either approach. For example, focusing on the build up of genetic associations in ecological time scales and their evolutionary effects in diverse communities would require explicitly modeling the population dynamics and demography of partner species. Agent-based simulations can yield important insight to this question. On the other hand, the integration of genetic association with other mechanisms for fitness feedbacks also remains to be fully explored (see Section 4.7.1), which can be done with models that treat population structure implicitly (McGlothlin et al. 2010, Akçay and Van Cleve 2012, Van Cleve and Akçay, 2014).

4.4 Modeling phenotypic feedbacks

In many mutualisms, the rewards from investing into a partner arise from the partner changing its investment in return. One classic example (described in more detail in Section 4.5.3, and in Boxes 4.2 and 5.2) is the legume-rhizobium symbiosis, where high nitrogen fixation by a nodule increases the resource allocation of the plant to that nodule, and low (or no) fixation elicits a reduction in the plant investment (Kiers et al. 2003, Simms et al., 2006). The term *phenotypic feedbacks* denotes feedbacks between the investments that happen within the time scale of the interaction, reflecting the fact that investments in such mutualisms can be thought of as flexible phenotypes of individuals. Phenotypic feedbacks are distinct from genetic associations and intergenerational feedbacks due to differential reproduction of partners. A huge amount of research effort in mutualism theory is focused on modeling phenotypic feedbacks. This section focuses on the simplest cases, feedbacks in pairwise interactions, or that are otherwise undirected to particular individuals. Section 4.5 will deal with phenotypic feedbacks directed at particular partners, e.g., as a function of their performance.

The theory of iterated games (Maynard Smith and Price 1973, Axelrod and Hamilton, 1981) has historically been a very popular way of modeling phenotypic feedbacks, starting from the first explicitly game-theoretic study in biology, Maynard Smith and Price (1973). Phenotypic feedbacks can be modeled as conditional strategies in iterated games, where behavior in one round depends on the partner's behavior in the previous round. This approach had a meteoric rise after the landmark paper by Axelrod and Hamilton (1981), who showed that a very simple conditional strategy that simply copies the opponent's previous move can outcompete a multitude of sophisticated strategies. This set off an explosion of studies that pitched a multitude of simple conditional strategies against each other in various agent-based simulations. While the modeling enterprise soon took on a life of its own with relatively little connection to biological questions, there have been several studies that have provided useful insight into biological questions, and mutualisms in particular.

In one such study, Doebeli and Knowlton (1998) ask how cooperation between species can evolve when an individual's investment in the mutualism is responsive to its partner's investment. Doebeli and Knowlton assume pairwise interactions on a lattice-structured population. Each pair plays a constant, finite number of rounds with each other before reproducing. At each round, each individual's investment to the mutualism is determined by the the outcome of the previous round. More precisely, if a focal individual invests x and its partner y in a given round, the payoff to the focal individual is given by $B(y)-C(x)$ (benefit from partner's investment minus the cost from own investment). The investment of the focal individual in the next round is then given by $a+b(B(y)-C(x))$, where a and b are assumed to be genetically encoded strategies (i.e., they are constant during the time scale of the interaction). The investments of the partner species are determined likewise using a linear response rule. The total fitness of an individual is the sum of its payoffs in each round; this total fitness determines the evolution of the investment strategies a and b.

Using individual-based simulations, Doebeli and Knowlton found that in a well-mixed population (with global dispersal), the only evolutionarily stable investment strategy is no investment at all ($a=b=0$). On the other hand, if dispersal is localized (so that genetic associations start building up within and between species), unconditional and conditional investments can evolve even in finitely repeated games. The intuition behind this result is that although a positive response slope is costly to a focal individual at the end of a finite interaction (when there is no possibility of positive feedbacks), population structure allows some of this cost to be "recouped" through effects on kin: offspring of a focal individual are more likely to interact with other mutualists as a consequence of the focal individual's positive response. Population structure and phenotypic feedback in general interact synergistically in supporting cooperation (see also Section 4.7.1).

One of the key assumptions of Doebeli and Knowlton (1998) is that the interaction lasts for a finite number of rounds. This differentiates their model from a set of related studies for within-species cooperation (e.g., McNamara et al. 1999, Taylor and Day 2004) that instead assumed that the number of rounds is very large and therefore concentrated on the equilibrium of the phenotypic dynamics (i.e., when from one round to the next, the investment levels do not change). Among these, Taylor and Day (2004) found that conditional investment and cooperative investment equilibria can indeed evolve even in unstructured populations. The intuition behind this result is sometimes called partner fidelity feedback (Bull and Rice 1991, Sachs et al., 2004): if one stays with a given partner long enough and the partner employs a conditional investment strategy, it does not pay to cheat it for short-term benefit.

Akçay et al. generalized the behavioral equilibrium-based approach and found that the effects of such feedbacks in continuous, symmetric games can in general be characterized by a local response coefficient ρ, which controls how much a partner of a focal individual increases (or decreases) their investments in response to a small change in the focal individual's investment Akçay et al. (2009), Akçay and Van Cleve (2012), and Van Cleve and Akçay (2014). For within-species interactions with population structure, we used the Price equation approach to show that the response coefficient ρ plays a role symmetric to relatedness r in its effect on selection (Akçay and Van Cleve, 2012). Moreover, relatedness and responsiveness interact with each other synergistically in promoting more investment, such that populations with both genetic structure and phenotypic responses can maintain significantly more cooperation than expected by their separate effects. Van Cleve and Akçay (2014) extended such analysis to a simple model of symbiotic mutualism, finding that the synergistic interaction between phenotypic response coefficients on the one hand and within- and between-species genetic associations on the other is a general prediction of these models.

Phenotypic feedbacks can also be modeled using the methodology of quantitative genetics, which refers to phenotypic feedbacks as indirect genetic effects (IGEs; Moore et al. 1997, McGlothlin et al. 2010). The IGE framework starts with a standard quantitative genetics framework but then incorporates the fact that in a social interaction, an individual's phenotype (a behavior or a morphological trait) is a function of not just its own genetic makeup, but those of its partners as well. In the IGE framework, each individual's phenotype is a linear, additive function of its own breeding value and the phenotypes of others it interacts with multiplied by some IGE coefficient (usually denoted by Ψ). In this way, IGE models bear a close resemblance to the linear response rule models, except that the IGE coefficient is usually taken to be a constant (with some exceptions, e.g., Kazancioğlu et al. 2012). Akçay and Van Cleve (2012) demonstrated that for a single phenotype, there is a one-to-one mapping between IGE models in structured populations (McGlothlin et al. 2010) and the behavioral response formulation, with ρ and Ψ invertible functions of each other. The advantage of IGE models is that they use the machinery of quantitative genetics, which can handle multiple traits and their phenotypic and genetic covariances, and provide a robust framework for empirical measurement of such responses (McGlothlin and Brodie III 2009). The intersection of IGE models and models of behavioral and proximate mechanisms mediating the phenotypic feedbacks is an exciting area of future research (Akçay and Simms 2011, Akçay and Van Cleve, 2012).

4.5 Partner choice and beyond: economic theory of mutualism

Cooperation between species can also evolve if individuals preferentially interact with a cooperative subset of partners, or selectively terminate interactions with non-cooperative ones; in other words, if they exert partner choice. The idea of choice by organisms as a selective force goes back at least to Darwin's sexual selection theory, and has also played a central role in social evolution theory

(Axelrod and Hamilton 1981, Bull and Rice 1991). It also has close parallels with consumer choice and related economic concepts and thus has been a focus for the application of ideas and techniques from economics to biological problems (Hammerstein and Hagen 2005). There are many important biological questions to ask about partner choice, such as when to choose, what to base partner choices on, how stringent partner choice should be, etc. Answering this diverse set of questions requires a somewhat eclectic set of approaches, which are the focus of this section.

4.5.1 Supply and demand: biological markets

One prominent application of an economic idea to biology has been the concept of biological markets (Noe and Hammerstein 1994, 1995). Biological market theory asserts that, just like in human markets, the supply and demand for partners of high quality will determine the evolutionary dynamics of a mutualism through partner choice. This idea was first modeled by Noe and Hammerstein (1994), who considered two species that trade in some "commodities" (e.g., resources such as nitrogen or carbon, or services such as pollination). One of the species (say, species B) can invest by producing commodities of high or low value to the other species, low-value commodities being less costly to produce. Furthermore, there is some noise in production of commodities by species B, such that a high investment might produce a low value, and vice versa. This last feature ensures that there is phenotypic variability in the population that makes partner choice worthwhile, even if the population is fixed for one investment type. Species A, on the other hand, has a fixed amount of investment, but it has the ability to choose its partner from the B species, but that comes at a cost, in terms of the effort spent sampling different partners. The model has two different kinds of equilibria depending on the relative abundance of the species: if Species A is abundant, A individuals will accept low-value partners, and the B species will invest into low-value commodities. On the other hand, if B individuals are abundant, A individuals will only trade with high-value partners and consequently the B population is fixed for high investment. These results make sense from a supply–demand perspective: when A individuals are relatively rare, B individuals are willing to pay the cost of high investment to be chosen by a partner. Interestingly, there is a range of relative abundances where both equilibria exist at the same time, meaning that two different populations with the same abundances can inhabit either regime, depending on the initial conditions.

Applying the consumer choice metaphor at the individual decision-making level, Kummel and Salant (2006) modeled a plant that can associate with multiple mycorrhizal partners. Each partner is characterized by an exchange function, which denotes how much nutrient it supplies for a given amount of carbon it receives. The plant then has to decide how much nutrient it "buys" from each "supplier." Kummel and Salant showed that the optimal allocation of carbon will equalize the marginal cost of nutrient acquisition from each partner. Interestingly, the exchange function characterization admits a much richer space of variation in partner quality than a dichotomy between mutualists and cheaters: a partner can be very beneficial at low exchange rates (the exchange function goes up steeply at first) but not at higher exchange rates (the exchange function levels off), and vice versa. Kummel and Salant showed that depending on the set of exchange functions available to the plant, complex patterns of partner choice can emerge. An extension of this model that incorporates mycorrhizal fitness shows that it can also generate context dependency in the benefits to the mycorrhizae (Akçay, unpublished results).

Yet another approach based on the idea of markets was proposed by Schwartz and Hoeksema (1998, Hoeksema and Schwartz 2003), who asked when trade between two species is beneficial to begin with. In their model, both species need two distinct resources (say, carbon and phosphorus). The two species can either take both resources from the environment or they can each take up only one and trade with each other for the other. This situation is directly analogous to comparative advantage models from international trade going back to Adam Smith and David Ricardo (Ricardo, 1891). The basic idea is that countries might do best if they focus on producing goods for which they have a comparative advantage (e.g., Portuguese wine is better than British), and rely on trade to obtain other goods that they need. Schwartz and Hoeksema (1998) showed that under some "exchange rates" of the two resources, it is optimal for each species to specialize in taking up only one resource and trade with each other for the other. The existence of such exchange rates depends on the ratio at which each species optimally consumes the two resources (i.e., the stoichiometry of their growth), and their relative efficiency in obtaining each resource directly from the environment. Specialization with trade is optimal when the two species are relatively more efficient in taking up different resources from the environment.

4.5.2 The paradox of the market

The mechanisms of partner choice and biological markets as means to select against cheaters create something of a paradox: if partner choice is effective, cooperative partners will increase in frequency, leading most individuals to interact only with cooperative partners, leading to lower benefits from choice. Eventually, the cost of partner choice will create selection against it, which opens the door for cheaters to invade again. This dynamic is well known in the study of sexual selection as the "paradox of the lek" (Kirkpatrick and Ryan 1991), and has attracted a persistent research effort over more than thirty years. Foster and Kokko (2006) tackled this issue with a model in which a host interacts with symbionts that are drawn from a distribution of mutualistic quality, but chooses the more beneficial ones. This choice depletes available variation in mutualist quality locally, but Foster and Kokko proposed that it is continually replenished by migration from a source population in which variation is held constant. With enough such migration, costly partner choice and a high level of average cooperation can be maintained. In a fundamental sense, though, this model begs the question, since it assumes some unspecified mechanism that maintains the variation in the source population. On this issue, Foster and Kokko point out that the benefit from choice in mutualisms does not require additive genetic variation in the partner quality (as it does in the paradox of the lek), so purely environmental or non-additive genetic variation (both of which are easy to maintain even under selection) can make partner choice beneficial in the absence of additive genetic variation in partner quality. This is consistent with empirical evidence that suggests that most variation in partner quality is not additive (with unconditionally beneficial and non-beneficial partners); rather, the costs and benefits are highly context dependent (see Chapters 6 and 10, as well as Section 4.7.2). However, such non-additive variation and mechanisms that generate and maintain variation in mutualist quality in general remain relatively unexplored in theory (Heath and Stinchcombe, 2014, see also Section 4.6).

4.5.3 Sanctions: partner choice in the legume-rhizobium symbiosis

One of the most important model empirical systems for studying partner choice has been the symbiosis between legume plants and the nitrogen-fixing soil bacteria collectively called rhizobia (see Boxes 4.2 and 5.2). Rhizobia are taken up from the soil and inhabit specialized plant root organs called nodules. A single plant can have hundreds of nodules, each inhabited by one or more rhizobium strains. Since rhizobia are not vertically transmitted, there is a conflict of interest between the plant and the rhizobia, and hence the prospect of cheating.

Partner choice in the legume-rhizobium system is frequently referred to as "sanctions," a term introduced by Denison (2000) who proposed that legume plants control cheating by selectively withholding resources to nodules that do not fix nitrogen. West et al. 2002a, b) have formalized Denison's ideas in two separate models. West et al. (2002b) considered a kin-selection model with local competition among a population of rhizobia that occupy the same plant, using the approach of Taylor and Frank (1996; see Section 4.3.1). To model partner choice by the plant, West et al. (2002b) assumed that a focal rhizobium strain's fitness is given by a continuous, increasing function of the focal strain's nitrogen fixation rate. They considered two different forms for this function: one where it depends only on the focal strain's fixation rate (which they termed fixed-rule sanctions) and one where it depends both on the focal strain's fixation rate and the average of all nodules (termed relative-rule sanctions). Both types of partner choice select for more nitrogen fixation compared to no partner choice. Increasing relatedness between rhizobia on different nodules increases the evolutionarily stable fixation rate under fixed-rule sanctions, but, interestingly, it has the opposite effect under relative-rule sanctions. This is because high relatedness increases the average nitrogen fixation rate on plants that contain a highly fixing focal strain, which decreases the benefit the focal strain obtains from the plant under relative-rule sanctions.

In the companion paper, West et al. (2002a) used the same modeling approach to ask when it is beneficial for the plant to engage in sanctions when interacting with rhizobium strains that vary in their nitrogen fixation rate. They modeled plant fitness as proportional to $R*N$, where R stands for the carbon the plant has available for its own growth and reproduction (after investing in the production of nodules) and N is the total nitrogen the plant obtains from the nodules and the soil. The amount of resources available for growth, R, is a function of how many nodules a plant makes initially, how many it sanctions, and what proportion of the resources invested into sanctioned nodules the plant can reclaim. Likewise, N is a function of how many nodules there are after sanctions, their average fixation rate, and the availability of nitrogen from the soil. The crucial assumption in the model is that the average fixation rate of the non-sanctioned nodules is increasing in the fraction of nodules sanctioned, although the mechanism by which this happens is not modeled. Under these assumptions, stricter sanctions are favored when more resources are recouped by sanctioning, there is a higher mean fixation rate among rhizobia before sanctions, and a higher variance in fixation rates.

These two models confirm our intuition that sanctions might favor cooperation and evolve through direct benefits. However, since they treat rhizobia and plants separately, they do not fully elucidate the patterns of fitness variation that sanctions would generate between plants and rhizobia. Further, West et al. (2002a) sidestepped two important facts: first, that the distribution of rhizobium strains of varying quality on a plant will determine how much benefit the plants gain by sanctioning, and second, the possibility of mutual control over partner choice. These factors are treated in the models discussed next.

4.5.4 Control, negotiation, and outside options in mutualisms

In the models in Sections 4.5.1, 4.5.2, and 4.5.3, partner choice is a unilateral affair: one party has complete control over who it interacts with and how much. When both parties have some control, richer dynamics ensue. Johnstone and Bshary (2002) considered a model inspired by the cleaner fish mutualism, where small cleaner fish remove ectoparasites from larger fish that visit their territories, but prefer to feed on healthy mucus tissue (see Box 6.2 for more on this mutualism). In the model of Johnstone and Bshary (2002), client–cleaner pairs gain mutual benefits as an increasing function of the time they spend together. The cleaners can, however, exploit the clients by biting healthy mucus tissue rather than removing parasites (Bshary and Grutter 2002), gaining an additional benefit per unit time at a cost to their clients. The evolutionarily stable exploitation level will be determined by the duration of the interaction at different exploitation levels, which in turn depends on which party controls the duration of the interaction. Johnstone and Bshary (2002) assumed that the interaction ends when the marginal cost of continuation to the cleaners exceeds the marginal benefit multiplied by a coefficient signifying the "power" cleaners have over the clients. The more power cleaners have, the closer the duration will be to their optimum. With increasing power to the cleaners, evolutionarily stable exploitation levels increase. This shows that the effectiveness of partner choice depends on whether it can be exercised free from interference from the partners.

In their initial model, Johnstone and Bshary (2002) did not account for the alternatives clients or cleaners have once they leave an interaction, in other words, their outside options. Johnstone and Bshary (2008) modeled outside options arising from population level availability of cleaners and clients, which is a function of their relative abundances, as well as how long they stay in each pairwise interaction. Johnstone and Bshary (2008) used a time-in/time-out approach, where individuals transition from being free-living (time in the "market") to being in an interaction (time out), at a rate that is proportional to the availability of partners. The time-out duration is determined as in Johnstone and Bshary

(2002), but with the additional consideration of the opportunity cost of leaving, which depends on how fast a new partner can be found. This opportunity cost emerges as a property of equilibrium, with only the overall abundance of clients and cleaners specified exogenously. The results support the biological market hypothesis: when clients are rare, low or no exploitation is expected at equilibrium, but when cleaners are rare, exploitation levels are expected to be high. In the former case, cleaners face a high opportunity cost of exploiting a client and causing it to leave, since replacing a partner takes a long time; in the latter case clients are facing the high opportunity cost. Thus, Johnstone and Bshary (2008) showed the importance of outside options in determining evolutionary outcomes of biological markets.

In interactions where fast partner switching is not an option, as in the legume-rhizobium mutualism, other mechanisms will determine the outcome of the mutualism. The models of West et al. (2002a, b) assume that the nitrogen fixation rates are "sealed bids," determined by the rhizobium genotype only, while the plant responds to these bids. Akçay and Roughgarden (2007) investigated an alternative where the fixation rates are the result of a "negotiation" process. The key idea is that once a nodule is established, there will be a range of carbon and nitrogen exchange rates with a focal nodule that make both the plant and the rhizobia in that nodule better off (grow more) relative to no exchange. But some of these exchange rates will favor the plant more, and some the rhizobia. Where along that continuum the outcome ends up can have important consequences for how partner choice works on a longer time scale. In the negotiation between bacteria and plants, the "offers" come in the form of stochastic fluctuations in resource fluxes. When these fluctuations increase one party's growth rate while decreasing the other's, Akçay and Roughgarden conjectured that there would be a period of costly conflict between the parties (modeled as a war of attrition), in which the party with more to gain (relative to giving in) wins. This simple process converges to the maximum of the product of the plant's and nodule's growth rates. This stable equilibrium point is called the Nash bargaining solution (Nash 1950, Rubinstein, 1982) in economics and provides a basis for quantitatively predicting the outcome of a negotiation process.

Akçay and Simms (2011) later extended the pairwise negotiation model to negotiation with two nodules, each providing an outside option to the plant vis-à-vis the other and allowing the "negotiation strategy" of the rhizobia to evolve. In particular, they considered rhizobia that have two evolving traits. The first was anthropomorphically termed "stubbornness," i.e., how hard it is to get a rhizobium strain to increase its nitrogen fixation or decrease carbon intake. A more stubborn strain fixes less nitrogen for more carbon, and thus is less cooperative. Second, a focal nodule could provide a good or bad outside option to the plant, by increasing its nitrogen fixation temporarily when the plant is negotiating with the other nodule. This outside option trait is nearly neutral for the focal nodule, since it is only a temporary increase in nitrogen fixation and does not affect the equilibrium fixation rate for the focal nodule directly. When the more stubborn (less cooperative) nodule has a lower growth rate than the more cooperative nodule, we can say the plant is exercising effective partner choice (or sanctions). It was shown that effective partner choice requires that the less stubborn strain also provides a better outside option to the plant. This finding provides a potential explanation for why some studies of the legume-rhizobium symbiosis show effective partner choice (Kiers et al. 2003, 2006, Simms et al. 2006), but others do not (Marco et al. 2009, Gubry-Rangin et al. 2010). Strikingly, regardless of whether partner choice is effective, selection for stubbornness (i.e., against cooperation) is always positive. This is because the plant has incomplete control over partner choice, and all else being equal, a more stubborn rhizobium strain will always extract more resources from the plant. To counteract this selection, a positive feedback at the whole-plant level is required, so that nodules on plants that grow better do better altogether. This finding shows that phenotypic feedbacks and selection at multiple levels might be required to overcome the limitations of partner choice when both parties have control over the interaction (see Box 4.2).

Grman et al. (2012) recently combined the Nash bargaining approach of Akçay and Roughgarden

(2007) with the comparative advantage model of Schwartz and Hoeksema (1998) to ask how a negotiated exchange rate affects the evolution of specialization in a plant–mycorrhizal mutualism. Mycorrhizal fungi infect the roots of a wide variety of plants, and trade nutrients they acquire from the soil with the plant for carbon (Smith and Read 2008). Grman et al. (2012) found that specialization is somewhat less likely with negotiated exchange, since it makes a party completely dependent on the partner (i.e., reduces the outside options) and means a worse bargaining outcome. Nevertheless, for empirically derived parameter regimes, specialization (especially of the fungus into nutrient uptake) does occur at the equilibrium. This study and Johnstone and Bshary (2008) together demonstrate the utility of combining models that apply at pair and population levels (see Box 4.2, and Section 4.7.1).

The general message from the set of models in this subsection is that the outcome of a mutualism depends strongly on the mechanisms of control each party has over partners and the relative outside options (e.g., alternative partners, ability to take up resources from the environment, etc.) each party has. This fact sets the stage for rich evolutionary and ecological dynamics that have not yet been fully explored.

4.5.5 Screening

An alternative to trying to stop a partner's exploitation during the interaction is to not associate with an exploiter in the first place. In the cleaner fish mutualism, for example, clients choose to interact with only non-exploiting partners, as determined by own experience or observing interactions with others (Bshary and Grutter 2002, 2006). As discussed in Section 4.5.1, such "market choice" can preclude the evolution of exploitation, but it requires that clients get reliable information in the face of incentives by the cleaners to misrepresent.

This is a problem that has direct analogies in economics: when a business wants to hire new employees, it will have imperfect information about the quality of the applicants. Obviously, applicants for the vacancy have an incentive to exaggerate their abilities. One way to resolve this problem is to make the applicants reliably signal their quality before the interaction, which usually involves costly signals (Spence 1973; the same idea was proposed independently in biology as the handicap principle by Zahavi 1975). But an entirely different way of separating high- and low-quality applicants is to let them sort themselves out. This is termed "screening" in economics (Rothschild and Stiglitz 1976, Riley 2001). For example, a firm can offer a job contract to a candidate that entails a low salary with a big bonus that is only paid if a certain performance criterion (e.g., sales volume) is satisfied. Candidates that know they cannot satisfy the criterion will have no incentive to accept the offer, and those that do accept it will automatically have the correct incentives to work hard, without the need for constant monitoring by their employer. The result mimics effective partner choice, but there is no explicit signaling of quality, or evaluation of partners by the chooser.

A series of recent papers proposed that screening theory can be applied to understand mutualisms (Archetti et al. 2011, Scheuring and Yu 2012). Archetti et al. (2011) used the symbiosis between the Hawaiian bobtail squid (*Euprymna scolopes*) and *Vibrio fischeri* bacteria (see Box 5.3) as an example. The bacteria colonize the light organ of the squid, where they emit light that reduces detection of the squid by predators. The luciferase enzyme that is responsible for bioluminescence also happens to remove reactive oxygen from the interior of the light organ, where it can damage the bacterial cells. Archetti et al. interpreted this as a screening mechanism: only those bacterial strains that can produce light can survive in this hostile environment. Therefore, for would-be cheaters (strains that don't produce light), colonizing the squid is a fitness dead end, and only light-producing strains will evolve the traits necessary to colonize the squid. This obviates the need to control cheating after the initial infection, similar to the job offer with bonus example in the previous paragraph.

The second paper, Scheuring and Yu (2012), proposed a different screening mechanism, based on competition between antibiotic-producing bacterial strains. Using a model originally developed by Mao-Jones et al. (2010) for the establishment of coral diseases, Scheuring and Yu argued that the host can screen for a beneficial microbiome by letting the antibiotics produced by beneficial symbionts create the hostile environment for the non-beneficial ones. Following Archetti et al. (2011), they call this mechanism "competitive screening." The crucial divergence of this model from economic screening models is that the symbiont dynamics within a host are bistable: if non-beneficial microbes initially colonize a host at high frequency, they persist at equilibrium. Thus, unlike standard screening models, the selection on cheaters for not colonizing the host requires some preinfection partner choice mechanism (or vertical transmission) on the part of the host that results in a high initial frequency of beneficial microbes.

4.5.6 Economics and mutualisms: a bright future

This section has illustrated how fruitful economic analogies and techniques have been in understanding partner choice and related phenomena in mutualisms. This is not a coincidence of course: mutualisms are exchanges of resources and services between different species, whereas economics studies the exchange of resources and services among humans. The future of economic theories of mutualisms seems very bright, with a great deal of work that remains to be done in exploring strategic questions related to mutualisms using game theory, market modeling, and related theories such as signaling and screening.

4.6 Evolution and maintenance of diversity in mutualisms

In nature, most mutualisms involve tremendous variation in how beneficial the different species or strains are to their partners. However, relatively few models have specifically focused on the emergence and maintenance of diversity (more specifically, variation in cooperative phenotypes) in mutualisms.

One of the most sustained theoretical contributions to this question has come from the adaptive dynamics (AD) approach. As described in Box 4.1, AD can be used to find conditions under which selection will cause a population to branch phenotypically, evolving in two different directions. Doebeli and Dieckmann (2000) analyzed an instance of such evolutionary branching in mutualisms as part of a larger compendium of models for adaptive speciation. They considered standard Lotka–Volterra dynamics of two mutualists, but let the interaction coefficients between the species evolve, with more benefits from the interaction if the trait values (e.g., flowering time for a flower and emergence time for its insect pollinator) match, with the trade-off that each species also has an independent (e.g., abiotic) optimum value for the trait. Doebeli and Dieckmann found that if these trait optima are far enough apart then evolutionary branching might occur, resulting in two species pairs that match each other well.

One potential objection to Doebeli and Dieckmann's approach is that in their model, the cost of mutualism is due to deviation from some external optimum, rather than costly investments that are beneficial to the partner. This issue is addressed by Ferrière et al. (2002), who considered the evolution of costly provisioning of some resource by partners in an obligate mutualism. Their model incorporates intraspecific competition in each species for both the resources provided by the partner species as well as for some other, mutualism-independent, resource. The ecological dynamics of the two populations are stable when rates of resource provision are positive but not too high. Cheating can destabilize mutualism evolutionarily (and lead populations to extinction) when individuals compete equally well for partner-provided resources regardless of their own provision. A competitive advantage for better mutualists—not surprisingly—counteracts cheating, but there is also the potential for evolutionary branching in the provision rate when the competitive advantage of mutualists is accelerating faster than the cost of providing commodities. Biologically, this means that mechanisms such as partner choice that create steep differences between the fitness of good and bad mutualist partners are more likely to create diversity in mutualistic quality.

Ferrière et al. (2007) extended this model to include a third, unrelated, species that only exploits its partner, without providing any benefits. They found that if an exploiter invades a system after high levels of cooperation have evolved, it can take over and cause the collapse of the mutualism. In contrast, if the exploiter invades the system early, the provision rates evolve to lower stable values, but are also immune to further invasion by exploiters. A similar effect was found by Golubski and Klausmeier (2010) in an "ultimatum game" model. In the same three-species setting, Jones et al. (2009) further found that the intensity of intraspecific competition among the mutualists and exploiters mediates their coexistence, with high intraspecific competition in both promoting coexistence, and low competition in one tending to drive the other to extinction.

Borrowing an approach from ecological coexistence theory, Hoeksema and Kummel (2003) used a patch occupancy model to explore how cooperative genotypes (in their case, of mycorrhizae) can coexist with non-cooperative genotypes. In contrast to the AD models, they assumed that the cooperative genotypes are actually competitively inferior to the non-cooperative types, due to their costly resource provisioning. Nonetheless, both can survive if cooperative mycorrhizae are better able to colonize empty root tips as they appear. This is one of the few models that has investigated the trade-offs between different traits affecting mutualist performance, which is likely to be of importance in maintaining mutualism variation.

4.7 Future directions

4.7.1 Integrating mechanisms and modeling approaches

The main message of this chapter is that there are many different mechanisms and selective pathways can resolve conflicts in mutualisms. Existing models tend to consider each selective pathway in isolation, but it is likely that multiple mechanisms frequently operate at the same time, and at different levels. For example, the model of Akçay and Simms (2011) for the legume-rhizobium mutualism suggests that the interaction of feedbacks at the nodule and whole-plant level determines whether there will be selection for cooperation or not. Box 4.2 by Simms and Porter discusses in more detail the multiple levels of selection in a hierarchically organized symbiosis. Using the legume-rhizobium mutualism as an example, they argue that phenotypic feedbacks and population structure at different scales generate different selection pressures at multiple levels of selection. The net outcome of selection on symbiosis traits is determined by the interaction of these feedbacks and levels. So far, there has been relatively little theoretical work that tries to integrate feedbacks and selection pressures at multiple levels.

In one of the only attempts to integrate multiple mechanisms, Foster and Wenseleers (2006) considered a model where the fitness of a mutualist of Species A depends on the mutualist's own investment, a "passive" phenotypic feedback (akin to byproducts) from the partner in response to investment, active partner choice, and the help received from the partner. They then decomposed the latter three feedbacks as a product of a total potential benefit, the realized fraction of the potential (due to the degree of fidelity between the partners), and the relatedness of Species A individuals that benefit from each feedback, and wrote a first-order condition for increased investment (i.e., the total derivative of fitness with respect to A's investment) in this expanded form. Interpreting this equation, Foster and Wenseleers argued that passive feedbacks and active partner choice should play a more important role in promoting mutualisms than genetic associations between species. Importantly, Foster and Wenseleers made explicit that different processes might operate at different scales, and therefore be subject to different population structures, summarized by their separate relatedness coefficients. On the other hand, the way they modeled phenotypic feedbacks precludes any interaction between these processes; in particular, each species' investment phenotype is assumed to be independent of these feedbacks. This would be inconsistent with what most people regard as active or passive phenotypic feedbacks, in which, say, a positive feedback to a good mutualist would most likely be higher mutualistic investment into that partner (e.g., Doebeli and Knowlton, 1998). When feedbacks occur over investments, a variety of models find that the genetic associations and phenotypic feedback coefficients interact both multiplicatively and additively (Lehmann and Keller 2006, McGlothlin et al. 2010, Akçay and Van Cleve 2012, Van Cleve and Akçay, 2014, see also Section 4.4), and Foster and Wenseleer's conclusions will likely change.

Box 4.2 Phenotypic feedbacks and multilevel selection in mutualisms

Ellen Simms and Stephanie Porter

Costly cooperative investments trade off with individual fitness and directly select for cheating [see earlier in this chapter; Figure 1(A)], but positive host responses to symbiont cooperation can counter selection for cheating (Figure 1(B); Sections 4.4, 4.5, and 4.7.3). These phenotypic feedbacks can align partners' fitness (Sachs et al., 2004) and produce positive net fitness covariance between partners [Figure 1(C)], despite the potential benefits of cheating.

Many mutualisms exhibit hierarchical spatial organization delimited by modules. Examples include flowers in the yucca–moth interaction, inflorescences in the wasp–fig interaction, and nodules in the legume-rhizobium interaction. Hierarchical spatial organization of symbionts creates the potential for multilevel selection (Okasha, 2006). Therefore, exclusively measuring selection and feedbacks at only one level—among modules, for example—would be inadequate for predicting the evolution of cooperation if multiple symbiont genotypes occur within a module or feedbacks and selection can occur at levels above and/or below the module (e.g., see the discussion of Akçay and Simms (2011) in Section 4.5.4).

The legume-rhizobium mutualism offers an especially interesting case for examining multilevel selection on symbionts. Soil-dwelling rhizobium bacteria infect legume roots, stimulate nodule formation, proliferate, and may differentiate to fix nitrogen in exchange for plant-derived photosynthates (Oldroyd et al., 2011). Nodules can contain one or more rhizobium genotypes (Friesen and Mathias, 2010). Reproductive rhizobium cells return to soil populations upon nodule senescence and symbiosis is reinitiated with each new nodule. Thus, the scales of selection relevant to the maintenance of cooperative rhizobia are: 1) among free-living rhizobium cells, 2) among rhizobium genotypes within a mixed nodule, 3) among nodules, and 4) among plants (Figure 2). Further, phenotypic feedbacks by hosts can modulate selection within each host generation.

Selection on cooperation in the legume-rhizobium symbiosis may differ in sign and magnitude among levels. Less-cooperative rhizobia might better survive in soil or compete for nodulation sites (Denison and Kiers 2011, Oldroyd et al., 2011), leading to selection against rhizobium cooperation, but phenotypic feedback via host choice prior to nodule formation can favor more cooperative rhizobium cells (Heath and Tiffin 2009). After infection, sanctions (Kiers et al. 2003) or partner fidelity feedback (Bull and Rice 1991, Sachs et al. 2004) can favor nodules with rhizobium subpopulations that express greater cooperation, leading to positive selection at the nodule level. If less-cooperative rhizobia could manipulate host responses or free-ride within mixed nodules then direct selection would favor less-cooperative rhizobia within nodule subpopulations (Friesen and Mathias 2010). However, rhizobium cheating might reduce host fitness and thereby favor cooperation via selection at the level of the entire host. Phenotypic feedback by hosts could also create cell-level selection by favoring more-cooperative rhizobium cells within a nodule. Thus, understanding the maintenance of cooperation in this symbiosis requires accounting for the levels, signs, and relative magnitudes of selection in the soil, in nodules, and among plants.

This accounting will be complicated by environmental variation that could dynamically shift the relative magnitudes of selection and host responses across scales. Increasing soil rhizobium density increases the number of mixed infection nodules on a host (Friesen and Mathias 2010, Sachs et al. 2010), which would increase within-nodule selection for cheating if uncooperative rhizobia were more fit than cooperative genotypes within mixed infection nodules. Host ontogeny affects the magnitude of selection at

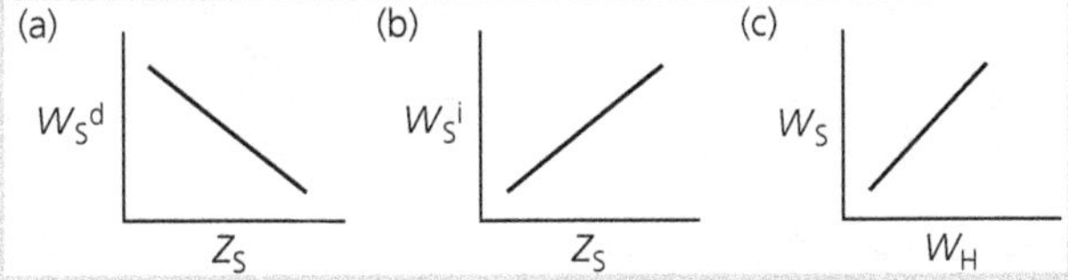

Figure 1 Direct symbiont fitness cost, W_S^d, as a function of resources allocated to cooperation phenotype, Z_S (a). Indirect symbiont fitness benet W_S^i, as a function of symbiont cooperation phenotype, resulting from feedback from the host (b). Net fitness alignment of host, W_H, and symbiont fitness W_S in stable mutualism (c).

continued

Box 4.2 *Continued*

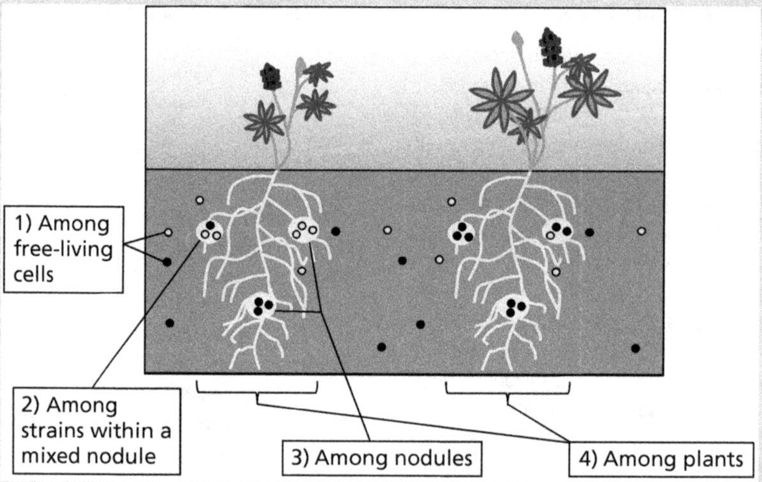

Figure 2 Levels of selection and phenotypic feedbacks relevant to the maintenance of cooperative rhizobia. Dark circles, cooperative rhizobia; light circles, uncooperative rhizobia.

the level of the whole host. Seedlings host few nodules and if early viability selection on seedlings dominates selection among plants, then plant-level selection for rhizobium cooperation could be strong. However, the number of nodules, and thus the potential diversity of symbionts sharing a host, increases as plants mature, which reduces whole-plant-level selection.

The availability of traded resources could also impact selection (Heath and Tiffin 2009, Akçay and Simms 2011). For example, shading and high external nitrogen both increase carbon limitation, leading plants to drive a harder bargain with rhizobia and thus favor more "stubborn" rhizobia that are poorer nitrogen fixers. However, if shading resulted from interplant competition, it could strengthen viability selection among plants, thereby increasing whole-plant-level selection in favor of rhizobial cooperation. High soil nitrate availability reduces nodulation rates via autoregulation of nodulation, which could also increase selection for cooperation if it strengthened preinfection partner choice. While we have highlighted levels of selection that we expect to be sensitive to external contexts, further development of this hypothesis requires more explicit theoretical models to clarify the implications of such selection, as well as empirical measurements of selection on symbionts at multiple scales in contrasting environmental conditions. This research will improve predictions about the evolution of cooperation within realistically rich patterns of environmental heterogeneity.

References

Okasha, S. (2006). *Evolution and the Levels of Selection.* Oxford: Oxford University Press.

Oldroyd, G.E., Murray, J.D., Poole, P.S., and Downie, J.A. (2011). The rules of engagement in the legume-rhizobial symbiosis. *Annual Review of Genetics*, **45**, 119–144.

Friesen, M.L. and Mathias, A. (2010). Mixed infections may promote diversification of mutualistic symbionts: why are there ineffective rhizobia? *Journal of Evolutionary Biology*, **23**, 323–334.

Denison, R.F. and Kiers, E.T. (2011). Life histories of symbiotic rhizobia and mycorrhizal fungi. *Current Biology*, **21**, R775–R785.

Heath, K.D. and Tiffin, P. (2009). Stabilizing mechanisms in a legume–rhizobium mutualism. *Evolution*, **63**, 652–662.

Sachs, J.L., Russell, J.E., Lii, Y.E., Black, K.C., Lopez, G., and Patil, A.S. (2010). Host control over infection and proliferation of a cheater symbiont. *Journal of Evolutionary Biology*, **23**, 1919–1927.

4.7.2 Context dependency

One of the major lessons from the past couple of decades in mutualism research is that the benefits from most mutualisms (or even whether the interaction is mutualistic or not) depend on both the biotic and abiotic context (see Chapter 10). Increasingly, it appears that most variation we see in the costs and benefits of mutualisms in nature are due to these context-dependent effects. Identifying cheaters unequivocally in nature has been challenging (see Chapter 6). One important source of context dependency is the evolved phenotypic feedbacks (partner fidelity or partner choice) that stabilize mutualisms. For example, in the relative sanctions model of West et al. (2002b), rhizobium fitness depends on the other strains infecting the same plant, because the plant's partner choice is relative to the mean nitrogen fixation rate. Models that explicitly focus on the causes and consequences of context dependency either due to environment (Pringle et al. 2013), life-history variation (Palmer et al. 2010), or phenotypic responses (Akçay and Simms 2011) will be useful in understanding the variation in the costs and benefits of mutualisms in the environment. Coupling these models with the ecological and resource dynamics of the mutualisms (e.g., soil nitrogen and plant interspecific competition in the legume-rhizobium case) will also help predict the responses of mutualistic interactions to global change (Kiers et al. 2010).

4.7.3 Alignment of evolutionary interests

With the focus on the cheater problem in the literature (see Chapter 6), it is easy to perceive mutualisms as a constant battleground. However, empirical research shows broad alignment of interest between parties in most, if not all, mutualisms (e.g. Friesen 2012, see also Chapter 6). Yet, most theory is based on the assumption that mutualisms must have an underlying conflict of interests. This assumption is certainly supported by what we know about "first principles" in many mutualisms, for example, the fact that nitrogen fixation *is* energetically costly for rhizobia. At face value, we seem to have a mismatch between theory and empirical findings. But a more nuanced reading of the theory suggests a different interpretation. All of the mechanisms discussed in this chapter that increase mutualistic genotypes' fitness relative to cheaters are implicitly aligning the evolutionary interests of the parties. In this view, mutualisms are rife with *potential* for conflicts of interests, but these are resolved through one or more mechanisms. Equivalently, one might say that mutualisms evolve only when the underlying conflicts of interests are sufficiently resolved.

Alignment of interests therefore provides a complementary perspective to the cheating problem. However, theoretical literature only recently started to focus explicitly on resolving conflicts of interests. Along these lines, Worden and Levin (2007) considered a population that starts out playing a prisoner's dilemma game, but is subject to mutations that change the payoff matrix of the game. They showed that eventually, the population evolves to play a game of perfectly aligned interests. The evolution toward mutual interest is unconstrained because Worden and Levin assume no cost to helping (or not hurting) one's partner. Later, Akçay and Roughgarden (2011) analyzed a related model where the mutation that gives an incentive to the partner to cooperate is costly to the focal individual. We showed that despite costs, such incentives can invade a population if they are sufficiently big to induce the partner to put up with cheating. Interestingly, under a diploid population genetic model, a stable polymorphism arises between the different alleles that modify the game. The result is a rich diversity of game structures that coexist in a population. Behaviorally, even though defecting individuals on average have higher fitness than cooperators, the system is nonetheless stable, because the genetic polymorphism is not in the behavior itself, but the games individuals play. These models have not yet been extended to interactions between species, which represents a fruitful avenue for future research. It will also be important to develop a formal framework to measure and quantify the alignment of fitness interests. Quantitative genetic theory is likely to provide a good basis for such a theory.

Alignment of interests is a harder problem to solve when there is asymmetric or private information (i.e., one party can condition its behavior on some information while the other party cannot), as in the signaling or screening models discussed in Section 4.5.5. The theory of mechanism design in economics (Myerson 1979, Myerson and Satterthwaite 1983, Akçay et al. 2012) deals with these situations and concerns itself with finding games that achieve the best outcome (defined according to some performance measure) given private information and conflicts of interests. In general, it is not always possible to achieve full efficiency (i.e., taking advantage of all the opportunities for mutual benefit) when parties are privately informed (Myerson and Satterthwaite 1983, Akçay et al. 2012). However, mechanism design provides techniques to compute game structures that achieve a second-best outcome given the constraints of private information. If one makes the working assumption that natural selection modifies games to achieve the most mutual benefit that is compatible with individual-level selection, these second-best mechanisms provide testable hypotheses about what we might expect to see in nature. More importantly, an evolutionary version of mechanism design theory can open the door to a deeper understanding of how natural selection structures biological organization for mutual benefit in both mutualisms and within-species cooperation (Akçay et al. 2010).

4.7.4 Conclusion

This chapter has presented a survey of the main ideas that have been influential in mutualism research. It is of course impossible to be all-inclusive (the existing literature is already too big for that), but it has tried to highlight the principal ideas in evolutionary theory of mutualisms and the modeling approaches used to address them. We are beginning to see a new generation of theoretical approaches to social evolution and mutualisms, and these hold the promise of both uncovering new theoretical principles but also bringing models closer to data. When this book is updated in a few decades we will surely see the fruits of that effort.

References

Akçay, E., Meirowitz, A., Ramsay, K.W., and Levin, S.A. (2012). Evolution of cooperation and skew under imperfect information. *Proceedings of the National Academy of Sciences USA*, **109**, 14936–14941.

Akçay, E., Roughgarden, J., Fearon, J.D., Ferejohn, J.A., and Weingast, B. (2010). Biological institutions: the political science of animal cooperation. Working Paper.

Akçay, E., and Roughgarden, J. (2007). Negotiation of mutualism: rhizobia and legumes. *Proceedings of the Royal Society, Series B: Biological Sciences*, **274**, 25–32.

Akçay, E., and Roughgarden, J. (2011). The evolution of payoff matrices: providing incentives to cooperate. *Proceedings of the Royal Society, Series B: Biological Sciences*, **278**, 2198–2206.

Akçay, E., and Simms, E.L. (2011). Negotiation, sanctions, and context dependency in the legume-Rhizobium mutualism. *The American Naturalist*, **178**, 1–14.

Akçay, E., and Van Cleve, J. (2012). Behavioral responses in structured populations pave the way to group optimality. *The American Naturalist*, **179**, 257–269.

Akçay, E., Van Cleve, J., Feldman, M.W., and Roughgarden, J. (2009). A theory for the evolution of other-regard integrating proximate and ultimate perspectives. *Proceedings of the National Academy of Sciences USA*, **106**, 19061–19066.

Archetti, M., Ubeda, F., Fudenberg, D., Green, J., Pierce, N.E., and Yu, D.W. (2011). Let the right one in: a microeconomic approach to partner choice in mutualisms. *The American Naturalist*, **177**, 75–85.

Axelrod, R., and Hamilton, W.D. (1981). The evolution of cooperation. *Science*, **211**, 1390–1396.

Bever, J.D., and Simms, E.L. (2000). Evolution of nitrogen fixation in spatially structured populations of rhizobium. *Heredity*, **85**, 366–372.

Bshary, R., and Grutter, A.S. (2002). Asymmetric cheating opportunities and partner control in a cleaner fish mutualism. *Animal Behavior*, **63**, 547–555.

Bshary, R., and Grutter, A.S. (2006). Image scoring and cooperation in a cleaner fish mutualism. *Nature*, **441**, 975–978.

Bull, J.J., and Rice, W.R. (1991). Distinguishing mechanisms for the evolution of co-operation. *Journal of Theoretical Biology*, **149**, 63–74.

Denison, R.F. (2000). Legume sanctions and the evolution of symbiotic cooperation by rhizobia. *American Naturalist* , **156**, 567–576.

Dieckmann, U., and Law, R. (1996). The dynamical theory of coevolution: a derivation from stochastic ecological processes. *Journal of Mathematical Biology*, **34**, 579–612.

Doebeli, M., and Dieckmann, U. (2000). Evolutionary branching and sympatric speciation caused by different types of ecological interactions. *The American Naturalist*, **156**, S77–S101.

Doebeli, M., and Knowlton, N. (1998). The evolution of interspecific mutualisms. *Proceedings of the National Academy of Sciences USA*, **95**, 8676–8680.

Ewald, P.W. (1987). Transmission modes and evolution of the parasitism-mutualism continuum. *Annals of the New York Academy of Sciences*, **503**, 295–306.

Ferdy, J.-B., and Godelle, B. (2005). Diversification of transmission modes and the evolution of mutualism. *American Naturalist*, **166**, 613–627.

Ferrière, R., Bronstein, J.L., Rinaldi, S., Law, R., and Gauduchon, M. (2002). Cheating and the evolutionary stability of mutualisms. *Proceedings of the Royal Society, Series B: Biological Sciences*, **269**, 773–780.

Ferrière, R., Gauduchon, M., and Bronstein, J.L. (2007). Evolution and persistence of obligate mutualists and exploiters: competition for partners and evolutionary immunization. *Ecology letters*, **10**, 115–126.

Fine, P.E. (1975). Vectors and vertical transmission: an epidemiologic perspective. *Annals of the New York Academy of Sciences*, **266**, 173–194.

Foster, K.R., and Kokko, H. (2006). Cheating can stabilize cooperation in mutualisms. *Proceedings of the Royal Society, Series B: Biological Sciences*, **273**, 2233–2239.

Foster, K.R., and Wenseleers, T. (2006). A general model for the evolution of mutualisms. *Journal of Evolutionary Biology*, **19**, 1283–1293.

Frank, S. (1994). Genetics of mutualism: the evolution of altruism between species. *Journal of Theoretical Biology*, **170**, 393–400.

Frank, S. (1998). *Foundations of Social Evolution*. Princeton: Princeton University Press.

Frank, S.A. (1997). The price equation, Fisher's fundamental theorem, kin selection, and causal analysis. *Evolution*, **51**, 1712–1729.

Friesen, M.L. (2012). Widespread fitness alignment in the legume-rhizobium symbiosis. *New Phytologist*, **194**, 1096–1111.

Golubski, A.J., and Klausmeier, C.A. (2010). Control in mutualisms: combined implications of partner choice and bargaining roles. *Journal of Theoretical Biology*, **267**, 535–545.

Grman, E., Robinson, T.M.P., and Klausmeier, C.A. (2012). Ecological specialization and trade affect the outcome of negotiations in mutualism. *The American Naturalist*, **179**, 567–581.

Gubry-Rangin, C., Garcia, M., and Béna, G. (2010). Partner choice in *Medicago truncatula–Sinorhizobium* symbiosis. *Proc. R. Soc. B*, **277**, 1947–1951.

Hamilton, W.D. (1964). The genetical evolution of social behaviour. *Journal of Theoretical Biology*, **7**, 1–16.

Hammerstein, P., and Hagen, E.H. (2005). The second wave of evolutionary economics in biology. *Trends in Ecology & Evolution*, **20**, 604–609.

Heath, K.D., and Stinchcombe, J.R. (2014). Explaining mutualism variation: A new evolutionary paradox? *Evolution*, **68**, 309–317.

Hoeksema, J.D., and Kummel, M. (2003). Ecological persistence of the plant–mycorrhizal mutualism: a hypothesis from species coexistence theory. *American Naturalist*, **162**, S40–S50.

Hoeksema, J.D., and Schwartz, M.W. (2003). Expanding comparative-advantage biological market models: contingency of mutualism on partner's resource requirements and acquisition trade-offs. *Proceedings of the Royal Society of London. Series B: Biological Sciences*, **270**, 913–919.

Johnstone, R.A., and Bshary, R. (2002). From parasitism to mutualism: partner control in asymmetric interactions. *Ecology Letters*, **5**, 634–639.

Johnstone, R.A., and Bshary, R. (2008). Mutualism, market effects and partner control. *Journal of Evolutionary Biology*, **21**, 879–888.

Jones, E.I., Ferrière, R., and Bronstein, J.L. (2009). Eco-evolutionary dynamics of mutualists and exploiters. *The American Naturalist*, **174**, 780–794.

Kazancioğlu, E., Klug, H., and Alonzo, S.H. (2012). The evolution of social interactions changes predictions about interacting phenotypes. *Evolution*, **66**, 2056–2064.

Kiers, E.T., Palmer, T.M., Ives, A.R., Bruno, J.F., and Bronstein, J.L. (2010). Mutualisms in a changing world: an evolutionary perspective. *Ecology Letters*, **13**, 1459–1474.

Kiers, E.T., Rousseau, R.A., and Denison, R.F. (2006). Measured sanctions: legume hosts detect quantitative variation in rhizobium cooperation and punish accordingly. *Evolutionary Ecology Research*, **8**, 1077–1086.

Kiers, E.T., Rousseau, R.A., West, S.A., and Denison, R.F. (2003). Host sanctions and the legume-rhizobium mutualism. *Nature*, **425**, 78–81.

Kirkpatrick, M., and Ryan, M.J. (1991). The evolution of mating preferences and the paradox of the lek. *Nature*, **350**, 33–38.

Kummel, M., and Salant, S.W. (2006). The economics of mutualisms: optimal utilization of mycorrhizal mutualistic partners by plants. *Ecology*, **87**, 892–902.

Lehmann, L., and Keller, L. (2006). The evolution of cooperation and altruism: a general framework and a classification of models. *Journal of Evolutionary Biology*, **19**, 1365–1376.

Lehmann, L., and Rousset, F. (2010). How life history and demography promote or inhibit the evolution of helping behaviours. *Philosophical Transactions of the Royal Society B: Biological Sciences*, **365**, 2599–2617.

Mao-Jones, J., Ritchie, K.B., Jones, L.E., and Ellner, S.P. (2010). How microbial community composition regulates coral disease development. *PLoS Biology*, **8**, e1000345.

Marco, D.E., Carbajal, J.P., Cannas, S., Pérez-Arnedo, R., Hidalgo-Perea, A., Olivares, J., Ruiz-Sainz, J.E., and Sanjuán, J. (2009). An experimental and modelling exploration of the host-sanction hypothesis in legume–rhizobia mutualism. *Journal of Theoretical Biology*, **259**, 423–433.

Maynard Smith, J., and Price, G.R. (1973). The logic of animal conflict. *Nature*, **246**, 15–18.

McGlothlin, J.W., and Brodie III, E.D. (2009). How to measure indirect genetic effects: the congruence of trait-based and variance-partitioning approaches. *Evolution*, **63**, 1785–1795.

McGlothlin, J.W., Moore, A.J., Wolf, J.B., and Brodie, E.D. (2010). Interacting phenotypes and the evolutionary process III. Social evolution. *Evolution*, **64**, 1–17.

McNamara, J.M., Gasson, C., and Houston, A.I. (1999). Incorporating rules for responding into evolutionary games. *Nature*, **401**, 368–371.

Moore, A.J., Brodie III, E.D., and Wolf, J.B. (1997). Interacting phenotypes and the evolutionary process I. Direct and indirect genetic effects of social interactions. *Evolution*, **51**, 1352–1362.

Myerson, R. (1979). Incentive compatibility and the bargaining problem. *Econometrica*, **47**, 61–74.

Myerson, R., and Satterthwaite, M. (1983). Efficient mechanisms for bilateral trading. *Journal of Economic Theory*, **29**, 265–281.

Nash, J. (1950). The bargaining problem. *Econometrica*, **18**, 155–162.

Noe, R., and Hammerstein, P. (1994). Biological markets: supply and demand determine the effect of partner choice in cooperation, mutualism and mating. *Behavioral Ecology and Sociobiology*, **35**, 1–11.

Noe, R., and Hammerstein, P. (1995). Biological markets. *Trends in Ecology & Evolution*, **10**, 336–339.

Palmer, T.M., Doak, D.F., Stanton, M.L., Bronstein, J.L., Kiers, E.T., Young, T.P., Goheen, J.R., and Pringle, R.M. (2010). Synergy of multiple partners, including freeloaders, increases host fitness in a multispecies mutualism. *Proceedings of the National Academy of Sciences USA*, **107**, 17234–17239.

Price, G.R. (1970). Selection and covariance. *Nature*, **227**, 520–521.

Price, G.R. (1972). Extension of covariance selection mathematics. *Annals of Human Genetics*, **35**, 485–490.

Pringle, E.G., Akçay, E., Raab, T.K., Dirzo, R., and Gordon, D.M. (2013). Water stress strengthens mutualism among ants, trees and scale insects. *PLoS Biology*, **11**, e1001705.

Ricardo, D. (1891). *Principles of political economy and taxation*. London: G. Bell and Sons.

Riley, J.G. (2001). Silver signals: twenty-five years of screening and signaling. *Journal of Economic Literature*, **39**, 432–478.

Rothschild, M., and Stiglitz, J. (1976). Equilibrium in competitive insurance markets: an essay on the economics of imperfect information. *The Quarterly Journal of Economics*, **90**, 629–649.

Rubinstein, A. (1982). Perfect equilibrium in a bargaining model. *Econometrica*, **50**, 97–109.

Sachs, J.L., Mueller, U.G., Wilcox, T.P., and Bull, J.J. (2004). The evolution of cooperation. *Quarterly Review of Biology*, **79**, 135–160.

Sachs, J.L., Skophammer, R.G., and Regus, J.U. (2011). Evolutionary transitions in bacterial symbiosis. *Proceedings of the National Academy of Sciences*, **108**, 10800–10807.

Scheuring, I., and Yu, D.W. (2012). How to assemble a beneficial microbiome in three easy steps. *Ecology letters*, **15**, 1300–1307.

Schwartz, M.W., and Hoeksema, J.D. (1998). Specialization and resource trade: biological markets as a model of mutualisms. *Ecology*, **79**, 1029–1038.

Simms, E.L., Taylor, D.L., Povich, J., Shefferson, R.P., Sachs, J.L., Urbina, M., and Tausczik, Y. (2006). An empirical test of partner choice mechanisms in a wild legume-rhizobium interaction. *Proceedings of the Royal Society, Series B: Biological Sciences*, **273**, 77–81.

Smith, S.E., and Read, D.J. (2008). *Mycorrhizal Symbiosis*. New York: Academic Press.

Spence, M. (1973). Job market signaling. *The Quarterly Journal of Economics*, **87**, 355–374.

Stewart, A.D., Logsdon, J.M., and Kelley, S.E. (2005). An empirical study of the evolution of virulence under both horizontal and vertical transmission. *Evolution*, **59**, 730–739.

Taylor, P.D. and Day, T. (2004). Stability in negotiation games and the emergence of cooperation. *Proceedings of the Royal Society of London, Series B: Biological Sciences*, **271**, 669–674.

Taylor, P.D., and Frank, S.A. (1996). How to make a kin selection model. *Journal of Theoretical Biology*, **180**, 27–37.

Van Cleve, J., and Akçay, E. (2014). Pathways to social evolution: reciprocity, relatedness, and synergy. *Evolution*, **68**, 2245–2258.

Weeks, A.R., Turelli, M., Harcombe, W.R., Reynolds, K.T., and Hoffmann, A.A. (2007). From parasite to mutualist: rapid evolution of wolbachia in natural populations of drosophila. *PLoS Biology*, **5**, e114.

West, S.A., Kiers, E.T., Pen, I., and Denison, R.F. (2002a). Sanctions and mutualism stability: when should less beneficial mutualists be tolerated. *Journal of Evolutionary Biology*, **15**, 830–837.

West, S.A., Kiers, E.T., Simms, E.L., and Denison, R.F. (2002b). Sanctions and mutualism stability: why do rhizobia fix nitrogen? *Proc. R. Soc. B*, **269**, 685–694.

Worden, L., and Levin, S.A. (2007). Evolutionary escape from the prisoner's dilemma. *Journal of Theoretical Biology*, **245**, 411–422.

Yamamura, N. (1993). Vertical transmission and evolution of mutualism from parasitism. *Theoretical Population Biology*, **44**, 95–109.

Yamamura, N., Higashi, M., Behera, N., and Yuichiro Wakano, J. (2004). Evolution of mutualism through spatial effects. *Journal of Theoretical Biology*, **226**, 421–428.

Zahavi, A. (1975). Mate selection: a selection for a handicap. *Journal of Theoretical Biology*, **53**, 205–214.

CHAPTER 5

Evolution and genetics of mutualism

Carina A. Baskett and Douglas W. Schemske

5.1 Introduction

In his book *Fertilisation of Orchids* (1862), Darwin used flowers as convenient systems to study adaptive evolution. That flowers are key traits in a mutualistic interaction was not the main focus of his research, but rather a useful feature that allowed him to explore how natural selection contributes to adaptation. His work on floral adaptation set the stage for much of the next hundred and fifty years of empirical research on evolution in mutualisms. It is only in the last several decades that researchers have asked evolutionary questions in other mutualistic systems, and begun to think about and study mutualism evolution itself.

In this chapter, we focus on patterns and processes of adaptation in mutualisms, including the genetic basis of mutualistic evolution. The overarching question is how mutualistic evolution differs from evolution driven by other selective agents, such as the abiotic environment or antagonistic interactions. For example, is selection stronger or weaker compared to other interactions? Do genes underlying mutualistic adaptations tend to have large or small effects on phenotype? Throughout our review, we provide a conceptual perspective on mutualism evolution, and, where possible, offer empirical examples. Although we have learned much from empirical research on mutualism evolution, researchers usually ask different questions in different systems. Therefore, we believe there is a lack of available empirical evidence to evaluate these general questions. The next major hurdle in understanding the evolution and genetics of mutualism is to ask questions across systems, so we can begin to understand the ways in which the evolution of mutualisms may be unique.

A comprehensive study of the genetics and evolution of a given mutualism should include phylogenetic evidence of its history, studies of mutualistic traits and their selective value, and an understanding of the genes that underlie mutualistic adaptations, particularly traits involved in adaptive transitions. At present, there are no systems for which all of this information is known. In particular, there are few systems that provide both comprehensive genetic resources and an opportunity to evaluate ecological interactions. We thus focus much of our attention on two mutualisms in which researchers have answered disparate questions, using different approaches: 1) plant–pollinator interactions, and 2) endosymbioses between bacteria and their sap-sucking insect hosts. Studies of plant–pollinator mutualisms provide information on the adaptive and genetic changes involved in the putative early steps of transitions between partner taxa, while studies of endosymbiosis illustrate the value of molecular genetic approaches to understand the genomic consequences of long-term mutualism evolution. We also highlight three systems in the chapter boxes that provide exciting opportunities for the study of mutualism evolution: attine ants and fungi (Box 5.1); legumes and rhizobia (Box 5.2); and squid and bioluminescent bacteria (Box 5.3). Our goal is to use the handful of mutualisms that are relatively well studied to suggest processes and patterns likely to hold more generally.

Mutualism. Edited by Judith L. Bronstein.

Box 5.1 The attine ant–fungus mutualism

Ulrich G. Mueller

The contemporary interaction

Fungus-growing (attine) ants obligately depend on mutualistic fungi for food. Attine ants cultivate their fungi on plant-derived substrate, such as freshly cut leaves in leafcutter ants, or a mixture of plant debris, seeds, and frass in primitive, lower-attine ants (Mueller et al. 2005, Mehdiabadi and Schultz 2010). The enormous sizes of leafcutter colonies (up to 5000 gardens/colony, 5–10 million workers) make them dominant Neotropical herbivores, whereas all other attines have small, inconspicuous colonies. The ants cultivate gardens of fungal monocultures (Mueller et al. 2010), but the fungus is immersed in complex microbiota that the ants engineer to optimize garden growth (Mueller et al. 2005, Mueller 2012).

In contrast to the ants' obligate dependency on fungi, associations of the corresponding fungi range from facultative to obligate. Only the fungi cultivated by a derived clade of higher-attine ants (including the leafcutters) appear obligately tied to mutualistic life. In contrast, lower-attine fungi are occasionally recruited from free-living populations (Mueller et al. 1998, Mikheyev et al. 2010), and these fungi can transition between free-living and symbiotic life, thus retaining the facultative nature of ancestral associations at the origin of the mutualism 55 million years ago (Mehdiabadi and Schultz 2010). Fungal cultivars are vertically transmitted by foundress queens from parent to offspring nests, but the fungi are also occasionally exchanged horizontally between ant lineages (Mueller et al. 1998, Mikheyev et al. 2010). It is unclear whether lower-attine fungi are coevolved (Mueller et al. 1998, Mehdiabadi et al. 2012), whereas higher-attine fungi have coevolved features in response to fungiculture, such as nutritive hyphal swellings (gongylidia) harvested by the ants.

Fungiculture liberates attine ants from hunting and scavenging, the typical modes of ant nourishment. Attines evolved a novel feeding niche for ants, tapping into abundant resources (plant debris, leaves) to support enormous colonies and population densities that are difficult to sustain for hunting ants. The ants' dependency on fungal crops also incurs costs, as loss of the fungus leads to death of the ant farmer. Benefits and costs of mutualistic life to the fungi are more complex. Sheltered and tended by the ants, the fungus is guaranteed a competition-free and protected existence, but the ants' fungicultural manipulations effectively enslave and sterilize the fungi, limiting them to clonal propagation (Mueller et al. 2010). The evolutionary interests of the *facultatively* symbiotic lower-attine fungi differ from the ants' interests, whereas the interests of the *obligately* symbiotic higher-attine fungi are tightly coupled to the ants'. Nevertheless, even the higher-attine mutualism is not completely free of host–symbiont conflict (e.g., sex-ratio conflict, because only females propagate the fungus between nests; Mueller 2002).

Phylogenetic context

Lower-attine fungi are closely related to free-living fungi, indicating that these fungi are occasionally recruited by the ants (Mueller et al. 1998). In contrast, higher-attine fungi are obligate mutualists that were evolutionarily derived from within one clade of lower-attine fungi (Mikheyev et al. 2010; Figure 1 in Box 5.1). Phylogenetic analyses contradict strict ant–fungus cocladogenesis (Mikheyev et al. 2010, Mehdiabadi et al. 2012; Figure 1 in Box 5.1), which is predicted if the fungi were propagated strictly vertically between ant generations. Instead, most attine mutualisms are diffusely organized, with many fungal lineages in a clade associating with many ant species in a corresponding clade. Phylogenetic dating indicates that the last common ancestor of leafcutter fungi is much younger than the ancestor of leafcutter ants (Mikheyev et al. 2010). This discord suggests that new fungi arise and spread by lateral transfer, displacing fungi of earlier origin.

Genetic basis of mutualistic traits

Little is known about the genetics underlying adaptations to the mutualism. The two leafcutter genomes sequenced so far revealed modifications related to nutrient acquisition (e.g., loss of detoxification genes; Suen et al. 2010, Nygaard et al. 2011). Leafcutter fungus evolved a novel laccase enzyme that aids detoxification of defensive plant chemicals (De Fine Licht et al. 2013). This laccase is overexpressed in the gongylidia consumed by the ants, passes undigested through the ant gut, is applied to the garden during fecal manuring, and thus helps to detoxify the leaf substrate. Comparative genomics revealed acceleration of evolutionary rates in strongly coevolving partners (e.g., leafcutter mutualism), but not in weakly coevolving partners (e.g., lower-attine mutualism; Mikheyev et al. in revision).

Early evolutionary steps

Attine fungiculture most likely evolved either from adventitious interactions with litter-decomposing fungi growing on walls of nests built in leaf litter, or from a system of fungus-vectoring by ants (myrmecochory) in which specialized fungi

Box 5.1 *Continued*

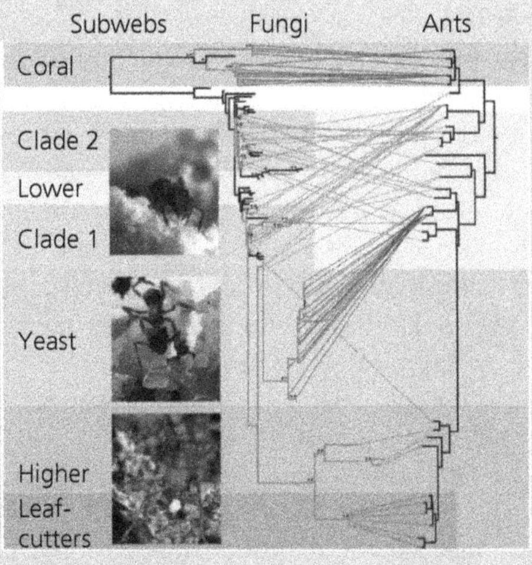

Figure 1 Phylogenetic associations between attine ants and fungal cultivars. Fungal cultivars (gray lines) were domesticated repeatedly from free-living fungi (black lines), followed by lateral spread of cultivars within defined ant clades. Phylogenetically primitive *Apterostigma* ants cultivate pterulaceous coral fungi, which are distantly related to the lepiotaceous fungi cultivated by all other attine ants. Ants in "lower" attine clades collect various plant debris for gardens, and cultivate fungi from at least two independently acquired cultivar clades (photo: *Cyphomyrmex wheeleri*). Cultivars in the "yeast" group are grown in single-celled form by the ants, whereas all other cultivars are grown as filamentous mycelia (photo: *Cyphomyrmex rimosus* attending yeast garden). Only the fungi cultivated by "higher" attine ants produce specialized nutritive structures (gongylidia) harvested by the ants, and these fungi therefore appear to have coevolved. Ants in the monophyletic "leafcutter" clade use predominantly fresh leaves for fungiculture (photo: *Atta cephalotes* harvesting gongylidia). Reprinted from Mikheyev et al. (2010); see phylogenetic methods therein. See Plate 2 for color version.

relied on ants for dispersal (Mueller 2002). The latter model contrasts with the traditional view in which attine fungi are passive symbionts that became domesticated by ants. Rather, the "myrmecochorous" model for the mutualism's origin assumes evolutionary modifications in the fungi that facilitated fungal vectoring before the ants became fungus farmers.

Future directions

Genomics of attine ants and fungi will further elucidate evolutionary transitions of the mutualism, for example functional-genomic traits underlying innovation at the origin of leafcutter fungiculture. Metagenomic analyses of garden microbiota should elucidate the engineering of microbiota by the ants to optimize nutrition and hygiene. Emerging fungicultural principles may help bio-inspire novel approaches for manipulating beneficial microbiota in agriculture or industry (Mueller et al. 2005).

References

De Fine Licht, H.H., Schiøtt, M., Rogowska-Wrzesinska, A., Nygaard, S., Roepstorff, P., and Boomsma, J.J. (2013). Laccase detoxification mediates the nutritional alliance between leaf-cutting ants and fungus-garden symbionts. *Proceedings of the National Academy of Sciences U.S.A.*, **110**, 583–587.

Mehdiabadi, N.J, Mueller, U.G., Brady, S.G., Himler, A. G., and Schultz, T.R. (2012). Symbiont fidelity and the origin of species in fungus-growing ants. *Nature Communications*, **3**, 840

Mehdiabadi, N.J., and Schultz, T.R. (2010). Natural history and phylogeny of the fungus-farming ants (Hymenoptera: Formicidae: Myrmicinae: Attini). *Myrmecological News*, **13**, 37–55.

Mikheyev, A.S., Mueller, U.G., and Abbott, P. (2010). Comparative dating of attine ant and lepiotaceous cultivar phylogenies reveals coevolutionary synchrony and discord. *American Naturalist*, **175**, E126–E133.

Mikheyev, A.S., Watanabe, Y., Affeldt, S., and Mueller, U.G. (2015). Molecular signatures of mutualistic coevolution in attine ants and their fungal cultivars. *Proceedings of the National Academy of Sciences U.S.A.*, in revision.

Mueller, U.G. (2002). Ant versus fungus versus mutualism: ant-cultivar conflict and the deconstruction of the attine ant-fungus symbiosis. *American Naturalist*, **160**, S67–S98.

Mueller, U.G. (2012). Symbiont recruitment versus ant-symbiont co-evolution in the attine ant-microbe symbiosis. *Current Opinion in Microbiology*, **15**, 269–277.

Mueller, U.G., Gerardo, N.M., Aanen, D.K., Six, D.L., and Schultz, T.R. (2005). The evolution of agriculture in insects. *Annual Review of Ecology, Evolution, and Systematics*, **36**, 563–595.

Mueller, U.G., Rehner, S.A., and Schultz, T.R. (1998). The evolution of agriculture in ants. *Science*, **281**, 2034–2038.

Mueller, U.G., Scott, J.J., Ishak, H.D., Cooper, M., and Rodrigues, A. (2010). Monoculture in leafcutter ant gardens. *PLoS One*, **5**, e12668.

Nygaard, S., Zhang, G., Schiøtt, M., et al. (2011). The genome of the leaf-cutting ant *Acromyrmex echinatior* suggests key adaptations to advanced social life and fungus farming. *Genome Research*, **21**, 1339–1348.

Suen, G., Teiling, C., Li, L., et al. (2010). The genome sequence of the leaf-cutter ant *Atta cephalotes* reveals insights into its obligate symbiotic lifestyle. *PLoS Genetics*, **7**, e1002007.

5.2 Natural selection in mutualisms

Does natural selection driven by mutualistic interactions differ from selection imposed by the abiotic environment or by antagonistic interactions? First, consider how adaptation to a biotic factor may differ from adaptation to an abiotic factor. In biotic interactions, adaptive peaks are "moving," owing to the coevolution of interacting species, whereas adaptive peaks may be more often "fixed" for an abiotic selective agent because the abiotic environment cannot evolve or coevolve (Schemske 2002, 2009). This may be particularly true in the tropics, where the abiotic environment is more consistent and predictable than in the temperate zone (Janzen 1967, Schemske 2009).

If biotic adaptive peaks are more mobile than abiotic peaks, then biotic selection may be stronger, more often directional than stabilizing (Rutter and Rausher 2004), and more spatially and temporally variable than abiotic selection (Siepielski et al. 2009, Thompson 2005). Both directional (e.g., Rutter and Rausher 2004) and stabilizing (e.g., Wright and Meagher 2004) selection gradients have been recorded for mutualistic traits in natural populations, but there is not enough data to evaluate whether one is more common, or to analyze the context in which one or the other may be expected. Several reviews have compiled estimates of the form and magnitude of selection from observational studies (Siepielski et al. 2009, Kingsolver and Diamond 2011), yet none has compared these metrics for different selective agents, e.g., biotic versus abiotic factors, or for mutualisms in particular. This may be due, in part, to the difficulty of determining selective agents in observational studies (Wade and Kalisz 1990, Siepielski et al. 2009).

Evidence from studies of experimental evolution shows that selection does differ when the adaptive peak is fixed versus moving, but exactly how it differs (in strength or variability) is unclear. In an antagonistic bacteria–phage system, when bacteria were allowed to coevolve, the phage populations experienced greater genetic and phenotypic divergence from the ancestor and greater between-population divergence, compared to populations in which bacteria were not allowed to coevolve (Paterson et al. 2010). Experimental evolution may be a powerful approach for revealing how selection is unique in coevolving systems (Hembry et al. 2014), but we are unaware of any mutualistic systems that have been used to address this question.

Thus far, we have focused on how selection may differ for biotic and abiotic selective agents. How might selection in an antagonistic interaction differ from selection in a mutualistic interaction? Both interactions may involve moving, coevolving adaptive peaks. However, an antagonism consists of one-way exploitation, while a mutualism consists of mutual exploitation. How this difference might affect selection is unclear.

Current observational and experimental evidence is insufficient to determine whether the form of natural selection in mutualisms is unique. In fact, we are aware of only four mutualisms in which selection gradients have been measured, all involving plants: pollination (e.g., Campbell et al. 1994, 1996), ant–plant protection (e.g., Rudgers and Strauss 2004, Rutter and Rausher 2004), seed dispersal (e.g., Siepielski and Benkman 2010), and nitrogen fixation in legumes (Porter and Simms 2014). Selection gradients in mutualisms have only been estimated for one side of the interaction (plants), except for the study by Porter and Simms (2014), which found selection for cheating in rhizobial bacteria but not in their legume hosts. To answer the question of whether selection in mutualisms differs in form or magnitude from selection by other types of selective agents, studies of additional mutualisms and estimates of selection gradients for both interacting species are needed.

5.3 The genetics of mutualism evolution

To understand the genetics underlying adaptation in a mutualism, it is important to consider the evolutionary history of the interaction. Is the mutualism in its early stages, such as a recent shift in pollinators between sister species of plants, or has the mutualism been in place for millennia? We expect that many fewer genes will be involved in the early stages of an interaction, but they may have relatively large phenotypic effects (Section 5.3.2).

Box 5.2 The legume–rhizobia mutualism

Maren L. Friesen

The contemporary interaction

Rhizobia are soil bacteria that form a facultative endosymbiosis with legume plants. Rhizobia fix plant-inaccessible atmospheric nitrogen into plant-accessible ammonium in exchange for plant carbon. The benefit to the host depends on both plant and bacterial genotype, and interactions between them. The molecular basis of this interaction can include rhizobium signal genes and host receptor genes, as well as host disease resistance genes.

Nodules, specialized plant organs that house rhizobia, develop following a sophisticated infection process. Almost all rhizobia produce "nod factors" in response to flavonoids secreted by the plant. These bacterial signals enable them to enter root hairs in "infection threads," which then triggers the root cortex to give rise to the nodule meristem. The penetrating infection threads ramify and follow the meristem into the developing nodule, where infection droplets break off to become "symbiosomes," bacterial cells encased in a plant membrane where bacteroids produce nitrogenase (Oldroyd et al. 2011).

Phylogenetic context

Nodulating symbioses between plants and nitrogen-fixing prokaryotes are both more recent and restricted than plant–mycorrhizal fungi symbioses, which are involved in nutrient uptake from the soil. All nodulating plants are in a subclade of the Eurosid I group of Angiosperms. This includes both actinorhizal plants that associate with *Frankia* and rhizobia-associating species in the legume family Fabaceae and the non-legume genus *Parasponia* (family Ulmaceae; Doyle 2011). Determinate nodules, formed by soy and lotus, contain bacteroids that remain viable and de-differentiate into free-living rhizobia; this state may be ancestral to indeterminate nodules, formed by *Medicago* and clover, in which plant peptides trigger terminal bacteroid differentiation.

In contrast to their phylogenetically narrow hosts, rhizobia are paraphyletic. Most rhizobia that interact with legumes in the Papilionoid subfamily belong to the Rhizobiaceae family of the alpha-proteobacteria, but distantly related beta-proteobacteria are the dominant symbiont of the Mimosid subfamily and have acquired nodulation genes via horizontal transfer (Chen et al. 2003). Rhizobiaceae contains non-rhizobial genera, such as *Agrobacterium*.

Genetic basis of mutualistic traits

Rhizobia are transmitted horizontally via soil, so unlike vertical intracellular symbionts, they do not show genome reduction, and not all rhizobia strains in the rhizosphere nodulate (Denison and Kiers 2004). Rhizobia possess symbiotic gene "islands" containing nodulation and nitrogen fixation genes, located on transmissible genetic elements. In one case, a field trial of introduced *Lotus corniculatus* led to symbiosis island transfer from an introduced inoculant *Mesorhizobium* strain into a native, previously non-nodulating *Mesorhizobium* strain (Sullivan and Ronson 1998). Comparative genomic analysis fails to identify core genes specific to nodulating bacteria, underscoring the diverse nature of symbiotic programs (Masson-Boivin et al. 2009).

Early evolutionary steps

Before becoming legume symbionts, rhizobia were likely commensal root endophytes—some are currently found in this role in *Arabidopsis thaliana* (Bulgarelli et al. 2012). Nodulation has evolved multiple times within Eurosid I. The ancestral state of the interaction was likely intercellular "crack entry" that still occurs during submerged conditions in *Sesbania rostrata* (Masson-Boivin et al. 2009). Rhizobia in these "infection pockets" sometimes form infection threads; thus, infection pockets may be ancestral. In the non-legume *Parasponia* and some early-branching legumes, rhizobia fix nitrogen within infection threads rather than symbiosomes; legume symbiosomes are thus likely derived from infection threads.

Several plant components of the nodulation program have been co-opted from other plant developmental processes (Szczyglowski 2003). Multiple genes in the signal transduction pathway downstream of nod factor perception are also required for mycorrhizal infection (Oldroyd et al. 2011). In the non-legume *Parasponia*, a single putative receptor is essential for both rhizobia and mycorrhiza to become intracellular (Op den Camp et al. 2011); this constraint has been overcome in Papilionoid legumes through gene duplication. In addition, genes for infection thread formation originated via tandem duplication of paralogs involved in pollen tube growth (Rodríguez-Llorente et al. 2004). Multiple observations support genetic overlap/crosstalk between nodulation and lateral root development, suggesting co-option of these pathways for symbiosis (Szczyglowski 2003). Importantly, host control of symbionts may be co-opted from host processes that facilitated the origin of the rhizobium symbiosis,

continued

Box 5.2 *Continued*

rather than evolutionary responses to "cheating" rhizobia, of which we currently lack examples (Friesen 2012).

Future directions

Over a century of agronomically motivated study has resulted in an exquisitely detailed understanding of early infection processes and catalyzed the development of substantial genomic resources for legumes and rhizobia. While molecular studies focus largely on a few species within Papilionoid legumes, a broader evolutionary perspective can leverage the multiple origins of nodulation (e.g., Op den Camp et al. 2011). Evolutionary ecological and quantitative genetic studies of natural populations of rhizobia and legumes can elucidate trade-offs and selection pressures shaping host and symbiont genomes. Outstanding problems include understanding why rhizobia performance varies across host genotypes and elucidating the molecular basis of host regulation of symbiosis. The rhizobia–legume system has the potential to greatly advance our understanding of mutualism coevolution and diversity.

References

Bulgarelli, D., Rott, M., Schlaeppi, K., et al. (2012). Revealing structure and assembly cues for Arabidopsis root-inhabiting bacterial microbiota. *Nature*, **488**, 91–95.

Chen, W., Moulin, L., Bontemps, C., Vandamme, P., Bena, G., and Boivin-Masson, C. (2003). Legume symbiotic nitrogen fixation by beta-proteobacteria is widespread in nature. *Journal of Bacteriology*, **185**, 7266–7272.

Denison, R., and Kiers, E. (2004). Lifestyle alternatives for rhizobia: mutualism, parasitism, and forgoing symbiosis. *FEMS Microbiology Letters*, **237**, 187–193.

Doyle, J.J. (2011). Phylogenetic perspectives on the origins of nodulation. *Molecular Plant-Microbe Interactions*, **24**, 1289–1295.

Friesen, M.L. (2012). Widespread fitness alignment in the legume-rhizobium symbiosis. *New Phytologist*, **194**, 1096–1111.

Masson-Boivin, C., Giraud, E., Perret, X., and Batut, J. (2009). Establishing nitrogen-fixing symbiosis with legumes: how many rhizobium recipes? *Trends in microbiology*, **17**, 458–466.

Oldroyd, G.E.D., Murray, J.D., Poole, P.S., and Downie, J.A. (2011). The rules of engagement in the legume-rhizobial symbiosis. *Annual Review of Genetics*, **45**, 119–144.

Op den Camp, R., Streng, A., De Mita, S., et al. (2011). LysM-Type mycorrhizal receptor recruited for rhizobium symbiosis in nonlegume *Parasponia*. *Science*, **331**, 909–912.

Rodríguez-Llorente, I.D., Pérez-Hormaeche, J., El Mounadi, K., et al. (2004). From pollen tubes to infection threads: recruitment of *Medicago* floral pectic genes for symbiosis. *Plant Journal*, **39**, 587–598.

Sullivan, J.T., and Ronson, C.W. (1998). Evolution of rhizobia by acquisition of a 500-kb symbiosis island that integrates into a phe-tRNA gene. *Proceedings of the National Academy of Sciences USA*. **95**, 5145–5149.

Szczyglowski, K. (2003). Symbiosis, inventiveness by recruitment? *Plant Physiology*, **131**, 935–940.

In addition, an interaction that has evolved only rarely, such as the attine ant-fungus mutualism (Box 5.1), may have a more complex genetic basis than one that has evolved independently many times, such as ant–plant protection mutualisms (Weber and Keeler 2013).

Most models of adaptive evolution have focused on fixed adaptive peaks (Section 5.3.2). How might a coevolving, moving adaptive peak (Section 5.2) affect the genetics of mutualism evolution? Here we explore this question with regard to both advanced and early stages of mutualism evolution.

5.3.1 Advanced stages

Once mutualisms are established, the interaction will be evolutionarily "fine-tuned" over hundreds or perhaps thousands of generations of reciprocal selection and coevolution, as each organism is selected to extract higher benefit from the interaction while providing less in return. This ongoing mutual exploitation provides an opportunity for various traits, pathways, and genes to be modified or lost, so we expect that the genetic basis of older mutualistic adaptations would involve many more genes than that of more recently derived adaptations.

In some cases, genes involved in long-established mutualisms become tightly linked, and are essentially inherited as a single unit. As Maren Friesen discusses in Box 5.2, the genes controlling many aspects of the mutualism between legumes and their nitrogen-fixing bacteria are found on transmissible genetic elements; they can be considered "islands" of symbiotic genes, hundreds of kilobases long (MacLean et al. 2007). The finding that the genomic regions involved in symbiotic functions of rhizobia are highly dynamic, with a large number of insertion sequence elements (MacLean et al. 2007), illustrates that there has been substantial molecular evolution related to the symbiosis.

Gene loss and genome reduction is commonly documented in symbiotic mutualisms, which can in turn affect the host genome (see also Chapter 2). For example, recent sequencing of the pea aphid (*Acyrthosiphon pisum*) genome has revealed twelve genes of bacterial origin and complementary amino acid synthesis capabilities with its obligate *Buchnera* symbiont due to gene loss from the bacterium (Richards et al. 2010). However, the extent of gene loss may depend on the mode of transmission of symbionts between hosts. Intracellular insect endosymbionts often show evidence of genome reduction compared to free-living relatives (McCutcheon and Moran 2012), but rhizobia, which are transmitted horizontally via soil (Box 5.2), do not. As Ulrich Mueller discusses in Box 5.1, phylogenetic analyses of the attine ant–fungus mutualism suggest that fungi are not strictly vertically propagated between ants, and that clades of symbiotic fungi are considerably younger than clades of their corresponding ant hosts (Figure 1 in Box 5.1). Genome sequencing has revealed the loss of genes associated with nutrient acquisition in leafcutter ants. Although detailed genomic data for the fungi are not yet available, we might hypothesize that the degree of gene loss will be related to the mode of transmission, with vertically transmitted fungi exhibiting greater gene loss.

Dramatic cases of genome size reduction have been recorded in insect intracellular endosymbiont genomes, which consist of a subset of genes found in their free-living relatives (McCutcheon and Moran 2012). In fact, several intracellular symbionts in sap-sucking insects have the smallest recorded genomes of any organism (McCutcheon and Moran 2012; see Box 2.3). This extreme genome reduction is hypothesized to be due to drift; with relaxed selection on non-essential genes, deletions are easily fixed due to extreme population bottlenecks during vertical transmission from insect parent to offspring (McCutcheon and Moran 2012). Loss of DNA repair genes is hypothesized to contribute to other patterns commonly seen in endosymbiont genomes (i.e., rapid sequence evolution and A + T bias; McCutcheon and Moran 2012).

Genome reduction in insect endosymbionts can have tangible impacts on the host's ecology. The glassy-winged sharpshooter (*Homalodisca vitripennis*) is a leafhopper (Cicadellidae) that feeds on xylem sap, which is a particularly nutrient-poor food source. The sharpshooter hosts two endosymbionts: *Sulcia muelleri* (Bacteroidetes) is an obligate endosymbiont acquired over 260 mya, which is present in many insects in the Auchenorrhyncha suborder and has one of the smallest recorded bacterial genomes with only 245 kilobases (McCutcheon and Moran 2007). *Baumannia cicadellinicola* (Gammaproteobacteria) is more recently associated with the sharpshooter, and is not found in other plant-sucking insects (Moran et al. 2008). Neither of the two bacteria is capable of synthesizing all the amino acids necessary for animals on its own; rather, they have complementary synthesis pathways (McCutcheon and Moran 2007). Presumably, when the sharpshooter ancestor was feeding on phloem, *Sulcia* would have lost the ability to synthesize nutrients that are present in phloem. This gene loss is unidirectional, so it is hypothesized that association with *Baumannia* was necessary for sharpshooters to feed on xylem, which contains fewer nutrients than phloem (Moran et al. 2008).

The long-term genetic consequences of mutualism, particularly obligate symbioses, can be dramatic. As genomic data become available for more organisms, we will be able to ask questions about how genetics and ecological context (e.g., transmission mode and degree of specialization) interact to influence mutualism evolution.

5.3.2 Early stages

One of the persistent controversies in the study of adaptation concerns the effect size of mutations that are fixed as populations adapt to a new environment (Barton and Keightley 2002, Rockman 2011). Do populations adapt by a large number of small "mutational steps," or are large mutational steps sometimes involved? As discussed above, coevolution in mutualisms may lead to mobile adaptive optima, which could influence the underlying genetics of mutualistic traits. However, the theory of genetic architecture has thus far ignored the agents of selection.

Fisher (1930) proposed that adaptations are comprised of a nearly infinite number of mutations, each with an infinitesimally small effect on the phenotype, based on the assumption that large effect mutations have negative, pleiotropic effects, and hence move populations away from their optimum phenotype. Kimura (1983) proposed a modification to Fisher's infinitesimal theory, suggesting that mutations of intermediate effect size play the major role in adaptation since these are less prone to stochastic loss via genetic drift than are small-effect mutations. Orr (1998) proposed that large-effect mutations could be favored when populations are far from their optimum, but that mutations of progressively smaller effect would contribute to adaptation as the population approached its optimum, resulting in a combination of a few large-and many small-effect mutations during the adaptive "walk." Although there is considerable evidence to support Orr's view that adaptations are often comprised of large-effect mutations (Barrett and Hoekstra 2011), there is still controversy regarding the importance of major genes in adaptive evolution (Rockman 2011).

A key assumption of the models of Fisher, Kimura, and Orr is that the optimum phenotype is stationary. If the optimum is constantly moving, as might be expected for coevolving mutualists or antagonists (Schemske 2002, 2009, Louthan and Kay 2011, Matuszewski et al. 2014), large-effect mutations might be favored more often than if the optimum was fixed (Orr 2005). It could also be argued that large mutational steps are required for a major ecological transition. At the origin of a mutualism, for example, the sign of an interaction between species may change from negative to positive (as in the case of a seed predator that becomes a pollinator). There is presently insufficient data to know if the early mutational steps involved in mutualism evolution differ in size or number from those involved in adaptation to antagonistic interactions or to abiotic sources of selection. At issue is the shape and stability of the fitness landscape. Are the valleys between fitness peaks wider (more ecologically divergent) or deeper (negative fitness consequences are greater for moving away from the optimum) on average for mutualism evolution? Are the peaks more mobile?

Frequent evolutionary transitions between modes of pollination make plant–pollinator mutualisms a fertile system for investigating early stages in mutualism evolution. Pollination can be observed or manipulated in the field and greenhouse, allowing researchers to study traits that are known to be ecologically relevant, such as structures involved in attraction and reward (e.g., petals and nectaries). The genetic basis of these traits can then be elucidated by mapping quantitative trait loci (QTL, see below), and hypotheses for the traits and genes that contribute to early stages of pollinator shifts can be tested.

In brief, QTL mapping uses genetic markers to identify genomic regions associated with phenotypic traits in a "mapping" population. The mapping population is produced by crossing populations or species that differ in the traits of interest, followed by one or more rounds of crossing or self-fertilization to produce a genetically and phenotypically diverse population with recombined parental genotypes. Compared to many animals, plants are ideal for such studies because crosses between populations or species are often fertile, and researchers can obtain the large number of progeny required to maximize statistical power. Thus far, QTL studies of pollination mutualisms have been "unilateral" (Bronstein 1994). While they identify genetic mechanisms of plant adaptation, they do not reciprocally elucidate the genetic basis of pollinator adaptations. Here we discuss examples from two different plant systems, *Mimulus* (Phrymaceae) and *Petunia* (Solanaceae), to highlight progress made to date.

Bradshaw et al. (1998) investigated the genetic basis of traits that contribute to floral divergence between bee-pollinated *Mimulus lewisii* and its close relative, hummingbird-pollinated *M. cardinalis*. *Mimulus lewisii* flowers are pink, produce little nectar, and have broad petals thrust forward as a landing platform for bees, and the anthers and stigma are inserted within the floral tube. *Mimulus cardinalis* flowers are red, produce large quantities of nectar, have reflexed petals, and the anthers and stigma protrude from the floral tube. Genetic mapping studies in this system identified

12 QTL for nine floral traits, including at least one QTL of moderate to large effect for each trait (Bradshaw et al. 1998). Field studies of pollinator visitation in arrays of F2 hybrids determined that petal carotenoid concentration and nectar volume had the greatest influence on pollinator visitation, with bees exhibiting a strong preference for pink-flowered plants with low petal carotenoid concentration, while hummingbirds preferred plants with high nectar volume (Schemske and Bradshaw 1999). That both of these traits are controlled by QTL of large effect in this system suggests that the transition from bee to hummingbird pollination may have involved relatively few genes of large effect.

A major role of the single QTL for petal carotenoid production was demonstrated in field experiments intended to mimic the effect of a mutation at the carotenoid locus. Near-isogenic lines (NILs) were produced by introgressing the QTL for petal carotenoids from *M. cardinalis* to *M. lewsii*, and vice versa (Bradshaw and Schemske 2003). Here we focus on the substitution of the *M. cardinalis* allele into *M. lewisii*, given phylogenetic analysis suggesting that bee pollination is ancestral and that hummingbird pollination is derived (Beardsley et al. 2003). Bee visitation to the pink-flowered, *M. lewsii* parent (Figure 5.1A) was much higher than that to the orange-red *M. lewisii* NIL (Figure 5.1B), while the converse was observed for hummingbirds, suggesting that increased petal carotenoid concentration was probably a major, early step in the shift to hummingbird pollination.

Figure 5.1 Flowers of *Mimulus lewisii* used in field experiments to examine the genetics of a pollinator switch from bee to hummingbird pollination. a. Wild-type (WT); b. Near-isogenic line (NIL) in which the genomic region containing the allele for petal carotenoids in the hummingbird-pollinated *M. cardinalis* was introgressed into *M. lewisii* (reprinted from Bradshaw and Schemske 2003). See Plate 3 for color version.

Petunia is a South American genus of approximately sixteen species of herbaceous perennials that has become an important model system for elucidating the genetics of floral traits and pollinator transitions (Venail et al. 2010). Bee pollination appears to be the ancestral state, but species pollinated by hawkmoths and by hummingbirds are also observed. For example, hummingbird-pollinated *P. exserta* is thought to be derived from the hawkmoth-pollinated *P. axillaris* (Klahre et al. 2011). Both of these species possess long floral tubes and produce a large volume of nectar, but *P. exserta* has unscented, red flowers, while *P. axillaris* has strongly scented, white flowers (Klahre et al. 2011). QTL mapping of floral scent identified two large-effect QTL (Klahre et al. 2011), including one gene region containing the myb transcription factor *ODORANT1*, which is known to regulate enzymes involved in scent production (Klahre et al. 2011). To examine the effects of scent and flower color on hawkmoth visitation in this system, Klahre et al. (2011) introgressed the major scent QTL into both white- and red-flowered plants, and carried out pollinator choice experiments in a wind tunnel with the hawkmoth *Manduca sexta*. Moths displayed a strong preference for scented over non-scented flowers of the same color, but when presented with both scented, red flowers and non-scented, white flowers, there was no clear preference. These results suggest that floral scent and color are both strong cues to pollinators, and that at least some of the early stages in the pollinator shift probably involved large mutational steps.

To conclude, the genetic architecture of adaptation in mutualisms is poorly understood, but there is evidence supporting the idea that genes of large effect are important in the transitions between different pollination syndromes. Despite limited power to generalize, the studies of *Mimulus* and *Petunia* both show evidence of QTL of large effect. More general support for the hypothesis of larger effect sizes involved in biotic adaptation was obtained by Louthan and Kay (2011),who found that QTL controlling floral traits involved in pollination mutualisms had significantly larger effect sizes than did those for non-flowering traits.

5.4 Emerging systems and tools

5.4.1 Moving beyond pollination

The available studies of natural selection in mutualisms are primarily restricted to plant–animal interactions—typically plant–pollinator mutualisms, and then almost exclusively from the plant perspective. To achieve a synthesis that addresses the major questions in mutualism evolution will require many studies covering a wider range of mutualisms. The accompanying boxes highlight three emerging systems for the study of mutualism evolution. In each, the authors describe the contemporary interaction, provide the phylogenetic context, discuss the genetic basis of mutualistic traits and early steps in the evolution of the mutualism, and present ideas for future research.

In the attine ant–fungus mutualism (Box 5.1), ants collect plant fragments that are used as a substrate for fungi cultured and consumed by ants in subterranean gardens. As Ulrich Mueller discusses, this interaction has existed for 55 million years, yet there is substantial variation in its specificity. The ants are obligately associated with fungi, but some of the fungi are facultative, and are close relatives of free-living taxa. A strong indication of coevolution in this mutualism is the finding from genomic studies that leafcutter fungi have evolved a novel enzyme for detoxification of host plant chemicals, while leafcutter ants have lost detoxification genes.

The mutualism between legumes and nitrogen-fixing bacteria discussed by Maren Friesen in Box 5.2 takes place in root nodules colonized by soil bacteria (*Rhizobium*) that gain access to the roots through a complex process of infection. The bacteria are facultative and horizontally transmitted; thus, there is considerable opportunity for gene exchange with free-living strains. Particularly fascinating is the finding that genes involved in various aspects of plant development have been co-opted for nodulation.

The marine mutualism between squids and bioluminescent bacteria described by Michele Nishiguchi in Box 5.3 offers an extraordinary example of how mutualistic interactions can expand the behavioral repertoire of the host. In this system, free-living bioluminescent bacteria (family Vibrionaceae) colonize the light organ of uninfected, juvenile squid, and there produce light that contributes to prey capture and predator avoidance. Phylogenetic evidence suggests that the mutualism has evolved twice, and is ancestral in the family Sepiolidae where it has been lost in three of seven genera. This may be a case where the formation of a mutualism is difficult, but loss is straightforward, and repeatedly favored. Of particular note is the evolution of host immune functions that create conditions favorable to the growth and reproduction of particular bacterial strains in the light organ.

The results that emerge from studies of these and other emerging systems will dramatically improve our understanding of mutualism evolution. An additional system with great promise is the mutualism between plants that provide extrafloral nectar to ants in exchange for protection against herbivores. Extrafloral nectaries have evolved independently many times (Weber and Keeler 2013), and there is a range of generalized to specialized ant × plant interactions. Thus, this system provides an interesting opportunity to investigate different stages of mutualism evolution and to compare the phenotypic and genetic steps involved in multiple independent clades. Cotton (*Gossypium*) may be of particular interest, in that some species produce leaf and extrafloral nectaries that are involved in an ant–plant mutualism (Rudgers and Strauss 2004), and genetic studies find that the loss of nectaries has a simple genetic basis (Meyer and Meyer 1961).

5.4.2 Genomic approaches

The extraordinary breakthroughs in gene sequencing technologies have provided new insights into the long-term evolutionary consequences of mutualism evolution. For example, the proliferation of available data on insect endosymbionts has prompted fascinating lines of inquiry into the causes and consequences of genome reduction (McCutcheon and Moran 2012), which we briefly described in Section 5.3.1. Yet in most studies, molecular work has far outpaced our knowledge of the natural history and ecology of most interactions. Work on the relationship between the pea aphid and its facultative endosymbiont *Hamiltonella defensa* exemplifies what we can learn from combining ecological and

Box 5.3 Bobtail squids light the way to study marine mutualisms

Michele K. Nishiguchi

The contemporary interaction

Beneath the sand in calm, tropical waters of the Indo-West Pacific lies a little-known benthic squid that alters its coloration to match down-welling moonlight in a "stealth-like" behavior known as counterillumination (Jones and Nishiguchi 2004). These bobtail or dumpling squids (Cephalopoda: Sepiolidae) harbor bioluminescent bacteria (family Vibrionaceae) in an epithelial lined cavity called the light organ (Nyholm and Nishiguchi 2008). These bacteria are found free-living in the surrounding seawater, where uninfected juvenile squids acquire them after hatching. Since vibrio bacteria infect their hosts environmentally with each generation, the identity, number, and relatedness of successfully colonizing bacterial isolates are influenced by a number of ecological factors (Jones et al. 2006, Soto et al. 2009, 2012).

Light produced by the bacteria is focused downward (ventrally) using both a lens and reflective ventral tissues. Since the squids are nocturnal, the production of bacteriogenic light may allow squids to locate prey and evade predators through silhouette reduction. The vibrio bacteria benefit because they can grow four times faster in the light organ than in normal seawater.

Phylogenetic context

Nearly all squids in the family Sepiolidae contain similar bacteriogenic light organs (seven genera), but the trait has been lost in at least three genera (Nishiguchi et al. 2004). Molecular phylogenetic evidence indicates that light organs in the Sepiolidae are an ancestral trait (Nishiguchi et al. 2004). The only other family of squids that contain bacteriogenic light organs are the Loliginidae, where it is also thought to be derived (Lindgren et al. 2012, Nishiguchi et al. 2004). Interestingly, bacteriogenic light organs have only evolved twice in the Cephalopoda, and the bacteria in both cases are in the family Vibrionaceae.

Genetic basis of mutualistic traits

Genetic mechanisms that drive host specificity have been extensively studied in symbiotic vibrios. The ability to culture and genetically manipulate vibrio bacteria in the lab has given researchers the capability to test which gene(s) are responsible for each component of symbiont competency. Colonization factors such as flagella, biofilm production, nitric oxide production, and peptidoglycan have been shown to influence the degree to which the mutualism can be established (Chavez-Dozal et al. 2012). Additionally, three genomes of *Vibrio fischeri* strains have been sequenced (Ruby et al. 2005), enabling researchers to compare various host-specific genes that affect colonization, host range, and whether the host is a squid or a fish (Mandel et al. 2009). No sepiolid squid genome has been sequenced, but genes that are up- or down-regulated have been documented. These complement *Vibrio* signals, and include such functions as neutralization of bacterial toxins, apoptosis, nitric oxide response, and the host daily rhythms of venting (Koropatnick et al. 2004, Weir et al. 2010). These specific host immune functions select the appropriate population of *V. fischeri* that can withstand the biochemical onslaught of the squid's immune system in order to colonize and persist in the light organ.

Early evolutionary steps

Morphological and molecular evidence has provided evidence regarding the early evolution of the squid–vibrio mutualism. The nearby accessory nidamental gland secretes the outermost covering of egg cases and also hosts bacteria that may provide protection against biofouling (Collins et al. 2012). In some species the gland and the light organ are connected, and it is hypothesized that this gland may have been both a precursor to the light organ and the source of its bacterial colonizers (Nishiguchi et al. 2004 and references therein). Additionally, the protein aldehyde dehydrogenase (ALDH), the main component of the cephalopod eye lens, is biochemically convergent to the sepiolid light organ lens (Montgomery and McFall-Ngai 1992). Thus, we are starting to understand how preexisting features in squid tissues were co-opted for symbiotic functions.

Future directions

The ability to manipulate and experimentally evolve different vibrio bacteria to accommodate their sepiolid squid hosts has provided insights from genes to ecosystems that have been instrumental in understanding how the squid–bacteria mutualism has evolved (Nyholm and Nishiguchi 2008). For example, symbiotic genes responsible for colonization and persistence can be genetically modified to examine subtle differences between closely related *V. fischeri* strains during colonization of different species of host squid. Similarly, these bacteria can be grown under a variety of conditions *in vitro* to determine whether factors in the abiotic environment may select key features that preadapt symbiotic bacteria for life in the host light organ. The ability to link

continued

Box 5.3 *Continued*

changes in the abiotic environment to adaptation and evolution in both partners is of special interest considering climate change and human disturbance. Understanding how mutualisms are affected by plastic and adaptive responses to both abiotic and biotic change will be the next challenge facing symbiosis researchers.

References

Chavez-Dozal, A.A., Hogan, D., Gorman, C., Quintanal-Villalonga, A., and Nishiguchi, M.K. (2012). Multiple *Vibrio fischeri* genes are involved in biofilm formation and host colonization. *FEMS Microbial Ecology*, **81**, 562–573.

Collins, A.J., LaBarre, B.A., Won, B.S.W., et al. (2012). Diversity and partioning of bacterial populations within the accessory nidamental gland of the squid *Euprymna scolopes*. *Applied and Environmental Microbiology*, **78**, 4200–4208.

Jones, B.W., Lopez, J.E., Huttenberg, J., and Nishiguchi, M.K. (2006). Population structure between environmentally transmitted *Vibrios* and bobtail squids using nested clade analysis. *Molecular Ecology*, **15**, 4317–4329.

Jones, B.W., and Nishiguchi, M.K. (2004). Counterillumination in the bobtail squid, *Euprymna scolopes* (Mollusca: Cephalopoda). *Marine Biology*, **144**, 1151–1155.

Koropatnick, T.A., Engle, J.T., Apicella, M.A., Stabb, E.V., Goldman, W.E., and McFall-Ngai, M.J. (2004). Microbial factor-mediated development in a host-bacterial mutualism. *Science*, **306**, 1186–1188.

Lindgren, A.R., Pankey, M.S., Hochberg, F.G., and Oakley, T.H. (2012). A multi-gene phylogeny of Cephalopoda supports convergent morphological evolution in association with multiple habitat shifts in the marine environment. *BMC Evolutionary Biology*, **12**, 129.

Mandel, M.J., Wollenberg, M.S., Stabb, E.V., Visick, K.L., and Ruby, E.G. (2009). A single regulatory gene is sufficient to alter bacterial host range. *Nature*, **458**, 215–218.

Montgomery, M.K., and McFall-Ngai, M.J. (1992). The muscle-derived lens of a squid bioluminescent organ is biochemically convergent with the ocular lens. *Journal of Biological Chemistry*, **267**, 20999–21003.

Nishiguchi, M.K., Lopez, J.E., and Boletzky, S.V. (2004). Enlightenment of old ideas from new investigations: the evolution of bacteriogenic light organs in squids. *Evolution and Development*, **6**, 41–49.

Nyholm, S.V., and Nishiguchi, M.K. (2008). The evolutionary ecology of a sepiolid squid-*Vibrio* association: from cell to environment. *Vie et Milieu*, **58**, 175–184.

Ruby, E.G., Urbanowski, M., Campbell, J., et al. (2005). Complete genome sequence of *Vibrio fischeri*: a symbiotic bacterium with pathogenic congeners. *Proceedings of the National Academy of Sciences USA*, **102**, 3004–3009.

Soto, W., Gutierrez, J., Remmenga, M.R., and Nishiguchi, M.K. (2009). Salinity and temperature effects on physiological responses of *Vibrio fischeri* from diverse ecological niches. *Microbial Ecology*, **57**, 140–150.

Soto, W., Punke, E.B., and Nishiguchi, M.K. (2012). Evolutionary perspectives in a mutualism of sepiolid squid and bioluminescent bacteria: Combined usage of microbial experimental evolution and temporal population genetics. *Evolution*, **66**, 1308–1321.

Weir, A.M., Nyholm, S.V., Mandel, M.J., et al. (2010). Transcriptional patterns in both host and bacterium underlie a daily rhythm of anatomical and metabolic change in a beneficial symbiosis. *Proceedings of the National Academy of Sciences USA*, **107**, 2259–2264.

genetic information. *Hamiltonella defensa* retains genes related to host invasion (Moran et al. 2005), and because it lacks genes for synthesizing essential amino acids, it appears that both *H. defensa* and the pea aphid rely on the host's obligate endosymbiont *Buchnera* to fulfill their nutritional needs from phloem sap (Degnan et al. 2009). The mutualistic function provided by *H. defensa* is defense against endoparasitic wasps by disrupting early development of wasp larvae (Oliver et al. 2003). The association is obligate for the bacteria but facultative for the aphid, as the proportion of infected aphids increases with higher parasitoid pressure, and declines when parasitoid pressure is low (Oliver et al. 2008). Since they are more tractable to manipulate compared to obligate relationships, facultative symbioses could be key for understanding the early steps and genetics of mutualism establishment and transitions, especially when studied in a phylogenetic context for both partners.

5.4.3 Experimental evolution

Evolution experiments in microbes may be a powerful approach to complement theory and studies in natural systems for elucidating the processes of mutualism evolution and for answering the question of how evolution in mutualisms may be unique compared to other selective agents. An astonishing example demonstrating possible early steps toward mutualism occurred in a laboratory setting in the late 1960s. After less than 1000 generations (five years), the D strain of *Amoeba proteus* became obligately dependent on a bacteria (*Candidatus Legionella jeonii*) that was initially pathogenic (Jeon 2004 and references therein). In later experimental introductions of the derived symbiotic bacteria into uninfected hosts, the amoebae evolved dependence in only 200 generations (18 months; Jeon 2004). The only known benefit for the amoebae is accelerated growth during initial infection. The infected strain performs worse than the uninfected ancestral strain under some conditions, such as at high temperatures (Jeon 2004). Perhaps given enough time (especially outside a laboratory setting), partner fidelity feedback (Chapter 4) would select for more obvious mutualistic function (Sachs et al. 2004). Although we are not aware of evolution experiments involving a microbial mutualism that occurs in nature, recent studies establishing novel mutualisms can test theoretical predictions of the contexts in which simple mutualisms establish and remain stable (Harcombe 2010, Hillesland et al. 2014).

5.5 Future directions

We presently lack answers to many fundamental questions regarding the evolution of mutualisms. However, the development of new phylogenetic and molecular approaches and the emergence of new model systems should contribute to substantial improvements in our understanding in the coming years. Here we highlight a few of the many open questions regarding the genetics and evolution of mutualisms that merit future consideration.

5.5.1 Comparison to antagonisms

One question that remains unexplored is how evolution in an antagonistic interaction is expected to differ from evolution in a mutualistic interaction. Much attention has been paid to how the two types of interactions grade into each other in terms of cheating in mutualisms (Chapter 6). However, to our knowledge the only general comparisons of coevolutionary dynamics in different types of interactions are theoretical (see Chapters 4 and 7). For example, Yoder and Nuismer (2010) predict that antagonistic coevolution promotes phenotypic diversification under certain conditions, whereas mutualistic coevolution restricts it. Their model did not account for cheating, or for how antagonism and mutualism may differ in their effects on reproductive isolation. Empirical evidence related to this question is wholly lacking.

5.5.2 Speciation

In comparison to other biotic and abiotic factors, how important are mutualisms for speciation? To answer this question will require comparative studies conducted at broad taxonomic levels to search for patterns (see Chapter 3), followed by detailed investigations of the ecological and evolutionary mechanisms whereby mutualisms might contribute to reproductive isolation. Phylogenetic approaches are critical for identifying patterns. For example, Dodd et al. (1999) found that animal-mediated pollination was associated with higher species diversity of plant families, but animal-mediated seed dispersal was not. In a diverse group of orchids, Waterman et al. (2011) found that pollination contributed to speciation, but that the fungal symbiosis contributed only to species coexistence. Plant–pollinator mutualisms obviously have great potential for influencing rates of speciation, in that evolutionary divergence between populations in floral traits and pollinators can often have direct effects on reproductive isolation (Bradshaw and Schemske 2003, Schemske and Bradshaw 1999, Kay and Sargent 2009). As discussed in Chapter 3, for example, the extraordinary diversity of figs (*Ficus* spp, > 600 species) must be related in part to the highly specialized nature of the pollination mutualism. Might mutualisms other than those based on pollination also contribute to speciation? The role of ecological divergence in speciation has experienced a revival, with the idea that adaptive divergence between populations may directly or indirectly lead to greater reproductive isolation (Nosil 2012). Thus, any factor or interaction that contributes to adaptive divergence can under some circumstances be viewed as a mechanism of speciation (Sobel et al. 2010).

5.5.3 The latitudinal diversity gradient

Along with many other biotic interactions studied, mutualisms often show a strong latitudinal gradient of interaction strength, prevalence, or importance (Schemske et al. 2009). For example, the incidence of animal-mediated pollination increases toward the equator (Regal 1982), as does the importance of ant–plant mutualisms related to plant defense (Schemske et al. 2009). With a better understanding of mutualism evolution, we can begin to address the question of whether gradients in the importance of species interactions are a cause or effect (or both) of the remarkable and as yet poorly understood latitudinal diversity gradient. Studies that examine the importance and evolution of mutualisms on latitudinal scales will require new model systems (especially in the tropics) with well-developed genetic resources in order to compare evolutionary processes between environments.

5.5.4 Anthropogenic environmental change

Understanding the evolution and genetics of mutualisms is critical to a growing number of applied problems. For example, as discussed in Chapters 13 and 14, mutualism evolution is relevant to conservation and restoration, such as declines in pollinator populations due to human pressures—might evolution of increased self-pollination mediate the impact of pollinator loss for plants (Eckert et al. 2009)? Climate change and invasive species may also disrupt mutualisms and cause novel interactions. In the case of climate change, these changes are expected to occur due to range shifts and phenological mismatches of mutualistic partners, both of which may be mitigated by adaptation to changing conditions (Hegland et al. 2009). We cannot accurately assess the potential for mutualistic interactions to adapt to climate change or invasive species without knowing details about the genetic basis of mutualisms, such as how many genes are involved in adaptation.

5.5.5 Conclusions

To understand the fundamental patterns and processes underlying mutualism evolution, and particularly to identify how evolution in mutualisms may be unique, questions about the evolutionary ecology and genetic basis of adaptive traits must be addressed on a broad taxonomic scale, and include studies of all species involved in the interaction. For most mutualisms (including those highlighted here), our understanding of the natural history, genes, traits, and fitness consequences of the interaction are limited and strikingly asymmetrical (Bronstein et al. 2006). Such bias severely limits our ability to assess the selective forces and evolutionary steps involved in the evolution of the interactions, and in particular, to determine the role of coevolution.

Fortunately, active evolutionary research in the systems highlighted here, along with others that we did not have space to review, is broadening our taxonomic coverage. These systems represent a range of life histories, specialization of the interaction, ease of lab culture and experimental manipulation, and molecular genetic and phylogenetic tools. This is an exciting time to study mutualism evolution, as the goal of finding empirical support for generalizations across systems may soon be within reach.

Acknowledgments

We would like to thank our box authors, U. Mueller, M. Friesen, and M. Nishiguchi, for their generous time and efforts. We would also like to thank J. Eilers and R. Atchison for assistance with formatting and references, and many friendly reviewers for their suggestions and comments, which greatly improved the manuscript: D. Brickley, J. Bronstein, E. Dittmar, A. Douglas, K. Kay and her lab members, J. Lau, N. Moran, U. Mueller, C. Oakley, S. Perez, and J. Rudgers.

References

Barrett, R.D.H., and Hoekstra, H.E. (2011). Molecular spandrels: tests of adaptation at the genetic level. *Nature Reviews Genetics*, **12**, 767–780.

Barton, N.H., and Keightley, P.D. (2002). Understanding quantitative genetic variation. *Nature Reviews Genetics*, **3**, 11–21.

Beardsley, P.M., Yen, A., and Olmstead, R.G. (2003). AFLP phylogeny of *Mimulus* section *Erythranthe* and the evolution of hummingbird pollination. *Evolution*, **57**, 1397–1410.

Bradshaw, H.D., Otto, K.G., Frewen, B.E., McKay, J.K., and Schemske, D.W. (1998). Quantitative trait loci affecting differences in floral morphology between two species of monkeyflower (*Mimulus*). *Genetics*, **149**, 367–382.

Bradshaw, H.D., and Schemske, D.W. (2003). Allele substitution at a flower colour locus produces a pollinator shift in monkeyflowers. *Nature*, **426**, 176–178.

Bronstein, J.L. (1994). Our current understanding of mutualism. *Quarterly Review of Biology*, **69**, 31–51.

Bronstein, J.L., Alarcon, R., and Geber, M. (2006). The evolution of plant-insect mutualisms. *New Phytologist*, **172**, 412–428.

Campbell, D.R., Waser, N.M., and Price, M.V. (1994). Indirect selection of stigma position in *Ipomopsis aggregata* via a genetically correlated trait. *Evolution*, **48**, 55–68.

Campbell, D.R., Waser, N.M., and Price, M.V. (1996). Mechanisms of hummingbird-mediated selection for flower width in *Ipomopsis aggregata*. *Ecology*, **77**, 1463–1472.

Darwin, C. (1862). *On the various contrivances by which British and foreign orchids are fertilised by insects*. London: John Murray.

Degnan, P.H., Yu, Y., Sisneros, N., Wing, R.A., and Moran, N.A. (2009). *Hamiltonella defensa*, genome evolution of protective bacterial endosymbiont from pathogenic ancestors. *Proceedings of the National Academy of Sciences of the U.S.A.*, **106**, 9063–9068.

Dodd, M.E., Silvertown, J., and Chase, M.W. (1999). Phylogenetic analysis of trait evolution and species diversity variation among angiosperm families. *Evolution*, **53**, 732–744.

Eckert, C.G., Ozimec, B., Herlihy, C.R., Griffin, C.A., and Routley, M.B. (2009). Floral morphology mediates temporal variation in the mating system of a self-compatible plant. *Ecology*, **90**, 1540–1548.

Fisher, R.A. (1930). *The genetical theory of natural selection*. Oxford: Clarendon Press.

Harcombe, W. (2010). Novel cooperation experimentally evolved between species. *Evolution*, **64**, 2166–2172.

Hegland, S.J., Nielsen, A., Lazaro, A., Bjerknes, A.L., and Totland, O. (2009). How does climate warming affect plant-pollinator interactions? *Ecology Letters*, **12**, 184–195.

Hembry, D.H., Yoder, J.B., and Goodman, K.R. (2014). Coevolution and the diversification of life. *American Naturalist*, **184**, 425–438.

Hillesland, K.L., Lim, S.J., Flowers, J.J., et al. (2014). Erosion of functional independence early in the evolution of a microbial mutualism. *Proceedings of the National Academy of Sciences of the U.S.A.*, **111**, 14822–14827.

Janzen, D.H. (1967). Why mountain passes are higher in the tropics. *American Naturalist*, **101**, 233–249.

Jeon, K.W. (2004). Genetic and physiological interactions in the amoeba-bacteria symbiosis. *Journal of Eukaryotic Microbiology*, **51**, 502–508.

Kay, K.M., and Sargent, R.D. (2009). The role of animal pollination in plant speciation: integrating ecology, geography, and genetics. *Annual Review of Ecology Evolution and Systematics*, **40**, 637–656.

Kimura, M. (1983). *The Neutral Theory of Molecular Evolution*. Cambridge, UK: Cambridge University Press.

Kingsolver, J.G., and Diamond, S.E. (2011). Phenotypic selection in natural populations: what limits directional selection? *American Naturalist*, **177**, 346–357.

Klahre, U., Gurba, A., Hermann, K., et al. (2011). Pollinator choice in *Petunia* depends on two major genetic loci for floral scent production. *Current Biology*, **21**, 730–739.

Louthan, A.M., and Kay, K.M. (2011). Comparing the adaptive landscape across trait types: larger QTL effect size in traits under biotic selection. *BMC Evolutionary Biology*, **11**.

MacLean, A.M., Finan, T.M., and Sadowsky, M.J. (2007). Genomes of the symbiotic nitrogen-fixing bacteria of legumes. *Plant Physiology*, **144**, 615–622.

Matuszewski, S., Hermisson, J., and Kopp, M. (2014). Fisher's geometric model with a moving optimum. *Evolution*, **68**, 2571–2588.

McCutcheon, J.P., and Moran, N.A. (2007). Parallel genomic evolution and metabolic interdependence in an ancient symbiosis. *Proceedings of the National Academy of Sciences of the U.S.A.*, **104**, 19392–19397.

McCutcheon, J.P., and Moran, N.A. (2012). Extreme genome reduction in symbiotic bacteria. *Nature Reviews Microbiology*, **10**, 13–26.

Meyer, J.R., and Meyer, V.G. (1961). Origin and inheritence of nectariless cotton. *Crop Science*, **1**, 167–169.

Moran, N.A., Degnan, P.H., Santos, S.R., Dunbar, H.E., and Ochman, H. (2005). The players in a mutualistic symbiosis: insects, bacteria, viruses, and virulence genes. *Proceedings of the National Academy of Sciences of the U.S.A.*, **102**, 16919–16926.

Moran, N.A., McCutcheon, J.P., and Nakabachi, A. (2008). Genomics and evolution of heritable bacterial symbionts. *Annual Review of Genetics*. Palo Alto: Annual Reviews.

Nosil, P. (2012). *Ecological Speciation*. New York: Oxford University Press.

Oliver, K.M., Campos, J., Moran, N.A., and Hunter, M.S. (2008). Population dynamics of defensive symbionts in aphids. *Proceedings of the Royal Society B-Biological Sciences*, **275**, 293–299.

Oliver, K.M., Russell, J.A., Moran, N.A., and Hunter, M.S. (2003). Facultative bacterial symbionts in aphids confer resistance to parasitic wasps. *Proceedings of the National Academy of Sciences of the U.S.A.*, **100**, 1803–1807.

Orr, H.A. (1998). The population genetics of adaptation: the distribution of factors fixed during adaptive evolution. *Evolution*, **52**, 935–949.

Orr, H.A. (2005). Theories of adaptation: what they do and don't say. *Genetica*, **123**, 3–13.

Paterson, S., Vogwill, T., Buckling, A., et al. (2010). Antagonistic coevolution accelerates molecular evolution. *Nature*, **464**, 275–U154.

Porter, S.S., and Simms, E.L. (2014). Selection for cheating across disparate environments in the legume-rhizobium mutualism. *Ecology Letters*, **17**, 1121–1129.

Regal, P.J. (1982). Pollination by wind and animals—ecology of geographic patterns. *Annual Review of Ecology and Systematics*, **13**, 497–524.

Richards, S., Gibbs, R.A., Gerardo, N.M., et al. (2010). Genome sequence of the pea aphid *Acyrthosiphon pisum*. *PLoS Biology*, **8**.

Rockman, M.V. (2011). The QTN program and the alleles that matter for evolution: all that's gold does not glitter. *Evolution*, **66**, 1–17.

Rudgers, J.A., and Strauss, S.Y. (2004). A selection mosaic in the facultative mutualism between ants and wild cotton. *Proceedings of the Royal Society B-Biological Sciences*, **271**, 2481–2488.

Rutter, M.T., and Rausher, M.D. (2004). Natural selection on extrafloral nectar production in *Chamaecrista fasciculata*: the costs and benefits of a mutualism trait. *Evolution*, **58**, 2657–2668.

Sachs, J.L., Mueller, U.G., Wilcox, T.P., and Bull, J.J. (2004). The evolution of cooperation. *Quarterly Review of Biology*, **79**, 135–160.

Schemske, D.W. (2002). Ecological and evolutionary perspectives on the origins of tropical diversity. In Chazdon, R.L. and Whitmore, T.C. (eds.) *Foundations of Tropical Forest Biology*. Chicago: University of Chicago Press.

Schemske, D.W. (2009). Biotic interactions and speciation in the tropics. In Butlin, R.K., Bridle, J.R., and Schluter, D. (eds.) *Speciation and Patterns of Diversity*. Cambridge, UK: Cambridge University Press.

Schemske, D.W., and Bradshaw, H.D. (1999). Pollinator preference and the evolution of floral traits in monkeyflowers (*Mimulus*). *Proceedings of the National Academy of Sciences of the U.S.A.*, **96**, 11910–11915.

Schemske, D.W., Mittelbach, G.G., Cornell, H.V., Sobel, J.M., and Roy, K. (2009). Is there a latitudinal gradient in the importance of biotic interactions? *Annual Review of Ecology Evolution and Systematics*, **40**, 245–269.

Siepielski, A.M., and Benkman, C.W. (2010). Conflicting selection from an antagonist and a mutualist enhances phenotypic variation in a plant. *Evolution*, **64**, 1120–1128.

Siepielski, A.M., DiBattista, J.D., and Carlson, S.M. (2009). It's about time: the temporal dynamics of phenotypic selection in the wild. *Ecology Letters*, **12**, 1261–1276.

Sobel, J.M., Chen, G.F., Watt, L.R., and Schemske, D.W. (2010). The biology of speciation. *Evolution*, **64**, 295–315.

Thompson, J.N. (2005). *The Geographic Mosaic of Coevolution*. Chicago: University of Chicago Press.

Venail, J., Dell'Olivo, A., and Kuhlemeier, C. (2010). Speciation genes in the genus *Petunia*. *Philosophical Transactions of the Royal Society B-Biological Sciences*, **365**, 461–468.

Wade, M.J., and Kalisz, S. (1990). The causes of natural selection. *Evolution*, **44**, 1947–1955.

Waterman, R.J., Bidartondo, M.I., Stofberg, J., et al. (2011). The effects of above- and below-ground mutualisms on orchid speciation and coexistence. *American Naturalist*, **177**, E54–E68.

Weber, M.G., and Keeler, K.H. (2013). The phylogenetic distribution of extrafloral nectaries in plants. *Annals of Botany*, **111**, 1251–1261.

Wright, J.W., and Meagher, T.R. (2004). Selection on floral characters in natural Spanish populations of Silene latifolia. *Journal of Evolutionary Biology*, **17**, 382–395.

Yoder, J.B., and Nuismer, S.L. (2010). When does coevolution promote diversification? *American Naturalist*, **176**, 802–817.

CHAPTER 6

The exploitation of mutualisms

Joel L. Sachs

6.1 Introduction

The evolution of mutualisms presents a paradox. Natural selection favors selfish traits rather than cooperative ones in most scenarios (Darwin 1859, Williams 1966). Thus, mutualisms are expected to evolve only under a restrictive set of conditions. Moreover, selection is predicted to favor the exploitation of mutualisms, a situation in which individuals gain fitness benefits from a mutualist service but do not reciprocate. Exploiters can be selectively superior to mutualists because they gain the benefits of a mutualism without the associated costs of being cooperative (Yu 2001), and thus exploitation can ultimately cause a mutualism to break down (Sachs and Simms 2006). Exploitation of mutualism can occur in two different ways. First, the exploiters can be "cheater mutants," erstwhile mutualists that have evolved to exploit their partners (Bronstein 2001, Bull and Rice 1991, Sachs et al. 2004). By definition, the evolution of cheater mutants requires an evolutionary reversal from a mutualist trait (Sachs and Simms 2006). Secondly, exploitation can occur when a species from outside a mutualism takes advantage of the resources or services that one mutualist species offers another, and does so without reciprocation. These "parasites of mutualisms" (Yu 2001) have also been called exploiter species (Bronstein 2001) and aprovechados ("advantage takers"; Soberon and Martinez del Rio 1985). Unlike cheater mutants, a parasite of mutualism might have no history of cooperation with the mutualist it exploits. The categorization of cheater mutants and parasites of mutualisms is imperfect. Some forms of exploitation reveal blurred lines between these definitions, especially in multipartner or diffuse mutualisms where it is difficult to reconstruct the past history of interactions.

In the past fifty years the study of mutualism has undergone several paradigmatic shifts. The perceived importance of exploitation has changed dramatically over this time. As late as the mid-1960s many biologists held the opinion that natural selection favors "the good of the species," and hence the evolutionary maintenance of cooperative interactions was not viewed critically. The work of W.D. Hamilton and George Williams fundamentally shifted the paradigm, by revealing the primacy of individual selection in shaping traits, and hence that selfishness rather than cooperation should be favored in most settings (Hamilton 1964, Williams 1966). Subsequently, Trivers, May, and Axelrod and Hamilton brought the study of cooperation between species into the limelight (Axelrod and Hamilton 1981, May 1976, Trivers 1971). These researchers introduced a theoretical framework predicting that mutualisms are prone to exploitation, and thus that mutualisms can be both ecologically and evolutionarily unstable. Over the next decades began an empirical focus on mutualism that led to the examination of diverse species interactions. Exploitation was a key focus in these studies (Bronstein 1994a, b). The paradox of mutualism persistence began to be resolved in a general way in the 1990s and 2000s in a series of papers that modeled conditions under which mutualists can be favored over exploiters (Sachs et al. 2004, Bull and Rice 1991, Denison 2000, Foster and Wenseleers 2006, West et al. 2002) or in which mutualists and exploiters can stably coexist (Ferriere et al. 2002, Jones et al. 2009).

Mutualism. Edited by Judith L. Bronstein.

This chapter begins by reviewing empirical work on exploitation and outlining the types of evidence used to define exploitation of mutualism. My goals are to examine the degree to which exploitation occurs in mutualisms, whether exploitation evolves within a mutualism (cheater mutants) or comes from outside the interaction (parasites of mutualism), and how exploitation affects mutualism stability. My review of available datasets suggests that the evolution of cheater mutants appears to occur rarely and only in obligate interactions. In contrast, parasites of mutualism are comparatively common. I then review theoretical work. I focus on models that predict a set of conditions in which mutualist traits are favored and exploitation is disfavored, and that show how exploitation can coexist with mutualism. Theoretical predictions are related back to empirical studies of mutualisms whenever possible. Here, the discussion reveals some disagreement between theory and the empirical data. Evolutionary models often suggest that cheater mutants can evolve and potentially destabilize mutualisms, but this prediction has not been borne out in the empirical data. Several hypotheses are described that can explain this paucity of empirical evidence for cheater mutants. Although this remains an open question, the evidence suggests that mutualisms are not often beset by conflicts of interest, as is often assumed in models.

6.2 Empirical evidence for the exploitation of mutualisms

Exploitation of mutualism can be defined empirically in terms of three characteristics of an interaction. Exploiters bear traits that (i) take advantage of a service or resource provided by mutualists, (ii) enhance the fitness of the exploiter, and (iii) decrease the fitness of the mutualist being exploited. Exploiters do not encompass predators, pathogens, and parasites that eat, infect, or otherwise inhabit mutualists, because these individuals do not take advantage of the mutualistic interaction. Moreover, exploiters do not include individuals that are somehow disabled from providing a benefit to a mutualist partner. This can happen if a potential mutualist is of insufficient size, or is otherwise unfit (and thus is unable to provide benefits), or because the potential mutualist is specialized to a different set of partners (i.e., maladapted mutualist). Poor and maladapted mutualists are different from exploiters in that they do not themselves gain benefit from their apparent exploitation (Sachs and Simms 2008).

To differentiate between the two types of exploiters—cheater mutants and parasites of mutualism—data must resolve whether the exploitative individuals are evolutionarily derived from mutualists (cheater mutants) or from outside the mutualist interaction (parasites of mutualism). To elucidate the origins of the exploitative trait requires a well-sampled and resolved phylogeny and good trait information for all the taxa involved. These data can be hard to come by. The question of origins can also become ambiguous in mutualisms that involve networks of interacting partners, wherein it is challenging to ascribe specificity to any set of partners, or the specificity might vary within or among populations. Most empirical studies of exploitation generally fall short of the ideal dataset, which would include information about the mutualist service being exploited, the fitness effects on the exploiters and the individuals being exploited, and the evolutionary origins of the exploitation trait. Below, I review empirical studies of exploitation. My review of empirical evidence is not exhaustive, but is intended to be a guide to a few well-studied mutualistic systems, and to a few key types of data that have been gathered.

6.2.1 Pollination mutualisms

In pollination mutualisms, vertebrates and insects transfer pollen for flowering plants in exchange for rewards, usually nectar. But pollinators and plants can experience a conflict of interest over the optimal amount of nectar provided (which is higher for the pollinator than the plant), versus the amount of pollination provided (which is higher for the plant; Bell 1986, Rathcke 1992). Mutualists of either partner species could conceivably exploit the other by evolving to reduce or negate their service (cheater mutants). Examples of cheater mutants would be if a nectar-producing plant evolved to eliminate nectar production or if a pollinating insect evolved to take nectar without pollination. Exploitation can also come from outside of a plant–pollinator interaction, if a third species can gain benefits from the mutualism without paying costs (parasites of mutualism). An example of a parasite of this mutualism would be if an insect with no pollinating behavior evolved to exploit a plant's nectar.

From the perspective of plant fitness, pollination can be exploited by offering little or no nectar. In an extreme example, about one-third of orchid species never produce nectar (Smithson and Gigord 2003). These orchids appear to be parasites of pollination mutualisms, as they rely on pollinators that get nectar from other plant species. Some nectarless orchids mimic nectar-bearing species (Dafni 1984), and others bear flowers that look and/or smell like female insects, thus attracting males to attempt copulation on the flower and transfer pollen in the process. In many cases nectarless orchids suffer from low reproductive success relative to plant species that do provide nectar, likely because pollinators learn to discriminate against the exploiter species (Smithson and Gigord 2003). In some plant taxa, different strategies of nectar production can coexist and thus be compared. In the honey mesquite, *Prosopsis glandulosa*, about half of individuals are nectarless. But as in the orchids, pollinators discriminate against the nectarless plants, leading to significant reduction in fruit set (Golubov et al. 1999). Plants can also exploit pollinators in more nuanced ways. For instance, Bell (1986) modeled a mixed strategy in which plants offered nectar in only a subset of their flowers, dependent on the cost borne by the pollinator for discriminating among flowers. Empirical work on the species *Cerinthe major* supports the model, and shows that this host plant bears many nectarless flowers, especially when it is costly for the insects to take the time to discriminate among flowers (Gilbert et al. 1991). Nectarless flowers have evolved in many plant lineages, and in some cases being nectarless can significantly enhance plant fitness (Smithson and Gigord 2003). In many cases, however, the pollinators appear to detect and avoid plants with poor nectar production. To further examine the potential for plants to exploit pollinators, more studies are needed to reconstruct the evolutionary gain and loss of nectarless flowers and to compare the relative fitness effects of these strategies in natural populations.

Pollinators can exploit host plants by eating nectar without providing pollination, which often occurs when an animal makes or uses a hole in the petal tissue to gain easy access to the nectar (Irwin et al. 2010). Among insects, many ants rob nectar and almost never pollinate flowers, and thus represent pure parasites of the mutualism (Irwin and Maloof 2002). A mixed strategy of pollination and robbing also appears common, wherein individual insects alternate strategies between bouts of visitation, or among plant species visited (Irwin et al. 2010). Nectar robbing is nearly ubiquitous upon some plant species, especially those with tubular flowers (Irwin and Maloof 2002). For many pollinator and pollen-robbing species it is difficult to estimate fitness benefits and costs of exploitation as the interactions are often diffuse (with many interacting partners) and because individual animals can vary strategies.

Pairwise pollination systems offer a simpler model of these interactions, such as occurs between figs and fig wasps, between yuccas and yucca moths, and between globeflowers (*Trollius*) and flies, wherein a single or a few insect species provides most or all of the pollination for a single plant species. In these obligate partnerships the insects generally oviposit into the ovaries of the plant, and then deposit pollen in those flowers, and the developing insect larvae subsequently feed on a subset of the developing seeds. Thus, there is potential conflict over the production of plant versus pollinator offspring. Insects that are exploiters lay their eggs into flowers without pollinating them, or oviposit directly into developing fruits (that were the products of pollination by another insect). This non-pollination strategy has evolved multiple times in yucca moths (Pellmyr et al. 1996), and at least once in fig wasps (Machado et al. 2000) and globeflower flies (Despres and Jaeger 1999). Non-pollination in the yucca moth and fig wasp appears to evolve only after an ancestrally pollinating species colonized a novel host that already had a pollinator. The host switching makes it difficult to resolve the exact history of how the exploitation evolved. In contrast, in the globeflower system, pollinating and non-pollinating flies all coexist on a single host species, and hence there is stronger evidence that cheater mutants have evolved and then speciated from mutualist ancestors (Despres and Jaeger 1999). More work is needed to discover and examine these exploitative traits before they become fixed in populations. For instance, work has uncovered the presence of rare, non-pollinating individuals within several species of fig wasps that normally pollinate their fig hosts (Jánder and Herre 2010). Fitness evidence suggests that these individuals get punished by hosts; hence the non-pollinators might not have the capacity to spread in the population (Jánder and Herre 2010).

6.2.2 Legume–rhizobium symbiosis

Rhizobia are soil bacteria that have the ability to form root nodules on legumes, within which the bacteria fix nitrogen for the host in exchange for photosynthates (see Box 5.2). Legumes gain substantial growth benefits from rhizobia. In turn, beneficial rhizobia attain high densities in soils nearby legume hosts, presumably due to release from senescing nodules (Sachs et al. 2009, 2010a, Denison 2000). Despite these mutual benefits, fitness interests can conflict between the partners. In particular, nitrogen fixation is metabolically costly (Trainer and Charles 2006), and models predict that selection can favor rhizobia that minimize fixation (West et al. 2002). Consistent with these predictions, legumes are commonly infected with rhizobia that provide zero or marginal amounts of fixed nitrogen to host (Sachs and Simms 2008). However, non-fixing rhizobia might represent maladapted mutualists rather than cheaters (i.e., the non-beneficial rhizobia gain no benefit from the interaction). Importantly, rhizobia that enhance fitness for one legume genotype often provide little or no benefits to other host populations or species, likely because specificity to a subset of partners leads to maladaptation to others (Burdon et al. 1999). This pattern suggests that rhizobia that have been identified as non-fixing (on one host) might be highly beneficial if inoculated onto a more suitable host. Moreover, non-fixing strains have been most frequently isolated from agricultural and pastoral sites, where transfer of plants, mixture of soils, and chemical soil amendments might enhance the probability of maladaptive interactions (Sachs et al. 2010a).

Most studies of the legume–rhizobium symbiosis measure the effect of different rhizobial strains on the host without examining the fitness of the rhizobial partner, a pattern that is common in studies of mutualism (Bronstein 1994a). But it is possible to simultaneously measure the fitness of both infected legume hosts (via growth or seed set) and rhizobia (via population size of rhizobia inside or released from the nodule). One wild rhizobial strain has been discovered to provide zero growth benefit to sympatric host plants, but nonetheless exhibits higher fitness within host root nodules than beneficial strains (Sachs et al. 2010a). But this rhizobium is not an unambiguous example of a cheater mutant. Phylogenetic analyses showed that this exploiter did not evolve from mutualist ancestors, but instead from a non-symbiotic lineage, and gained the ability to infect hosts via a horizontal gene transfer event (Sachs et al. 2010a). Moreover, this exploiter rhizobium exhibits very low fitness when it must compete with other rhizobia (Sachs et al. 2010b). A meta-analysis of diverse legume–rhizobium interactions supports the idea that cheating might be absent or rare in this system. The study, which analyzed the fitness effects (upon legume hosts) of diverse rhizobia, found a strong positive correlation between benefit to hosts and the rhizobial fitness upon those hosts (Friesen 2012). Thus, rhizobia that caused low host fitness also exhibited low fitness themselves. No study has yet unambiguously demonstrated that rhizobia have the capacity to exploit hosts to their own benefit. Instead, most non-fixing rhizobial infections are caused by host–symbiont mismatch, as opposed to the rhizobial exploitation of the host.

6.2.3 Invertebrate–algal symbiosis

Dinoflagellate algae in the genus *Symbiodinium* infect a broad array of marine invertebrate hosts including bivalves, corals, hydras, and jellyfish (Trench 1993; see Boxes 9.1 and 10.3). These algae provide their hosts with sugars derived from photosynthesis (Balderston and Claus 1969) in exchange for nitrogen and other nutrients (Muscatine 1990). *Symbiodinium* encompasses at least nine diverse lineages. Two of the lineages, named clades "A" and "D," often provide poor or negligible benefits to invertebrate hosts relative to other *Symbiodinium* (Stat et al. 2008, Lesser et al. 2013), but these algal strains can spread to dominate the hosts under thermally stressful conditions that are increasingly common in reef communities (Stat and Gates 2011, Stat et al. 2008). In general, the clade A and D symbionts do not appear to be their hosts' typical partners. Moreover, clade A holds an ancestral position within the *Symbiodinium* lineage and is closely related to pathogenic algae, leading to the hypothesis that clade A exhibits no evolutionary history of mutualism with invertebrate hosts (Lesser et al. 2013). *Symbiodinium* clades A and D might be best described as parasites of the mutualism, as they infect hosts that typically harbor other clades of algal symbionts. But the fitness outcomes of these interactions are still unclear both for algal and animal partners. Studies must discern whether hosts suffer reduced reproductive output from clade A or D infections and also test if these algae gain a significant fitness advantage over more beneficial taxa within hosts.

Individual animal hosts can harbor a diversity of *Symbiodinium* strains, suggesting the possibility that a range of beneficial and harmful symbionts can coexist. The mode of symbiont transmission among hosts can have important selective effects on the relative success of mutualist versus parasite symbiont strains. For instance, depending on the host biology, *Symbiodinium* can be transmitted vertically among hosts (i.e., from host parent to offspring), horizontally (among unrelated hosts), or both. Vertical transmission couples the interest of symbiont and host and should favor mutualist symbionts, whereas horizontal transmission decouples the fitness interest of the symbiont from the host, and allows parasitic algae to gain advantage in the populations (Sachs et al. 2004). Experiments on jellyfish–algal symbioses have supported these predictions: experimental populations of *Symbiodinium* that were enforced to undergo repeated rounds of horizontal transmission evolved to provide lower fitness benefits to jellyfish hosts while evolving to grow more rapidly within host tissue (Sachs and Wilcox 2006). But importantly, hosts that were infected with the horizontally transmitted algae were quite stunted in their growth, such that the total fitness of the harmful algae that escaped from these hosts was also poor (Sachs and Wilcox 2006). These data suggest that *Symbiodinium* populations within hosts can range from beneficial to harmful, but that partner fidelity feedback between host and symbiont fitness (see Section 6.3) can be an important factor that selects against exploitation.

6.2.4 Phylogenetic analyses of mutualist lineages

Phylogenetic and population-genetic analyses can be powerful tools to investigate the degree to which exploitative traits emerge in mutualist lineages (Sachs and Simms 2006; see also Box 7.2). Among studies that have adopted these tools, there are few examples that are consistent with the evolution of cheater mutants from mutualist ancestors. For instance, the mutualism between large blue butterflies (Lycaenidae) and defensive ants has generated cheater species that have evolved from mutualist ancestors (Als et al. 2004, Pierce et al. 2002). The larvae of blue butterfly provide nutritious secretions to ant partners in exchange for protection (Pierce et al. 2002). The cheater traits are expressed by the larvae of the lycaenid genera *Phengaris* and *Maculinea* that exhibit a range of parasitic behaviors on their ant hosts (Als et al. 2004). Only one origin of parasitism from mutualism has been confidently resolved, but the broad distribution of exploitative traits in the lycaenid and riodinid butterflies suggests that there might be frequent origins of cheating (Pierce et al. 2002). Another origin of exploitation is exemplified by the evolutionary loss of photosynthetic traits in plants. The plants associate with ectomycorrhizal mutualists, soil fungi that attach to plant roots and provide nutrients and or protection in exchange for resources from the plant. By losing photosynthetic capacity, the plant partners have undergone a transition from mutualist to parasite on their fungal partners (Bidartondo and Bruns 2001, Taylor and Bruns 1997). The histories of the specific plant–fungal partnerships are not fully resolved, and it is not clear if partner switches or multipartner interaction networks occurred concurrent with the evolutionary loss of photosynthetic capability. Thus, the plants have evolved to exploit mutualism, but it is not clear if these lineages evolved as parasites of the mutualism or are cheater mutants. Finally, the evolutionary stability of mutualist lineages has also been studied at much deeper phylogenetic levels, in particular by examining mutualist traits in ancient lineages of bacteria. When traits of mutualism, parasitism, and non-association were reconstructed on bacterial phylogenies of a whole phylum or kingdom, a few transitions from mutualism to exploitation were uncovered (Sachs et al. 2011b, 2014). Thus, even over the course of a billion-plus years of bacterial evolution, evidence of the origins of cheaters is scant.

6.2.5 Summary of empirical evidence

In the mutualisms highlighted above, non-beneficial or harmful partners that interact with mutualists are common. But the failure to provide a benefit to a mutualist can be caused by multiple factors other than exploitation, including maladaptation and poor partner quality. It can be challenging empirically to demonstrate that exploitation of mutualisms represents an adaptive strategy. Among the examples of exploitation that appear adaptive, most are consistent with parasites of mutualism rather than cheater mutants.

6.3 Theoretical predictions for the conditions that stabilize mutualism

Theoreticians have formulated three types of models under which mutualistic traits are stably maintained in populations. These models have been termed byproducts, partner fidelity feedback, and partner-choice sanctions (Sachs et al. 2004). Mutualism models are also discussed elsewhere in this volume (see Chapter 4) so the discussion here is brief. The simplest models predict that mutualist phenotypes are cost-free byproducts of other traits (Sachs et al. 2004, Foster and Wenseleers 2006). Such "byproduct mutualisms" (see also Chapter 2) are interactions characterized by a producer species, one that automatically generates a byproduct (most often a secretion or excretion), and a receiver species that feeds upon or otherwise benefits from the byproduct. The byproducts can flow one way in the partnership, which thus could be defined as a commensalism, or byproducts can be exchanged. The mutualistic trait of the producer is favored by natural selection, but for reasons unrelated to the mutualism. For example, bacteria must rid themselves of waste compounds into the environment to maintain cell function, but these byproducts are then made available to and useful for other bacterial species in their community. Some bacteria cannot live without the byproducts of other bacterial species (Morris et al. 2012). Byproduct models are unique among models of interspecific cooperation in that no conflicts of interest are predicted, since there are no costs of association, and thus selection for exploitation should be wholly absent (Foster and Wenseleers 2006). Other examples of byproduct mutualism include predators that necessarily leave portions of their kills behind and thus provide food for carrion-eating species, and ungulates that generate excreta, which happen to be key food resources for dung beetles (Sachs et al. 2004).

In partner fidelity feedback models, mutualistic traits are maintained via feedbacks from partners that reciprocate fitness benefits (Trivers 1971, Bull and Rice 1991, Sachs et al. 2004, Foster and Wenseleers 2006). Mutualisms between ants and plant hosts are a classic case of partner fidelity. Plant hosts such as *Acacia* offer housing and food to ants in exchange for protection from herbivory (Janzen 1966, Chamberlain and Holland 2009, Frederickson and Gordon 2009). These plant services promote the health and growth of the ant colony; in turn, the increased number of ants defending the plant allows the plant to grow and return even more benefits to the ants. But partner fidelity does not operate across generations in this interaction. New seedlings must be colonized by ants, and those ants need not come from the parent plant. Thus, partner fidelity can support cooperation within the life span of the plant, but breaks down between plant generations, leading to potential conflict. In Box 6.1, Megan Frederickson describes empirical work that studies exploitation in this mutualism.

Organelles and their eukaryote hosts experience an extreme form of partner fidelity. Vertical transmission of the organelle from host parent to offspring links the fitness interests of both partners over millennia. Furthermore, any microbe that is vertically transmitted in a host lineage, or that interacts with a host for many cell generations, can

Box 6.1 Some ants sterilize their host plants, but are they "cheaters"?

Megan E. Frederickson

Ant–plant associations are a textbook example of mutualism, but these interactions also reveal conflict that can threaten the maintenance of mutualism. In these interactions, the ants defend their host trees against herbivory in exchange for food and housing from the plant in the form of hollow structures called domatia (Chamberlain and Holland 2009, Janzen 1966, Frederickson and Gordon 2009). Initially, an ant queen colonizes a domatia-bearing plant when the plant is small, and the plant and the ant colony grow together for many years, during which time the relationship is exclusive (i.e., a single ant colony nests in one host plant). Partner fidelity feedback is expected to occur between ant colony and plant growth. Ant workers defend their plant from herbivores, increasing plant growth, and as the plant grows it provides more domatia (and frequently more extrafloral nectar or food bodies) to its resident ant colony, increasing ant colony growth (Frederickson and Gordon 2009). Thus, natural selection favors mutualism and not cheating because a partner that fails to cooperate reduces its own fitness. For example, if an ant colony does not protect its plant from herbivores, the plant does not produce the domatia necessary for the ant colony to expand and produce reproductives. However, partner fidelity feedback is not always faithfully maintained between ant and plant partners, since new seedlings are colonized by ants that do not come from the parent plant. There is no guarantee that the offspring of either partner will associate in the next generation and thus neither partner gains from the reproduction of the other.

Some of these same ant species exhibit a curious behavior: they destroy flowers on their host trees, severely reducing plant fecundity, even while their nestmates may patrol and defend leaves against herbivores, promoting plant growth (Yu and Pierce 1998; Figure 1 of Box 6.1). Yu and Pierce proposed that sterilizing ants manipulate the trade-off between

Figure 1 Left: *Allomerus octoarticulatus* workers patrol and defend a young *Cordia nodosa* shoot against herbivores. Top right: workers of this ant species also destroy *C. nodosa* inflorescences by gnawing on the delicate floral buds. Bottom right: undamaged *C. nodosa* floral buds, for comparison. Photos originally appeared in Frederickson (2009): Conflict over reproduction in an ant-plant symbiosis: why Allomerus octoarticulatus ants sterilize Cordia nodosa trees, *The American Naturalist*, 173, 675–681. © 2009 by The University of Chicago. See Plate 4 for color version.

continued

Box 6.1 *Continued*

growth and reproduction in their host plants. The prediction is that ants attack flowers to promote plant growth, and by corollary ant colony fitness at the expense of plant reproduction, from which ants derive no fitness benefit (Yu and Pierce 1998). Empirical work supports this prediction, showing that artificially sterilized trees produce more domatia than control trees and can thus house larger, more fecund ant colonies (Frederickson 2009). Flower-attacking behavior may therefore be adaptive because ant colony fitness depends only on plant growth, but not on plant reproduction. Moreover, sterilizing ants could impose selection on plants for floral defenses against ants, possibly leading to coevolutionary arms races between flower-attacking ants and flower-defending plants. However, this fundamentally depends on whether sterilizing ants decrease plant fitness. Sterilizing ants substantially reduce fruit set, but most ant-plants are long-lived, making direct measurements of total lifetime fitness difficult (Yu and Pierce 1998, Orivel et al. 2011, Palmer et al. 2010). Furthermore, because ant-plants often outlive their ant colonies, they can be repeatedly colonized by ants throughout their lives—often by different species of ants, not all of which sterilize trees (Palmer et al. 2010, Frederickson 2009). A recent demographic model showed that the ontogenetic timing of ant partners has a major effect on the fitness of *Acacia drepanolobium* trees in East Africa (Palmer et al. 2010). In this model, the sterilizing ant *C. nigriceps* actually increases lifetime plant fitness because it tends to colonize young *A. drepanolobium* trees and strongly promotes their survival; later, these trees are colonized by *C. mimosae*, which does not sterilize, permitting fruit set. If, like *C. nigriceps*, other sterilizing ants increase the lifetime fitness of their host plants, we cannot regard them as "cheaters" and they will not give rise to antagonistic coevolution between ant and plant partners.

References

Chamberlain, S.A., and Holland, J.N. (2009). Quantitative synthesis of context dependency in ant-plant protection mutualisms. *Ecology*, **90**, 2384–2392.

Frederickson, M.E. (2009). Conflict over reproduction in an ant-plant symbiosis: why *Allomerus octoarticulatus* ants sterilize *Cordia nodosa* trees. *The American Naturalist*, **173**, 675–681.

Frederickson, M.E., and Gordon, D.M. (2009). The intertwined population biology of two Amazonian myrmecophytes and their symbiotic ants. *Ecology*, **90**, 1595–1607.

Janzen, D.H. (1966) Coevolution of mutualism between ants and acacias in Central America. *Evolution*, **20**, 249–275.

Orivel, J., Lambs, L., Malé, P.J.G., Leroy, C., Grangier, J., Otto, T., Quilichini, A., and Dejean, A.. (2011). Dynamics of the association between a long-lived understory myrmecophyte and its specific associated ants. *Oecologia*, **165**, 365–376.

Palmer, T.M., Doak, D.F., Stanton, M.L., Bronstein, J.L., Kiers, E.T., Young, T.P., Goheen, J.R., and Pringle, R.M. (2010). Synergy of multiple partners, including freeloaders, increases host fitness in a multispecies mutualism. *P. Natl. Acad. Sci, USA*, **107**, 17234–17239.

Yu, D.W., and Pierce, N.E. (1998). A castration parasite of an ant-plant mutualism. *Proceedings of the Royal Society of London B*, **265**, 375–382.

benefit by actively enhancing the fitness of the host, and the enhanced host health and growth feeds back to benefit the microbe. Under partner fidelity, the interdependence of fitness selects for cooperation, and partner fidelity feedback is predicted to scale with the intimacy and interdependency of an interaction (Foster and Wenseleers 2006). The experiments on jellyfish–algal symbioses described in this chapter (Section 6.2.3) are consistent with this prediction. The populations of *Symbiodinium* enforced to undergo repeated rounds of horizontal transmission, and thus reduced partner fidelity with the host, evolved to provide lower fitness benefits to those hosts (Sachs and Wilcox 2006). Thus partner fidelity systems are not immune to cheaters or to parasites of mutualism. But fitness feedbacks can strongly limit the benefits that exploiters reap from these interactions (Foster and Wenseleers 2006).

In partner choice and sanctions models, one mutualist partner exhibits some degree of control such that it can select for more cooperative members of the interacting species (Bull and Rice 1991, Denison 2000, Foster and Wenseleers 2006, Sachs et al. 2004). The twin ability of a mutualist to choose among partners (partner choice) and/or to modulate investment or punishment of partners based on quality (sanctions) can generate a powerful selective force against cheater mutants as well as parasites of mutualism (Sachs et al. 2004). In Box 6.2, Redouan Bshary describes cleaning mutualisms, wherein fish

Box 6.2 Mechanisms and examples of cheating in cleaning mutualisms

Redouan Bshary

In marine cleaning mutualisms, "client" fish visit so-called "cleaners" to have ectoparasites removed. Over 100 species of fishes and shrimps have been observed to clean, but only a few species are highly dependent on cleaning interactions for food (Cote 2000). The mutual benefits of cleaning interactions appear to be obvious, as removing the parasites yields the cleaner a meal. In line with this view, studies have provided evidence that regular access to the cleaner wrasse *Labroides dimidiatus* can enhance the growth, body condition, and stress levels of clients (Clague et al. 2011, Ros et al. 2011). Nevertheless, three sources of conflict have been described: (i) cleaners are at risk of being eaten when client species are piscivorous; (ii) clients are at risk of being exploited if cleaners eat a client's protective layer of mucus instead of ectoparasites; and (iii) clients can get exploited by parasites of the cleaning mutualism, such as sabre-tooth blennies that mimic cleaner wrasse coloration but that eat scales and mucus off clients instead of cleaning them.

A cleaner's risk of being eaten

Client species regularly eat small fish the size of cleaners, but cases in which cleaners have been observed to get eaten are very rare and apparently occur outside of cleaning interactions (Cote 2000). In the lab, predation may more readily occur (Grutter 2004), but its ecological relevance is questionable. In general, predatory clients show reduced hunting activity in the presence of cleaner wrasse *L. dimidiatus* (Cheney et al 2008). As to why predatory clients refrain from eating cleaners, even though the latter even enter their mouths, virtually nothing is known. When predator species do attack cleaners, it is unclear if they are actively seeking cleaning stations for hunting, and thereby inflicting costs on both cleaners and small clients.

A client's risk of being cheated

A comparative study of wrasses suggests that cleaning mutualism started as a byproduct mutualism and that only more dependent cleaner species evolved a taste for client mucus (Barbu et al. 2011). A preference for mucus may have evolved only once in the common ancestor of the two genera *Labroides* and *Larabicus*. Clients of *L. dimidiatus* show a variety of control mechanisms that enhance the service quality provided by the cleaners. The mechanisms depend on a client's strategic options. "Resident" clients with access to only one cleaning station punish (i.e., sanction) mucus-feeding cleaners through aggressive chasing (Bshary and Grutter 2002). In contrast, "visitor" clients with access to several cleaning stations switch to a different station for their next inspection (i.e. partner choice) if the cleaner cheated them (Bshary and Schaffer 2002). It should be noted that despite the clear evidence for conflict in the *L. dimidiatus* system, this system has also provided the clearest evidence for the mutualistic outcome of cleaning interaction. From a theoretical viewpoint, the reasons are probably that clients are in control of whether an interaction takes place and interactions are iterated (i.e., both partner choice and partner fidelity feedback can stabilize the mutualism). If these two factors do not hold, the system should drift further toward parasitism.

Parasites of the cleaning mutualism

Only cleaners of the genus *Labroides* have been described as models for sabre-tooth blennies. The fangblenny, *P. rhynorhynchus*, regularly bites clients of cleaners. The mimics actively seek the presence of cleaners and apparently benefit from increased access to victims and reduced punishment (Cote and Cheney 2004). The blennies' activity not only reduces the payoffs for clients, but also cleaners with a mimic neighbor have reduced access to clients (Cote and Cheney 2004). Just as is common for other mutualisms, parasites of the mutualism are a common source of exploitation, but there is relatively little evidence of cheater mutants.

References

Barbu, L., Guinand, C., Alvarez, N., Bergmüller, R., and Bshary, R. (2011). Cleaning wrasse species vary with respect to dependency on the mutualism and behavioural adaptations in interactions. *Animal Behaviour*, **82**, 1067–1074.

Bshary, R., and Grutter, A.S. (2002). Asymmetric cheating opportunities and partner control in a cleaner fish mutualism. *Animal Behaviour*, **63**, 547–555.

Bshary, R., and Schäffer, D. (2002). Choosy reef fish select cleaner fish that provide high service quality. *Animal Behaviour*, **63**, 557–564.

Cheney, K.L., Bshary, R., and Grutter, A.S. (2008). Cleaner fish cause predators to reduce aggression toward bystanders at cleaning stations. *Behavioural Ecology*, **19**, 1063–1067.

Clague, G.E., Cheney, K.L., Goldizen, A.W., McCormick, M.I., Waldie, P.A., and Grutter, A.S. (2011). Long-term cleaner fish presence affects growth of a coral reef fish. *Biological Letters*, **7**, 863–865.

continued

Box 6.2 *Continued*

Côté, I.M. (2000). Evolution and ecology of cleaning symbioses in the sea. *Oceanography and Marine Biology: An Annual Review*, **38**, 311–355.

Côté, I.M., and Cheney, K.L. (2004). Distance-dependent costs and benefits of aggressive mimicry in a cleaning symbiosis. *Proceedings of the Royal Society B*, **271**, 2627–2630.

Grutter, A.S. (2004). Cleaner fish use tactile dancing behavior as a preconflict management strategy. *Current Biology*, **14**, 1080–1083.

Ros, A.F.H., Lusa, J., Meyer, M., et al. (2011). Does access to the bluestreak cleaner wrasse *Labroides dimidiatus* affect indicators of stress and health in resident reef fishes in the Red Sea? *Hormones and Behaviour*, **59**, 151–158.

visit other species to have ectoparasites removed, and where there appears to be both exploitation and means to punish uncooperative individuals.

Well-studied examples of choice and sanction systems also occur in microbial mutualisms in which beneficial bacteria infect an animal or plant host (Sachs et al. 2011b). For example, Sepilioid squids host bioluminescent bacteria in a specialized light organ that the squid employs to evade predators (Ruby 1996; see Box 5.3). Non-luminescent bacteria mutants have been generated (that are otherwise isogenic with cooperative strains), but host squids can discriminate against these mutants. Within the squid light organ, non-luminescent strains of bacteria do not proliferate because these are selectively poisoned (Visick et al. 2000). Parallel experiments have also been performed on legumes and rhizobia, whereby non-fixing rhizobial mutants and isogenic nitrogen-fixing rhizobia were coinfected into host legumes. In these experiments, the non-fixing strain always gets efficiently sanctioned by the host (Singleton and Stockinger 1983).

Partner fidelity feedback and partner-choice sanctions are imperfect, and thus exploiters are almost never eradicated. For instance, many pollinators can discriminate among flowers and select for nectar production (partner choice), but this is apparently not so efficient as to eliminate nectarless flowers. Moreover, even when bacteria are vertically transmitted from host parent to offspring (enforcing partner fidelity), the bacteria can evolve to exploit their hosts, as occurs when mitochondria or endosymbiotic bacteria manipulate their host's sex ratio (Douglas 2008). Nonetheless, some models suggest that mutualisms can persist even in the presence of exploiters. Inter- and intraspecific competition appears to be key features that can stabilize interactions of mutualists and exploiters, allowing mutualist and exploiter lineages to diverge enough to coexist (Ferriere et al. 2002, Jones et al. 2009).

6.4 A new paradox for mutualism?

To summarize, exploitation of mutualism is common, but most cases involve parasites of mutualisms rather than cheater mutants. Given that natural selection can favor the exploitation of mutualisms, why are cheater mutants so rare? The paucity of evidence for cheating can be explained in three main ways, each of which is explored below.

The simplest explanation is that cheater mutants are in fact common, but the types of datasets that would reveal their existence are lacking. Evidence of cheaters requires data on the fitness effects of an interaction on multiple partners, as well as phylogenetic data that resolve the origins of the cheating trait. It could be argued that an insufficient number of mutualisms have been examined in this depth.

A second explanation is that cheating strategies are themselves evolutionarily unstable. Once a cheater begins to spread in a population of mutualists, mutualism can rapidly dissolve (Sachs and Simms 2006). Dissolution of mutualism can occur if cheaters drive local extinction of their interacting species or cheater mutants cause the interacting species to abandon the interaction entirely. However, detecting these short-lived transitions is difficult in population genetic studies because the researcher would need to capture just the right moment in time when cheaters have emerged but have not yet driven a dissolution event. Moreover, ancestral state reconstruction would not uncover transitions that drive extinction, since these would represent missing taxa. But new data from bacterial–eukaryote mutualisms contradict the idea that cheating causes instability. A phylogenetic reconstruction of mutualist traits in proteobacteria suggests that mutualist bacteria experience similar extinction rates to other taxa and rarely transition from mutualism into other types of interactions. Hence there is little evidence that cheating causes the dissolution of mutualisms (Sachs et al. 2014).

A final and compelling idea is that mutualisms originate only under restrictive conditions that disfavor the emergence of cheating. For instance, the classic "mutualistic environment" hypothesis predicts that mutualist partners do not experience conflicts of interest. Under this model, mutualisms are stabilized at their origin and thus the traits that modulate the interaction exhibit evolutionary stasis (Law and Lewis 1983). In contrast to this idea, most frameworks of mutualism described above predict conflicts between partners over control of the dispensation of fitness benefits, and expect mutualists to be selected to constantly maximize the benefits and minimize the costs of the interaction through mechanisms of cheating, defense, and counterdefense (Herre et al. 1999, Egger and Hibbett 2004, Douglas 2008). Thus, a key question is to resolve the degree to which conflicts of interest are inherent to mutualisms.

Evolutionary genetic analyses can examine evidence for conflicts of interests in species interactions by studying the molecular evolution of genes that modulate mutualisms (Sachs et al. 2011a). For instance, evidence of intense arms race evolution that occurs between hosts and microbial parasites has been elegantly revealed through the rapid molecular evolution of pathogenicity loci (in microbes) and immune loci (in hosts). Molecular analysis of "mutualism loci" (i.e., genes that modulate beneficial species interactions) can resolve whether there is evidence of stabilizing selection on these genes (purifying selection—consistent with stable mutualism) or positive selection (selection for novel mutations—consistent with arms race dynamics). Such molecular evolution studies have now been performed in several mutualisms. Much of these data reveal that mutualism loci exhibit evolutionary stasis compared to homologous loci in parasitic species (Kimbrel et al. 2013, Jiggins et al. 2002, Brownlie et al. 2007), though see Sachs and colleagues for counterexamples (Sachs et al. 2011a). These data suggest that conflicts of interest are often minimal or lacking among mutualists.

6.5 Conclusion

Mutualisms are defined as interactions in which members of different species provide each other with net fitness benefits. But theoretical work since the 1970s has modeled mutualisms as a balance of cooperation and conflict between partners. In the dominant theoretical framework for mutualism, conflicts of interest are predicted to arise such that each partner is selected to gain the maximum benefit from the interaction while paying the minimum possible costs (Trivers 1971, Axelrod and Hamilton 1981, Bull and Rice 1991, Foster and Wenseleers 2006). These conflicts can lead to cheating, and to a degradation of the mutualist benefits via reciprocal evolution of defense and counterdefense traits between mutualists (Sachs et al. 2011a). Review of empirical work reveals that mutualisms are often rife with conflict, but that most exploitation of mutualisms comes from parasites of mutualism, i.e., third-party species that capitalize on the fitness benefit that one mutualist gives to another. In contrast, the evolution of exploitation within a mutualism, via the emergence of cheating from a mutualist ancestor, appears to be relatively rare. Examples of "cheater mutants" are only found in obligate mutualisms, wherein an exploited partner might have few options to choose among partners, abandon one partner for another, or abandon the interaction entirely. Data from molecular evolutionary analyses of mutualist traits are inconsistent with arms race dynamics between mutualists, and instead support the hypothesis that mutualisms are characterized by coadaptation and evolutionary stasis. Hence, there may be little or no selective pressure to cheat in most mutualist populations. This could be because one partner has complete control over the other through partner choice and sanctions, or because partner fidelity feedback maintains the mutualisms. Conflicts of interest must arise in mutualisms to some degree, since it is impossible for the fitness interests of two species to completely overlap. But in most mutualisms, under most scenarios it appears that the fitness of interests between partners is aligned. Without conflicts of interests mutualisms are easily maintained over evolutionary time, because cheater mutants exhibit poor fitness compared to mutualists.

References

Als, T.D., Vila, R., Kandul, N.P., Nash, D.R., Yen, S.H., Hsu, Y.F., Mignault, A.A., Boomsma, J.J., and Pierce, N.E. (2004). The evolution of alternative parasitic life histories in large blue butterflies. *Nature*, **432**, 386–390.

Axelrod, R., and Hamilton, W.D. (1981). The evolution of cooperation. *Science*, **211**, 1390–1396.

Balderston, W.L., and Claus, G. (1969). A study of the symbiotic relationships between *Symbiodinium microadriaticum* Freudenthal, a zooxanthella and the upside down jellyfish, *Cassiopea* sp. *Nova Hedwigia*, **17**, 373–382.

Bell, G. (1986). The evolution of empty flowers. *Journal of Theoretical Biology*, **118**, 253–258.

Bidartondo, M.I., and Bruns, T.D. (2001). Extreme specificity in epiparasitic Monotropoideae (Ericaceae): widespread phylogenetic and geographical structure. *Molecular Ecology*, **10**, 2285–2295.

Bronstein, J.L. (1994a). Conditional outcomes in mutualistic interactions. *Trends in Ecology & Evolution*, **9**, 214–217.

Bronstein, J.L. (1994b). Our current understanding of mutualism. *Quarterly Review of Biology*, **69**, 31–51.

Bronstein, J.L. (2001). The exploitation of mutualisms. *Ecology Letters*, **4**, 277–287.

Brownlie, J.C., Adamski, M., Slatko, B., and McGraw, E.A. (2007). Diversifying selection and host adaptation in two endosymbiont genomes. *BMC Evolutionary Biology*, **7**, 68.

Bull, J.J., and Rice, W.R. (1991). Distinguishing mechanisms for the evolution of cooperation. *Journal of Theoretical Biology*, **149**, 63–74.

Burdon, J.J., Gibson, A.H., Searle, S.D., Woods, M.J., and Brockwell, J. (1999). Variation in the effectiveness of symbiotic associations between native rhizobia and temperate Australian *Acacia*: within species interactions. *Journal of Applied Ecology*, **36**, 398–408.

Chamberlain, S.A., and Holland, J.N. 2009. Quantitative synthesis of context dependency in ant-plant protection mutualisms. *Ecology*, **90**, 2384–2392.

Dafni, A. (1984). Mimicry and deception in pollination. *Annual Review of Ecology and Systematics*, **15**, 259–278.

Darwin, C. (1859). *The Origin of Species*. New York: Gramercy Books.

Denison, R.F. (2000). Legume sanctions and the evolution of symbiotic cooperation by rhizobia. *American Naturalist*, **6**, 567–576.

Despres, L., and Jaeger, N. (1999). Evolution of oviposition strategies and speciation in the globeflower flies *Chiastocheta* spp. (Anthomyiidae). *Journal of Evolutionary Biology*, **12**, 822–831.

Douglas, A.E. (2008). Conflict, cheats and the persistence of symbioses. *New Phytologist*, **177**, 849–858.

Egger, K.N., and Hibbett, D.S. (2004). The evolutionary implications of exploitation in mycorrhizas. *Canadian Journal of Botany-Revue Canadienne De Botanique*, **82**, 1110–1121.

Ferrière, R., Bronstein, J.L., Rinaldi, S., Law, R., and Gauduchon, M. (2002). Cheating and the evolutionary stability of mutualisms. *Proceedings of the Royal Society B-Biological Sciences*, **269**, 773–780.

Foster, K.R., and Wenseleers, T. (2006). A general model for the evolution of mutualisms. *Journal of Evolutionary Biology*, **19**, 1283–1293.

Frederickson, M.E., and Gordon, D.M. (2009). The intertwined population biology of two Amazonian myrmecophytes and their symbiotic ants. *Ecology*, **90**, 1595–1607.

Friesen, M.L. (2012). Widespread fitness alignment in the legume-rhizobium symbiosis. *New Phytologist*, **194**, 1096–1111.

Gilbert, F.S., Haines, N., and Dickson, K. (1991). Empty flowers. *Functional Ecology*, **5**, 29–39.

Golubov, J., Eguiarte, L.E., Mandujano, M.C., Lopez-Portillo, J., and Montana, C. (1999). Why be a honeyless honey mesquite? Reproduction and mating system of nectarful and nectarless individuals. *American Journal of Botany*, **86**, 955–963.

Hamilton, W.D. (1964). Genetical evolution of social behaviour I. *Journal of Theoretical Biology*, **7**, 1–16.

Herre, E.A., Knowlton, N., Mueller, U.G., and Rehner, S.A. (1999). The evolution of mutualisms: exploring the paths between conflict and cooperation. *Trends in Ecology and Evolution*, **14**, 49–53.

Irwin, R.E., Bronstein, J.L., Manson, J.S., and Richardson, L. (2010). Nectar robbing: ecological and evolutionary perspectives. *Annual Review of Ecology, Evolution, and Systematics*, **41**, 271–292.

Irwin, R.E., and Maloof, J.E. (2002). Variation in nectar robbing over time, space, and species. *Oecologia*, **133**, 525–533.

Jánder, K.C., and Herre, E.A. (2010). Host sanctions and pollinator cheating in the fig tree-fig wasp mutualism. *Proceedings of the Royal Society B-Biological Sciences*, **277**, 1481–1488.

Janzen, D.H. (1966). Coevolution of mutualism between ants and acacias in Central America. *Evolution*, **20**, 249–275.

Jiggins, F.M., Hurst, G.D.D., and Yang, Z.H. (2002). Host-symbiont conflicts: positive selection on an outer membrane protein of parasitic but not mutualistic Rickettsiaceae. *Molecular Biology and Evolution*, **19**, 1341–1349.

Jones, E.I., Ferrière, R., and Bronstein, J.L. (2009). Eco-evolutionary dynamics of mutualists and exploiters. *American Naturalist*, **174**, 780–794.

Kimbrel, J.A., Thomas, W.J., Jiang, Y., Creason, A.L., Thireault, C.A., Sachs, J.L., and Chang, J.H. (2013). Mutualistic co-evolution of Type III effector genes in *Sinorhizobium fredii* and *Bradyrhizobium japonicum*. *PLoS Pathogens*, **9**, e1003204.

Law, R., and Lewis, D.H. (1983). Biotic environments and the maintenance of sex—some evidence from mutualistic symbioses. *Biological Journal of the Linnean Society*, **20**, 249–276.

Lesser, M.P., Stat, M., and Gates, R.D. (2013). The endosymbiotic dinoflagellates (*Symbiodinium* sp.) of corals are parasites and mutualists. *Coral Reefs*, **32**, 603–611.

Machado, C.A., Jousselin, E., Kjellberg, F., Compton, S.G., and Herre, E.A. (2000). Phylogenetic relationships, historical biogeography and character evolution of fig pollinating wasps. *Proceedings of the Royal Society of London, B.*, **268**, 685–694.

May, R.M. (1976). Models for two interacting populations. In May, R.M. (ed.) *Theoretical Ecology, Principles and Applications*. Philadelphia: W.B. Saunders.

Morris, J.J., Lenski, R.E., and Zinser, E.R. (2012). The black queen hypothesis: evolution of dependencies through adaptive gene loss. *Mbio*, **3**, e00036–12.

Muscatine, L. (1990). The role of symbiotic algae in carbon and energy flux in coral reefs. In Z. Dubinsky (ed.) *Ecosystems of the World: Coral Reefs*. Amsterdam: Elsevier.

Pellmyr, O., LeebensMack, J., and Huth, C.J. (1996). Non-mutualistic yucca moths and their evolutionary consequences. *Nature*, **380**, 155–156.

Pierce, N.E., Braby, M.F., Heath, A., Lohman, D.J., Mathew, J., Rand, D.B., and Travassos, M.A. (2002). The ecology and evolution of ant association in the Lycaenidae (Lepidoptera). *Annual Review of Entomology*, **47**, 733–771.

Rathcke, B.J. (1992). Nectar distributions, pollinator behavior, and reproductive success. In M.D. Hunter, T. Ohgashi, and P.W. Price (ed.) *Effects of Resource Distribution on Animal–Plant Interactions*. New York: Academic Press.

Ruby, E.G. (1996). Lessons from a cooperative, bacterial-animal association: the *Vibrio fischeri Euprymna scolopes* light organ symbiosis. *Annual Review of Microbiology*, **50**, 591–624.

Sachs, J.L., Ehinger, M.O., and Simms, E.L. (2010a). Origins of cheating and loss of symbiosis in wild *Bradyrhizobium*. *Journal of Evolutionary Biology*, **23**, 1075–1089.

Sachs, J.L., Essenberg, C.E., and Turcotte, M.M. (2011a). New paradigms for the evolution of beneficial infections. *Trends in Ecology & Evolution*, **26**: 202–209.

Sachs, J.L., Kembel, S.W., Lau, A.H., and Simms, E.L. 2009. In situ phylogenetic structure and diversity of wild *Bradyrhizobium* communities. *Applied and Environmental Microbiology*, **75**, 4727–4735.

Sachs, J.L., Mueller, U.G., Wilcox, T.P., and Bull, J.J. 2004. The evolution of cooperation. *Quarterly Review of Biology*, **79**, 135–160.

Sachs, J.L., Russell, J.E., Lii, Y.E., Black, K.C., Lopez, G., and Patil, A.S. (2010b). Host control over infection and proliferation of a cheater symbiont. *Journal of Evolutionary Biology*, **23**, 1919–1927.

Sachs, J.L., and Simms, E.L. (2006). Pathways to mutualism breakdown. *Trends in Ecology & Evolution*, **21**, 585–592.

Sachs, J.L., and Simms, E.L. (2008). The origins of uncooperative rhizobia. *Oikos*, **117**, 961–966.

Sachs, J.L., Skophammer, R.G., Bansal, N., and Stajich, J.E. (2014). Evolutionary origins and diversification of proteobacterial mutualists. *Proceedings of the Royal Society B: Biological Sciences*, **281**, 20132146.

Sachs, J.L., Skophammer, R.G., and Regus, J.U. (2011b). Evolutionary transitions in bacterial symbiosis. *Proceedings of the National Academy of Sciences*, **108**, 10800–10807.

Sachs, J.L., and Wilcox, T.P. (2006). A shift to parasitism in the jellyfish symbiont *Symbiodinium microadriaticum*. *Proceedings of the Royal Society B-Biological Sciences*, **273**, 425–429.

Singleton, P.W., and Stockinger, K.R. (1983). Compensation against ineffective nodulation in soybean. *Crop Science*, **23**, 69–72.

Smithson, A., and Gigord, L.D.B. (2003). The evolution of empty flowers revisited. *American Naturalist*, **161**, 537–552.

Soberon, J.M., and Martinez del Rio, C. (1985). Cheating and taking advantage in mutualistic interactions. In D.H. Boucher (ed.) *The Biology of Mutualisms*. London: Croom Helm.

Stat, M., and Gates, R.D. (2011). Clade D Symbiodinium in Scleractinian corals: a "nugget" of hope, a selfish opportunist, an ominous sign, or all of the above? *Journal of Marine Biology*, **2011**, 1–9.

Stat, M., Morris, E. and Gates, R.D. (2008). Functional diversity in coral-dinoflagellate symbiosis. *Proceedings of the National Academy of Sciences of the United States of America*, **105**, 9256–9261.

Taylor, D.L., and Bruns, T.D. (1997). Independent, specialized invasions of ectomycorrhizal mutualism by two nonphotosynthetic orchids. *Proceedings of the National Academy of Sciences of the United States of America*, **94**, 4510–4515.

Trainer, M.A., and Charles, T.C. (2006). The role of PHB metabolism in the symbiosis of rhizobia with legumes. *Applied Microbiology and Biotechnology*, **71**, 377–386.

Trench, R.K. (1993). Microalgal-invertebrate symbioses—a review. *Endocytobiosis and Cell Research*, **9**, 135–175.

Trivers, R.L. (1971). Evolution of reciprocal altruism. *Quarterly Review of Biology*, **46**, 35–57.

Visick, K.G., Foster, J., Doino, J., McFall-Ngai, M., Ruby, E.G. (2000). *Vibrio fischeri* lux genes play an important role in colonization and development of the host light organ. *Journal of Bacteriology*, **182**, 4578–4586.

West, S.A., Kiers, E.T., Simms, E.L., and Denison, R.F. (2002). Sanctions and mutualism stability: why do rhizobia fix nitrogen? *Proceedings of the Royal Society of London, B*, **269**, 685–694.

Williams, G.C. (1966). *Adaptation and Natural Selection*. Princeton, NJ: Princeton University Press.

Yu, D.W. (2001). Parasites of mutualisms. *Biological Journal of the Linnean Society*, **72**, 529–546.

CHAPTER 7

Coevolution in mutualisms

Bruce Anderson

7.1 Introduction

Coevolution can be defined as the reciprocal evolutionary change of interacting species, driven by natural selection. A frequently cited example of coevolution stems from the early writings of Darwin (1862): Having examined the extraordinarily long nectar spurs of the Madagascan star orchid *Angraecum sesquipedale*, Darwin successfully predicted the existence of a seemingly implausible hawkmoth pollinator with a tongue measuring close to 30 cm in length. In explaining his prediction, Darwin described the first mechanistic model of the coevolutionary process, hypothesizing that selection induced by the orchid's long nectar spurs drove the evolution of even longer moth tongues, enabling the moths to access nectar deeper within the flower. Simultaneously, he posited that the evolution of longer moth tongues exerted selection for longer nectar spurs so that pollen transfer would be more efficient, resulting in "a race in gaining length" between moth tongues and floral nectaries (Darwin 1862). Darwin's prediction was confirmed some forty years later when a hawkmoth with an extremely long proboscis was found in Madagascar.

However, as Arditti et al. (2012) point out, coevolution is not confirmed by the mere interaction between a moth with a long proboscis and a flower with a long spur. Long spurs and proboscides could potentially be adaptations to other unrelated selective pressures. For example, it has been proposed that the long moth proboscides did not evolve to sip nectar from long-tubed flowers, but rather to distance the moths from the attractive parts of short-tubed flowers where predatory spiders lie in wait (Wasserthall 1997). Such debates speak to the historical controversy surrounding the coevolutionary concept, and the consequent oscillations between general acceptance of coevolution as a powerful selective force, and general scepticism of its importance. Perhaps one problem with the hawkmoth-orchid exemplar is that it gives the impression that reciprocal selection can only be very narrowly applied to two reciprocally interacting traits such as proboscides and spurs, involved in an apparently positive feedback cycle. However, coevolution can also be used more broadly to encompass situations where two organisms (rather than just two interacting traits) undergo some form of evolutionary change or adaptive response to one another (as used by Thompson 1994, 2005). Importantly, adaptations should only be defined as coevolutionary when they are a result of a shared evolutionary history, not simply a chance conjoining of two species with preadapted or preexisting complimentary traits (also see Section 3.3.1 for what constitutes adaptation).

In this chapter, I will provide a brief history of coevolutionary studies, demonstrating that the concept can be equally applied to mutualisms and antagonistic relationships. Next I shall discuss some of the challenges in demonstrating coevolution, highlighting studies that provide powerful evidence for reciprocal selection and evolutionary change. Following this, I review some of the patterns associated with phenotypic traits and the phylogenies of interacting species that are often taken as partial evidence for coevolution. Here I show that the usefulness of pattern-based studies is not in demonstrating coevolution, but in testing predictions

Mutualism. Edited by Judith L. Bronstein.

about how coevolution is expected to influence the evolution of adaptive traits, specialization, or speciation. While the main focus of this chapter is on coevolution in mutualisms, I have attempted to highlight differences and similarities in which mutualisms and antagonisms are expected to evolve, potentially giving rise to differences in the patterns that they produce. In some cases I turn to antagonisms as exemplars because empirically based exemplars of mutualism frequently do not exist for some coevolutionary concepts.

7.2 A short history of coevolution

Although Darwin frequently used terms such as "mutually adapted" and "coadaptations" to describe coevolved interactions, the father of evolution never actually used the term "coevolution." The inception of the term can be traced back to Charles Mode (1958), who developed the first mathematical model of coevolution. This model described gene-for-gene interactions in parasites and their hosts, in which each gene for host resistance is matched by a gene for parasite virulence. The variable outcomes of interactions between flax plants and flax rust fungus were first attributed to this gene-for-gene matching or mismatching of parasite virulence and host resistance by Flor (1942).

Antagonistic interactions were also the focus of a landmark study by Ehrlich and Raven (1964). In a study that popularized the term "coevolution," they examined the dietary patterns of butterfly larvae and found that they were often specialized on unrelated plants with similar chemical defenses. They envisaged that the evolution of a novel insect counterdefense would allow that insect to gain access to a new suite of potential plant hosts with similar defenses, facilitating diversification and speciation. In response, plants would continually evolve new chemicals to counter the evolution of insect resistance, which may also result in further plant radiation, a process called "escape-and-radiate coevolution" (Thompson 1989). From the pollination of the Madagascan star orchid to the two antagonisms described thus far, these contrasting examples demonstrate the variability in the kinds of interactions and mechanisms which can give rise to coevolution.

The 1970s saw a proliferation of mutualism studies, many of which assumed that apparent matches between interacting partners were based on coevolutionary processes. This prompted Janzen (1980) to appeal for more rigorous usage of the term based upon a clear definition, stressing the point that coevolution requires reciprocal evolutionary change. Perhaps because of the difficulty in demonstrating coevolution in accordance with Janzen's strict criteria, many researchers only appeared to consider it as a major evolutionary driving force in intimate and highly specialized relationships, where complete codependence makes it difficult to ignore its probable importance. Central among these relationships were the classic examples of coevolved mutualisms such as the highly specialized brood site mutualisms of yuccas and figs. Because of their simplicity and pairwise nature, obligately specialized mutualisms and highly specialized parasitisms often served as the primary models for studying the coevolutionary process. Many researchers (e.g. Janzen 1981, Howe 1984) subsequently began to perceive coevolution as unlikely or unimportant if interactions were not highly specialized, obligate, and symmetrical.

The critical thinking of the 1980s led to the recognition that most mutualisms (other than some brood site mutualisms, such as those between figs and fig wasps) involved several interacting species, rather than just a single interacting pair. Recent analytical and visualization techniques also confirmed the early observations that interaction webs tend to be highly asymmetrical (see Chapter 11). For example, plants may form guilds in which several species are pollinated or dispersed by a small number of vectors (sometimes only one). However, each vector itself may visit a wide range of plant species (Faegri and van der Pijl 1979). Though this relative complexity complicates our ability to quantify, model, or visualize the reciprocal effects of coevolution on individual populations that interact within multispecific or asymmetric networks, it does not negate the possibility that coevolution is taking place within complex or asymmetrical interactions.

7.2.1 Geographic mosaic of coevolution

After the 1980s, the multispecific and geographically variable nature of many interactions was cleverly taken advantage of to study coevolution in a geographic context. This idea, developed by John Thompson and his colleagues, is usually referred to as the geographic mosaic of coevolution. Three of Thompson's books document a plethora of studies on the many ways in which geography influences coevolution (Thompson 1994, 2005, 2013). Because of the inherently heterogeneous nature of the earth's biotic and abiotic surface, all species are faced with a complex mosaic of geographically and temporally divergent selective pressures (Thompson 1997). This geographic context provides us with a beautifully replicated framework with which biologists have been able to study coevolution (see Thompson

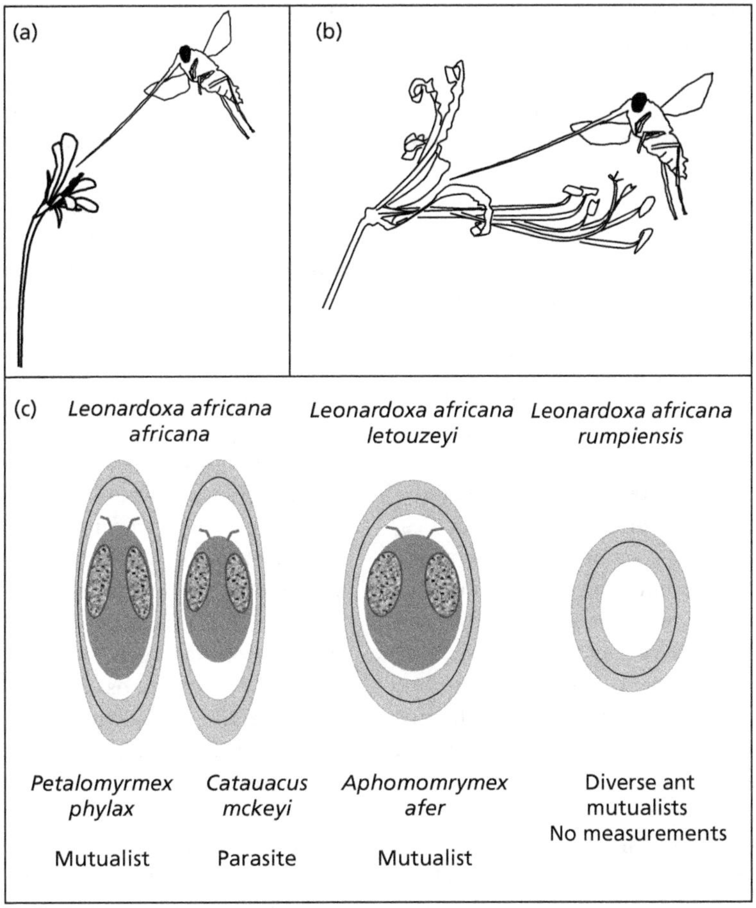

Figure 7.1 Few studies have attempted to demonstrate whether traits facilitating specialization have evolved through processes of filtration versus exploitation. a) In order to access the nectar at the bottom of the floral tube, the long proboscid fly will be required to insert its proboscis deep into the tube. A tube such as this could have evolved to filter out less effective pollinators with shorter proboscides or alternatively it could have evolved to better exploit the morphological characteristics of certain pollinators. The long mouthparts of the pollinators are most likely to have evolved to exploit certain long-tubed flowers more efficiently, suggesting that the matching of traits could have been the result of coevolution. b) The elongated floral parts of *Nerine humilis* could only have evolved to exploit the unusual body parts of its primary pollinator (*Proseoca ganglbaueri*) more efficiently by placing pollen on the fly's abdomen. The filtering out of less efficient pollinators can be ruled out because the nectaries of the flower can be reached by long- and short-tongued pollinators (there is no corolla tube). The flies show no specific adaptations to forage from these flowers, suggesting a unilateral rather than a coevolutionary relationship. c) The differing prostoma shapes (means in black and standard errors in grey) of three subspecies of *Leonardoxa africana* plants effectively filter out ants with larger head sizes, allowing only a certain subset to colonize the plants. One subspecies (*Leonardoxa africana rumpiensis*) is colonized by a diversity of ant species and consequently exhibits a more variable size of entrance hole. In contrast, if the head shapes of the ants have evolved in response to the plants, they could only have evolved to exploit certain species better. Adapted and printed with permission from Brouat et al. (2001).

1999). The geographic mosaic of coevolution recognizes that organisms are unlikely to interact with exactly the same communities throughout their range. Thus, in one population reciprocal selection between two species may be strong, while in another, it may be weaker because it occurs in a different community context. In other populations the same two species may not even co-occur, and organisms may interact with entirely different suites of partners. Under these biotically variable conditions we expect that the magnitude and symmetry of reciprocal selection should be geographically and temporally variable, resulting in a broad mosaic of outcomes that include coevolution in some populations but one-sided evolution and the lack of evolution in others, or mutualism in some populations but antagonism in others (Thompson 1997).

An example of a relationship with geographically variable outcomes due to changes of interacting community members are woodland stars (*Lithophragma parviflorum*) and their pollinators. Woodland stars are very effectively pollinated by the moth *Greya pollitella* (Thompson and Pellmyr 1992). However, this is a brood site mutualism: the moths also lay eggs in some of the plant's ovules, and their larvae consume a small proportion of the developing seeds upon hatching (Thompson and Pellmyr 1992, Pellmyr and Thompson 1996). In some populations, effective copollinators such as bombyliid flies are abundant. They pollinate but do not lay eggs in the flowers (Pellmyr and Thompson 1996). Although both bombyliid flies and *Greya* moths provide benefits to the plants, a plant that is well visited by bombyliids alone should produce a greater seed yield than one equally pollinated by *Greya* moths alone, because the bombyliids do not parasitize any of the seeds. Thompson and Cunningham (2002) showed that the relationship between *Greya* and *Lithophragma* is geographically and temporally variable: when bombyliids were abundant, the relationship between *Lithofragma* and *Greya* is antagonistic, but when bombyliids were scarce, *Greya* and *Lithofragma* are mutualistic.

These data (and others, e.g. Schemske and Horwitz 1984) suggest that when a single species interacts with several others, the participants within those multispecific assemblages are seldom equal in their contributions. In particular, certain species are likely to offer better or more consistent rewards than others. While inefficient partners may still provide a service when considered in isolation, they may have a net antagonistic effect if they inefficiently use resources which would otherwise have attracted more efficient partners. As a result, specialization may evolve in mutualisms, either to better exploit a particular partner (e.g. Stebbins 1970) or to prevent interactions with less efficient partners (e.g. Johnson et al. 2006, Brouat et al. 2001). Figure 7.1 illustrates the subtle differences between these two concepts using an ant–plant mutualism and a pollination mutualism. The ant–plant relationship (Figure 7.1), and the pollination of *Greya* both demonstrate that the geographic mosaic of coevolution can potentially generate a wide array of interaction outcomes. This makes it easy to envisage how coevolution may translate into one of the most important diversifying forces on the planet.

7.3 Demonstrating coevolution

Modeling approaches have demonstrated that coevolution can, at least in theory, affect the traits of organisms interacting in both reciprocally specialized contexts (e.g. Gandon et al. 1996) and highly asymmetrically specialized contexts (e.g. Nuismer at al 2013, Guimarães et al. 2011). Furthermore, coevolution has been invoked to explain the outcomes of interactions that include both mutualisms and antagonistic relationships (Nuismer et al. 1999, Yoder and Nuismer 2010). However, because the kinds of patterns associated with coevolutionary processes can often also be produced by non-coevolutionary processes (Nuismer et al. 2010; see also Section 6.4), demonstrating patterns consistent with the coevolutionary process typically only constitutes partial evidence for its existence. Incontrovertible empirical evidence for the importance or existence of coevolution has been difficult to obtain. To demonstrate coevolution between two organisms, a good starting point would be to identify putatively coevolved traits which are variable, and then show that there is a predictable adaptive fit between the traits of the interacting mutualists. While this sounds like a simple step, it may frequently be the most difficult hurdle. For example, it is hard to pinpoint and quantify coevolving traits in some of the most tightly associated mutualisms for which many would not hesitate to implicate a history of coevolution (e.g. lichens, coral/algae mutualisms). The next step would be showing that there is selection operating on the identified traits of each partner, and that in each case, the selective agent acting on the trait is the putatively coevolving partner. Finally, one needs to demonstrate that traits in each species have changed as a result of the interaction. While coevolutionary studies typically concentrate on phenotypic change, it is also possible to study the reciprocal change in the genotypes of interacting organisms. Similar to phenotypic studies, most studies linking adaptation and genetics in mutualisms have typically concentrated on a single mutualist partner (see Chapter 5). However, Noah Whiteman (Box 7.1) describes an exciting future for coevolutionary genomics, in which the changes in the genotypes of both interacting organisms can be studied.

Box 7.1 Coevolutionary genomics

Noah Whiteman

In the same way that one can quantify reciprocal phenotypic change to demonstrate coevolution, it may also be possible to demonstrate coevolution through the reciprocal changes in allele frequencies driven by natural selection. Advances in the field of genomics, which encompasses classical genetics, quantitative genetics, and population and comparative genomics, are allowing for the discovery of interacting genes in pairs or communities of interacting species that mediate coevolutionary outcomes. Furthermore, the evolution of traits can be facilitated or constrained by their underlying genetic architecture, which includes how many loci and alleles code for a trait, which other traits those loci code for, and how those loci interact with other loci. It is clear that the outcomes of coevolutionary interactions (e.g. winners and losers of coevolutionary races) could depend on the underlying genetic architecture of those traits. Genomics approaches are likely to bear the most fruit when considering relatively specific interactions between hosts and symbionts that infect their hosts, whether the outcome is mutualistic or parasitic. This is principally because of the difficulties associated with identifying clear agents of selection when coevolution is diffuse. The field of coevolutionary genomics is most likely to be advanced by focusing on experimentally tractable, genome-enabled species with well-matched generation times.

Potential signatures of coevolutionary interactions have frequently been studied by observing change or patterns in the phenotypic traits of interacting organisms, but the complementary approach of studying coevolving genes is still in its infancy. However, most genomic studies have been on antagonistic relationships and have only identified the putatively coevolving genes in one of the interacting species. For example, Dobler et al. (2012) demonstrated that the same agent of natural selection (plant toxins that block the sodium pump, an essential transmembrane carrier) appears to have driven the fixation of identical amino acid substitutions in the sodium pump enzyme (Na,K-ATPase) of distantly related but specialized insects that attack milkweeds. It is less clear whether the insects have in turn caused the host plants to evolve new resistance strategies in response to the evolution of these traits, making coevolution a possible but not definite driver of the genetic changes in the insects. This pattern of convergence at the molecular level is very similar to that observed in sodium channels of snakes that have evolved resistance to tetrodotoxins produced by prey species (Feldman et al. 2012). Although the evidence for coevolution is, perhaps, stronger in the latter case than in the milkweed–insect example, the genomic targets of snake-driven selection have not been characterized in newts that serve as prey for snakes (Brodie et al. 2002, 2005). Nonetheless, progress in identifying the actual genes underlying variation in traits that underlie species interactions has been slow. Why is this the case? Interactions between host and symbiont genotypes give rise to new phenotypic outcomes that are emergent properties of interacting alleles (Wade 2007). Thus, the genes that underlie coevolving traits require the identification of at least two loci (one to several loci per species). This is a difficult task even in one species. However, a good example of how whole-genome sequences of two interacting species can reveal potential patterns of coevolution (that can then be tested by further study) is the metabolic complementarity between aphids and their *Buchnera* spp. endosymbionts (International Aphid Genomics 2010). Furthermore, analogies with epistatic interactions (when expression of one gene depends on expression of another) between genes within the genome of one species can be

continued

Box 7.1 *Continued*

drawn between genes that interact in two different organisms. Wade (2007) has proposed that cotransmission of host and symbiont alleles across generations can eventually lead to mutualistic interactions between species because the fate of one allele in the host depends on the fate of the other in the symbiont, and potentially vice versa. A similar argument has been proposed for the attenuation of virulence in parasites with vertical transmission (Clayton and Tompkins 1994). The potential for such interactions can be measured in many systems in which genetic markers are available by estimating a cotransmission parameter theta (Wade 2007).

A different conceptual approach comes from the field of quantitative genetics. Interactions between host and symbiont genomes are often ignored in typical quantitative genetic models. Consequently, important genetic variation underlying infection outcomes is often overlooked. A different gap exists in coevolutionary biology. Studies aiming to identify the genetic bases of host–symbiont interactions have focused on long-standing gene-for-gene models in which a few loci of major effect interact (such as the examples involving predator–prey and plant–herbivore interactions) in two interacting species to mediate infection or associational outcome (Flor 1971). However, the heritability of most organismal traits is due to quantitative genetic variation, in which many genes of small effect explain phenotypic variation (Yang et al. 2010; Chapter 5).

Although quantitative genetics has been applied to understand antagonistic coevolved systems, the actual genetic polymorphisms underlying infection or associational outcomes have seldom been identified (Lambrechts et al. 2005, 2006, 2009). Progress in this field requires application of new quantitative models to host–symbiont systems in which the genomic architecture of the association can be dissected in each species. An elegant quantitative genetic model has been proposed by Lambrechts (2010) that could facilitate the identification of interacting genetic polymorphisms in the genomes of coevolving host and symbiont species. In this framework (Box 7.1, Figure 1), infection outcomes are determined by three main terms: (1) additive genetic variation in the host (G_H), (2) additive genetic variation in the parasite (G_P), and (3) variation from interactions between host and parasite genotypes (G_{HP}). However, partitioning of variance in this manner still does not yield the expected genetic polymorphisms. This approach still requires more in-depth

Lambrechts et al. (2006) extends the classic framework in which phenotypes are the summed effects of genes (G), the environment (E), and the interaction between genes and environment (G × E). For simplicity, the G × E and G × G × E interactions are not considered. This is given by :

$$y = \mu + g + \varepsilon$$

where y is the phenotype of an individual, μ is the average population value, g is the genotypic value and ε is the environmental value. If an interaction occurs between symbiont and host, the model is extended to:

$$y = \mu + g_H + g_P + g_{HP} + \varepsilon$$

where g_H is the host genotypic value, g_P is the symbiont's genotypic value and g_{HP} is the genotypic value of the G x G interaction between host and symbiont. Additive and epistatic interactions allow simultaneous consideration of additive and epistatic effects of both host and symbiont loci:

$$y = \mu + \sum \alpha_{HP} + \sum \beta_{HP} + \varepsilon$$

where $\sum \alpha_{HP} = \sum \alpha_H + \sum \alpha_P$ is the sum of the additive effects of loci in host and symbiont, and $\sum \beta_{HP} = \sum \beta_H + \sum \beta_P + g_{HP}$ is the sum of the interaction effects between loci within host and symbiont genomes and interaction effects between host and symbiont genomes.

Figure 1 A quantitative, coevolutionary genetics framework for host–parasite interactions.

Box 7.1 *Continued*

studies through genome-wide association (GWA) studies, classic mapping strategies, and functional genetic studies.

Pea aphids and their endosymbiotic *Buchnera* bacteria (Vogel and Moran 2011), as well as *Medicago* spp. plants and their rhizosphere bacteria (Gorton et al. 2012), are both excellent genome-enabled systems in which to explore the genetic bases of mutualisms. However, coevolutionary studies in these systems are challenging to realistically carry out, principally because of the mismatch in generation time between host and symbiont. Conversely, from a practical perspective, studies in which host and symbiont are more evenly matched in terms of life cycle are likely to hold the most promise in the context of identifying gene × gene interactions that mediate coevolutionary outcomes. The study of interactions between microbial species is potentially one way to cope with this issue. For example, communities of microbes living in extreme environments, in which mutualistic interactions may be rampant, are already the focus of intensive study (Stams and Plugge 2009).

In conclusion, the future is bright for those interested in closing the phenotype-genotype-fitness loop for coevolved species and subtending traits that are the targets and agents of reciprocal natural selection. This enthusiasm is due to the rapid progress in the development of new genomics tools, new theory, particularly in quantitative genetics, and the availability of new experimental model systems for the study of coevolution that involve complex eukaryotic species rather than microbes, which have received most attention by researchers (Gloss et al. 2013). However, because the field relies on historical associations between organisms, illuminating the processes that led to current patterns is likely to remain extremely challenging. Clever approaches, such as leveraging of quiescent life stages that can be resurrected (e.g., *Daphnia* and its microparasites), could lead to breakthroughs because the genomes of previous generations can be directly interrogated and could help give insight into current genomic patterns (Decaestecker et al. 2007). Similarly evolve and resequence approaches (Turner et al. 2011) that involve host–symbiont or predator–prey interactions are likely to be extremely illuminating because loci detected as being subject to selection can be screened in natural populations. Thus, the pairing of experimental genomics and landscape genomics could be a powerful approach to validate results generated in the laboratory. The challenge, however, is in determining the nature of criteria required to prove that coevolution has occurred, and the availability of genomic data have not made this task any easier.

References

Brodie, E.D. 3rd., Feldman, C.R., Hanifin, C.T., Motychak, J.E., Mulcahy, D.G, Williams, B.L., and Brodie, E.D. Jr. (2005). Parallel arms races between garter snakes and newts involving tetrodotoxin as the phenotypic interface of coevolution. *Journal of Chemical Ecology*, **31**, 343–356.

Brodie, E.D. Jr., Ridenhour, B.J., and Brodie, E.D. 3rd. (2002). The evolutionary response of predators to dangerous prey: hotspots and coldspots in the geographic mosaic of coevolution between garter snakes and newts. *Evolution*, **56**, 2067–2082.

Clayton, D.H., and Tompkins, D.M. (1994). Ectoparasite virulence is linked to mode of transmission. *Proceedings of the Royal Society of London B*, **256**, 211–217.

Decaestecker, E., Gaba, S., Raeymaekers, J.A., Stoks, R., Van Kerckhoven, L., Ebert, D., and De Meester, L. (2007). Host-parasite "Red Queen" dynamics archived in pond sediment. *Nature*, **450**, 870–873.

Dobler, S., Dalla, S., Wagschal, V., and Agrawal, A.A. (2012). Community-wide convergent evolution in insect adaptation to toxic cardenolides by substitutions in the Na,K-ATPase. *Proceedings of the National Academy of Sciences USA*, **109**, 13040–13045.

Feldman, C.R., Brodie, E.D. Jr., Brodie, E.D. 3rd., and Pfrender, M.E. (2012). Constraint shapes convergence in tetrodotoxin-resistant sodium channels of snakes. *Proceedings of the National Academy of Sciences USA*, **109**, 4556–4561.

Flor, H.H. (1971). Current status of the gene-for-gene concept. *Annual Review of Phytopathology*, **9**, 275–296.

Gloss, A.D., Nelson Dittrich, A.C., Goldman-Huertas, B., and Whiteman, N.K. (2013). Maintenance of genetic diversity through plant–herbivore interactions. *Current Opinion in Plant Biology*, **16**, 443–450.

Gorton, A.J., Heath, K.D., Pilet-Nayel, M.L., Baranger, A., and Stinchcombe, J.R. (2012). Mapping the genetic basis of symbiotic variation in legume–rhizobium interactions in *Medicago truncatula*. *G3: Genes, Genomes, Genetics*, **2**, 1291–1303.

International Aphid Genomics, C. (2010). Genome sequence of the pea aphid *Acyrthosiphon pisum*. *PLoS Biology*, **8**, e1000313.

Lambrechts, L. 2010. Dissecting the genetic architecture of host-pathogen specificity. *PLoS Pathogens*, **6**, e1001019.

Lambrechts, L., Chevillon, C., Albright, R.G., Thaisomboonsuk, B., Richardson, J.H., Jarman, R.G., and Scott, T.W. (2009). Genetic specificity and potential for local adaptation between dengue viruses and mosquito vectors. *BMC Evolutionary Biology*, **9**, 160.

Lambrechts, L., Fellous, S., and Koella, J.C. (2006). Coevolutionary interactions between host and parasite genotypes. *Trends in Parasitology*, **22**, 12–16.

Lambrechts, L., Halbert, J., Durand, P., Gouagna, L.C., and Koella, J.C. (2005). Host genotype by parasite genotype interactions

continued

Box 7.1 *Continued*

underlying the resistance of anopheline mosquitoes to *Plasmodium falciparum*. *Malaria Journal*, **4**, 3.
Stams, A.J.M., and Plugge, C.M. (2009). Electron transfer in syntrophic communities of anaerobic bacteria and archaea. *Nature Reviews Microbiology*, **7**, 568–577.
Turner, T.L., Stewart, A.D., Fields, A.T., Rice, W.R., and Tarone, A.M. (2011). Population-based resequencing of experimentally evolved populations reveals the genetic basis of body size variation in *Drosophila melanogaster*. *PLoS Genetics*, **7**, e1001336.
Vogel, K.J., and Moran, N.A. (2011). Sources of variation in dietary requirements in an obligate nutritional symbiosis. *Proceedings of the Royal Society of London B*, **278**, 115–121.
Wade, M.J. (2007). The coevolutionary genetics of ecological communities. *Nature Reviews Genetics*, **8**, 185–195.
Yang, J., Benyamin, B., McEvoy, B.P., Gordon, S., Henders, A.K., Nyholt, D.R., Madden, P.A., Heath, A.C., Martin, N.C., Montgomery, G.W., Goddard, M.E., and Visscher P.M. (2010). Common SNPs explain a large proportion of the heritability for human height. *Nature Genetics*, **42**, 565–569.

Some phenotypic studies of coevolution in mutualisms satisfy a few of these criteria, but none of them do a good job of satisfying all of them. Based on Darwin's ideas on coevolving plant and pollinator traits, the quantification of pollinator foraging apparatus (e.g. pollinator proboscis length) and the tubular structures of flowers (e.g. corolla tubes) have been well studied in many systems and several studies suggest instances where the traits of pollinators and plants appear to be predictably matched (e.g. Steiner and Whitehead 1991, Anderson and Johnson 2008, Figures 7.2 and 7.3). Similarly, in some African fig pollination mutualisms, large figs tend to be associated with female fig wasp pollinators with elongated heads, while smaller figs are pollinated by wasps with shorter, more square-shaped heads. The elongated heads are thought to be an adaptation for pushing through the longer ostioles of the large figs. Because male wasps do not enter fig ostioles, the head size–fig size relationship is not strong for the male wasps. However, it is strong for unrelated, non-pollinating wasp taxa which enter the figs as parasites (van Noort and Compton 1996). Ovipositor lengths of fig wasps are also frequently closely matched to the style lengths of the figs (Weiblen 2004).

Several studies have also demonstrated selection on traits that are putatively coevolving; however, few studies have quantified selection acting on the traits of both organisms. For example, many studies have documented positive directional selection on floral tube or spur lengths, suggesting partial evidence in support of Darwin's coevolutionary race (e.g. Nilsson 1988, Alexandersson and Johnson 2002, Muchhala and Thomson 2009, Anderson et al. 2010a, Ellis and Johnson 2010). However, none of these studies investigated selection on the proboscis lengths of their pollinators. Because it is often hard to estimate reproductive success (i.e. to measure fitness) from the fleeting encounters of free-living partners such as pollinators or seed dispersers, quantifying selection gradients on them is a major hurdle. Furthermore, demonstrating correlations between fitness and trait magnitudes usually does not show explicitly that the agent of selection is the putatively coevolving partner. Instead, selection studies often show only that a measure of fitness is associated with the magnitude of a particular trait (e.g. Anderson et al. 2010a). Using a pollination mutualism, Pauw et al. (2009) elegantly circumnavigated some of these problems by measuring the effects of floral tube length on the efficiency of pollen receipt after visitation by pollinators with a known proboscis length. Simultaneously, the authors were able to measure the amount of nectar that pollinators with known proboscis lengths were able to consume from flowers of known tube length. This enabled them to show that tube lengths and proboscis lengths were both the agents as well as the targets of reciprocal, directional selection. Left open to debate is whether pollen receipt and nectar consumption are accurate measures of fitness. In fact, the difficulty of quantifying total fitness measurements is a weakness of many of the aforementioned selection studies.

Another problem with selection studies is that even if reciprocal selection is demonstrated, it is still

Plate 1 Selected paleontological reconstructions of mutualism (Labandeira 2010). (a) The long-proboscate mesopsychid scorpionfly *Lichnomesopsyche gloriae* from the Middle Jurassic of Inner Mongolia, China, a probable pollinator of some gymnospermous plants.
(b) Reconstructions of two mid-Mesozoic pollinator associations between mesopsychid scorpionflies and their inferred host plants. At left, *L. gloriae* feeds on the pollen drops of *Caytonia sewardi* via an integumental tube under the lower lip of the cupule. At right is another mesopsychid, *Vitimopsyche kozlovi*, from the mid Early Cretaceous of Liaoning, China, feeding on catchment funnel nectaries from the ovulate cone of a cheirolepidiaceous conifer and achieving pollination in the process. Original art by Mary Parrish (Figure 1 in Box 3.1).

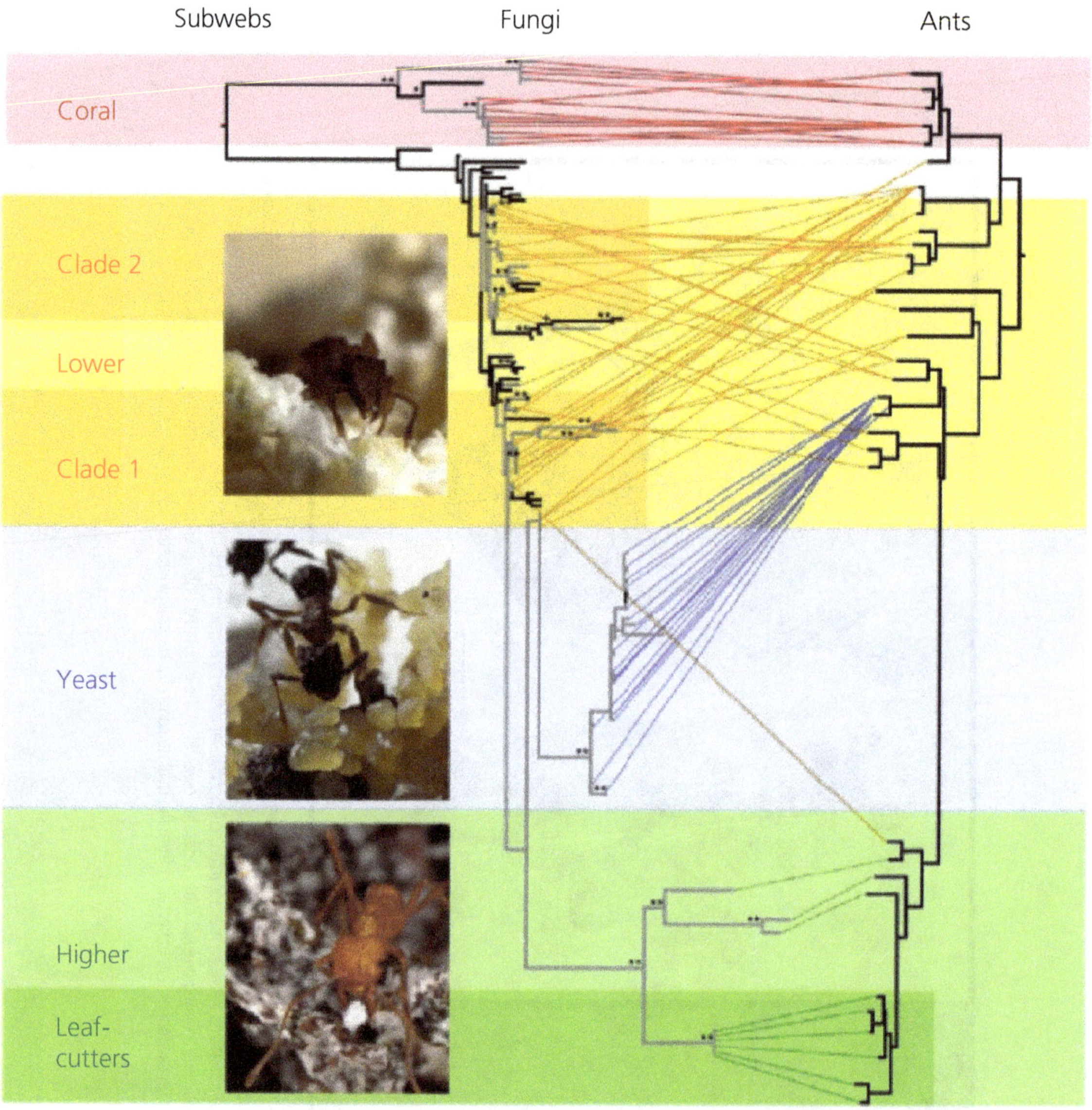

Plate 2 Phylogenetic associations between attine ants and fungal cultivars. Fungal cultivars (gray lines) were domesticated repeatedly from free-living fungi (black lines), followed by lateral spread of cultivars within defined ant clades. Phylogenetically primitive *Apterostigma* ants cultivate pterulaceous coral fungi, which are distantly related to the lepiotaceous fungi cultivated by all other attine ants. Ants in "lower" attine clades collect various plant debris for gardens, and cultivate fungi from at least two independently acquired cultivar clades (photo: *Cyphomyrmex wheeleri*). Cultivars in the "yeast" group are grown in single-celled form by the ants, whereas all other cultivars are grown as filamentous mycelia (photo: *Cyphomyrmex rimosus* attending yeast garden). Only the fungi cultivated by "higher" attine ants produce specialized nutritive structures (gongylidia) harvested by the ants, and these fungi therefore appear to have coevolved. Ants in the monophyletic "leafcutter" clade use predominantly fresh leaves for fungiculture (photo: *Atta cephalotes* harvesting gongylidia). Reprinted from Mikheyev et al. (2010) with permission from Chicago University Press. Comparative dating of attine ant and lepiotaceous cultivar phylogenies reveals coevolutionary synchrony and discord, *American Naturalist*, 175, E126–E133; see phylogenetic methods therein (See Figure 1 of Box 5.1).

Plate 3 Flowers of *Mimulus lewisii* used in field experiments to examine the genetics of a pollinator switch from bee to hummingbird pollination. a. Wild-type (WT); b. Near-isogenic line (NIL) in which the genomic region containing the allele for petal carotenoids in the hummingbird-pollinated *M. cardinalis* was introgressed into *M. lewisii* (reprinted from Bradshaw and Schemske 2003 with permission from Nature Publishing Group), Allele substitution at a flower colour locus produces a pollinator shift in monkeyflowers, *Nature*, 426, 176–178.) (See Figure 5.1).

Plate 4 Left: *Allomerus octoarticulatus* workers patrol and defend a young *Cordia nodosa* shoot against herbivores. Top right: workers of this ant species also destroy *C. nodosa* inflorescences by gnawing on the delicate floral buds. Bottom right: undamaged *C. nodosa* floral buds, for comparison. Photos originally appeared in Frederickson, M.E. (2009), Conflict over reproduction in an ant-plant symbiosis: why *Allomerus octoarticulatus* ants sterilize *Cordia nodosa* trees, *The American Naturalist*, 173, 675–681 (© 2009 by The University of Chicago) (Figure 1 of Box 6.1).

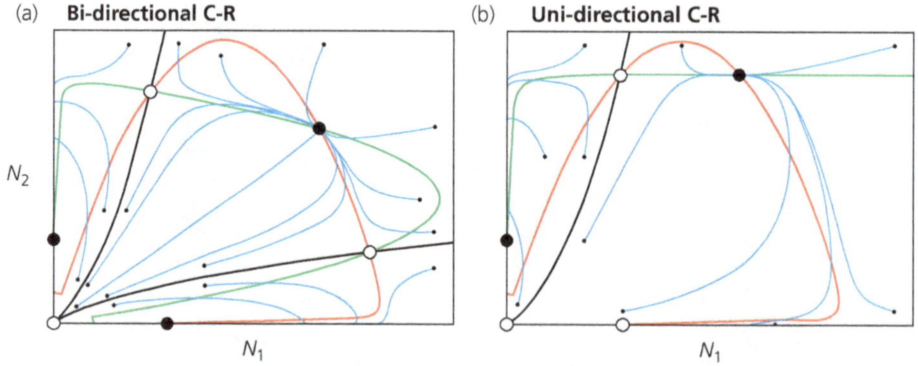

Plate 5 Phase-plane diagrams for the population dynamics of bidirectional (a) and unidirectional (b) consumer–resource mutualisms between facultative–facultative species with population densities of N_1 and N_2. Red and green lines are zero-growth isoclines ($dN_i/dt = 0$) for N_1 and N_2, respectively. Stable and unstable equilibria (nodes) are identified by filled and open circles, respectively. Saddle points have a black line (separatrix) passing through them to the origin, subdividing the phase-plane space into different basins of attraction that correspond with a particular node. Starting with the origin and moving clockwise, the equilibria for panel (a) are unstable node, stable node, saddle point, stable node, saddle point, stable node; for panel (b), they are unstable node, stable node, saddle point, stable node, saddle point. Transient dynamics of some population trajectories in phase-plane space are depicted by blue lines, for which the initial conditions for N_1 and N_2 of each trajectory are represented by filled black circles (See Figure 8.5).

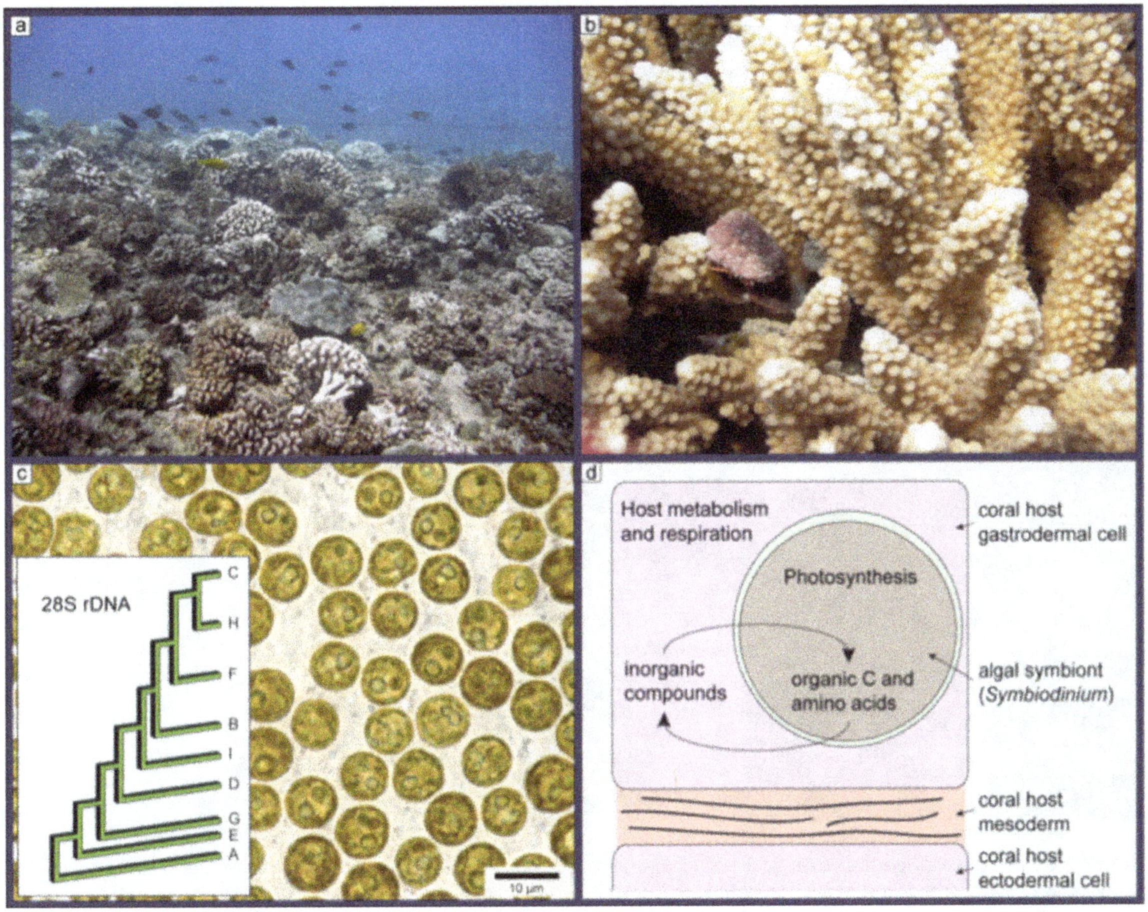

Plate 6 (a) The corals on the forereef of Moorea, French Polynesia, that grow and thrive as a result of the mutualistic association between corals and algae belonging to the genus *Symbiodinium*. (b) A close-up of a coral colony of a branching species of *Acropora*. (c) A bright-field microscope view of dinoflagellate algae in the genus *Symbiodinium* that form a mutualistic symbiosis with corals (image courtesy of Peta Clode, CMCA, The University of Western Australia). The inset shows a cartoon representation of the phylogeny of *Symbiodinium* and the evolutionary relationship of the nine lineages (clades A–I) within the genus based on the sequence of partial nuclear ribosomal DNA. (d) The localization of the algal symbionts that are housed within the gastrodermal cells of the coral host and the cycling of nutrients between the partners (See Box 9.1 Figure 1).

Plate 7 The brown color visible across various coral taxa (a) comes from the intracellular symbionts that are present in large densities throughout the coral polyp and the connective tissues [(b), *Stylophora pistillata*]. These single-celled endosymbionts are dinoflagellates of the genus *Symbiodinium* (c). [(a) ©George Roff; (b) ©Eugenia Sampayo; (c) ©Natalia Winkler.] (See Box 13.2 Figure 1).

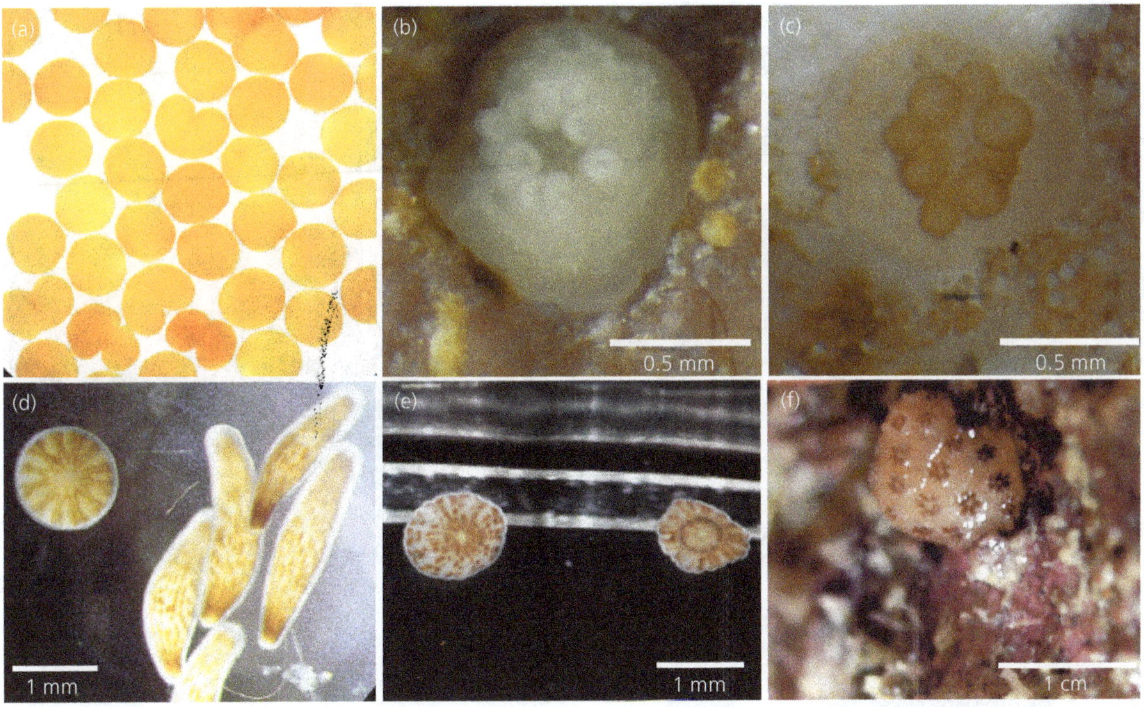

Plate 8 Symbiont acquisition occurs during the larval stages either from (a)–(c) the environment or (d)–(f) the maternal colony. The aposymbiotic egg/sperm bundles of the broadcast spawning coral *Acropora millepora* go through (a) various developmental stages before becoming swimming larvae that take up symbionts from the environment. Newly settled recruits (b) quickly develop a calcium carbonate skeleton (four-day old recruit shown) and (c) within ten days significant densities of symbionts are visible with the naked eye (starting in the tentacles). In contrast, (d) the swimming larvae released by the brooding coral *Stylophora pistillata* already contain a full complement of the maternal symbionts. Maternal transfer is linked to high levels of specificity, and likely improves settlement success to the local environment from (e) settlement into (f) adulthood (six-month-old juvenile colony). [(a), (d), (e) ©Eugenia Sampayo; (b), (c), (f) ©Natalia Winkler.] (See Box 13.2 Figure 2).

Plate 9 Images of restorations involving mutualisms. (1) Removal of black locust trees and their nitrogen-fixing symbionts restores natural nutrient cycling. Malcolm, et al. (2008). Soil nitrogen conditions approach preinvasion levels following restoration of nitrogen-fixing black locust (*Robinia pseudoacacia*) stands in a pine-oak ecosystem. *Restoration Ecology*, 16, 70–78. (2) Identifying effective mycorrhizal partner species can be critical for successfully culturing rare plants such as these orchid seedlings. Swarts, N.D., and Dixon, K.W. (2009). Terrestrial orchid conservation in the age of extinction. *Annals of Botany*, 104, 543–556. (3) Corals are cultivated for future transplanting in reef restorations. Shaish, L., et al. (2010). Employing a highly fragmented, weedy coral species in reef restoration. *Ecological Engineering*, 36, 1424–1432. (4) Bird perches installed in isolated restoration patches can encourage seed dispersal from source fragments. Bustamante-Sanchez, M.A., and Armesto, J.J. (2012). Seed limitation during early forest succession in a rural landscape on Chiloe Island, Chile: implications for temperate forest restoration. *Journal of Applied Ecology*, 49, 1103–1112. (5) Importing non-native giant tortoises to islands where endemic tortoise species are extinct can restore important dispersal function to endemic large-seeded plants. Griffiths, C. J., et al. (2011). Resurrecting extinct interactions with extant substitutes. *Current Biology*, 21, 762–765. (6) An experiment to assess the effectiveness of pollinator restoration plantings (MacLeod et al. *In preparation*). Winfree lab photo. Photo credits: (1) Spotlight News. (2) Nontachaiyapoom et al. (2011). Symbiotic seed germination of *Grammatophyllum speciosum* Blume and *Dendrobium draconis* Rchb. f., native orchids of Thailand. *Scientia Horticulturae*, 130:1, 303–308. Elsevier. (3) Shaish et al. 2010. Elsevier. (4) Orangutan Information Centre (OIC) <http://orangutancentre.org/>. (5) Hansen et al. (2010). Ecological history and latent conservation potential: large and giant tortoises as a model for taxon substitutions. *Ecography*, 33, 272–284. John Wiley & Sons (permission via RightsLink®). (6) Rachael Winfree. <http://winfreelab.com> (See Figure 14.1).

difficult to ascertain how important coevolution was as a force driving the evolution of a particular trait. For example, the hummingbird-pollinated plant *Nicotiana glauca* has long-tubed flowers, which are probably adapted to hummingbird beak dimensions (Nattero and Cocucci 2007). Although native to Argentina, the plant is a successful invader in many parts of the world, including southern Africa, where unrelated but functionally similar sunbirds probe the flowers with their long beaks (Geerts and Pauw 2009). The long, tubular corolla is a useful preadaptation to sunbird pollination, and may well be under present selection by the sunbirds. However, selection by sunbirds clearly did not play a role in the evolution of long tubes in *Nicotiana*. Sunbirds may, however, select for slight modifications in the tube dimensions. Consequently, the interpretation of selection studies needs to be carefully considered within the context of the interaction. This example also demonstrates the importance of showing that traits have changed in response to their interacting partners.

Among-population trait divergence and phylogenetically controlled analyses of trait divergence have both been used to demonstrate evolutionary change (e.g. Weiblen 2004, Chapter 3). However, it is still difficult with this approach to ascertain whether one or both species have undergone evolutionary change in response to one another. Some studies have attempted to combine trait variation approaches with selection or local adaptation studies in order to demonstrate which traits are likely to have changed in response to one another (e.g. Anderson and Johnson 2008, Pauw et al. 2009). A series of powerful studies document sexual dimorphism in the bill shapes of purple-throated carib hummingbirds. Two species of *Heliconia* are important nectar sources for the hummingbirds, and in sympatry they usually have different tube shapes which are advertised with different floral colours (Temeles et al. 2000). Each hummingbird sex preferentially feeds from *Heliconia* flowers that fit their bill morphology, and it appears that the fit between bill and flower morphology also affects the foraging efficiency of the birds (Temeles et al. 2000, Temeles and Kress 2003). This has been interpreted as evidence for evolutionary change by the birds in response to the different feeding niches presented by *Heliconia* (Temeles et al. 2000). Each species of *Heliconia* is, however, polymorphic for floral colour and corolla morphology, and the relative prevalence of each morph is dependent on variations in the floral community context (Temeles and Kress 2003). Floral polymorphisms in the two *Heliconia* have been interpreted as evolutionary responses to the two pollinator niches. Consequently, coevolution in this system is supported by evidence for evolutionary responses of hummingbirds to *Heliconia* flowers as well as evolutionary responses of *Heliconia* flowers to humingbirds (Temeles and Kress 2003).

Another powerful tool in studying morphological change is the documentation of phenotypic responses of interacting organisms over short time scales. Galetti et al. (2013) demonstrated that a hundred years after the loss of large-gaped seed dispersers, palm populations had evolved smaller seeds compared to populations which had intact seed dispersing fauna (a case of unilateral evolutionary change). Although the nature of this study made it impossible to look at reciprocal changes in bird morphology, putatively reciprocal change has been demonstrated in interactions where both interacting partners have fast generation times. Mostly such studies have been on host–parasite systems involving annual plants or rapidly producing invertebrates and their pathogens (e.g. Thrall et al. 2012, Koskella and Lively 2007). An advantage of these fast-generation systems is that it is often easy to estimate fitness in both partners through measurements of mortality and virulence. Resurrection ecology, where it is possible to "travel back in time" by resurrecting organisms trapped in old strata, has also produced some good evidence for reciprocal evolutionary change (Decaestecker et al. 2007). Here it is possible to trace temporal changes of resistance and virulence phenotypes, as well as to infect ancient hosts with more recent parasites and vice versa (e.g. Decaestecker et al. 2007). Presently, real-time coevolution and resurrection studies typically focus on antagonistic relationships. However, Weeks et al. (2007) examined *Wolbachia* infections of *Drosophila* in which *Wolbachia* appear to have rapidly evolved a mutualistic relationship from a previously parasitic one (see Box 2.1 and Chapter 3). Initially *Drosophila* fecundity was reduced by *Wolbachia* infections, but after twenty years of evolution, the fecundity of infected *Drosophila* was higher than for uninfected *Drosophila*. So far, evolutionary changes only appear to have been made on the part of the *Wolbachia* (i.e. unilateral evolution rather than coevolution has taken place). However, this does not preclude the possibility that future evolutionary changes in the *Drosophila* may facilitate infection by the mutualistic *Wolbachia*.

Although unequivocal evidence for coevolution is rare, partial evidence consistent with coevolution in both mutualisms and antagonisms is so widespread that, taken together, it may be hard to imagine a world in which coevolution had no role to play (Thompson 1994, 2005). In situations where the phenotypic variability of organisms affects each other's fitness on a routine basis, coevolution is most likely at work, either making fine or major adjustments to the genotypes and phenotypes of those interacting organisms. In many cases, perhaps, the questions we need to ask are not whether coevolution has occurred, but whether the products of coevolution are consequential, and whether coevolution can lead to quantifiable trait change. To do this we need to have a clear idea of what patterns in nature could conceivably be generated by coevolution.

7.4 Coevolutionary patterns

Coevolutionary patterns include the ways in which matches or mismatches between putatively coevolving traits occur in time and space, as well as the phylogenetic patterns of coevolving clades. Some of the earliest putative examples of coevolution appear to be based on the close phenotypic matching of traits between interacting organisms. As we have seen, Darwin (1862) himself invoked the idea of coevolution as the potential mechanism behind the matching lengths of hawkmoth tongues and orchid spurs. However, while this kind of matching may well be the result of coevolution, trait-matching cannot prove coevolution; similarly, a mismatch cannot disprove coevolution (Janzen 1980, Nuismer et al. 2010). In fact, broad-scale patterns of matching traits may be expected whenever it is advantageous for at least one partner (not necessarily both) to adapt to the traits of the other. This is exemplified in Figure 7.2, where similar patterns of trait matching are demonstrated in a putatively coevolved species pair and in a pair of species where the traits are unlikely to have evolved through coevolution. Trait matching can consequently be expected in mutualisms, antagonisms, and commensalisms (Nuismer et al. 2010). However, the type of relationship or the kind of coevolutionary mechanism may affect the degree or likelihood of matching (Nuismer et al. 2010). In this section I will highlight some of the patterns frequently attributed to coevolution, and explain some of the coevolutionary and other mechanisms that may produce them. I will also draw attention to the expected similarities and differences between patterns produced by mutualistic versus antagonistic coevolution.

7.4.1 Matching of traits

7.4.1.1 Coevolutionary complimentarity

Despite potential conflicts of interest between mutualists (see Chapter 6), traits may frequently coevolve to facilitate mutual access to rewards and services. Consequently, there is often selection for a certain degree of trait complimentarity between partners. Mutual recognition systems between nitrogen-fixing rhizobial bacteria and the plants they colonize (Miklasheviches et al. 2001) provide an example of trait complimentarity in mutualism. The plants involved in this interaction (described in more detail in Box 5.2) control the pathway for the production of a key amino acid used by the bacteria, while bacteria control the pathway for the production of a key amino acid used by the plants (Lodwig et al. 2003). It is thought that the mutualism is stabilized because neither the plant nor the bacteria can dominate the relationship without having an inverse effect on fitness. Another example of trait complimentarity can be found in the mutualistic interaction between the tree *Leonardoxa africana* and certain ants that protect it from herbivory (Figure 7.1). The tree produces modified thorns in which the ants live (domatia), but the ants need to chew through a weak, unlignified spot called a prostomium in order to gain entry (Brouat et al. 2001). Different populations of *L. africana* are associated with different species of ants with different head shapes. The shapes and sizes of the associated ants are closely matched to the shapes and sizes of the prostomia. It is evident that the prostomia act as a filter allowing a small subset of ants to interact with the plant (Brouat et al. 2001). Phylogenetic analyses support the idea that the head shapes of some ant species are highly derived, suggesting coevolutionary origins. While trait matching can also be observed in antagonistic relationships, its root causes are usually different.

7.4.1.2 Coevolutionary races

Unlike mutualisms, antagonisms involve individuals of one species that usually experience a net loss in fitness as a result of the interaction, while individuals of the other species usually gain fitness. Consequently one may expect the "losing" species to evolve mechanisms enabling it to escape the interaction through the mismatching of traits, while the "gaining" species should evolve mechanisms to match the traits of the "losing" species in order to continue benefiting from the interaction (Thompson 2013). This dynamic engenders an "escape-and-chase" scenario. The oscillatory cycles of coevolving polymorphisms are a good example of escape-and-chase coevolution. Gene-for-gene coevolution takes the format of oscillating polymorphisms for resistance genes in the host, and virulence genes in the parasite. Fluctuations in resistance and virulence gene frequencies are underpinned by negative frequency dependent selection, in which rare parasite phenotypes are not easily recognized by hosts, making them more infectious than more common phenotypes. However, as they become more common, selection on the host for defense against that phenotype increases. If each infection phenotype is coded for by a specific genotype, and each resistance phenotype is coded for by a specific resistance genotype, the expectation is that different genotypes for infection and resistance will cycle through time, sometimes with close infection-resistance phenotype matching (e.g. Thrall and Burdon 2003), but occasionally with host and parasite phenotypes being completely mismatched and out of phase (e.g. Thrall et al. 2012). In these types of relationships, the frequency of mismatched versus matched traits could be dictated by how fast the parasite evolves compared to the host. Frequent mismatches favoring the host may be expected with a relatively quickly evolving host, but close matching of resistance to virulence phenotypes (favoring the parasite) would be expected if the parasite evolved relatively quickly compared to the host.

While the oscillatory races of gene-for-gene coevolution have never been documented for mutualisms, the antagonistic aspects of mutualisms may give rise to escalatory races between coevolved traits. The most famous example of this is Darwin's race (described in Section 7.1), where a positive feedback system is generated by increases in floral tube length selecting for increases in pollinator tongue length which in turn selects for increases in tube length. This leads to a race in which traits become larger or more extreme. Clearly, such a race of exaggeration cannot continue indefinitely: the race is halted when an opposing selection force equals and stabilizes the selection force for further exaggeration in the traits of either partner. For example, trait elongation on hawkmoth proboscides may stop if it becomes too difficult for the hawkmoth to maneuver a proboscis in excess of a certain length. We would then expect the plants to evolve spurs which are marginally longer than the moth tongues, forcing the moths to make contact with the reproductive parts of the flowers whenever they visit. In putatively coevolving mutualisms, traits that are expected to coevolve under this kind of model have frequently been found to vary geographically. For example, three-fold variation in the proboscis length of a long-proboscid fly species was demonstrated across its range (Anderson and Johnson 2008). Similar variation has been observed for other species of long-proboscid fly (Pauw et al. 2009, Newman et al. 2014) as well as for the forelegs of *Rediviva* bees, which use their legs to mop up oil from the deep, twin spurs of oil-producing *Diascias* (Steiner and Whitehead 1991). In each of these systems the lengths of the floral parts correlate with the lengths of the pollinator parts (see Figure 7.2 for correlations between pollinator foraging traits and floral tube length). Similar correlations have been found in antagonistic relationships where escalatory coevolution has been invoked to explain the covariation in toxicity and resistance of newts and their garter snake predators respectively (e.g. Hanifin et al. 2008), as well as the evolution of extremely extended mouthparts in weevils and the correspondingly thick pericarps of their host plants (e.g. Toju and Sota 2006). While correlated patterns of trait exaggeration may be a frequent outcome of coevolution, they alone do not prove that coevolution is or has been occurring. This is because correlated traits can also be an outcome of unilateral evolution where one species adapts to another but not vice versa (Figure 7.2). Trait correlations could also conceivably occur if abiotic selective pressures select on the traits of both interacting species (see Nuismer et al. 2010 for further discussion on this topic). Patterns of trait matching may, however, be used to test whether matching is stronger in mutualisms versus parasitisms or commensalisms (e.g. Anderson et al. 2010b; see Section 7.4.2).

7.4.2 Mismatching of traits

7.4.2.1 Asymmetries in escalatory races

Even though escalatory races may produce correlated traits, correlations do not necessarily mean that the traits are well matched. Two kinds of mismatches can be distinguished in trait matching studies. The first is that regressions between traits are never perfect. There are always some populations of interacting individuals that appear to match better than others (Anderson et al. 2010b). Variability in the degree of mismatch between populations of interacting species may be the result of variability in the way in which populations respond to each

Figure 7.2 Similar patterns of trait matching between mutualistic, coevolved species and antagonistic, unilaterally evolved species.
a) *Zaluzianskya microsiphon* is the principal nectar source for the pollinator (*Prosoeca ganglbaueri*), whose probosicis length is closely matched to the tube lengths of the plants in each population. b) *P. ganglebaueri* flies are deceitfully attracted to the mimetic orchid *Disa nivea* which has no nectar. Even though the tube length of the floral spur does not select on fly tongue length, the tongue lengths of the flies select on the tube lengths of the plants, generating patterns of trait matching. Long tubes facilitate precise pollen placement at the base of the fly proboscis (see arrow). Reprinted with permission from Anderson and Johnson (2009).

other. One of the pillars of the geographic mosaic of coevolution is that the strength and symmetry of selection between interacting organisms is likely to vary among populations, leading to coevolutionary hotspots and coevolutionary coldspots (Thompson 1994, 1999, 2005).

Regressions of matching traits may also differ in their slopes (Anderson et al. 2010b). Theoretically, we could conceive of a situation where the average trait measurements of interacting populations are perfectly matched, so that on average, neither population has a functional advantage over the other. If these traits were geographically variable but perfectly matched at each site, we could draw a theoretical line through the data points, which we could call the expected line for perfect trait matching. We could then collect real data on trait matching from multiple populations and determine whether the line for perfect trait matching is a good fit of the real data or not. Deviations from the line for perfect trait matching suggest that one species may have the functional "upper hand" in the relationship. The best examples of this are from the parasite and predator–prey literature on coevolution where the values for perfect trait matching were experimentally calculated (e.g. Toju and Sota 2006, Hanifin et al. 2008). Camellia weevils have evolved an extraordinarily long rostrum with chewing mouthparts on the end, used to bore through the thick woody pericarp of Japanese camellias, so that their eggs can be laid on the seeds within. Perfect balance or trait matching is when the rostrum is long enough that half of the beetle's boring attempts within a population of Japanese camellias are successful and the seeds of the camellia are reached (Toju and Sota 2006). Toju and Sota (2006) and Hanifin et al. (2008) identified consistent mismatches in the traits of interacting species, suggesting that winners and losers of coevolutionary races may be predictable.

Ultimately, the dimensions of interacting traits are governed by the net outcome of several selection pressures acting on them, not just the reciprocal selection pressures generated by the interaction. In other words, the traits of interacting organisms are not only shaped by selection driven by each other, but also by constraints selecting against further trait exaggeration. Often, these constraints are generated from the abiotic environment in which the interaction takes place, and these constraints are likely to be quite different for the interacting organisms. For example, the rostrums of camellia weevils and the pericarps of Japanese camellias exhibit predictable patterns of phenotype mismatching within a strongly correlated pattern of phenotype matching (Toju and Sota 2006). The exaggeration of pericarp thickness is thought to be constrained by resource availability, while the exaggeration of weevil mouthparts is thought to be constrained by allometry (Toju et al. 2011). Consequently, when resources are scarce, pericarp thickness is the first trait to be constrained, and the traits of the weevils exceed those of the plants. At resource-poor sites, the arms race does not proceed very far, and both pericarp thickness and rostrum length are poorly developed. However, when resources are not limiting, the race can continue to much greater lengths. In these populations the increased exaggeration of weevil mouthparts is eventually allometrically constrained, allowing the pericarps to become relatively more exaggerated and the plants to dominate the race (Toju et al. 2011). Increased exaggeration of traits may also be genetically constrained (Box 7.1). For example, Hanifin et al. (2008) found that although newt toxicity and snake resistance were closely correlated, snakes were consistently more resistant to newt toxicity than predicted by a scenario of perfect trait matching. It was hypothesized that the relative success of the snakes was due to the fact that high snake resistance in some populations can be explained by variation of a single nucleotide substitution (Geffeney et al. 2005), making it potentially easier to evolve than newt toxicity.

These advances in the literature on host–parasite coevolution lead to questions about whether similar trait mismatches occur in mutualisms, and whether mismatches in mutualisms tend to be larger or smaller than in antagonisms. We may also start asking questions about which factors contribute most toward trait mismatches. The answers to these questions may enable us to make predictions about the direction of trait mismatches in interacting species and about which species are expected to win coevolutionary races.

While models are an obvious place to start answering such questions, one empirical study examined patterns of trait matching using multiple published data sets on the traits of interacting plants and insects (Anderson et al. 2010b). They found that correlations between traits frequently occurred irrespective of whether the relationships were mutualistic or antagonistic. Although their analyses were crude in comparison to the host–parasite examples cited in this section, they nevertheless found similar mismatches in traits, comparable to those found in the antagonistic systems. Typically, the larger the traits, the more extreme the absolute mismatch; a case in point is that the most extreme mismatch was found for Darwin's famous orchid, with a spur length of approximately 30 cm, and its primary hawkmoth pollinator, with a proboscis length of only about 23 cm. More intriguing was the observation that the direction of trait mismatch evident in Darwin's orchid and hawkmoth appeared to hold for most interacting plant–insect partners across the spectrum of traits examined. In other words, the traits of plants tended to be more exaggerated than the traits of the interacting insects (Figure 7.3). The authors speculate that asymmetries in the constraints or selective pressures acting on insect versus plant traits were the primary reasons for the patterns of mismatch. However, modeling approaches may prove more useful in dissecting out the real causes of trait mismatches. In Box 7.2, Scott Nuismer models the matching and mismatching of traits. He demonstrates that inherent differences in the selection pressures against further trait exaggeration (i.e. constraints resulting in stabilizing selection) in insects versus plants may be extremely important in generating trait mismatches. Furthermore, he also shows that the relative population sizes of plants and insects can also play an important role in generating trait mismatches.

7.4.3 Trait divergence between sites

The divergence of traits through coevolution is classically associated with competition. For example, two of Darwin's finches competing for food may both diverge in their beak characteristics as they specialize on slightly different resources (e.g. large versus small seeds, *sensu* Grant and Grant 2006). Consequently Darwin's finches in populations

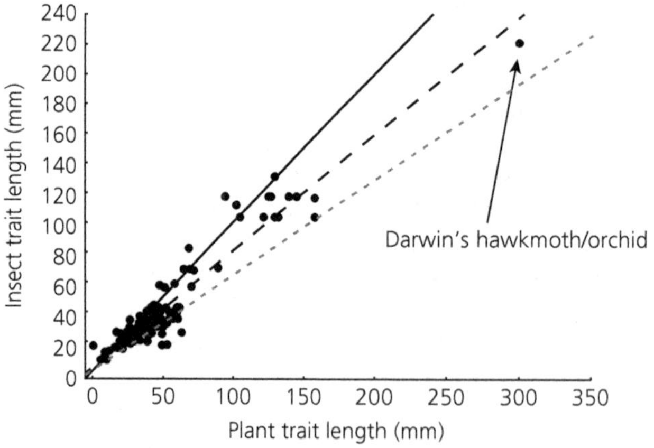

Figure 7.3 Scatterplot showing the relationship (ordinary least squares regression) between putatively complimentary traits for interacting plants and insects. Each data point represents a species mean for an interacting pair of plants and insects where at least one species (usually the plant) is highly specialized on the other. The solid line is the slope of unity or the line where traits are "perfectly matched" and it is hypothesized that neither the plant nor the insect has a functional "upper hand." The black, dashed line is the line of best fit using ordinary least squares regression and suggests that plant traits are typically more exaggerated than insect traits. The gray, dotted line describes a model adjusted for insect phylogeny, effectively magnifying the asymmetries of trait exaggeration. To illustrate the point using an extreme example, Darwin's Madagascan star orchid has a spur length of 300 mm while its principal pollinator has a tongue length of only 230 mm. Reprinted with permission from Anderson et al. (2010b).

Box 7.2 When do trait mismatches coevolve?

Scott L. Nuismer

Work exploring the relationship between traits of interacting mutualist species demonstrates that although traits of mutualists tend to be correlated, trait mismatching appears to be common. Specifically, by estimating the slope of the relationship between plant and animal traits for published data sets, Anderson et al. (2010) showed that the traits of plants tended to be more exaggerated than the traits of the interacting insects (Figure 7.3). To explore the possible causes of this pervasive pattern of trait mismatching, I develop a simple model of mutualistic coevolution and use it to predict the slope of the relationship between plant and insect traits.

I model mutualistic coevolution between a pollinating insect species, x, and a plant species which it pollinates, y. I assume the species live within a metapopulation consisting of N individual populations, each of which is inhabited by n_x and n_y insect and plant individuals, respectively. The mutualistic interaction is mediated by a single quantitative trait in each species, with coevolutionary selection favoring phenotype matching with strength S_x and S_y, respectively. In addition, insect and plant species experience stabilizing selection of strength ψ_x and ψ_y favoring trait values θ_x and θ_y respectively. As long as selection is weak and additive genetic variance fixed, this model can be used to study the evolution of trait means (Nuismer et al. 2010, Yoder and Nuismer 2010).

In order to tie the predictions of our model to empirical studies, I calculate the slope of a regression of insect trait on plant trait using reduced major axis (RMA) regression:

$$\beta_{xy} = \frac{\sigma_x}{\sigma_y} \tag{B1}$$

where σ_x and σ_y measure the standard deviations of trait means across the metapopulation for insect and plant, respectively. As long as a dynamic equilibrium between selection and drift has been reached, our model reveals that the slope of this relationship is given by:

$$\beta_{xy} = \frac{\sqrt{G_y n_y (S_y + \psi_y)^2 + G_x (n_x S_x^2 + n_y (S_y \psi_x + S_x \psi_y + \psi_x \psi_y)}}{\sqrt{G_x n_x (S_x + \psi_x)^2 + G_y (n_y S_y^2 + n_x (S_y \psi_x + S_x \psi_y + \psi_x \psi_y)}} \tag{B2}$$

Inspection of (B2) reveals several important insights. First, a slope of 1 is the exception rather than the rule. Only for very specific combinations of parameter values should a 1:1 relationship among traits of the coevolving species occur, the most obvious being when both species are identical with respect to S, G, n, and ψ. Second, in addition to the strength of coevolutionary selection, S, and stabilizing selection ψ, which have been previously hypothesized to shape the slope (Anderson et al. 2010), population sizes n and additive genetic variances G also appear to play an important role.

To gain further insight, I evaluated (B2) numerically. These analyses show that, in general, slopes greater than one are favored when species x has smaller population size than species y, experiences weaker stabilizing and coevolutionary selection than species y, and has less additive genetic variance than species y. Surprisingly, these analyses also reveal that the slope is much more sensitive to asymmetries in local population sizes and strengths of stabilizing selection than it is to asymmetries in the strength of coevolutionary selection (Box 7.2, Figure 1).

Intuitively, the effect of each evolutionary force on the slope can be understood by considering its influence on spatial variability among trait means, σ_i. Specifically, forces reducing spatial variability in species x decrease the slope whereas forces reducing spatial variability in species y increase the slope. Within our simple model, the evolutionary forces that reduce spatial variability include stabilizing selection (which I assume is spatially homogenous), coevolutionary selection, and additive genetic variance (because it facilitates the response to selection). Only random genetic drift increases spatial variability, with this effect greatest for small population sizes.

These results may help to explain why recent studies have observed such frequent mismatches in traits as interpreted by their slopes (Anderson et al. 2010). Specifically, slope deviations in these studies could have been caused by differences in the effective population sizes of plants and insects, differences in the constraints acting on plant and insect traits, or differences in the strength of selection for exaggeration of plant and insect traits. Of these possibilities, the first two seem the most plausible since only very extreme asymmetries in coevolutionary selection could generate the slopes observed in empirical studies whereas modest asymmetries in local population sizes and levels of constraint are consistent with the data (e.g., Box 7.2, Figure 1).

Taken together, the results of this model illustrate that the slope inferred from studies of trait matching is influenced by many evolutionary forces, among which coevolution is but one. The failure of the slope to provide unique and compelling evidence for coevolution raises an obvious question: should we abandon studies of trait matching as a

continued

Box 7.2 *Continued*

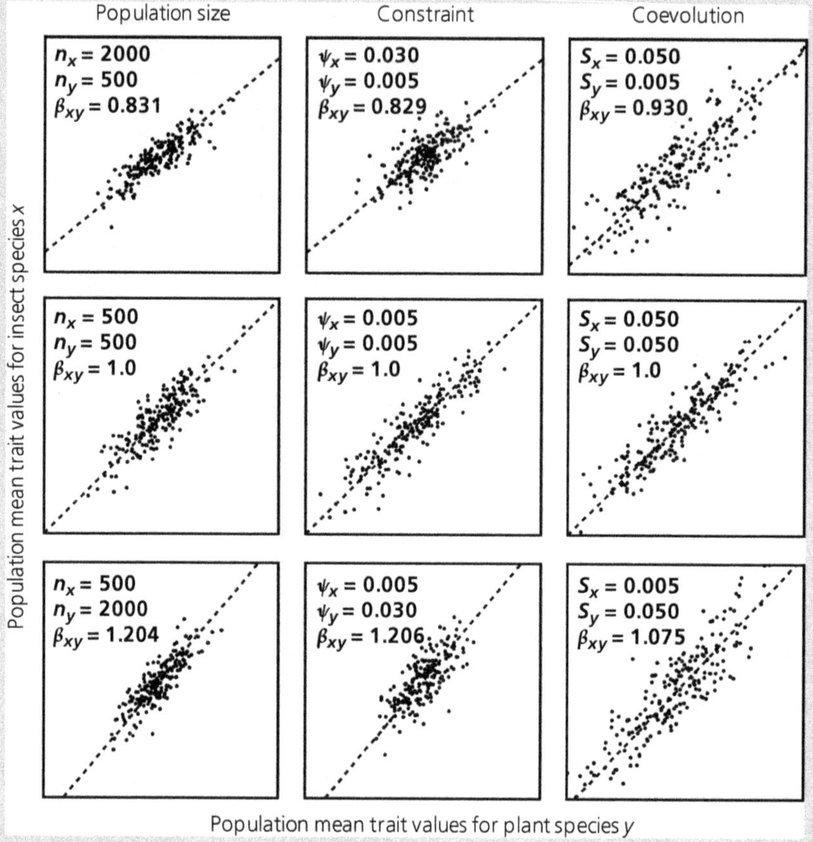

Figure 1 The slope of the reduced major axis (RMA) regression predicted by the analytical model (B2) shown as a dashed line. The dots represent trait values for individual populations as predicted by a stochastic simulation of a metapopulation consisting of 200 individual populations. Key parameters are shown within the upper left-hand corner of each figure, as is the predicted slope, with the first column demonstrating the effects of relative population size on trait matching slopes, the second column demonstrating the effects of relative constraints upon trait matching slopes, and the third column demonstrating the effects of the relative strengths of coevolutionary selection on trait matching slopes.

failed endeavor? My opinion is that this would be premature. In addition to identifying patterns ripe for mechanistic explanation, studies of trait matching have identified potentially important general trends and serve as a means to study differences in the patterns of mismatching between mutualisms and antagonisms. Identifying the mechanisms generating these patterns will require the development of model-based statistical approaches that can wring the most possible information from existing data sets. If such approaches can generate even coarse estimates for the strength of coevolutionary selection within natural systems then our understanding of evolving species interactions could be vastly improved.

References

Anderson, B., Terblanche, J.S., and Ellis, A.G. (2010). Predictable patterns of trait mismatches between interacting plants and insects. *BMC Evolutionary Biology*, **10**, 1471–2148.

Nuismer, S.L., Gomulkiewicz, R., and Ridenhour, B.J. (2010). When is correlation coevolution? *American Naturalist*, **175**, 525–537.

Yoder, J.B., and Nuismer, S.L. (2010). When does coevolution promote diversification? *American Naturalist*, **176**, 802–817.

without competitors may have different beak dimensions than in populations of the same species which do have to compete with other finches for food (character displacement). Importantly, intraspecific divergence is the product of a heterogeneous landscape in which populations coexist with mosaics of different competitors.

While coevolving mutualisms are seldom the examples that people turn to when they think about trait divergence, environmental heterogeneity could drive trait divergence of mutualist traits in a similar way. The environments experienced by coevolving mutualists are heterogeneous in terms of the biotic components (communities) in which the mutualisms are embedded, as well as in terms of their abiotic components (Thompson 2005). *Greya politella* visits and pollinates several species of woodland stars across its range. Multiple traits of these flowers (e.g. ovary position, stigma length, floral depth) differ between species (Thompson et al. 2013). The *Greya* moths, now recognized as a species complex consisting of several lineages (Rich et al. 2008), exhibit multiple traits (e.g. ovipositor length, proboscis length, and the length of abdominal segments) that covary with the traits of the plants they are found on. The covariation and divergence of moth and plant reproductive traits appear to maintain the effectiveness of both oviposition and pollination across a variable biotic mosaic (Thompson et al. 2013). Similarly intraspecific covariation of pollinator and plant traits described for escalatory races (Section 7.4.1.2) also appear to vary geographically in response to one another (e.g. Steiner and Whitehead 1991, Anderson and Johnson 2008, Pauw et al. 2009, Newman et al. 2014, Figure 7.2). However, these studies raise the question of why traits are geographically divergent and why the escalatory race does not stop at the same place in all populations, leading to a lack of trait divergence across populations. Once again, studies on antagonisms offer the best empirical evidence for why the escalatory traits vary geographically. For Japanese camellias, it appears as if variability in resource availability plays an important role in halting the escalatory race between the weevils and camellias at different endpoints (Toju et al. 2011). We can borrow from these ideas and hypothesize that different abiotic conditions could stop escalatory races at different end points for pollination mutualisms too. Perhaps strong winds prevent insects from evolving very long proboscides in certain areas, whereas lower wind speeds allow the escalatory race to proceed much further in other areas. We could also invoke differences in plant communities visited by pollinators as a factor which may affect the end point of an escalatory race (Anderson et al. 2010b).

7.4.4 Convergence of traits

7.4.4.1 Trait convergence within sites

Mutualisms usually involve many species. In multispecific mutualisms, the traits of many species may converge in response to the traits of a species (or a group of functionally similar species) that all of them are utilizing (Thompson 2005, Nuismer et al. 2013). For example, the traits of plants may converge to form a functional match in response to the traits of a shared pollinator (Figure 7.4). Groups of organisms with traits that have converged in response to a common selective force are called guilds (e.g. Faegri and van der Pijl 1979). Familiar to many is the hummingbird-pollinated guild of unscented, red-flowered plants with long corollas and exerted reproductive parts (Faegri and van der Pijl 1979), and the sweetly scented, pale, long-tubed flowers of plants pollinated by moths (Faegri and van der Pijl 1979). In some cases plant guilds may interact with just a single pollinator. However in many cases cases a plant guild will revolve around multiple pollinators that have similar traits (Johnson 2010). Similarly, fruits also display adaptations to groups of seed dispersal vectors (van der Pijl 1982). However, we still need to question whether coevolution or unilateral evolution is the driver of trait convergence to form guilds in these asymmetrically specialized mutualisms.

Asymmetries in specialization are likely to translate into asymmetries in the reciprocity of selection: a highly specialized flower may be obligately dependent on one pollinator and consequently the pollinator may select strongly upon the floral traits of that plant. However, if that pollinator visits several plant species for nectar, it is unlikely to depend

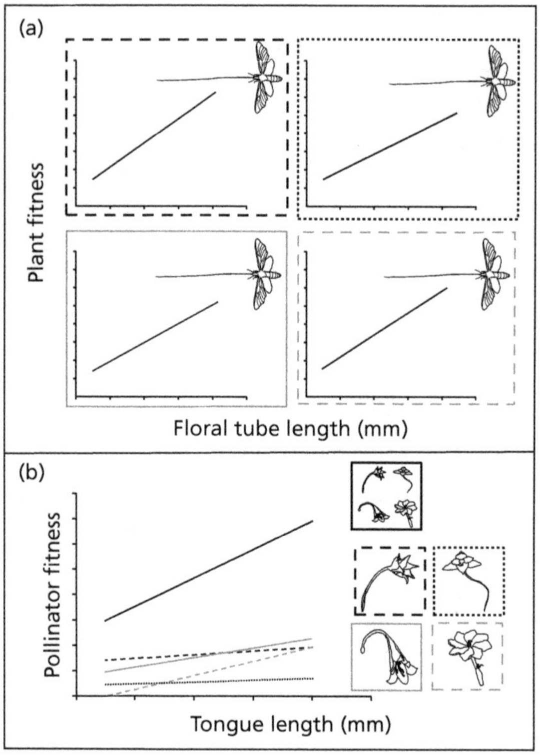

Figure 7.4 A hypothetical data set depicting the selective pressures on plant and pollinator traits (selective agents pictured for each figure) in an asymmetrically specialized pollination guild. The plant guild consists of four species: *Gladiolus longicollis* (dashed black line), *Crinum bulbispermum* (solid gray line), *Aerangis muscicola* (dotted black line), and *Gardenia thunbergia* (dashed gray line), all pollinated primarily by a single long-tongued moth pollinator, *Agrius convulvuli*. a) Because each plant is highly specialized on the moth pollinator, the moths exert a strong, positive directional selective pressure (measured by the steepness of the slope) on the tube lengths of each plant species. b) Since moths are not entirely dependent on any one plant species, each plant species will exert a relatively small selective pressure on moth proboscis length (shallow slopes). However, the cumulative effect of the entire guild (solid black line) exerts a strong selective pressure on pollinator tongue length, suggesting that coevolution in asymmetrically specialized guilds may play an important role in shaping the morphologies of all interacting partners. The relative selection strength imposed by each plant species is likely to be influenced by reward quality and quantity as well as the trait dimensions found in each population.

obligately on any one plant species; consequently, selection by one plant species on the pollinator is likely to be comparatively weak. So can such asymmetric selection lead to coevolution? Since the selection strength on a set of pollinator traits would be influenced by the entire plant guild and not just a single species, we should take into consideration the combined selective force exerted by a guild of functionally similar plant species, rather than the selection strength of each guild member individually. Since the traits of guild members often appear to be similar, each member may be generating a small but similar selective force on the pollinator. The strength of selection imposed by a guild should be strongly influenced by the similarity in the traits of the guild members. If guild members share similar traits, the combined or net selective pressure on the pollinator traits should be strong. However, if the traits of guild members are highly variable, they are likely to exert conflicting selective pressures. The resultant combined selection on the traits of the pollinator may be expected to be much weaker. In pollination guilds, the putatively coevolved traits of the floral guild members (e.g. corolla length) can show powerful convergence at the site level, such that different plants within a site are more similar to each other than are different populations of single species (Pauw et al. 2009, Anderson and Johnson 2009). This suggests that the traits of pollination guild members are frequently very similar at the site level and that they should impose similar selection pressures on the pollinator. Consequently, generalization by the pollinator is not necessarily synonymous with weak selection from the guild, because the additive selection pressures of all guild members is likely to be powerful (Figure 7.4). This way of thinking is supported by models of interacting mutualist communities, which predict that coevolution may play a very important role in the convergence and complimentarity of traits in multispecific interactions (Nuismer et al. 2013).

Due to differences in their abundance or in the quality of the commodities they provide to mutualists, some species within a guild may also play more important roles in driving coevolutionary dynamics than others. For example the traits of a rare guild member are less likely to affect the fitness of its pollinator than will a common member from the same guild, which is likely to constitute a more important food source. Nevertheless, coevolution between one interacting species pair has the ability to influence the traits of other, non-coevolving or weakly coevolving guild members. An extreme example of this phenomenon is the coevolutionary race between the proboscides of fly pollinators and the corolla tubes of their most important nectar plants (*Zaluzianskya microsiphon*, Figure 7.2), which are closely matched at each site (Anderson and Johnson 2008). *Z. microsiphon* plants are mimicked by a rewardless orchid (*Disa nivea*) which attracts long-proboscid fly pollinators due to their close resemblance to *Z. microsiphon* (Anderson et al. 2005). Clearly, the length of the orchid's nectarless spur exerts no selection on the tongue length of the pollinator. However, because spur length affects the efficiency of pollen placement, the orchid spurs track the coevolutionary race between flies and rewarding plants. Similarly, extremely rare guild members, which exert very weak selective forces on pollinators, are also expected to track, rather than drive, coevolutionary races between pollinators and more common guild members. Consequently, if relative abundances of guild members change through space and time, the relative importance of different guild members is likely to change accordingly. As a result, different guild members may be the primary drivers of coevolution at different times or sites. In this sense Arditti et al. (2012) was probably correct in questioning the idea of coevolution between Darwin's hawkmoth and the exceptionally uncommon Madagascan Star orchid.

7.4.5 Matching of phylogenies

A common and enduring misconception about coevolution is that it should give rise to identical patterns of phylogenetic divergence (matching phylogenies) for closely interacting species (e.g. Hafner and Nadler 1988, Cuthill and Charleston 2012). Coevolution is reciprocal evolutionary change, whereas cospeciation, codiversification, and parallel cladogenesis all refer to matching between the phylogenies of interacting taxa, such that divergence events in one taxon correspond with divergences in the other (Thompson 1994). Patterns of cospeciation have been proposed across a wide range of highly specialized mutualisms (e.g. Cruaud et al. 2012, figs and wasps; Nishiguchi et al. 1998, squid and bioluminescent bacteria; Mikheyev et al. 2010, leafcutter ants and fungi; Smith et al. 2008, yuccas and moths), as well as parasitisms (Hafner and Nadler 1988, gophers and lice; Bruyndonckx et al. 2009, bats and mites). However, almost all cophylogenies have also found host or partner switching in addition to cospeciation events. Interestingly, comparative studies on brood site mutualisms suggest that mutualists are more specialized than commensalists (Althoff et al. 2012, for interactions with yucca). Whether or not cospeciation is more prevalent in mutualist lineages than in lineages of non-pollinators that cohabit figs (Weiblen and Bush 2002) remains a matter of debate (Chapter 3). In any event, it is clear that cospeciation patterns may be observed irrespective of whether or not a relationship is coevolved (see Figure 1 in Box 7.3). In this box Kari Segraves describes how phylogenetic patterns of interacting species may be affected by patterns of gene flow which may often not be linked to coevolution, and the many ways in which phylogeny can be used to study coevolution. In fact, the parameter space for a direct causal link between coevolution and cospeciation is small because it requires cospeciation to act on the reproductive isolation of one or both species. While this is possibly the case for interactions in which one species controls or influences the mating patterns of another species, evidence for this link is limited (Moe and Weiblen 2012), or to the contrary (Althoff et al. 2012). Phylogeny is in fact a particularly poor tool for inferring most kinds of coevolution. However, it can still be a very powerful tool for coevolutionary studies. It is essential if, for example, one wishes to trace historical and contemporary patterns of host use, movement through geographic landscapes, rates of diversification, or the directions of morphological change through character state reconstructions (Chapter 3 and Box 7.3). In addition to using molecular tools to generate phylogenies, new advances in molecular technologies may enable biologists to identify the genes involved in coevolution. In Box 7.1, Noah Whiteman describes some of the methods and difficulties in identifying coevolutionary genes and how their identification may enable us to explore the links between reciprocal selection and genomic changes.

Box 7.3 Using phylogenetics to study coevolution

Kari A. Segraves

Since Ehrlich and Raven's seminal paper, coevolution has often been considered an important diversifying force in interacting taxa (Ehrlich and Raven 1964, Thompson 1994). Locally coevolving taxa may become highly specialized, potentially reducing the subsequent success of migrants moving among populations differing in coevolutionary history. This pattern of increasing specialization and reduction of gene flow among populations is the fuel for geographic variation in coevolved traits and speciation (Thompson 2005). Although coevolution may facilitate speciation in some types of interspecific interactions, recent theoretical work suggests that the direct role of coevolution in the diversification of mutualistic lineages is limited (Yoder and Nuismer 2010). Thus, there is a compelling need to rigorously assess how coevolution affects divergence in mutualistic taxa.

Phylogenetic tools offer a valuable approach to study the linkage between coevolution and diversification. Unfortunately, misinterpretation of cospeciation analyses (comparisons of the phylogenies of interacting taxa to assess the number of concordant speciation events) has incorrectly promoted the idea that matching phylogenies are sufficient evidence of coevolution (Box 7.3, Figure 1a). Phylogenetic comparisons cannot provide evidence of reciprocal evolutionary change, and like other correlative approaches, they should be combined with experimental data to demonstrate coevolution (Nuismer et al. 2010). Although concordant trees may be a result of coevolution, they may also be the result of other commonalities between two lineages, such as a shared biogeographic history that resulted in covicariance (Box 7.3, Figure 1b). Thus, matching phylogenies may occur in taxa that have no coevolutionary history and, likewise, tightly coevolved taxa may have significant mismatches in speciation events (Box 7.3, Figure 1c; e.g., Segraves 2010, this chapter). Consequently, phylogenetic information in the absence of good natural history, ecological, and biogeographic data can be misleading at best, and at worst, may result in erroneous conclusions about the importance of coevolution in the speciation process.

Since coevolution cannot be inferred directly from phylogenies, what role can phylogenetics play in coevolutionary studies? A phylogenetic framework can provide the basis for testing specific hypotheses about the importance of coevolution in shaping the evolutionary histories of putatively coevolving mutualists. Carefully crafted hypotheses combined with ecological or experimental data may distinguish coevolution from alternative explanations of the

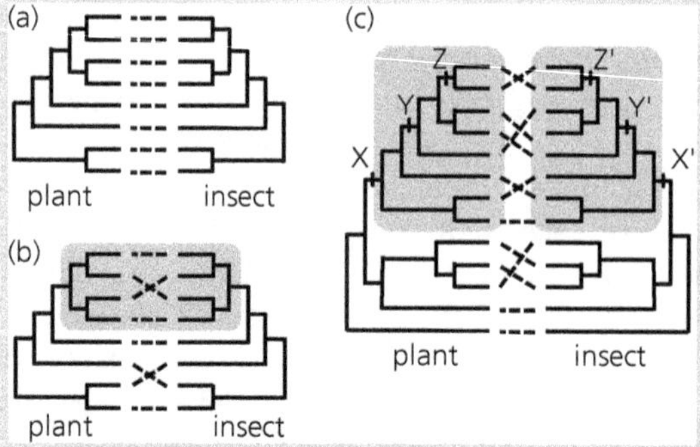

Figure 1 Examples of hypothetical phylogenetic patterns resulting from cospeciation analyses of interacting plant and insect lineages. Dotted lines connecting plants and insects indicate an interaction. a) Perfect tracking and significant concordance of the insect and plant speciation events suggests strong specialization or vertical transmission; however, there is no evidence of coevolution. b) General concordance of plant and insect phylogenies indicating cospeciation with a few host shifts. Clades highlighted by the box originated at the same time as major biogeographic events, suggesting that speciation was driven by vicariance and not coevolution. c) Discordant phylogenies showing substantial host shifting. Putatively coevolved clades are indicated by the boxes and coevolved traits are shown by the letters X/X′, Y/Y′, and Z/Z′. Coevolution is suggested by correlated shifts in traits between the plant and insect lineages. There is also enhanced speciation in the coevolved clades as compared to sister groups.

phylogenetic patterns. Together, these collective datasets allow one to ask whether coevolutionary selection has led to the formation of new species (e.g., Box 7.3, Figure 1c) or if non-coevolutionary explanations are a more likely cause of divergence (e.g., Box 7.3, Figure 1b). For systems in which there are data indicating reciprocal evolutionary change, modern comparative methods can take this analysis a step further to assess whether the coevolving traits are correlated with enhanced net speciation rates (e.g., Maddison et al. 2007).

An excellent illustration of the use of phylogenetics in coevolutionary studies is the recent work on the mutualism between the charismatic yucca, the Joshua tree, and its pollinating yucca moths. The Joshua tree occurs in the North American Mojave Desert, which was historically divided by the Bouse Embayment when rising sea level extended the Sea of Cortez northward. Both the Joshua tree (Lenz 2007) and pollinator moths (Pellmyr and Segraves 2003) consist of pairs of sister species, suggesting that coevolution may be causing speciation in both plants and insects. The key

Box 7.3 *Continued*

question to address is whether diversification was driven by coevolution or vicariance caused by historical changes in sea level.

Using phylogenetic approaches, Smith et al. (2008) showed that while Joshua trees seem to have diverged concordantly with the timing of the Bouse Embayment, moth divergence is significantly more recent—so recent, that simultaneous vicariant speciation of Joshua trees and moths is highly unlikely. Floral traits important to the pollination mutualism are significantly more divergent than vegetative traits which are independent of the interaction, and differences in floral traits are also correlated with shifts in moth morphology, a pattern consistent with pollinator-mediated selection on floral traits (Godsoe et al. 2008). Ancestral state reconstruction indicates that the pollinators diversified while interacting with the Joshua tree and not due to a host shift from another yucca species. Combined, these results suggest that coevolutionary changes in traits led to reproductive isolation and speciation in both plants and insects. Although the evidence points toward coevolution as a speciation mechanism, additional ecological work is required to assess alternative explanations. For example, since the Joshua tree appears to have diversified earlier than the moths, moth divergence may have been unilaterally driven by existing morphological differences in the plant rather than by reciprocal, coevolutionary change (Smith et al. 2008).

Clearly, there are a number of important factors to consider when using phylogenetics to study coevolution. First and foremost is a need for explicit, testable phylogenetic hypotheses that distinguish coevolution from alternative explanations. Like the Joshua tree example, these hypotheses will often be taxon specific and should be tailored for the study system. Second, under no circumstance should evidence of matching phylogenies be interpreted as coevolution. More often than not, concordant phylogenies are an indicator of a non-coevolutionary interpretation. Finally, phylogenetic studies need to be coupled with sound ecological work that shows reciprocal evolutionary change. The most powerful approach will combine clues from a diverse dataset to demonstrate a link between reproductive isolation and traits under coevolutionary selection.

References

Ehrlich, P.R., and Raven, P.H. (1964). Butterflies and plants: a study in coevolution. *Evolution*, **18**, 586–608.

Godsoe, W., Yoder, J.B., Smith, C.I., and Pellmyr, O. (2008). Coevolution and divergence in the Joshua tree/yucca moth mutualism. *American Naturalist*, **171**, 816–823.

Lenz, L.W. (2007). Reassessment of *Yucca brevifolia* and recognition of *Y. jaegariana* as a distinct species. Aliso, **24**, 97–104.

Maddison, W.P., Midford, P.E., and Otto, S.P. (2007). Estimating a binary character's effect on speciation and extinction. *Systematic Biology*, **56**, 701–710.

Nuismer, S.L., Gomulkiewicz, R., and Ridenhour, B.J. (2010). When is correlation coevolution? *American Naturalist*, **175**, 525–537.

Pellmyr, O., and Segraves, K.A. (2003). Pollinator divergence within an obligate mutualism: two yucca moth species (Lepidoptera; Prodoxidae: *Tegeticula*) on the Joshua tree (*Yucca brevifolia*; Agavaceae). *Annals of the Entomological Society of America*, **96**, 716–722.

Segraves, K.A. (2010). Branching out with coevolutionary trees. *Evolution: Education and Outreach*, **3**, 62–70.

Smith, C.I., Godsoe, W.K., Tank, S., Yoder, J.B., and Pellmyr, O. (2008). Distinguishing coevolution from covicariance in an obligate pollination mutualism: asynchronous divergence in Joshua tree and its pollinators. *Evolution*, **62**, 2676–2687.

Thompson, J.N. (1994). *The Coevolutionary Process*. Chicago, IL: University of Chicago Press.

Thompson, J.N. (2005). *The Geographic Mosaic of Coevolution*. Chicago, IL: University of Chicago Press.

Yoder, J.B., and Nuismer, S.L. (2010). When does coevolution promote diversification? *American Naturalist*, **176**, 802–817.

7.5 Conclusion

Many studies have excelled at showing patterns that could be attributed to coevolution (Thompson 2005). However, many of these patterns tend not to be mutually exclusive to the coevolutionary process. Because of this, it has been difficult to ascertain the relative importance of coevolution compared to other evolutionary processes in driving phenotypic change. Frequently, it has been easier to reason the importance of coevolution by arguing that certain interacting organisms are codependent and that they share long, intimate evolutionary histories. From there it is possible to ask questions about putatively coevolved relationships. For example, this chapter has highlighted some of the ways in which we expect the patterns of coevolving traits and phylogenies to differ under different interaction scenarios (e.g. the type or symmetry of the relationship). Some empirical tests of our predictions include phylogenetic data that have demonstrated that coevolution does not lead to a greater expectation of cospeciation than unilateral evolution. Similarly, mutualisms, antagonisms, and commensalisms all exhibit patterns of cospeciation. Recent studies also suggest that mutualisms and antagonisms can generate matching and mismatching of traits, although the jury is still out on whether the degree of matching/mismatching is strongly related to the kind of interaction, or whether it is related to other inherent qualities of the interacting species.

While theoreticians are hard at work formulating hypotheses for how the traits of mutualists coevolve, there remain few good empirical tests of these hypotheses. Furthermore, empirical studies on mutualisms have lagged behind work on antagonistic relationships. While it is not necessary to prove or disprove the importance of coevolution in every putative example, it is hard to gain a good understanding of coevolutionary mechanics without directly measuring the strengths and directions of selection gradients on interacting species pairs or communities of species. Although a few attempts have been made for antagonistic interactions (e.g. Benkman et al. 2003), even fewer are available for mutualisms, and measuring selection forces exerted by and received by multiple members of interaction webs is presently absent from the literature. This underlies the important question of how important coevolution is in the trait evolution of interacting communities versus interacting species pairs. While most coevolutionary models have been generated specifically for interacting pairs of species, it is evident that most mutualisms involve webs of interacting species (see Chapter 11). Future efforts need to be made in generating new models or adapting existing models on species-specific interactions so that they may be used to make predictions about how coevolution is expected to proceed in multispecific communities and what patterns we expect it to produce. So far, the expectation is that coevolution will play an important role in the evolution of traits in multispecific mutualisms (Nuismer et al. 2013).

Acknowledgments

I wish to thank U. Mueller, S. Nuismer, K. Segraves, and G. Weiblen for their very helpful comments on the manuscript. In addition, I would like to thank the authors of the three boxes, N. Whiteman, S. Nuismer, and K. Segraves, for their efforts in making this a much more well-rounded chapter. Finally I would like to thank J. Bronstein for her time and careful editing.

References

Alexandersson, R., and Johnson, S.D. (2002). Pollinator mediated selection on flower-tube length in a hawkmoth pollinated *Gladiolus* (Iridaceae). *Proceedings of the Royal Society B*, **269**, 631–636.

Althoff, D.M., Segraves, K.A., Smith, C.I., Leebens-Mack, J., and Pellmyr. O. (2012). Geographic isolation trumps coevolution as a driver of yucca and yucca moth diversification. *Molecular Phylogenetics and Evolution*, **62**, 898–906.

Anderson, B., Alexandersson, R., and Johnson, S.D. (2010a). Evolution and coexistence of pollination ecotypes in an African Gladiolus (Iridaceae). *Evolution*, **64**, 960–972.

Anderson, B., and Johnson, S.D. (2008). The geographical mosaic of coevolution in a plant-pollinator mutualism. *Evolution*, **62**, 220–225.

Anderson, B., and Johnson, S.D. (2009). Geographical covariation and local convergence of flower depth in a guild of fly-pollinated plants. *New Phytologist*, **182**, 533–540.

Anderson, B., Johnson, S.D., and Carbutt, C. (2005). Exploitation of a specialized mutualism by a deceptive orchid. *American Journal of Botany*, **92**, 1342–1349.

Anderson, B., Terblanche, J.S., and Ellis, A.G. (2010b). Predictable patterns of trait mismatches between interacting plants and insects. *BMC Evolutionary Biology*, **10**, 1471–2148.

Arditti, J., Elliott, J., Kitching, I.J., and Wasserthal, L.T. (2012). "Good Heavens what insect can suck it" – Charles Darwin, *Angraecum sesquipedale* and *Xanthopan morganii praedicta*. *Botanical Journal of the Linnean Society*, **169**, 403–432.

Benkman, C.W., Parchman, T.L., Favis, A., and Siepielski, A.M. (2003). Reciprocal selection causes a coevolutionary arms race between crossbills and lodgepole pine. *American Naturalist*, **162**, 182–194.

Brouat, C., Garcia, N., Andary, C., and McKey, D. (2001). Plant lock and key: Pairwise coevolution of an exclusion filter in an ant-plant mutualism. *Proceedings of the Royal Society of London B*, **268**, 2131–2141.

Bruyndonckx, N., Dubey, S., Ruedi, M., and Christe, P. (2009). Molecular cophylogenetic relationships between European bats and their ectoparasitic mites (Acari, Spinturnicidae). *Molecular Phylogenetics and Evolution*, **51**, 227–237.

Cruaud, A., Ronsted, N., Chantarasuwan B., et al. (2012). An extreme case of plant–insect codiversification: Figs and fig-pollinating wasps. *Systematic Biology*, **61**, 1029–1047.

Cuthill, J.H., and Charleston, M. (2012). Phylogenetic codivergence supports coevolution of mimetic *Heliconius* butterflies. *PLoS ONE*, **7**, e36464.

Darwin, C.R. (1862). *On the Various Contrivances By which British and Foreign Orchids Are Fertilized by Insects*, facsimile edn. London: John Murray.

Decaestecker, E., Gaba, E., Raeymaekers, J.A.M., Stoks, R., Van Kerckhoven, L., Ebert, D., and De Meester, L. (2007). Host–parasite "Red Queen" dynamics archived in pond sediment. *Nature*, **450**, 870–874.

Ehrlich, P.R., and Raven, P.H. (1964). Butterflies and plants: a study in coevolution. *Evolution*, **18**, 586–608.

Ellis, A.G., and Johnson, S.D. (2010). Gender differences in the effects of floral spur length manipulation on fitness in a hermaphrodite orchid. *International Journal of Plant Sciences*, **3**, 1010–1019.

Faegri, K., and van der Pijl, L. (1979). *The Principles of Pollination Ecology*, 3rd rev edn. Oxford: Pergamon Press.

Flor, H.H. (1942). Inheritance of pathogenicity in *Melampsora lini*. *Phytopathology*, **32**, 653–669.

Galetti, M., Guevara, R., Côrtes, M.C., et al. (2013). Functional extinction of birds drives rapid evolutionary changes in seed size. *Science*, **340**, 1086–1090.

Gandon, S., Capowiez, Y., Dubois, Y., Michalakis, Y., and Olivieri, I. (1996). Local adaptation and gene-for-gene coevolution in a metapopulation model. *Proceedings of the Royal Society of London B*, **263**, 1003–1009.

Geerts, S., and Pauw A. (2009). African sunbirds hover to pollinate an invasive hummingbird-pollinated plant. *Oikos*, **118**, 573–579.

Geffeney, S.L., Fujimoto, E., Brodie, E.D. III, Brodie, E.D. Jr., and Ruben, P.C. (2005). Evolutionary diversification of TTX-resistant sodium channels in a predator-prey interaction. *Nature*, **434**, 759–763.

Grant, P.R., and Grant, R.B. (2006). Evolution of character displacement in Darwin's finches. *Science*, **313**, 224–226.

Guimarães, P.R., Jordano, P., and Thompson, J.R. (2011). Evolution and coevolution in mutualistic networks. *Ecology Letters*, **14**, 877–885.

Hafner, M.S., and Nadler, S.A. (1988). Phylogenetic trees support the coevolution of parasites and their hosts. *Nature*, **332**, 258–259.

Hanifin, C.T., Brodie, E.D. Jr., and Brodie, E.D. II. (2008). Phenotypic mismatches reveal escape from arms-race coevolution. *PLoS Biology*, **6**, 471–482.

Howe, H.F. (1984). Constraints on the evolution of mutualisms. *American Naturalist*, **123**, 764–777.

Janzen, D.H. (1980). When is it coevolution? *Evolution*, **34**, 611–612.

Janzen, D.H. (1981). Patterns of herbivory in a tropical deciduous forest. *Biotropica*, **13**, 271–282.

Johnson, S.D. (2010). The pollination niche and its role in the diversification and maintenance of the southern African flora. *Philosophical Transactions of the Royal Society*, **365**, 499–516.

Johnson, S.D., Hargreaves, A.L., and Brown, M. (2006). Dark, bitter-tasting nectar functions as a filter of flower visitors in a bird-pollinated plant. *Ecology*, **87**, 2709–2716.

Koskella, B., and Lively, C.M. (2007). Advice of the rose: experimental coevolution of a trematode parasite and its snail host. *Evolution*, **61**, 152–159.

Lodwig, E.M., Hosie, A.H.F., Bourdès, A., et al. (2003). Amino-acid cycling drives nitrogen fixation in the legume-Rhizobium symbiosis. *Nature*, **422**, 722–726.

Mikheyev, A.S., Mueller, U.G., and Abbott, P. (2010). Comparative dating of attine ant and lepiotaceous cultivar phylogenies reveals coevolutionary synchrony and discord. *American Naturalist*, **175**, E126–E133.

Miklasheviches, E., Röhrig, H., Schell, J., and Schmidt, J. (2001). Perception and signal transduction of rhizobial NOD factors. *Critical Reviews in Plant Sciences*, **20**, 373–394.

Mode, C.J. (1958). A mathematical model for the coevolution of obligate parasites and their hosts. *Evolution*, **12**, 158–165.

Moe, A.M., and Weiblen, G.D. (2012). Pollinator-mediated reproductive isolation among dioecious fig species (Ficus, Moraceae). *Evolution*, **66**, 3710–3721.

Muchhala, N., and Thomson, J.D. (2009). Going to great lengths: selection for long corolla tubes in an extremely specialized bat-flower mutualism. *Proceedings of the Royal Society B*, **276**, 2147–2152.

Nattero, J., and Cocucci, A.A. (2007). Geographical variation in floral traits of the tree tobacco in relation to its hummingbird pollinator fauna. *Biological Journal of the Linnean Society*, **90**, 657–667.

Newman, E., Manning, J., and Anderson, B. (2014). Fine scale syndrome dynamics within a guild: geographic trait matching between long proboscid flies and the flowers that they pollinate. *Annals of Botany*, **113**, 373–384.

Nilsson, L.A. (1988). The evolution of flowers with deep corolla tubes. *Nature*, **334**, 147–149.

Nishiguchi, M.K., Ruby, E.G., and McFall-Ngai, M.J. (1998). Competitive dominance among strains of luminous bacteria provides an unusual form of evidence for parallel evolution in sepiolid squid-vibrio symbioses. *Applied and Environmental Microbiology*, **64**, 3209–3213.

Nuismer, S.L., Gomulkiewicz, R., and Ridenhour, B.J. (2010). When is correlation coevolution? *American Naturalist*, **175**, 525–537.

Nuismer, S.L., Jordano, P., and Bascompte, J. (2013). Coevolution and the architecture of mutualistic networks. *Evolution*, **67**, 338–354.

Nuismer, S.L., Thompson, J.N., and Gomulkiewicz, R. (1999). Gene flow and geographically structured coevolution. *Proceedings of the Royal Society of London B*, **266**, 605–609.

Pauw, A., Stofberg, J., and Waterman, R.J. (2009). Flies and flowers in Darwin's race. *Evolution*, **63**, 268–279.

Pellmyr, O., and Thompson, J.N. (1996). Sources of variation in pollinator contribution within a guild: the effects of plant and pollinator factors. *Oecologia*, **107**, 595–604.

Rich, K.A., Thompson, J.N., and Fernandez, C.C. (2008) Diverse historical processes shape deep phylogeographic divergence in a pollinating seed parasite. *Molecular Ecology*, **17**, 2430–2448.

Schemske, D.W., and Horwitz, C.C. (1984). Variation among floral visitors in pollination ability: a precondition for mutualism specialization. *Science*, **225**, 519–521.

Smith, C. I., Godsoe, W.K.W., Tank, S., Yoder, J.B., and Pellmyr, O. (2008). Distinguishing coevolution from coviariance in an obligate pollination mutualism: asynchronous divergence in Joshua tree and its pollinators. *Evolution*, **62**, 2676–2687.

Stebbins, L. (1970). Adaptive radiation of reproductive characteristics in angiosperms. I. Pollination mechanisms. *Annual Review of Ecology and Systematics*, **1**, 307–326.

Steiner, K.E., and Whitehead, V.B. (1991). Oil flowers and oil bees: further evidence for pollinator adaptation. *Evolution*, **45**, 1493–1501.

Temeles, E.J., and Kress, W.J. (2003). Adaptation in a hummingbird association. *Science*, **300**, 630–633.

Temeles, E.J., Pan, I.L., Brennan, J.L., and Hotwitt, J.N. (2000). Evidence for ecological causation of sexual dimorphism in a hummingbird. *Science*, **289**, 441–443.

Thompson, J.N, (1989). Concepts of coevolution. *Trends in Ecology and Evolution*, **4**, 179–183.

Thompson, J.N. (1994). *The Coevolutionary Process*. Chicago: The University of Chicago Press.

Thompson, J.N. (1997). Evaluating the dynamics of coevolution among geographically structured populations. *Ecology*, **78**, 1619–1623.

Thompson, J.N. (1999). Specific hypotheses on the geographic mosaic of coevolution. *American Naturalist*, **153**, S1–S14.

Thompson, J.N. (2005). *The Geographic Mosaic of Coevolution*. Chicago: The University of Chicago Press.

Thompson, J.N. (2013). *Relentless Evolution*. Chicago: The University of Chicago Press.

Thompson, J.N., and Cunningham, B.M. (2002). Geographic structure and dynamics of coevolutionary selection. *Nature*, **417**, 735–738.

Thompson, J.N., and Pellmyr, O. (1992). Mutualism with pollinating seed parasites amid co-pollinators: constraints on specialization. *Ecology*, **73**, 1780–1791.

Thompson, J.N., Schwind, C., Guimarães, G.B Jr., and Friberg M (2013). Diversification through multitrait evolution in a coevolving interaction. *Proceedings of the National Academy of Sciences*, **110**, 11487–11492.

Thrall, P.H., and Burdon, J.J. (2003). Evolution of virulence in a plant host-pathogen metapopulation. *Science*, **299**, 1735–1737.

Thrall, P.H., Laine, A-L., Ravensdale, M., Nemri, A., Dodds, P.N., Barrett, L.G., and Burdon, J.J. (2012). Rapid genetic change underpins antagonistic coevolution in a natural host-pathogen metapopulation. *Ecology Letters*, **15**, 425–435.

Toju, H., Abe, H., Ueno, S., Miyazawa, Y., Taniguchi, F., Sota, T., and Yahara, T. (2011). Gradients of arms race coevolution. *American Naturalist*, **177**, 562–573.

Toju, H., and Sota, T. (2006). Imbalance of predator and prey armament: Geographic clines in phenotypic interface and natural selection. *American Naturalist*, **167**, 105–117.

van der Pijl, L. (1982). *Principles of Dispersal in Higher Plants*. New York: Springer-Verlag.

van Noort, S., and Compton, S.G. (1996). Convergent evolution of agaonine and sycoecine (Agaonidae, Chalcidoidea) head shape in response to the constraints of host fig morphology. *Journal of Biogeogr.*, **23**, 415–424.

Wasserthal, L.T. (1997). The pollinators of the Malagasy star orchids *Angraecum sesquipedale, A. sororium*, and *A. compactum* and the evolution of extremely long spurs by pollinator shift. *Botanica Acta*, **110**, 343–359.

Weeks, A.R., Turelli, M., Harcombe, W.R., Reynolds, T., and Hoffmann, A.A. (2007). From parasite to mutualist: rapid evolution of *Wolbachia* in natural populations of *Drosophila*. *PloS Biology*, **5**, e114.

Weiblen, G.D. (2004). Correlated evolution in fig pollination. *Systematic Biology*, **53**, 128–139.

Weiblen, G.D., and Bush, G.L. (2002). Speciation in fig pollinators and parasites. *Molecular Ecology*, **11**, 1573–1578.

Yoder, J.B., and Nuismer, S.L. (2010). When does coevolution promote diversification? *American Naturalist*, **176**, 802–817.

SECTION 3

Introduction to Section 3

Judith L. Bronstein

Human awareness of mutualisms long predates their scientific study. In part this is related to their importance in agriculture, but it also reflects our simple curiosity about nature. As discussed in Chapter 1, a wealth of detail on particular mutualisms—many of them featured in the boxed highlights in this volume (see Table 1.1)—had already accumulated by the time ecology began to crystallize as a science. Although conceptual frameworks for the population, community, and ecosystem ecology of mutualism have taken quite some time to emerge, they have been solidly grounded in natural history from the start.

Population ecology emerged in the early twentieth century in an effort to understand fluctuations in animal population sizes, particularly species whose control had economic implications. Perhaps as a consequence, an interest in the population dynamics of mutualism was slow to develop and biologically unrealistic when it did. However, as J. Nathaniel Holland makes clear in Chapter 8, the field is now thriving. Theoretical development is proceeding rapidly, due in part, as Holland stresses, to increasing recognition that models of consumer/resource dynamics and functional responses are exceptionally useful for understanding (most) mutualisms. This field will continue to develop as we incorporate yet more theoretical approaches and concepts emerging from the study of antagonism, including metapopulation dynamics (Box 8.1, by Anna-Liisa Laine), phase-plane analysis (Box 8.2, by Kevin McCann), and interaction strengths (Box 8.3, by Tim Wootton). Population modeling has clearly been applied with great success to understanding many other aspects of mutualism. Examples highlighted in this volume include how mutualists and exploiters coexist (Box 9.3 by Charlotte Lee, Brian Inouye, and Tom Miller), how diversity can be maintained in mutualistic assemblages (Box 9.2 by Daniel Doak), and how resilient mutualisms might be in the face of anthropogenic change (Box 13.3 by Anthony Ives).

Perhaps surprisingly, empirical data have not been central in research on the population ecology of mutualism. The same cannot be said at the community level. Here, there is clearly a wealth of information on mutualistic communities to draw upon, as one can see, for example, in Box 9.1 on the coral–*Symbiodinium* symbiosis by Michael Stat. As in community ecology as a whole, the problem instead has always been a temptation to interpret the underlying causes of well-documented patterns a little too freely. Chapter 9, by Todd Palmer, Elizabeth Pringle, Adrien Stier, and Robert Holt, shows how far the study of mutualism's community ecology has progressed in the past twenty years or so, as well as how it can be enriched with ideas emerging from evolutionary biology. The field is crystallizing around central questions such as these: what are the causes and consequences of diversity within mutualistic assemblages? How do mutualisms and antagonisms interact to generate and potentially to stabilize larger-scale patterns? To what extent are range limits determined by mutualist distributions in space? Answers from model systems are building up the information necessary to predict patterns in communities we have not yet studied, as well as to recognize communities that may be most threatened by anthropogenic disruption of mutualism (Chapter 13).

One of the clearest and most significant patterns to emerge from community studies in the past twenty years is that mutualisms are not always mutualistic. In different community settings, mutualisms can swing between strong and weak, but also from mutualistic to commensal and even to antagonistic. As discussed in Chapter 9 and treated in depth by Jason Hoeksema and Emilio Bruna in Chapter 10, context-dependent outcomes of mutualisms have ecological and evolutionary effects at scales ranging from the individual to the ecosystem. For example, outcomes of defensive endosymbioses vary with a diversity of biotic and abiotic variables, with consequences for the intensity of pathogen attack (Box 10.1 by John Jaenicke); temperature-dependent success of corals and their symbionts are key to understanding when and where these mutualisms—which form the "backbone" of coral reef ecosystems worldwide—will break down (Box 10.3 by Tamar Goulet). Context dependency has been studied very unevenly across mutualisms; for example, in Box 10.2 Silvia Lomáscolo highlights context-dependent dynamics of seed dispersal mutualisms. Hoeksema and Bruna outline eight critical unanswered questions about where and when context dependency will have the strongest effects on mutualisms. Context dependency is likely a widespread phenomenon in non-mutualistic interactions as well. Although our understanding is clearly incomplete, studies of mutualisms are nevertheless at the forefront in this rapidly developing field.

As in other subfields of mutualism ecology and evolution, new tools are emerging with which to study community-scale phenomena. Some are illuminating patterns in nature that were not previously recognized. This is exceptionally clear in the rapidly emerging field of network dynamics. Emerging out of food web theory, network analysis has been applied with resounding success to mutualism, as Chapter 11 by Jordi Bascompte and Jens Olesen documents. This approach has been particularly informative with regard to questions highlighted in Chapter 9 about patterns of partner diversity within mutualisms. While there are undoubtedly *mutualists* that are generalists or specialists, network analysis is making clear that we can no longer conceive of *mutualisms* as being generalized or specialized: the level of specialization between partners is generally asymmetrical, with major consequences for both the ecology and the evolution of mutualism. A new community-level unit, the interaction module (given a theoretical basis in Box 11.2 by Roger Guimará), can now be recognized.

It is critical to look beyond the currently popular community scale, to ecosystems as a whole. Certain mutualisms have been well studied in ecosystem contexts already, as Nancy Johnson summarizes in Chapter 12. In marine environments, the coral–*Symbiodinium* symbiosis is the critical, threatened engineer of coral reef ecosystems (see Boxes 9.1, 10.3, and 13.2). The premier terrestrial examples involve soil microbial communities, reviewed by Thomas Whitham and Louis Lamit in Box 12.2 Among these are belowground mutualistic symbioses involving nitrogen-fixing bacteria (see also Box 4.2 by Ellen Simms and Stephanie Porter, and Box 5.2 by Maren Friesen) and mycorrhizal fungi (highlighted by Jonathan Leake in Box 12.1). Less well known are biological soil crust communities, discussed in Box 12.3 by Buck Sanford. Indeed, mutualisms are so essential to atmospheric and soil composition and to plant function that, Johnson argues, one could rightly say that mutualisms structure the world. They certainly are critical in agriculture, as she documents, and thus to human population growth. Once one recognizes their central roles in the provision of critical ecosystem services, the importance of managing mutualisms wisely, conserving them, and restoring them once they have been disrupted becomes startlingly clear. These themes are elaborated upon in Section 4 of this volume.

CHAPTER 8

Population ecology of mutualism

J. Nathaniel Holland

8.1 Population ecology and interspecific interactions

A population is a group of individual organisms of the same species living together in an area of sufficient size that population dynamics—changes in population size over time—are largely driven by births and deaths (Turchin 2003). Population ecology aims to understand the relationships of populations with individuals of the same population (intraspecific interactions), with individuals of populations of different species (interspecific interactions), and with the abiotic environment. Population dynamics arise from these intraspecific, interspecific, and environmental influences on demographic processes of birth (*B*), death (*D*), immigration (*I*), and emigration (*E*). A change in population size or density (*N*) at one point in time (*t*) over some time interval (*t* + 1) is expressed in the simplest form as,

$$N_{t+1} = N_t + B - D + I - E. \qquad (8.1)$$

Births and immigration increase population size, while deaths and emigration decrease population size. Groups of organisms living together in a smaller area are local populations, whose metapopulation dynamics are shaped mostly by dispersal and migration (see Box 8.1 for an explanation of metapopulation dynamics of mutualism). The basic principles of population ecology apply to organisms broadly, whether they be a predator, mutualist, competitor, or otherwise. For this reason, we first present some basic principles of population ecology, which should serve as a foundation for inquiry into the population ecology of any particular species interaction, mutualism included.

The study of population dynamics is concerned with the underlying causes and consequences of the short- and long-term changes in population size, which tend to fall into two broad categories, density-independent and density-dependent population dynamics. Density-independent population dynamics occur when population growth shows no relationship with population size. A key example of density-independent population dynamics is the Malthusian law of exponential population growth, which states that a population increases or decreases exponentially as long as all individuals experience the same constant biotic and abiotic environment (Turchin 2003). Exponential growth is formalized by assuming that a population is in a sufficient area that its size is determined by births and deaths. Equation (8.1) can be expressed as a differential equation for changes in population size (*dN*) over time (*dt*):

$$\frac{dN}{dt} = bN - dN = (b - d)N = rN, \qquad (8.2)$$

where *b* and *d* are the number of births or deaths per individual per unit of time, and for which *bN* = *B*, *dN* = *D*, and *r* is the intrinsic rate of population growth. Population density (*N*) through time follows a J-shaped curve, whereby a population's size increases exponentially. The per capita population growth rate (*dN/Ndt*) is invariant with density (*N*), having a y-intercept of *r* and a slope

Mutualism. Edited by Judith L. Bronstein.

Box 8.1 Metapopulation ecology and mutualism

Anna-Liisa Laine

A metapopulation is a group of local populations of the same species that are discrete entities in space, interacting with one another through the movement (emigration, immigration) of individuals and genes among the local habitat patches that occur within some regional area. Like birth and death rates shaping the dynamics of populations, extinction and (re-)colonization rates of local habitat patches govern the dynamics of metapopulations. Levins' (1970) classic model of metapopulation dynamics (Box 8.1 Figure 1a) assumes that (1) local populations (habitat patches) are of equal size and isolation, (2) local populations of the metapopulation have independent (uncorrelated) dynamics, and (3) individual movement among local populations is sufficiently low that local dynamics of extant populations occur on a fast time scale with no great effect compared with the dynamics of the metapopulation as a whole. More contemporary source–sink models (Box 8.1 Figure 1b) take into account local habitat-dependent demography, in which sink populations have a negative growth rate and are maintained by immigrants from source populations with a positive growth rate (Hanski 1999). More spatially realistic metapopulation theory (Box 8.1 Figure 1c) now identifies how the area and isolation of local habitat patches influences their extinction and colonization rates, which, in consideration of the influence of landscape structures on these probabilities, has enabled the application of metapopulation theory to the design of reserves in conservation biology (Hanski and Gaggiotti 2004).

Empirical studies have shown metapopulation dynamics with a frequent turnover of local populations in systems with no large and permanent source population (Hanski and Gaggiotti 2004). Moreover, any given species' population is embedded in a community of interacting species sharing the same network of suitable habitat. Such interspecific interactions can affect rates of migration, extinction, and colonization, thereby modulating metapopulation dynamics of the interacting species. Importantly, metapopulation dynamics interacting species can create and maintain spatially aggregated distributions of species even in the absence of environmental heterogeneity (Hanski 1999).

Indeed, spatial ecology and metapopulation processes extend beyond the population to the community, shaping the structure and dynamics of interspecific interactions (Hanski 1999). While we are only just beginning to consider the ecology and evolution of mutualism in the context of metapopulation dynamics, mutualistic interactions are inherently

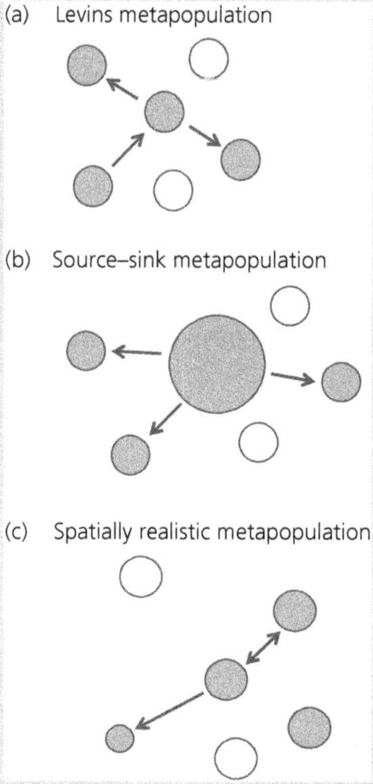

Figure 1 Conceptual depictions of metapopulations in which white circles are unoccupied habitat patches, gray are occupied patches, and arrows indicate movement between patches. (a) In the classic Levins metapopulation, colonization and extinction rates are the same for all patches, while (b) source–sink metapopulations are defined by asymmetry in movement from source to sink patches. (c) In spatially realistic metapopulations, the dynamics depend on the size and location of the habitat patches.

tied to processes of metapopulation dynamics. For example, mutualisms may be the defining feature of suitable habitat, and its quality, for a wide range of species, such as mycorrhizal fungi and nitrogen-fixing bacteria that are needed by most vascular plants. The dispersal services of mutualistic interactions link local populations to one another, as with frugivorous birds dispersing seeds or pollinators required for sexual reproduction of plants (Herre et al. 1999). Mutualisms may well be the key to the dispersal and migration and hence the metapopulation dynamics of many species.

Recent theoretical and empirical advances have demonstrated that metapopulation dynamics can have profound consequences for the stability and dynamics of mutualism

Box 8.1 *Continued*

(Saikkonen et al. 2002, Amarasekare 2004). In mutualistic interactions characterized by asymmetry in mobility of the interaction partners, the dispersing mutualist can rescue sink communities from extinction, provided there is at least one source community in which both species have abundances above the extinction threshold. If the benefit of the rescue effect to sink populations is outweighed by the cost of dispersal to sources, mutualistic communities linked by dispersal may experience a greater loss of diversity than communities that are isolated (Amarasekare 2004). In general, the consequences of immigration and emigration for the dynamics of mutualists depend strongly on which demographic rate is influenced by mutualism. Empirical studies suggest that mutualistic interactions in fragmented landscapes are likely to suffer from Allee effects. Plant species relying on animal vectors for pollination and seed dispersal suffer a reduction in reproductive success at low abundances because small patch sizes and increased isolation limit pollen transfer and prevent seed dispersal (Aguilar et al. 2008). Finally, though not expounded upon here, metapopulation dynamics of mutualism can have fundamental influences on eco-evolutionary trajectories of species and their community interactions through dispersal behavior, genetic structure, and evolutionary dynamics (Parker 1999, Yamamura et al. 2004). As with the study of metapopulation ecology as a whole, future studies of mutualism and metapopulation dynamics should include processes of migration, gene flow, spatially correlated dynamics, local extinction, genetic drift, local adaptation, and so forth (Hanski and Gaggiotti 2004).

References

Aguilar, R., Quesada, M., Ashworth, L., Herrerias-Diego, Y., and Lobo, J. (2008). Genetic consequences of habitat fragmentation in plant populations: susceptible signals in plant traits and methodological approaches. *Molecular Ecology*, **17**, 5177–5188.

Amarasekare, P. (2004). Spatial dynamics of mutualistic interactions. *Journal of Animal Ecology*, **73**, 128–142.

Hanski, I. (1999). *Metapopulation Ecology*. Oxford: Oxford University Press.

Hanski, I., and Gaggiotti, O.E. (2004). Metapopulation biology: past, present and future. In I. Hanski and O.E. Gaggiotti (eds.), *Ecology, Genetics and Evolution of Metapopulations*, pp. 3–22. San Diego: Elsevier.

Herre, E.A., Knowlton, N., Mueller, U.G., and Rehner, S.A. (1999). The evolution of mutualisms: exploring the paths between conflict and cooperation. *Trends in Ecology and Evolution*, **14**, 49–53.

Levins, R. (1970). Extinction. *Lecture Notes in Mathematics*, **2**, 75–107.

Parker, M.A. (1999). Mutualism in metapopulations of legumes and rhizobia. *American Naturalist*, **153**, S48–S60.

Yamamura, N., Higashi, M., Behera, N., and Wakano, J.Y. (2004). Evolution of mutualism through spatial effects. *Journal of Theoretical Biology*, **226**, 421–428.

of zero. Exponential growth is a key source of instability for populations, as they grow to unbounded sizes or decline to extinction, with positive or negative growth rates (r), respectively.

The instability of exponential population growth arises from density independence. It is certainly feasible for density-independent factors such as climatic and environmental events (e.g., hurricanes, floods, tornadoes) to suppress populations and affect their fluctuations. Yet, populations routinely experience some form of density dependence in which demographic rates vary with density, resulting in functional relationships between per capita population growth rates (dN/Ndt) and density (N). Birth (b) and death (d) rates typically decrease and increase, respectively, with increases in population density (N), so that dN/Ndt decreases with N, but under some circumstances dN/Ndt can increase with N (i.e., Allee effects; Morales et al. 2007). The instability of density-independent exponential population growth is resolved with negative density dependence of intraspecific self-limitation, which can be formalized with two similar models of logistic population growth:

$$\frac{dN}{dt} = rN\left(\frac{K-N}{K}\right) \text{ or } \frac{dN}{dt} = rN - dN^2. \qquad (8.3)$$

K is the carrying capacity of the population and d sets the death rate of self-limitation. With either formulation of logistic population growth, density through time follows an S-shaped curve, whereby populations reach a maximum density.

The per-capita population growth rate (dN/Ndt) is a negative function of density (N), having a y-intercept of r and a slope of $-r/K$ or $-d$. Self-limitation produces negative density dependence, whether due to overcrowding, space constraints, or limiting resources. See Case (2000) for further analyses of exponential and logistic growth, along with other basic principles of population ecology.

In addition to self-limitation, another key source of density dependence for populations is their interactions with other species. Indeed, populations do not occur in isolation of other species. Interspecific interactions occur when the actions, traits, or density of one population result in a positive (+) or negative (−) change in an attribute of another species' population, such as demographic rates (B, D, I, E), population growth, density, fitness, or trait values (Abrams 1987). In this way, species interactions are differentiated into discrete forms given their pairwise interaction outcomes, including predation (+ −), competition (− −), and mutualism (+ +). Species interactions are central to density-dependent population dynamics, as variation in one species' density can alter the birth and death rates of the species with which it interacts. Such density dependence can shape the short-term transient dynamics and long-term stability properties of interacting populations, including monotonic approach of density toward an equilibrium, damped oscillations in which fluctuations in density decrease as an equilibrium is approached, and stable limit cycles in which density exhibits repeating oscillations (Box 8.2 describes basic population dynamic theory that gives rise to these stability properties).

Great progress has been made in understanding the density-dependent population dynamics of two particular forms of interspecific interaction, namely predation and competition. Our understanding of them progressed through the early establishment of Lotka–Volterra (L–V) theory, followed by the development of mechanistic models and decades of interplay between theory and empiricism. Even though L–V models of predation and competition were largely phenomenological, they still provided direction for developing mechanistic theory with density dependence (Rosenzweig and MacArthur 1963, Leon and Tumpson 1975, Hsu et al. 1978). See Case (2000) for analyses of L–V and other models of predation and competition.

In contrast, mutualism has lagged as a third pillar in the study of the population ecology of species interactions. This is in part because initial theoretical models suggested they were unstable due to unbounded population growth arising from never-ending positive feedback in population densities (Gause and Witt 1935, May 1976). Plenty of models since then have included various intrinsic and extrinsic factors that can lead to stability, though empirical studies of them are surprisingly limited. This chapter summarizes shortcomings of historical models that stifled theory for decades, reviews the new generation of population dynamic theory for mutualism, synthesizes empirical studies of the density dependence of mutualism, and proposes future research directions. In doing so, contemporary advances in theory are emphasized, including mechanisms for the density dependence of mutualism for which research at the interface of theory and empiricism will be critical to future progress. The focus in this chapter is on direct mutualistic interactions, rather than indirect mutualisms in which the effects of the two interacting species on one another are mediated entirely by the density or traits of a third species (see Chapter 9). Also, this chapter focuses on the population and ecological dynamics of mutualism, and as such does not cover issues on the evolutionary dynamics of mutualism highlighted in previous chapters.

8.2 Historic Lotka–Volterra theory

A prominent shortcoming of the theory for the population dynamics of mutualism is the Lotka–Volterra (L–V) model (May 1976, Holland et al. 2006). Instead of providing a foundation for more mechanistic models, the L–V model of mutualism has stifled theory. The L–V model has been reinvented time and again, and studies continue to build on it (e.g., Ringel et al. 1996, Bascompte et al. 2006, Iwata et al. 2011). Largely due to biologically inaccurate results of the L–V model, mutualism continues to be perceived as an interesting interaction, but one less important to population dynamics than predation or competition (Turchin 2003). To move beyond the

L–V model, it is important to understand its historical development and deficiencies.

Initial theory for the population dynamics of mutualism was developed by Gause and Witt (1935) nearly a decade after the Lotka–Volterra theory for predation and competition. As the negative signs of L–V competition equations were replaced with positive signs to denote the benefits of mutualism, the following set of equations emerged, now referred to as the L–V model of mutualism:

$$\frac{dN_1}{dt} = r_1 N_1 \left(\frac{K_1 - \alpha_{11} N_1 + \alpha_{12} N_2}{K_1} \right) \text{ and}$$
$$\frac{dN_2}{dt} = r_2 N_2 \left(\frac{K_2 - \alpha_{22} N_2 + \alpha_{21} N_1}{K_2} \right) \tag{8.4}$$

N_i is the population density of species i, r_i, and K_i are the intrinsic growth rate and carrying capacity of species i [cf. equation (8.3)], and α_{ii} is the self-limitation or intraspecific competition parameter of N_i, which is often set to unity so that the parameter can be omitted (Vandermeer and Boucher 1978, Goh 1979). Mutualism is represented by + $\alpha_{ij}N_j$, where α_{ij} is the mutualistic effect of N_j on the population growth of N_i, or its interaction strength (a concept presented and discussed in Box 8.3). If $\alpha_{ij}N_j$ is set to zero, then equations (8.4) reduce to (8.3) and populations grow logistically.

We can examine the behavior and outcome of these dynamic equations using phase-plane analyses (Box 8.2 describes the basic theory of phase-plane analyses). Setting equations (8.4) equal to zero (with α_{ii} = 1.0) and solving for N_1 and N_2, we obtain the zero population growth isoclines, $dN_i/dt = 0$, for the two mutualistic species:

$$N_1 = K_1 + \alpha_{12} N_2 \text{ and } N_2 = K_2 + \alpha_{21} N_1. \tag{8.5}$$

These are linear lines with intercepts of K_i and positive slopes of α_{ij}. Figure 8.1 plots two key scenarios of the isoclines of equations (8.5) in (N_1, N_2)-coordinate phase-plane space, for which more detailed analyses occur elsewhere (Vandermeer and Boucher 1978, Goh 1979). For the L–V model here, the density of N_i decreases when above and increases when below the zero population growth isocline of N_i. In addition to the origin ($N_1 = N_2 = 0$), equilibria occur where the zero population growth isoclines intersect one another or the N_1- and N_2-axes. As α_{ij} is the slope of the linear isocline, increasing or decreasing α_{12} and α_{21} contributes to whether the two isoclines intersect and form an equilibrium, though the actual community matrix analysis is a bit more complicated (Box 8.2; Goh 1979, Case 2000).

Three main results emerge from phase-plane analyses of the L–V model for mutualism. First, if the product of interaction strengths of the mutualistic species is greater than unity ($\alpha_{12}\alpha_{21} > 1$) or the product of intraspecific self-limitation ($\alpha_{12}\alpha_{21} > \alpha_{11}\alpha_{22}$), then the zero-growth isoclines of N_1 and N_2 do not intersect (Figure 8.1a). Consequently, an equilibrium does not occur for positive densities of both N_1 and N_2, but instead unbounded population growth ensues in what May (1976) called an "orgy of mutual benefication," an empirically infeasible result akin to never-ending exponential population growth [equation (8.2)]. The two equilibria on the axes of N_1 and N_2 are saddle points of K_1 and K_2, the density of each species absent of interactions with the other. Introducing individuals of the other species leads dynamics away from these saddle points, and populations grow unbounded.

Second, if $\alpha_{12}\alpha_{21} < 1$ or $\alpha_{12}\alpha_{21} < \alpha_{11}\alpha_{22}$, then the zero-growth isoclines of N_1 and N_2 do intersect, forming a stable equilibrium of mutualistic coexistence (Figure 8.1b). Densities of N_1 and N_2 at this equilibrium are greater than the K_1 and K_2 saddle points when interactions do not occur, reflecting the principle that mutualism benefits interacting species by increasing their population densities (see Chapter 1). However, the stability of the equilibrium arises from the negative density dependence of self-limitation overriding the positive feedback of the benefits of mutualism, as weak or asymmetric interaction strengths are required for $\alpha_{12}\alpha_{21} < 1$ or $\alpha_{12}\alpha_{21} < \alpha_{11}\alpha_{22}$. While weak interaction strengths of L–V models can prevent unbounded growth, they suggest that mutualism itself has little effect on population dynamics. Yet, mutualisms do contribute to the dynamics of populations in nature (Addicott 1979, Morales 2000).

Third, though not well recognized (cf. Kot 2001), L–V theory suggests that mutualistic interactions lack population oscillations, as populations either grow unbounded or asymptotically approach an equilibrium (Figure 8.1). L–V models predict that

Box 8.2 Population dynamics and phase-plane analysis of interspecific interactions

Kevin McCann

Population dynamics arising from interspecific interactions include both short-term transient behaviors and long-term stability properties. Long-term stability properties include equilibrium and non-equilibrium steady states in population size or density, N_i, whereas short-term transient behavior refers to the time series of N_i dynamics leading to the long-term steady states. While population dynamic models of interacting species can appear overwhelming, they produce a surprisingly small set of dynamic behaviors and outcomes. In particular, three types of qualitative behaviors occur in population dynamic models. First, a long-term stable equilibrium of an interacting population, N_i, may be attained through transient behavior characterized by monotonic dynamics, whereby the equilibrium steady state is approached asymptotically (Box 8.2 Figure 1a). Second, a long-term stable equilibrium may be reached through transient behavior described by oscillatory decays, whereby the equilibrium steady state is approached through damped oscillations (Box 8.2 Figure 1b). While their transient dynamics differ, these are the only two forms of stable population dynamics whereby a steady-state equilibrium is reached in which all rates of change in N_i are zero. Third, a non-equilibrium steady state may occur in which the average rate of change in N_i over time is zero. Population dynamics associated with such an unstable equilibrium can be characterized by periodic limit cycles or complex oscillations in N_i. For example, a population can depart from an unstable equilibrium and move to stable non-equilibrium dynamics described by limit cycles (Box 8.2 Figure 1c), in which the population's trajectory repeats itself through time. Non-equilibrium steady states can also be characterized by complex oscillations in which the population's trajectory does not repeat itself, but has almost repeatable structure, such as chaos and quasi-periodic population dynamics.

The theoretical framework for studying such population dynamic behaviors is through phase-plane analysis (Case 2000, Edelstein-Keshet 2005, Henson 2012), a quantitative technique for the graphical and visual assessment of the dynamics of two state variables in a second-order system of equations such as $dN_1/dt = f(N_1,N_2)$ and $dN_2/dt = f(N_1,N_2)$. In this case the abundance or density of two interacting species, N_1 and N_2, are the state variables that vary dynamically with one another. In turn, the phase-plane portrait is defined by the two state variables with a coordinate system of N_1 and N_2. Each dynamic equation has a zero population growth isocline, the combination of values of the state variables (N_1 and N_2) that yield solutions of zero for the dynamical equation. For example, those values of N_1 and N_2 that give $dN_1/dt = 0$ yield N_1's isocline. Phase-plane analysis is a mathematical technique that reveals relationships between the underlying geometry of the zero population

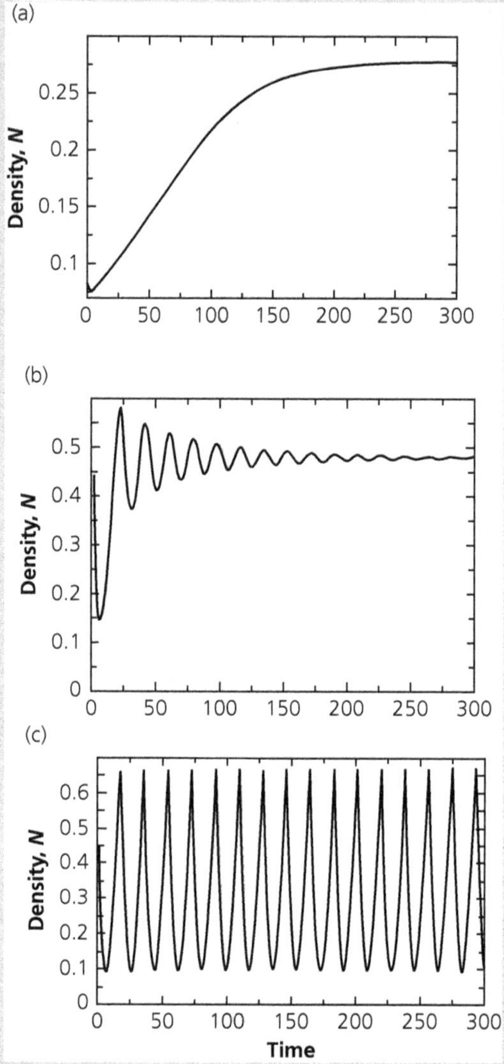

Figure 1 Three types of qualitative behaviors in the dynamics of population density, N, through time, namely
(a) monotonic dynamics of a long-term stable equilibrium,
(b) oscillatory decays in transient behavior to a long-term stable equilibrium, and (c) limit cycles of non-equilibrium dynamics.

Box 8.2 *Continued*

growth isoclines ($dN_i/dt = 0$) of the interacting species through phase-plane portraits and the biology of the equations and their dynamic outcomes. The curves defining the zero population growth isoclines represent the densities of that population that produce zero population growth (Box 8.2 Table 1). Any point on the $dN_1/dt = 0$ isocline (i.e., N_1-isocline) means that the population N_1 does not grow when at this density. Note, though, that the population of N_2 can grow on the other population's isocline. This is not true, however, where the two isoclines of the interacting species (N_1, N_2) intersect, whereby an equilibrium occurs as neither population changes with time.

To illustrate this graphical technique, consider the classic Rosenzweig–MacArthur consumer–resource model of predator–prey dynamics (Rosenzweig and MacArthur 1963). Changes in the isoclines of the two species produce predictable dynamical outcomes (Box 8.2 Figure 2a–c) based on the biology (e.g., attack rate) described by the equations. The Rosenzweig–MacArthur predator–prey equations are:

$$\frac{dN_1}{dt} = rN_1\left(\frac{K-N_1}{K}\right) - \frac{aN_1N_2}{b+N_1} \quad \text{B.1}$$

$$\frac{dN_2}{dt} = \frac{eaN_1N_2}{b+N_1} - mN_2 \quad \text{B.2}$$

where r is the prey (N_1) per capita rate of increase, K is the carrying capacity, a is the maximum attack rate, b is the half-saturation rate, e is the conversion efficiency of biomass consumed into consumer (predator) biomass (N_2), and m is the

Table 1 Definitions of some key terminology of phase-plane analyses of the population dynamics of species interactions.

State variables	The variables that are defined dynamically in the equations. In equations (B.1) and (B.2), these are N_1 and N_2.
Parameters	The variables that are independent of the equations. In equations (B.1) and (B.2) these are a, r, K, e, and m.
Phase space	A space defined by the state variables, e.g., N_1 and N_2 in equations (B.1) and (B.2).
Isoclines	The combination of values of the state variables (N_1 and N_2 here) that yield zero solutions for a given dynamical equation (e.g., those values of N_1 and N_2 that give $dN_1/dt = 0$ yield the N_1-isocline).
Initial value	The choice of values of all state variables at time t = 0.
Equilibrium steady state	A point in the phase space where all rates of change are zero (i.e., isoclines intersect) such that if the trajectory landed exactly on that point, it would remain there forever. An equilibrium can be an attractor, neutral, repellor, or saddle. Stable equilibrium steady states are represented as a time series by examples in Box 8.2 Figure 1(a) and (b).
Non-equilibrium steady state	A set of points in the phase space where the average rate of change over the set of points is zero such that if the trajectory landed exactly on one of these points, it would remain within this set forever. A non-equilibrium steady state can be an attractor, neutral, repellor, or saddle. We define two different manifestations of non-equilbrium states:
Periodic oscillation	A trajectory in phase space that repeats itself (e.g. consumer–resource limit cycle). A periodic oscillation can be an attractor, neutral, repellor, or saddle. Represented as a time series by example Box 8.2 Figure 1(c).
Complex oscillations	A trajectory in phase space that does not repeat itself but has almost repeatable structure (e.g., chaos, quasi-periodic dynamics). Such non-periodic dynamics can also be attractors, neutral, repellors, or saddles.
Transient	The portion of the time series before the dynamics settle down on the attractor. As an example, in Box 8.2 Figure 1(a) and (b) the transient portion is all the dynamics prior to the point where the solution reaches the equilibrium and then flatlines (i.e. stays the same for all time).
Attractor	A point (equilibrium steady state) or set of points (e.g. non-equilibrium steady state) in the phase space that a dynamical system, once given an initial value, ends up at and then remains. Different initial values may end up at different attractors in a given dynamical system.

continued

Box 8.2 *Continued*

predator's density-independent mortality rate. In equations (B.1) and (B.2), the state variables are N_1 and N_2, while the parameters that are independent of the equations are a, r, K, e, and m.

The zero-growth isoclines of N_1 and N_2 are obtained by setting equations (B.1) and (B.2) equal to zero and solving for N_1 and N_2's isoclines, respectively:

$$N_2 = r\left(\frac{K - N_1}{K}\right) \Big/ (b + N_1) \quad \text{B.3}$$

$$N_1 = mb/(ea - m) \quad \text{B.4}$$

The phase-plane technique proceeds by plotting the two zero-growth isoclines in (N_1,N_2)-coordinate space to examine the relationships of the isoclines with one another. Once the isoclines are plotted, the equilibrium solution can be determined for a given set of parameter values for equations (B.1) and (B.2). The equilibrium occurs where the two isoclines intersect. An equilibrium, therefore, is the combination of predator and prey densities that simultaneously attain zero population growth rates for both species. Biologically, this means that if two interacting populations occur on this equilibrium, they would not change again in density in time, as the population growth rates are zero.

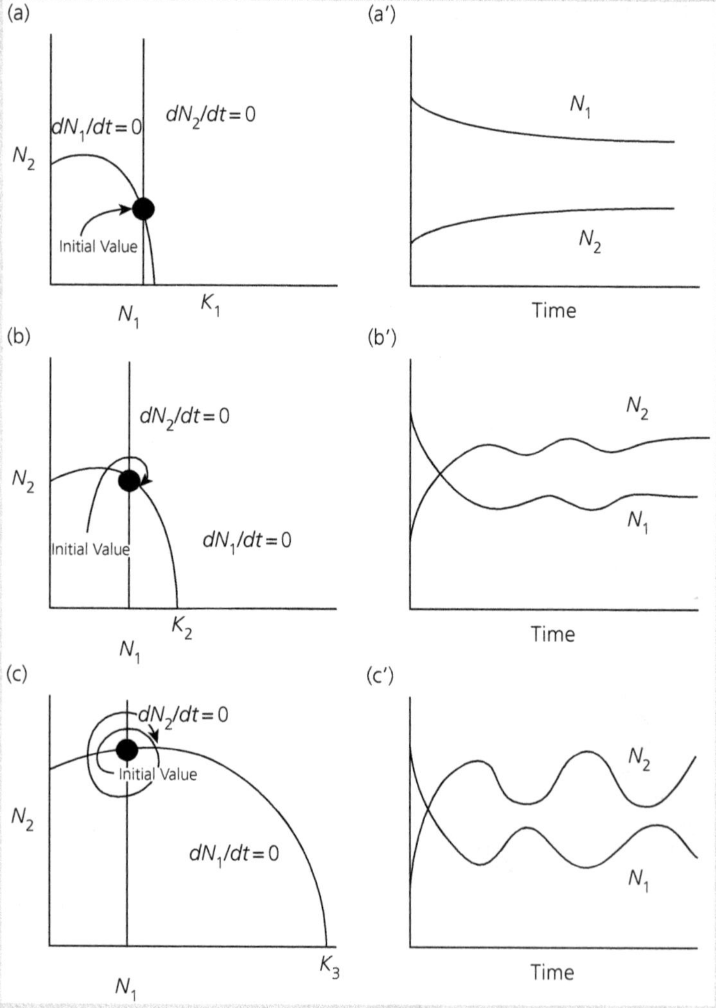

Figure 2 Phase-plane diagrams (a), (b), (c) of N_2 versus N_1, and N_1, N_2 temporal dynamics (a'), (b'), (c') of the classic Rosenzweig–MacArthur model of predator–prey dynamics. The zero-growth isoclines are $dN_1/dt = 0$ and $dN_2/dt = 0$, the equilibrium is marked with a filled circle, and transient dynamics are depicted by curves ending with an arrow.

Box 8.2 *Continued*

As discussed above, there are really only two common states to an equilibrium. An equilibrium either tends to be stable (Box 8.2 Figure 2a,b) or unstable (Box 8.2 Figure 2c). A stable equilibrium means that if we start a numerical experiment near the equilibrium then the dynamics of the system "attract" toward the equilibrium. This model experiment is depicted in the phase-plane diagram of Box 8.2 Figures 2a and b, as well as the time series associated with each phase-plane shown in Box 8.2 Figures 2a′ and b′. Note, time is implicit in the phase plane but effectively one should be able to see how the trajectory in Box 8.2 Figures 2a or b are the same as the time series in Box 8.2 Figures 2a′ or b′, respectively. An unstable equilibrium means that if we start a numerical experiment near the equilibrium then the dynamics of the system "repel" away from the equilibrium. We depict this model experiment with the phase-plane portrait of Box 8.2 Figure 2c, as well as the time series shown in Box 8.2 Figure 2c′. In this case, the dynamics move away from the equilibrium and instead "attract" to a non-equilibrium attractor called a limit cycle. An unstable equilibrium always repels away toward some other attractor. This new attractor must be either an equilibrium or some non-equilibrium attractor (limit cycle, chaos).

While the intersection of the isocline depicts the equilibrium, the shapes of the isocline near the equilibrium determine whether the equilibrium is stable or unstable (see McCann 2012 for details). For the Rosenzweig–MacArthur model, any intersection of the predator-isocline, $dN_2/dt = 0$, that lies on the right side of the hump of the prey-isocline, $dN_1/dt = 0$ (Box 8.2 Figure 2a or b), produces a stable equilibrium. An intersection on the left side yields an unstable equilibrium with limit cycles (Box 8.2 Figure 2c). These relationships between the predator and prey isoclines always lead to these population dynamics, such that any change in a parameter that moves the predator isocline across the hump from right to left always destabilizes the predator–prey interaction from monotonic population dynamics and oscillatory decays to limit cycles. For example, if we start with a low K and then increase K (prey carrying capacity, a surrogate for production) then we move from a stable to an increasingly unstable interaction (Box 8.2 Figures 2a–c). This robust result is the famous paradox of enrichment (Rosenzweig 1971). A similar general result can be used to show that increasing attack rates drives the same sequence (coined the principle of interaction strength, McCann 2012).

References

Case, T.J. (2000). *An Illustrated Guide to Theoretical Ecology*. New York: Oxford University Press.

Edelstein-Keshet, L. (2005). *Mathematical Models in Biology*. SIAM Classics in Applied Mathematics 46. Philadelphia: Society for Industrial and Applied Mathematics. (Reprinting of original 1988 publication by McGraw Hill).

Henson, S.M. (2012). Phase plane analysis. In A. Hastings and L.J. Gross (eds.), *Encyclopedia of Theoretical Ecology*, pp. 538–545. Los Angeles: University of California Press.

McCann, K.S. (2012). *Food Webs*. New Jersey: Princeton University Press.

Rosenzweig, M.L. (1971). Paradox of enrichment: destabilization of exploitation ecosystems in ecological time. *Science*, **171**, 385–387.

Rosenzweig, M.L., and MacArthur, R. (1963) Graphical representation and stability conditions of predator-prey interactions. *American Naturalist*, **97**, 209–223.

fluctuations in density typical of other species interactions (damped oscillations, limit cycles) do not occur for mutualistic interactions. This too is inconsistent with nature, as mutualistic populations do fluctuate in space and time (Addicott 1979, Morales 2000). These three primary results indicate that the L–V model is inadequate even as a first step in studying the population dynamics of mutualism, as it produces silly solutions with biologically inaccurate predictions (May 1976, Holland et al. 2006).

To rectify the inadequate results of L–V theory, the prevailing idea has been that, at least when populations are large, the positive density dependence of mutualism must be countered by negative density dependence for populations to have bounded growth and some stability. Indeed, negative feedback tends to be dynamically stabilizing, while positive feedback, despite being common in nature, tends to be destabilizing (DeAngelis et al. 1986). The 1970s and 1980s were replete with models that incorporated various forms of negative feedback, including factors intrinsic and extrinsic to pairwise mutualisms such as self-limitation, age structure, patch dynamics, handling time,

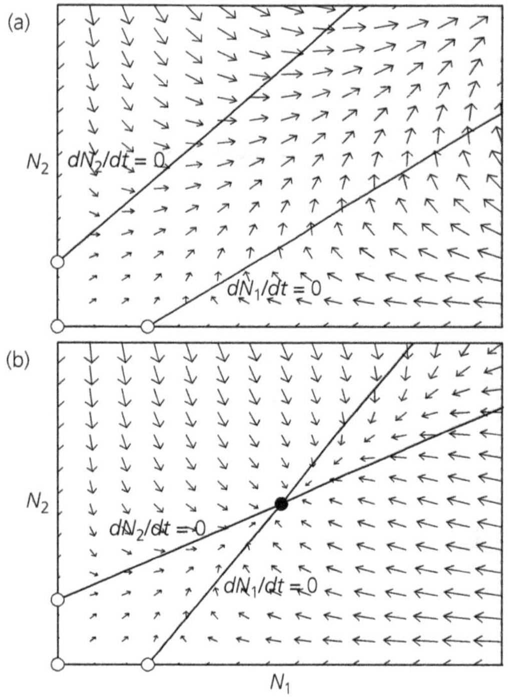

Figure 8.1 Phase-plane portraits of the Lotka–Volterra model [equations (8.4)] for the population dynamics of mutualism. Zero-growth isoclines ($dN_i/dt = 0$) of equations (8.5 are plotted for the population densities of N_1 and N_2. Stable and unstable equilibria (nodes) are shown with solid and filled circles, respectively. The small directional arrows depict vector fields in phase-plane space, showing the trajectories of transient dynamics for population densities of N_1 and N_2. The origin is an unstable equilibrium. Equilibria on N_1- and N_2-axes are saddle points, representing the carrying capacity of each species in the absence of interactions with the other. In panel (a), the product of the interaction strengths is greater than unity ($\alpha_{12}\alpha_{21} > 1.0$); zero-growth isoclines do not intersect, resulting in unbounded population growth as shown by the vector field of arrows leading to ever greater densities of N_1 and N_2. In panel (b), the product of the interaction strengths is less than unity ($\alpha_{12}\alpha_{21} < 1.0$); zero-growth isoclines intersect, forming a stable equilibrium where all transient dynamics converge, as shown by the vector field of arrows leading to the equilibrium. Parameter values for these solutions of equations (8.4) are $r_1 = r_2 = 1.0$ and $K_1 = K_2 = 30$ for panels (a) and (b); $\alpha_{12} = \alpha_{21} = 1.2$ for panel (a) and $\alpha_{12} = \alpha_{21} = 0.6$ for panel (b).

spatial heterogeneity, extrinsic resources, interspecific competition, and interspecific predation (Goh 1979, DeAngelis et al. 1979, 1986, Heithaus et al. 1980, Addicott 1981, Dean 1983, Wolin and Lawlor 1984, Pierce and Young 1986, Wright 1989). These theoretical efforts have been largely divorced from empirical studies, which have focused on documenting the many diverse benefits (and costs) exchanged in mutualisms. Models continue to show that self-limitation, competition, and predation can stabilize mutualism (Ringel et al. 1996, Johnson and Amarasekare 2013). While negative density dependence can stabilize mutualism, most theory through the 1990s did not explicitly modify the feature of the positive feedback of the L–V model that gives rise to its unrealistic results, namely linear functional responses ($+\alpha_{ij}N_jN_i$) for interspecific density dependence.

8.3 Interspecific density dependence and saturating functional responses

Early progress beyond L–V models was made by considering the *intra*-specific density dependence of mutualism, i.e., the benefits of mutualism to a recipient population decrease (and saturate) with the density of that population receiving the benefits (Soberon and Martinez del Rio 1981, Dean 1983, Wolin and Lawlor 1984). While this is a reasonable approach, an approach more consistent with species interactions in general is the consideration of *inter*-specific density dependence of nonlinear functional responses. As originally described for predation, a functional response represents how consumer exploitation of a resource varies with resource density, including linear, hyperbolic, and sigmoidal functions of Holling's Type I, II, and III functional responses.

Incorporating nonlinear functional responses into theory has been paramount to advancing studies of the population dynamics of species interactions. For example, mechanistic theory for predation was advanced over historic L–V models by including nonlinear functional responses, i.e., consumption rates of predators saturating with prey density (Rosenzweig and MacArthur 1963). Likewise, theory for exploitative competition was advanced over L–V models by incorporating density dependence between competitors and their shared resources, i.e., hyperbolic functional responses between a competitor's exploitation of a resource and resource supply (Leon and Tumpson 1975, Hsu et al. 1978). Indeed, a general feature of interspecific interactions is a hyperbolic functional response describing

Box 8.3 Interaction strengths and mutualism

J. Timothy Wootton

The strengths of interactions among species are key to understanding the structure and dynamics of ecological communities. Interaction strength is a diffuse concept, however, so care must be taken to identify precisely what it means in any given study, how it relates to ecological theory, and what motivates its empirical quantification. Recent reviews of the study of interaction strengths (Laska and Wootton 1998, Wootton and Emmerson 2005) identify four theoretical concepts that are aligned with most ecological studies: (1) the per capita effect of an individual of one species on an individual of another species; (2) the effect of one individual of one species on the population of another species near equilibrium; (3) the effect of a sustained input of individuals of one species on the equilibrium abundance of another; and (4) the effect of removing one species on the abundance of another. The first concept, which is often treated as a constant but can also be a nonlinear function that may include other species in the community, describes the parameters of functions in ecological models that govern changes in species abundances. Because of this strong link to theory, and its underlying influence on all other measures of interaction strength, it is often the most desirable to estimate empirically. The second concept estimates the direct effect of an individual of one species on the population of another, which is important in considering stability properties of communities. The third incorporates both direct and indirect effects of a species. The fourth includes both direct and indirect effects of a species, but also accounts for changes in the interaction network generated by a species. The latter measure has related to the most common experimental studies of species interactions. In general, generating empirical indices of interaction strength is most useful when explicitly linked to mathematical theory (Laska and Wootton 1998, Novak and Wootton 2008).

Quantifying interaction strengths has been largely targeted toward consumer–resource interactions of predation and competition (Wootton and Emmerson 2005). Estimating interaction strengths of consumer–resource interactions has included analysis of experimental manipulations of consumer density, observing consumption rates of individual consumers, inference based on fitting models to the dynamics of consumer and resource populations through time, and estimates based on energetic constraints and relationships with body size. In general, generating empirical estimates of interaction strengths in ecological communities has proven to be challenging, with many approaches needing further experimental tests and validation.

For the study of interaction strengths of mutualistic interactions between species, two key questions arise: (1) Can methods designed for consumer–resource interactions be extended to mutualism? (2) Are alternative approaches better suited to mutualism? One complicating factor in transferring these methods to mutualism is that consumer–resource interactions have a very well-defined functional basis—the transfer of energy and/or nutrients from one organism to another. While most mutualisms involve the movement of energy and nutrients, mutualisms also involve other underlying mechanisms, ranging from ant defense against herbivores to propagule and gamete dispersal. Some of these components share key elements with consumer–resource interactions, notably the provision and consumption of food rewards, so commonalities exist (Holland and DeAngelis 2010), but in some other cases new empirical approaches may be required. Furthermore, some mutualisms turn out to be indirect interactions arising from linked consumer–resource interactions, so common approaches may well apply here as well (Holland and DeAngelis 2010).

Although still rare, studies of mutualism have begun to estimate interaction strength involving a number of species pairs. The focus has largely been on plant–pollinator (Vázquez et al. 2006, 2012) and plant–seed disperser interactions (Jordano 1987, Bascompte et al. 2006). These interactions involve active mobile species, sessile species, and often strong signaling among them, features that in combination facilitate direct observation of interactions and their longer-term consequences. Commensurate with this advantage, investigators have used an observational approach based on estimates of the (relative) frequency of recorded interactions (flower visits, fruits removed) among species pairs. In general, the ratio of interactions of species *i* to total interactions observed (i.e., relative frequency) has served as an index of the interaction strength of species *i*, although Vázquez et al. (2012) used independent data on reproductive output to generate indices of both per visit and estimated impacts of all visits to flowers to generate metrics in the spirit of per capita and total impact interaction strengths.

Relative frequency of observed interactions seems a reasonable first index to get a handle on relative interaction strengths among mutualistic species, as it likely contributes to net impacts of species on one another, and interesting patterns in the distribution of this metric across species pairs

continued

Box 8.3 *Continued*

and networks have arisen. For example, relative frequencies of species interactions exhibit skewed distributions (Bascompte et al. 2006, Vázquez et al. 2012), similar to those of consumer–resource interactions (Wootton and Emmerson 2005). Nevertheless, we ultimately want to compare interaction strengths of mutualism with other species interactions, and to use them in dynamic models of ecological communities. In this regard, further advances are needed to measure strengths of mutualism in the currency of population size or biomass. Pollination and seed dispersal involve only part of the plant life cycle, and these components of reproduction may not correspond, at least linearly, with plant population dynamics when stage- and density-dependent survival play important roles. The demographic consequences of dispersal and gene flow are poorly known. Hence, studies integrating these mutualistic effects within full demographic models may provide more insights. Relative frequencies also contain no information on the population size or biomass of interacting species, and as a result may not yield per capita effects and may well vary with population sizes. For example, a species with a high frequency of interaction may have traits that make each individual more effective at an interaction, or alternatively may simply have a high population size. Additionally, observed interactions could each involve a different individual, or could be the same individual returning to a plant multiple times, so per interaction effects may not mirror per capita effects. For species receiving food resources, the impact of the interaction will depend on the contribution of alternative food resources, which may not be quantified in the context of studies of mutualisms. Including more information on population sizes of interacting species and the dependence of population growth on the target life stages will be useful in future work. Despite these limitations, there is some evidence that relative frequencies of interactions sometimes, but not always, correspond to important demographic components (Vázquez et al. 2012).

Experimental approaches have also been used to study mutualism, although they have generally not quantified interaction strength for the purpose of modeling, and they have been limited to small modules of interacting species (Stanton and Palmer 2011). Conceivably, these could be expanded to estimate interaction strength more explicitly, although logistical issues limit the number of interactions that can be handled. These studies also usually focus on a portion of the life cycle of species, so efforts to incorporate these into a full demographic analysis are also needed.

Other approaches used in consumer–resource studies, such as energetic constraint approaches and community dynamics modeling, have not been widely applied to the study of mutualism. The former might be especially applicable to components of mutualistic interactions involving energetic resource transfer (Holland and DeAngelis 2010), although it needs more rigorous field experimental testing. The latter may also be powerful because it is a fairly flexible approach that can address multiple mechanisms of mutualism, though it will be necessary to generate detailed time series of mutualists' densities in communities to implement it. In any event, there appears to be substantial scope for transfer of methods from traditional studies of consumer–resource interactions to studies of mutualism, just as food-web studies have inspired new insights into networks of mutualistic interactions (Bascompte et al. 2006).

References

Bascompte, J., Jordano, P., and Olesen, J.M. (2006). Asymmetric coevolutionary networks facilitate biodiversity maintenance. *Science*, **312**, 431–433.

Holland, J.N., and DeAngelis, D.L. (2010). A consumer-resource approach to the density-dependent population dynamics of mutualism. *Ecology*, **91**, 1286–1295.

Jordano, P. (1987). Patterns of mutualistic interactions in pollination and seed dispersal: connectance, dependence asymmetries and coevolution. *American Naturalist*, **129**, 657–677.

Laska, M.S., and Wootton, J.T. (1998). Theoretical concepts and empirical approaches to measuring interaction strength. *Ecology*, **79**, 461–476.

Novak, M., and Wootton, J.T. (2008). Estimating the nonlinear interaction strengths: an observational method for species-rich food webs. *Ecology*, **89**, 2083–2089.

Stanton, M.L., and Palmer, T.M. (2011). The high cost of mutualism: effects of four species of East African ant symbionts on their myrmecophyte host tree. *Ecology*, **92**, 1073–1082.

Vázquez, D.P., Lomáscolo, S.B., Maldonado, M.B., Chacoff, N.P., Dorado, J., Stevani, E.L., and Vitale, N.L. (2012). The strength of plant-pollinator interactions. *Ecology*, **93**, 719–725.

Vázquez, D.P., Morris, W.F., and Jordano, P. (2005). Interaction frequency as a surrogate for the total effect of animal mutualists on plants. *Ecology Letters*, **8**, 1088–1094.

Wootton, J.T., and Emmerson, M. (2005). Measurement of interaction strength in nature. *Annual Review of Ecology, Evolution and Systematics*, **36**, 419–444.

consumer–resource interactions (Turchin 2003). No matter how high resource density becomes, there is an upper limit to a functional response set by a species' biology. Actually, linear functional responses are only thought to occur in filter feeders (Jeschke et al. 2004).

Unrealistic results of the L–V model of mutualism are circumvented by recognizing that there is a maximum upper limit to the benefits and positive feedback of mutualistic interactions. In the L–V model, $+\alpha_{ij}N_jN_i$ represents a linear functional response in which the benefits and positive feedback of mutualism to N_i increase as a linear function of the density of N_j. Accordingly, increases in N_j lead to never-ending increases in N_i and vice versa, which lead to the artificial inequality of the L–V model ($\alpha_{12}\alpha_{21} < 1$) necessary for stability. As was the case with predation and competition, advances over L–V models of mutualism toward realistic and mechanistic theory began with interspecific density dependence, i.e., saturation in the benefits and positive feedback to one species as a function of the density of the species with which it interacts (May 1976, Holland et al. 2006).

To demonstrate the population dynamic consequences of hyperbolic functional responses for the benefits of mutualism, a third term can be included in the second equation of (8.3) to represent mutualistic interactions. Rather than a linear functional response which would produce L–V dynamics (DeAngelis et al. 1986), benefits of species j to species i are depicted by a hyperbolic Michaelis–Menten function, $\alpha_{ij}N_j/(b_j + N_j)$, where b_j is the half-saturation constant:

$$\frac{dN_1}{dt} = r_1N_1 + \frac{\alpha_{12}N_2N_1}{b_2 + N_2} - d_1N_1^2 \text{ and}$$
$$\frac{dN_2}{dt} = r_2N_2 + \frac{\alpha_{21}N_1N_2}{b_1 + N_1} - d_2N_2^2. \tag{8.6}$$

Other formulations of hyperbolic functional responses (e.g., Holling Type II) produce similar dynamics (Wright 1989, Kot 2001, Holland et al. 2006). Setting equations (8.6) equal to zero and solving for zero-growth isoclines ($dN_i/dt = 0$) yields:

$$N_1 = \left(r_1 + \frac{\alpha_{12}N_2}{b_2 + N_2}\right)\Big/ d_1 \text{ and}$$
$$N_2 = \left(r_2 + \frac{\alpha_{21}N_1}{b_1 + N_1}\right)\Big/ d_2. \tag{8.7}$$

The curvilinear isoclines have intercepts of r_i/d_i in (N_1, N_2) phase-plane space (Figure 8.2). The intercepts on the N_1 and N_2 axes are saddle points, which represent the equilibrium of each species (N_i) in the absence of interactions with the other ($N_j = 0$). In addition to the saddle points and unstable equilibrium at the origin ($N_1 = N_2 = 0$), the two isoclines intersect once forming a single stable equilibrium of mutualistic coexistence at which N_1 and N_2 densities are greater than their respective saddle point densities (Figure 8.2). This result is consistent with the concept that mutualism has positive (beneficial) effects on interacting populations by increasing their densities and population growth rates.

Depending on whether N_i is a facultative or obligate species, saddle points may intersect at the origin or a positive value on the N_i axis. The first term (r_iN_i) in equations (8.6) is the population growth rate of N_i in the absence of interactions with another mutualistic species, for which r_i can be greater than or equal to zero for facultative and obligate species, respectively. For a facultative species, population persistence ($N_i > 0$) occurs even when the density of a mutualistic population with which it interacts goes to zero ($N_j \rightarrow 0$). For facultative interactions, as in Figure 8.2 in which $r_i > 0$, if N_j in the dN_i/dt equation goes to zero, then logistic population growth ensues and the N_i population persists at its saddle point. For an obligate species, population extinction ($N_i = 0$) ensues when the density of a mutualistic population with which it interacts goes to zero ($N_j \rightarrow 0$). Setting $r_i = 0$ in equation (8.6) for an obligate species leads the saddle point for N_i on its axis in Figure 8.2 to go to the origin, such that if the density of N_j goes to zero then N_i also goes to extinction. Thus, the facultative/obligate nature of mutualistic interactions has important consequences for their population dynamics. While mutualisms were once thought to be largely species-specific obligate interactions, we now know that they are most commonly generalized facultative interactions given the nested structure of mutualistic networks in which specialized pairwise interactions are rare (Bascompte et al. 2003, Chapter 11).

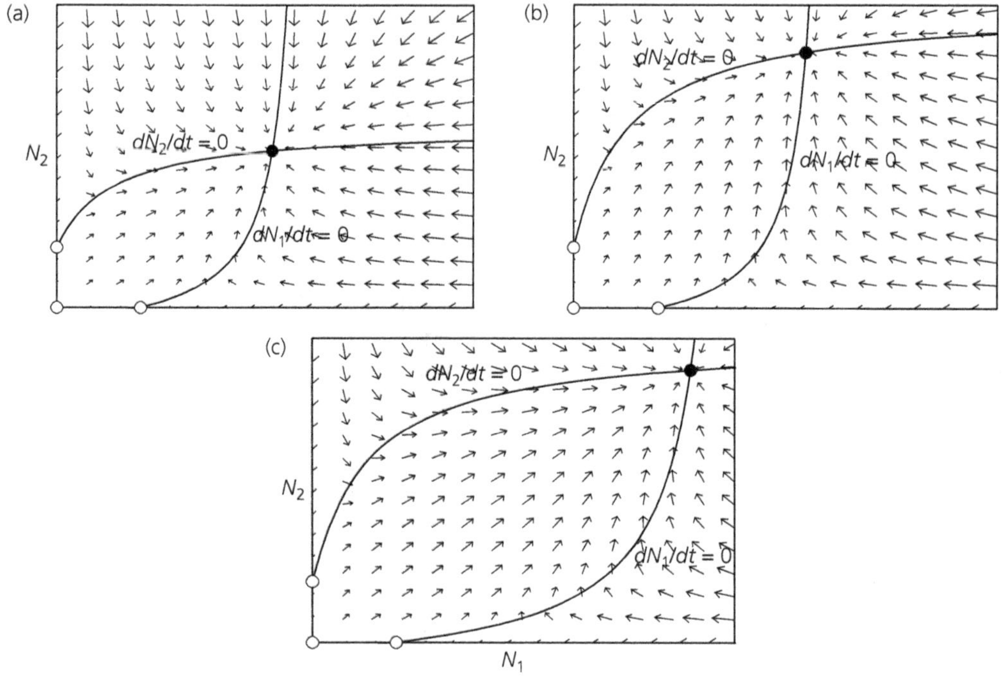

Figure 8.2 Phase-plane portraits of the saturating benefits model [equations (8.6)] for population dynamics of mutualism. Zero-growth isoclines ($dN_i/dt = 0$) of equations (8.7) are plotted for the population densities of N_1 and N_2. Stable and unstable equilibria (nodes) are shown with solid and filled circles, respectively. The small directional arrows depict vector fields in phase-plane space, showing the trajectories of transient dynamics for population densities of N_1 and N_2. The origin is an unstable equilibrium. Equilibria on N_1- and N_2-axes are saddle points, representing the carrying capacity of each species in the absence of interactions with the other. In panel (a), interaction strengths are mutually weak, specifically $\alpha_{12} = \alpha_{21} = 0.6$, such that the product of interaction strengths is less than unity ($\alpha_{12}\alpha_{21} < 1.0$). In panel (b), the interaction strengths are asymmetric, whereby one interaction strength is weak and one strong, specifically $\alpha_{12} = 0.6$ and $\alpha_{21} = 1.2$, but the product of interaction strengths is less than unity ($\alpha_{12}\alpha_{21} < 1.0$). In panel (c), interaction strengths are mutually strong, specifically $\alpha_{12} = \alpha_{21} = 1.2$, such that the product of interaction strengths is greater than unity ($\alpha_{12}\alpha_{21} > 1.0$). Stability of the stable equilibrium does not change among panels (a–c) with increasing interaction strengths, but rather equilibrium densities of N_1 and N_2 increase as expected with greater interaction strengths. Other parameter values for these solutions of equations (8.6) are $r_1 = r_2 = 1.0$, $b_1 = b_2 = 20$, and $d_1 = d_2 = 0.01$.

The hyperbolic functional response contributes to the stable equilibrium of mutualistic coexistence (Figure 8.2). Without identifying any particular biological mechanism for the density dependence, such saturating functional responses for the benefits and positive feedback of mutualism overcome the limitations of L–V models. First, mutualism does not lead to unbounded population growth (Figure 8.1a versus 8.2). Rather, all transient dynamics lead to a stable equilibrium of mutualistic coexistence (Figure 8.2). Second, the equilibrium does not depend on inequalities ($\alpha_{12}\alpha_{21} < 1$) of interaction strengths. Mutualistic coexistence occurs whether pairwise interaction strengths are weak (Figure 8.2a), asymmetric (Figure 8.2b), or strong (Figure 8.2c). Increasing interaction strengths simply increases equilibrium densities (Figure 8.2). Third, stability does not require invoking other biological processes of negative density dependence, such as increasing intraspecific density dependence with mutualism (Addicott 1981, Wolin and Lawlor 1984), extrinsic limitation of resources (Dean 1983), and competition or predation (Heithaus et al. 1980, Ringel et al. 1996, Johnson and Amarasekare 2013).

A saturating functional response for the positive feedback of mutualism is a general property that yields stability of pairwise interactions without invoking some other intrinsic and extrinsic source of negative density dependence. A growing number of studies are providing new insights into the ecological dynamics of mutualism by incorporating saturating functional responses into population dynamic models. For example, Ingvarsson and Lundberg (1995) showed that a hyperbolic functional response of a generalist pollinator can lead to stable equilibria and the regulation of plant populations as pollination saturates with the handling time of plant flowers by pollinators. They went on to show that when pollinator density is governed by floral resources of the plant population, functional responses lead to an Allee effect, such that below a certain pollinator abundance both plant and pollinator go extinct, but otherwise both species go to a stable equilibrium (Lundberg and Ingvarsson 1998). This approach of saturating functional responses has been extended from pairwise interactions to food web modules and large multispecies networks to examine their consequences for community dynamics (Okuyama and Holland 2008, Lee and Inouye 2010, Holland et al. 2013).

8.4 Contemporary theory for density dependence and population dynamics

While nonlinear functional responses are a minimal starting point for theory on the ecological dynamics of mutualism, equations (8.6) only entail monotonic damping in density toward a stable equilibrium. Theoretical predictions are still lacking for the most basic empirical phenomena of mutualism's population dynamics, including damped oscillations, population fluctuations, and density-dependent overexploitation by one mutualist driving another extinct. Incorporating the biological mechanisms responsible for the density-dependent functional responses of mutualism can help bridge theory and empiricism. Until recently there have been few mechanistic theories for the population dynamics of mutualism, in part due to the paucity of general concepts and principles that can cut across the many diverse mutualistic systems which differ so tremendously in their natural histories. What biological mechanisms and demographic currencies are most useful for studying the density-dependent population dynamics of mutualism? The following sections describe two contemporary approaches to the mechanisms underlying the density-dependent population dynamics of mutualism, in particular their benefits and costs and their consumer–resource interactions.

8.4.1 Benefits and costs of mutualism

Because mutualisms are commonly defined as mutually beneficial interactions between species, conceptual frameworks for theory have centered around the different types of benefits exchanged between species (Chapter 1). Benefits are grouped into three common forms: (1) nutritional resources (e.g., N, P, plant carbohydrates, nectar) obtained through biosynthesis or harvesting from the environment (e.g., plant–rhizobia, plant–mycorrhizal mutualisms), including the digestive breakdown of otherwise unusable resources (e.g., ruminant or termite gut flora symbioses); (2) transportation of gametes or progeny (e.g., pollination, seed dispersal mutualisms); and (3) defense from natural enemies (e.g., ant–plant mutualisms). Attention is also beginning to be given to the costs of mutualism. Costs may include (1) resources produced by a mutualist to "reward" another (e.g., photosynthates, nutrients, amino acids, nectar, extrafloral nectar, fruit); (2) structures produced by a mutualist to attract another (e.g., floral display, domatia); and (3) energetic costs of foraging and handling times spent obtaining those resources.

Benefits and costs have been argued for decades to be essential to understanding the growth and dynamics of mutualistic populations (May 1976, Addicott 1979, Inouye and Taylor 1979, Soberon and Martinez del Rio 1981, Wolin and Lawlor 1984, Pierce and Young 1986, Cushman and Whitham 1989, Breton and Addicott 1992). Benefits and costs must increase and decrease demographic rates in ecological time for them to have consequences for population dynamics, whereby benefits must exceed costs for net benefits to increase the population growth rates of the interacting species. For decades now empirical studies have been dedicated to documenting the many diverse benefits and costs of mutualism. Extension of economic theory to mutualism as a biological market has been useful in understanding benefits as "commodities of trade," including when species may invest or divest in their acquisition and exchange (Schwartz and Hoeksema 1998).

Functional responses in terms of the benefits and costs of mutualism represent key biological mechanisms for how the density of one mutualistic population may shape the population growth and dynamics of another. Studies have begun to explicitly include benefit and sometimes cost functional responses into population dynamic models. For example, with an application to pollinating seed-consuming mutualisms, Holland, DeAngelis, and colleagues showed that such mechanistic formulations of density-dependent functional responses can reveal new insights into the ecological and evolutionary dynamics of mutualism, including damped oscillations, multiple stable equilibria, interspecific population regulation, and the evolution of traits to limit overexploitation (Holland et al. 2002, 2004, Holland and DeAngelis 2006). Morris et al. (2003) employed density-dependent functional responses, showing how various ecological dynamics and stabilizing mechanisms can arise for pollinating seed-consuming systems faced with third-party exploiter species that do not pollinate (such as fig/fig wasp and yucca/yucca moth pollination mutualisms). Thompson et al. (2006) employed saturating functional responses for benefits of mutualism in models of fish/cleaner-shrimp interactions and showed that the consequences of immigration and emigration for the population dynamics of mutualists can depend on which demographic rate is influenced by mutualism.

Central to these studies, along with the more recent work of Morris et al. (2010), is that the difference between benefit and cost functional responses can lead to a variety of net effect functional responses, including not only Type I, II, and III response curves, but also unimodal functions (Figure 8.3). Contrary to common thought, more mutualists are not necessarily better than fewer. With saturating net effect functional responses, the benefits to N_i increase with the density of N_j to a point of saturation in benefits. More is better to that point, at which positive effects do not continue to increase with density. With a unimodal net effect functional response, maximum net benefits arise at intermediate densities of the interacting species. Net benefits to N_i increase to the modal density of N_j, but diminish and may even become negative with further increases in the density of N_j. More is better to an intermediate point, after which more is not better and may even become worse. Thus, the shape of net effect functional responses arising from density-dependent benefits and costs is central to the dynamical properties of mutualism (Holland et al. 2002, Morris et al. 2010). In addition to the shape, the particular formulation of a saturating functional response can be key to the dynamics. Fishman and Hadany (2010) showed that, rather than standard Holling Type II or Michaelis–Menten functional responses, Beddington–DeAngelis functions (i.e., consumer interference) may arise for the benefits of particular mutualisms leading to yet different dynamical properties. Yoo and Holway (2011) showed that there is a unimodal growth rate of aphids defended by ants as a function of the ratio of ants to aphids, indicating that ratio-dependent functional responses may be important for mutualism (Holland et al. 2004).

8.4.2 Consumer–resource interactions of mutualism

Advances in theoretical and empirical studies of the density-dependent population dynamics of predation and competition have been made through the study of their consumer–resource (C–R) interactions (Murdoch et al. 2003, Turchin 2003). Though often equated with the outcome of predator–prey interactions, the C–R interaction is a mechanism (not outcome) of interspecific interactions, that is the actions, ways, or means by which individuals of different species interact with one another (Abrams 1987). The C–R mechanism of interaction relates the process of energy and nutrient transfer between the consumer and resource, from which changes in consumer and resource biomass or density arise from the interaction. Resources are biotic or abiotic factors that increase the biomass density of the consumer over some range of availability of the resource. Consumers reduce the biomass density or supply of the resource. In this way, density or biomass are the biological currencies that translate consumer exploitation of resources (Box 8.2). Predation is a direct C–R interaction in which a predator is a consumer that eats a prey resource. Exploitative competition is an indirect C–R interaction in which two competitive consumers exploit a shared basal

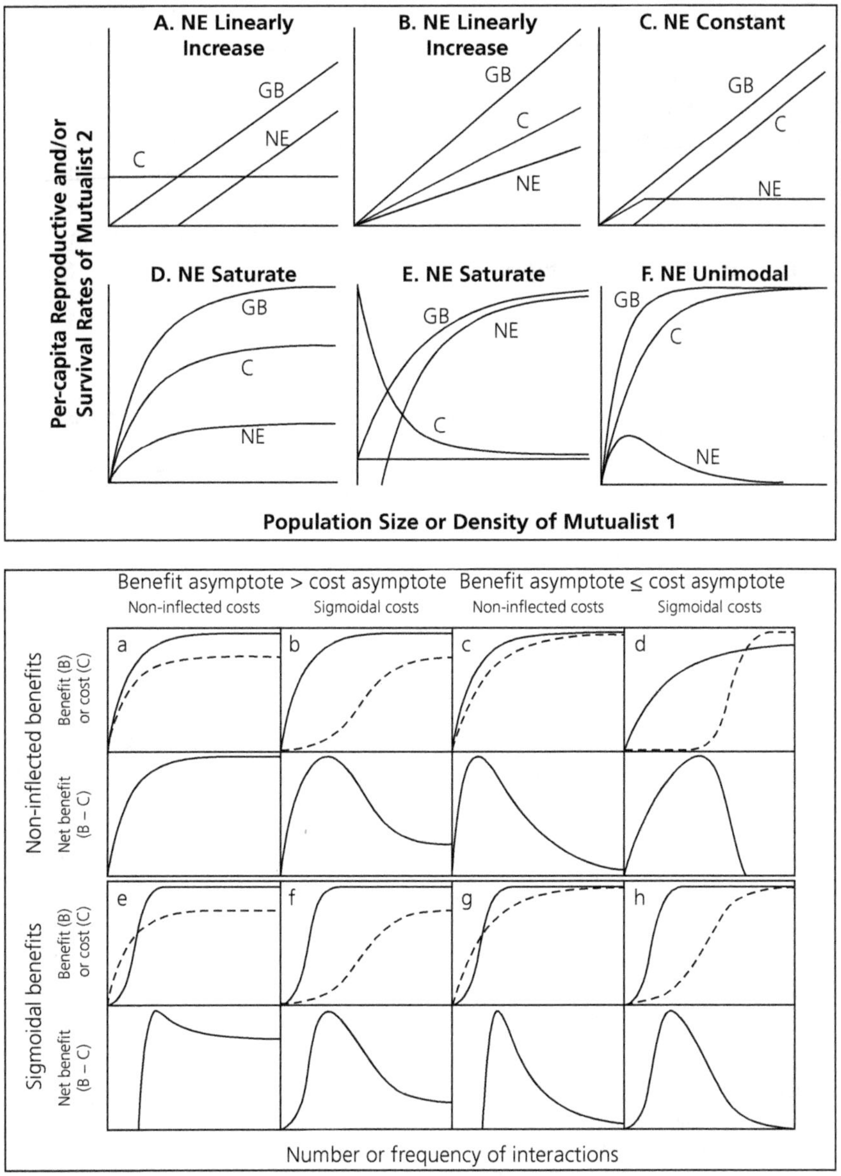

Figure 8.3 Examples of relationships between the benefit and cost functional response curves as shown in the top panel by Holland et al. (2002), in which net effects (NE) equal gross benefits (GB) minus costs (C), and in the bottom panel by Morris et al. (2010), in which net benefits equal benefits (B, solid line) minus costs (C, dashed line).

resource, whether it be an abiotic nutrient or biotic prey species. Mutualism is now recognized as a class of C–R interactions (Holland and DeAngelis 2010). A consumer mutualist (such as a pollinator) exploits a resource supplied by another mutualistic species (such as plant nectar) and in the process provisions that species with another resource or a non-trophic service of dispersal or defense (such as pollen transfer).

Nearly all mutualisms can be classified as bidirectional, unidirectional, or indirect C–R mutualisms (Holland and DeAngelis 2010). In bidirectional C–R mutualisms, each species functions as both a consumer and a resource (Table 8.1, Figure 8.4a); one mutualist (N_1) exploits resources supplied by another mutualist (N_2) and the latter also consumes resources supplied by the former. Unidirectional C–R mutualisms entail one species functioning as a consumer and the other as a resource, but neither functions as both (Table 8.1). Resources produced by a mutualist (N_1) attract a consumer mutualist (N_2), which, in the process of exploiting the resource, provisions the former with a service of dispersal (Figure 8.4b) or defense (Figure 8.4c). Indirect C–R mutualisms occur when the effects of species (N_1, N_2) on one another are mediated by the density or traits of a third species that is a consumer (P) or resource (R) of one or both of them (Table 8.1, Figures 8.4d,e,f).

Table 8.1 Examples of bidirectional, unidirectional, and indirect consumer–resource mutualisms, with N_1 and N_2 corresponding to the topological models of Figure 8.4. (Modified from Holland and DeAngelis 2010)

Mutualism	Species	Consumer	Resource
Bidirectional consumer–resource mutualisms			
Lichen	Fungi, algae	Fungi (N_2) Algae (N_1)	Algal photosynthates Nutrients, water
Coral	Coral, zooxanthellae	Coral (N_2) Zooxanthellae (N_1)	Algal photosynthates Nutrients, nitrogen
Mycorrhizal	Plant, mycorrhizal Fungi	Plant (N_1) Mycorrhizae (N_2)	Nutrient, phosphorus Root exudates, carbon
Nitrogen fixation	Plant, rhizobial bacteria	Plant (N_1) Rhizobium (N_2)	Nitrogen Root exudates, carbon
Myrmecotrophy	Plants, ants	Plant (N_1) Ant (N_2)	Nitrogen in debris plant food rewards
Ant agriculture	Ant, fungus	Ant (N_2) Fungus (N_1)	Fungus Ant-collected leaves
Digestive	Aphid, bacteria	Aphid (N_1) Bacteria (N_2)	Amino acids Aphid-ingested food
	Termite, protozoa	Termite (N_1) Protozoa (N_2)	Protozoa-digested food Termite-ingested cellulose
Unidirectional consumer–resource mutualisms			
Dispersal	Plant (pollen), animal Fungus (spores), beetle Plant (seed), vertebrate Plant (seed), ant	Animal (N_2) Beetle (N_2) Vertebrate (N_2) Ant (N_2)	Plant (N_1) nectar/pollen Tissue of fungus (N_1) Plant (N_1) fruit Plant (N_1) seed elaiosome
Defense	Plant, ant Ant, Lycaenid caterpillar Ant, Homopteran Plant, fungal endophytes	Ant (N_2) Ant (N_2) Ant (N_2) Fungi (N_2)	Plant (N_1) nectar/food body Caterpillar (N_1) secretions Homopteran (N_1) excretions Plant (N_1) carbon
Indirect consumer–resource mutualisms via a third species			
Müllerian mimicry	Two or more mimics	Mimics' predator (P)	Mimics (N_1, N_2)
Mixed foraging	Two or more vertebrates	Predator (P)	Foraging species (N_1, N_2)
Honey guide	Honey guide bird, badger	Bird (N_1), badger (N_2)	Bee larvae and honey (R)
Cleaning	Cleaner fish, client fish	Cleaner fish (N_2)	Client (N_1) ectoparasites (P)

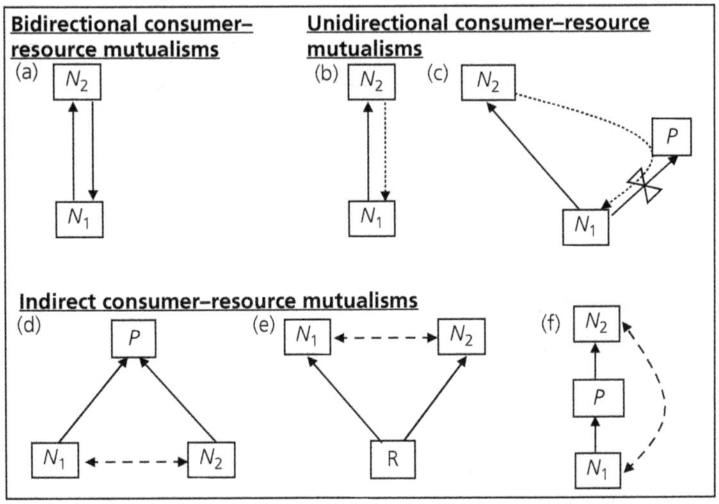

Figure 8.4 Conceptual models of bidirectional (a), unidirectional (b, c), and indirect (d, e, f) consumer–resource mutualisms involving interactions between two populations, N_1 and N_2, and in some cases a third population that is a natural enemy (*P*) or resource (*R*) of one or both of the mutualists. Trophic levels are depicted based on energy flow, such that if N_2 is above N_1, then it is on a higher trophic level. Solid, unidirectional arrows depict C–R interactions, directed from the resource to the consumer in the direction of energy flow or nutrient movement. Dotted, unidirectional arrows depict a non-trophic service of dispersal or defense, from the population providing the service to the one receiving it. Bidirectional dashed arrows depict indirect C–R interactions between N_1 and N_2. Joined triangles represent a rate modifier, in this case N_2 modifies the rate at which *P* attacks N_1 in panel (c). (Modified from Holland and DeAngelis 2010)

Central to the C–R interaction is the density-dependent functional response relating the interspecific coupling of resource exploitation by a consumer with the density of the resource. Based on C–R biology of mutualism, Holland and DeAngelis (2010) developed a general set of equations for the population dynamics of interacting mutualists that include both the positive and negative effects on population change associated with the supply and consumption of resources. For bidirectional C–R mutualisms, these equations are:

$$\frac{dN_1}{dt} = N_1[r_1 + c_1 f_1(N_1, N_2) - q_1 g_1(N_1, N_2) - d_1 N_1] \quad (8.8)$$

$$\frac{dN_2}{dt} = N_2[r_2 + c_2 f_2(N_1, N_2) - q_2 g_2(N_1, N_2) - d_2 N_2] \quad (8.9)$$

The first and fourth terms affect population change separate from mutualism. The first term is population growth independent of the interaction and the fourth term is density-dependent self-limitation. The second terms, $cifi(N_i,N_j)$, describe the positive effects on population change of species i due to its resource consumption or service acquisition from species j. $f_i(N_i,N_j)$ is the functional response of N_i, describing how resources or services acquired by N_i vary with N_j (and/or N_i), with currencies of biomass, energy, or nutrients consumed per consumer per unit of time. In the case of mutualistic services, such as pollination or defense, these are dimensionless quantities which translate into reproduction or survival. The third term, $q_i g_i(N_i,N_j)$, describes how resource supply to species j may reduce the biomass, nutrients, or energy of species i as a function of N_i and/or N_j, resulting in negative effects on population change of species i. Parameters c_i and q_i convert f_i and g_i into numerical responses, rates of change in the number or biomass density of species i. Equations (8.8) and (8.9) are readily modified to unidirectional C–R mutualisms by dropping the third term of equation (8.9), as the N_2 consumer does not provide a resource to N_1, as N_1 does to N_2:

$$\frac{dN_1}{dt} = N_1[r_1 + c_1 f_1(N_1, N_2) - q_1 g_1(N_1, N_2) - d_1 N_1] \quad (8.10)$$

$$\frac{dN_2}{dt} = N_2\left[r_2 + c_2 f_2(N_1, N_2) - d_2 N_2\right] \quad (8.11)$$

These models of C–R mutualisms resemble predator–prey interactions with the interspecific linkages of C–R interactions coupled between $f_1(N_1,N_2)$ and $g_2(N_1,N_2)$ and between $f_2(N_1,N_2)$ and $g_1(N_1,N_2)$. For example, in bidirectional C–R mutualisms involving mycorrhizal interactions, plants produce photosynthates (g_1) which are exploited by mycorrhizal fungi (f_2), while the fungi harvest a nutrient (g_2) such as phosphorus from the soil which is supplied as a resource to plants (f_1) (Table 8.1, Figure 8.4a). For unidirectional C–R mutualisms such as plant–pollinator interactions, plants produce floral nectar (g_1) which is consumed by animal pollinators (f_2), while pollinators provide a non-trophic service of pollination and pollen dispersal (f_1) but not a resource per se to plants (Table 8.1, Figure 8.4b). Likewise, lycaenid larvae produce food secretions as food resources (g_1) which are consumed by ants (f_2) which in turn provide the larvae with the non-trophic service of defense (f_1) that increases larval survival by decreasing rates of parasitism (Table 8.1, Figure 8.4c). The lack of the third term in the consumer equation (8.11) does not imply that there are never "costs" to N_2 for provisioning a service of dispersal or defense to N_1. Costs are likely in the form of time and energy associated with dispersal and defense, which manifest in the consumer's functional response.

Phase-plane analyses (Box 8.2) of these models parameterized after the Rosenzweig–MacArthur model of predation are revealing new insights into the transient dynamics and stability properties of mutualism (Holland and DeAngelis 2010). For example, Figure 8.5 shows how the inclusion of mechanistic consumer–resource interactions leads to transient dynamics and stability properties of mutualism that differ from Figure 8.2 with phenomenological representations of saturating functional responses. In the absence of mutualistic interactions, that is $N_i > 0$ when $N_j = 0$, each respective population grows logistically and persists at its equilibrium density of the positive node on its axis in phase-plane space. A central stable node of mutualistic coexistence occurs in the middle, upper-right of phase-plane space. At this stable equilibrium of coexistence, population densities (N_1, N_2) are greater than those of each species at its node in the absence of mutualistic interactions. Interactions lead to the stable node of mutualistic coexistence, but a variety of dynamical properties may occur, including stable and unstable nodes and saddle and spiral points (Figure 8.5).

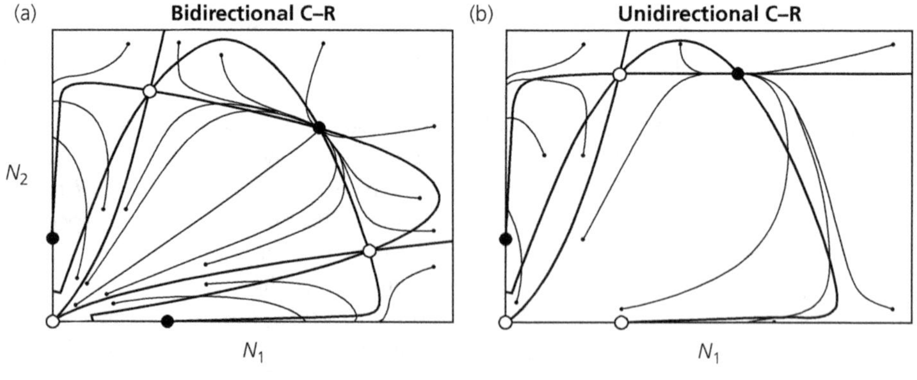

Figure 8.5 Phase-plane diagrams for the population dynamics of bidirectional (a) and unidirectional (b) consumer–resource mutualisms between facultative–facultative species with population densities of N_1 and N_2. The solid lines are zero-growth isoclines ($dN_i/dt = 0$) for N_1 and N_2, respectively. Stable and unstable equilibria (nodes) are identified by filled and open circles, respectively. Saddle points have a line (separatrix) passing through them to the origin, subdividing the phase-plane space into different basins of attraction that correspond with a particular node. Starting with the origin and moving clockwise, the equilibria for panel (a) are unstable node, stable node, saddle point, stable node, saddle point, stable node; for panel (b), they are unstable node, stable node, saddle point, stable node, saddle point. Transient dynamics of some population trajectories in phase-plane space are depicted by thinner lines, for which initial conditions for N_1 and N_2 of each trajectory are represented by filled circles. See Plate 5 for color version.

Associated with each saddle point is a separatrix running from the origin through the saddle point (Fig. 8.5). Separatrices subdivide phase-plane space into different basins of attraction. The central basin of attraction corresponds with a stable node of mutualistic coexistence, while peripheral basins of attraction correspond with stable nodes on the N_1 or N_2 axes ($N_i > 0$, $N_j = 0$). Peripheral basins of attraction shape the dynamics when the density of one species, N_i, is large relative to that of the other, N_j, and when the density of the latter is low, resulting in N_i overexploiting and driving N_j to extinction. N_i persists at the stable node on its axis in the absence of interactions with N_j, but importantly at a lower density than that of the stable node of mutualistic coexistence. When density-dependent conditions of the peripheral basins occur, population dynamics and interaction outcomes more closely resemble those of predation. Density-dependent overexploitation in unidirectional C–R mutualisms may only occur by N_2, as N_1 cannot overexploit N_2 as N_2 does not produce resources exploited by N_1. Also, the coexistence equilibrium for unidirectional C–R mutualisms may be a stable node or spiral point depending on the point of intersection of N_2's isocline with that of N_1. As a stable spiral point, dynamics can involve population cycles and damped oscillations toward the equilibrium.

Theoretical investigations into mutualism, and its relation with other species interactions, are building on the C–R approach. For example, de Mazancourt and Schwartz (2010) examined the eco-evolutionary dynamics of bidirectional C–R mutualisms, showing that basal resource limitation and the precise characterizations of resource trade between the mutualists can shape the stability and coexistence of mutualistic species. Employing saturating functional responses for two mutualistic species competing for a commonly shared resource provisioned by another, Lee and Inouye (2010) showed that mutualism can also facilitate competitive coexistence. Qualitative analyses by Jones et al. (2012) explored a variety of ways in which mutualism and competition may influence the dynamics of one another through the various structures that may shape the consumer–resource interactions of their food web modules. In sum, the C–R mechanism of interaction reveals population dynamic phenomena long recognized to occur in nature, including coexistence, population fluctuations, and extinction ensuing from density-dependent overexploitation.

8.5 Empirical studies of density dependence and population dynamics

Decades of interplay between theory and empiricism have been central to advancing our understanding of the ecological dynamics of predation and competition (Murdoch et al. 2003, Turchin 2003). While mutualism is not under-studied compared with other species interactions, empirical work is limited in its tests of the interspecific density dependence and population dynamics of mutualism. Only in recent years have mechanistic theories been developed for the density-dependent population dynamics of mutualism that can guide empiricism and for which empiricism can in turn validate, refute, or lead to their refinement. We are just beginning to conduct research at the interface of theory and empiricism (Morris et al. 2010, Kang et al. 2011, Morales 2011).

How do changes in the population growth or density of one mutualistic species vary with the density of another? While empirical studies are limited in their explicit tests of the density dependence of the benefits, costs, or consumer–resource interactions of mutualism, we do know that they are not static aspects of mutualistic populations, but rather vary with the abundance of the interacting species (Addicott 1979, Cushman and Whitham 1989, Breton and Addicott 1992, Nefdt and Compton 1996, Morales 2000). Although not always formulated as such, density-dependent functional responses have been shown for a variety of mutualisms (Fonseca 1999, Gange and Ayers 1999, Vazquez et al. 2005, Holland and DeAngelis 2006, Ness et al. 2006, Frederickson and Gordon 2007, Harmon and Andow 2007, Morris et al. 2010, Yoo and Holway 2011, Geib and Galen 2012). For example, both Gange and Ayers (1999) and Vannette and Hunter (2011) showed that plant benefits varied as hyperbolic or unimodal functions of mycorrhizal colonization density. Morris et al. (2010) used information-criterion statistics to examine functional responses of two pollination mutualisms, showing that hyperbolic and unimodal functions may occur. In yet other studies not testing functional responses per se, density dependence of mutualism's consumer–resource interactions is beginning to be demonstrated (Agrawal and Fordyce 2000, Chamberlain and Holland 2008).

While some important studies have examined the density dependence of mutualism, few empirical studies have tested the underlying factors governing various parameters of population dynamics, such as population growth rates, fluctuations, and oscillations (Morales 2011, Yoo and Holway 2011). One example of such a study is that by Morales (2011), in which information-criterion statistics were employed to assess the fit of various models entailing different formulations of linear and saturating functional responses to empirical data on ant interactions with treehoppers. Results not only show that mutualism can shape population dynamics, but that nonlinear density dependence of mutualistic interactions can occur between ants and treehoppers. The time is right for an increase in theoretical and empirical studies that give explicit consideration to density dependence in the population ecology of mutualism. While we know that linear functional responses lead us astray, empirical studies have only scratched the surface of testing the appropriate functional responses for different mutualism.

8.6 Conclusions

Historically, studies of the population ecology of mutualism have been largely divorced from central principles that have shaped our understanding of the population dynamics of other forms of species interactions. Unlike predation and competition, Lotka–Volterra models of mutualism with linear functional responses lead to artificial and unrealistic results, providing little to no foundation from which to build more mechanistic models and guide empirical studies. While the 1960s and 1970s yielded great progress in mechanistic theory for predation and competition, it has only been in recent times that advances over historic Lotka–Volterra models have been evident for mutualisms. It is clear that at a minimum nonlinear saturating functional responses resolve the mathematically artificial conclusions drawn from linear Lotka–Volterra models.

Great promise can be foreseen in the further development of mechanistic theory for and empirical studies of functional responses in terms of mutualism's benefits, costs, and consumer–resource interactions. Importantly, if the benefits, costs, and consumer–resource interactions of mutualism have consequences for population dynamics, then they must manifest in demographic rates of births, deaths, immigration, and emigration. While we have decades of studies of the various benefits of mutualisms for many different study systems, we know little of whether costs and resource provision are discernible in terms of demographic rates in ecological time. Studies are needed that explicitly examine which demographic rate is influenced by mutualism and how it varies with density-dependent functional responses.

The benefit and cost mechanism of interaction is leading to new insights into the density dependence and population dynamics of mutualism, but this approach has some shortcomings for theoretical and empirical studies. First, studies of the benefits of mutualism tend to emphasize only one species involved in an interaction. For example, plant–pollinator interactions are classic transportation mutualisms, yet this classification is phytocentric, emphasizing plant gains in reproduction and transportation of gametes while neglecting pollinator gains in food resources (nectar, pollen). Second, because the biological currencies of benefits and costs are often particular to one or a few specific types of mutualisms, general theory commonly shifts to system-specific models rather than mechanistic theory for mutualism in general. Third, benefits and especially costs may not be measured directly in the demographic currencies (births, deaths) that shape the density dependence of interspecific interactions. Benefits and costs are currencies that continue to differentiate mutualism from other species interactions. Fourth, and of particular importance, nearly all mutualisms entail the production and exchange of resources. In this way, mutualism, like predation and competition, inherently entails the consumer–resource interaction. In fact, the C–R mechanism is similar to that of the benefit and cost mechanism of interaction, as the benefits and costs typically equate with the consumption and supply of resources. Yet the C–R approach is more readily generalizable among mutualisms and species interactions overall, and it utilizes key currencies of biomass or density. In fact, the C–R approach unifies the population ecology of mutualism as a general mechanism for understanding density-dependent population dynamics of their interactions. Moreover, because predation, competition, and mutualism can all be unified under the common ecological framework of consumer–resource theory, the various types of interspecific interactions may have more in common than previously thought.

The C–R approach to mutualism is in its infancy, but can readily borrow from its long history and application in the study of the population dynamics of predation and competition. Such studies should greatly facilitate the direction of future interplay between theory and empiricism in the investigation of the ecological dynamics of mutualistic interactions. Indeed, there has been little to no quantitative interplay between theory and empiricism in the study of mutualism, with the exception of a few critical studies in recent years. For example, Morris et al. (2010) examine a series of theoretically feasible functional responses for the benefits and costs of pollination mutualisms, and then using empirical data and information-theoretic methods they assessed which theoretical functional responses best describe the empirical data of some particular study systems. In a study of mutualistic ant–treehopper interactions, Morales (2011) also used model selection methods to test among different models which best describe population dynamics arising from density-dependent functional responses of benefits. Methodological approaches (Bolker 2008) of studies such as these are key to providing the crucial interplay between theory and empiricism that is necessary to advance our understanding of the population ecology of mutualism.

Some important first steps have been taken in recent years for understanding the density dependence and population dynamics of mutualism, yet our knowledge of many prominent topics of study in population ecology remains sparse for mutualism, including population regulation (Lundberg and Ingvarsson 1998, Holland et al. 2004), interaction strengths (Box 8.3; Vazquez et al. 2012), and trait-mediated density dependence (Harmon and Andow 2007). Theory for limitations of mutualism's positive feedback and its influence on the population dynamics of interacting species centers around contrasting roles of intrinsic density-dependent functional responses, intraspecific self-limitation, interspecific competition, and extrinsic predation. With these competing hypotheses for the stabilization and dynamics of mutualism, empirical studies are needed that explicitly test these mechanisms and their relative influences on the density-dependent population dynamics of mutualism. This is one of many key areas of future research at the interface of theory and empiricism. Another example is that of the metapopulation ecology of mutualism (Box 8.1). While some initial theory and empirical studies have begun to be undertaken on the metapopulation ecology of mutualism, this is a major area needing further theoretical and empirical developments. Greater consideration needs to be given to the metapopulation dynamics of open systems of mutualistic interactions whereby migration and dispersal among local populations may be key to ecological dynamics (Box 8.1; Thompson 2005, Thompson et al. 2006). This is particularly important given that many mutualisms involve non-trophic services of dispersal (Table 8.1) that can mitigate colonization and extinction rates of metapopulations. As a final example, classic studies of the energy budgets of interacting species are just beginning to take place for mutualism with the recognition of its consumer–resource interactions as the means for biomass accumulation (Muller et al. 2009). While the study of the population ecology of mutualism has come a very long way since Gause and Witt (1935), and rapid progress has been made in the field in recent years, we still have a long way to go to catch up with the base of empirical and theoretical knowledge that now serves as a foundation for the study of predation and competition. We are well poised to make these leaps with further rapidity as mutualism declines in being perceived as an idiosyncratic interaction with little influence on the ecological dynamics of interacting species.

References

Abrams, P.A. (1987). On classifying interactions between populations. *Oecologia*, **73**, 272–281.

Addicott, J.F. (1979). A multispecies aphid-ant association: density dependence and species-specific effects. *Canadian Journal of Zoology*, **57**, 558–569.

Addicott, J.F. (1981). Stability properties of 2-species models of mutualism: simulation studies. *Oecologia*, **49**, 42–49.

Agrawal, A.A., and Fordyce, J.A. (2000). Induced indirect defence in a lycaenid-ant association: the regulation of a resource in a mutualism. *Proceedings of the Royal Society B: Biological Sciences*, **267**, 1857–1861.

Bascompte, J., Jordano, P., Melian, C.J., and Olesen, J.M. (2003). The nested assembly of plant-animal mutualistic networks. *Proceedings of the National Academy of Sciences of the United States of America*, **100**, 9383–9387.

Bascompte, J., Jordano, P., and Olesen, J.M. (2006) Asymmetric coevolutionary networks facilitate biodiversity maintenance. *Science*, **312**, 431–433.

Bolker, B.M. (2008). *Ecological Models and Data in R*. Princeton: Princeton University Press.

Breton, L.M., and Addicott, J.F. (1992). Density-dependent mutualism in an aphid-ant interaction. *Ecology*, **73**, 2175–2180.

Case, T.J. (2000). *An Illustrated Guide to Theoretical Ecology*. New York: Oxford University Press.

Chamberlain, S.A., and Holland, J.N. (2008). Density-mediated, context-dependent consumer resource interactions between ants and extrafloral nectar plants. *Ecology*, **89**, 1364–1374.

Cushman, J.H., and Whitham, T.G. (1989). Conditional mutualism in a membracid-ant association: temporal, age-specific, and density-dependent effects. *Ecology*, **70**, 1040–1047.

Dean, A.M. (1983). A simple model of mutualism. *American Naturalist*, **121**, 409–417.

DeAngelis, D.L., Post, W.M., and Travis, C.C. (1986). *Positive Feedback in Natural Systems*. New York: Springer-Verlag.

DeAngelis, D.L., Travis, C.C., and Post, W.M. (1979). Persistence and stability of seed-dispersed species in a patchy environment. *Theoretical Population Biology*, **16**, 107–125.

de Mazancourt, C., and Schwartz, M.W. (2010). A resource ratio theory of cooperation. *Ecology Letters*, **13**, 349–359.

Fishman, M.A., and Hadany, L. (2010). Plant-pollinator population dynamics. *Theoretical Population Biology*, **78**, 270–277.

Fonseca, C.R. (1999). Amazonian ant-plant interactions and the nesting space limitation hypothesis. *Journal of Tropical Ecology*, **15**, 807–825.

Frederickson, M.E., and Gordon, D.M. (2007). The devil to pay: a cost of mutualism with *Myrmelachista schumanni* ants in "devil's gardens" is increased herbivory on *Duroia hirsuta* trees. *Proceedings of the Royal Society B: Biological Sciences*, **274**, 1117–1123.

Gange, A.C., and Ayres, R.L. (1999). On the relation between arbuscular mycorrhizal colonization and plant "benefit." *Oikos*, **87**, 615–621.

Gause, G., and Witt, A. (1935). Behavior of mixed populations and the problem of natural selection. *American Naturalist*, **69**, 596–609.

Geib, J.C., and Galen, C. (2012). Tracing impacts of partner abundance in facultative pollination mutualisms: from individuals to populations. *Ecology*, **93**, 1581–1592.

Goh, B.S. (1979). Stability in models of mutualism. *American Naturalist*, **113**, 261–275.

Harmon, J.P., and Andow, D.A. (2007). Behavioral mechanisms underlying ants? Density-dependent deterrence of aphid-eating predators. *Oikos*, **116**, 1030–1036.

Heithaus, E.R., Culver, D.C., and Beattie, A.J. (1980). Models of some ant-plant mutualisms. *American Naturalist*, **116**, 347–361.

Holland, J.N., and DeAngelis, D.L. (2006). Interspecific population regulation and the stability of mutualism: fruit abortion and density-dependent mortality of pollinating seed-eating insects. *Oikos*, **113**, 563–571.

Holland, J.N., and DeAngelis, D.L. (2010). A consumer-resource approach to the density-dependent population dynamics of mutualism. *Ecology*, **91**, 1286–1295.

Holland, J.N., DeAngelis, D.L., and Bronstein, J.L. (2002). Population dynamics and mutualism: functional responses of benefits and costs. *American Naturalist*, **159**, 231–244.

Holland, J.N., DeAngelis, D.L., and Schultz, S. (2004). Evolutionary stability of mutualism: interspecific population regulation as an evolutionarily stable strategy. *Proceedings of the Royal Society of London Series B-Biological Sciences*, **271**, 1807–1814.

Holland, J.N., Okuyama, T., and DeAngelis, D.L. (2006). Comment on "Asymmetric coevolutionary networks facilitate biodiversity maintenance". *Science*, **313**, 1887–1887.

Holland, J.N., Wang, Y., Sun, S., and DeAngelis, D.L. (2013). Consumer-resource dynamics of indirect interactions in a mutualism-parasitism food web module. *Theoretical Ecology*, **6**, 475–493.

Hsu, S.B., Hubbell, S.P., and Waltman. P. (1978). A contribution to the theory of competing predators. *Ecological Monographs*, **48**, 337–349.

Ingvarsson, P.K., and Lundberg, S. (1995). Pollinator functional response and plant population dynamics: pollinators as a limiting resource. *Evolutionary Ecology*, **9**, 421–428.

Inouye, D.W., and Taylor, O.R. (1979). A temperate region plant-ant-seed predator system: consequences of extra floral nectar secretion by *Helianthella quinquenervis*. *Ecology*, **60**, 1–7.

Iwata, S., Kobayashi, K., Higa, S., Yoshimura, J., and Tainaka, K. (2011). A simple population theory for mutualism by the use of lattice gas model. *Ecological Modelling*, **222**, 2042–2048.

Jeschke, J.M., Kopp, M., and Tollrian, R. (2004). Consumer-food systems: why type I functional responses are exclusive to filter feeders. *Biological Reviews*, **79**, 337–349.

Johnson, C.A., and Amarasekare, P. (2013). Competition for benefits can promote the persistence of mutualistic interactions. *Journal of Theoretical Biology*, **328**, 54–64.

Jones, E.I., Bronstein, J.L., and Ferrière, R. (2012). The fundamental role of competition in the ecology and evolution of mutualisms. *Annals of the New York Academy of Sciences*, **1256**, 66–88.

Kang, Y., Clark, R., Makiyama, M., and Fewell, J. (2011). Mathematical modeling on obligate mutualism interactions between leaf-cutter ants and their fungus garden. *Journal of Theoretical Biology*, **289**, 116–127.

Kot, M. (2001). *Elements of Mathematical Ecology*. New York: Cambridge University Press.

Lee, C.T., and Inouye, B.D. (2010). Mutualism between consumers and their shared resource can promote competitive coexistence. *American Naturalist*, **175**, 277–288.

Leon, J.A., and Tumpson, D.B. (1975). Competition between two species for two complementary or substitutable resources. *Journal of Theoretical Biology*, **50**, 185–201.

Lundberg, S., and Ingvarsson, P.K. (1998). Population dynamics of resource limited plants and their pollinators. *Theoretical Population Biology*, **54**, 44–49.

May, R.M. (1976). Models of two interacting populations. In R.M. May (ed.), *Theoretical Ecology*. Boston: Blackwell Scientific Publications.

Morales, M.A. (2000). Mechanisms and density dependence of benefit in an ant-membracid mutualism. *Ecology*, **81**, 482–489.

Morales, M.A. (2011). Model selection analysis of temporal variation in benefit for an ant-tended treehopper. *Ecology*, **92**, 709–719.

Morales, M.A., Morris, W.F., and Wilson, W.G. (2007). Allee dynamics generated by protection mutualisms can drive oscillations in trophic cascades. *Theoretical Ecology*, **1**, 77–88.

Morris, W.F., Bronstein, J.L., and Wilson, W.G. (2003). Three-way coexistence in obligate mutualist-exploiter interactions: the potential role of competition. *American Naturalist*, **161**, 860–875.

Morris, W.F., Vazquez, D.P., and Chacoff, N.P. (2010). Benefit and cost curves for typical pollination mutualisms. *Ecology*, **91**, 1276–1285.

Muller, E.B., Kooijman, S.A.L.M., Edmunds, P.J., Doyle, F.J., and Nisbet, R.M. (2009). Dynamic energy budgets in syntrophic symbiotic relationships between heterotrophic hosts and photoautotrophic symbionts. *Journal of Theoretical Biology*, **259**, 44–57.

Murdoch, W.M., Briggs, C.J., and Nisbet, R.M. (2003). *Consumer-Resource Dynamics*. New Jersey: Princeton University Press.

Nefdt, R.J., and Compton, S.G. (1996). Regulation of seed and pollinator production in the fig-fig wasp mutualism. *Journal of Animal Ecology*, **65**, 170–182.

Ness, J.H., Morris, W.F., and Bronstein, J.L. (2006). Integrating quality and quantity of mutualistic service to contrast ant species protecting *Ferocactus wislizeni*. *Ecology*, **87**, 912–921.

Okuyama, T., and Holland, J.N. (2008). Network structural properties mediate the stability of mutualistic communities. *Ecology Letters*, **11**, 208–216.

Pierce, N.E., and Young, W.R. (1986). Lycaenid butterflies and ants: two-species stable equilibria in mutualistic, commensal, and parasitic interactions. *American Naturalist*, **128**, 216–227.

Ringel, M.S., Hu, H.H., and Anderson, G. (1996). The stability and persistence of mutualisms embedded in community interactions. *Theoretical Population Biology*, **50**, 281–297.

Rosenzweig, M.L., and MacArthur, R. (1963). Graphical representation and stability conditions of predator-prey interactions. *American Naturalist*, **97**, 209–223.

Schwartz, M.W., and Hoeksema, J.D. (1998). Specialization and resource trade: biological markets as a model of mutualisms. *Ecology*, **79**, 1029–1038.

Soberon, J.M., and Martinez del Rio, C. (1981). The dynamics of a plant-pollinator interaction. *Journal of Theoretical Biology*, **91**, 363–378.

Thompson, A.R. (2005). Dynamics of demographically open mutualists: immigration, intraspecific competition, and predation impact goby populations. *Oecologia*, **143**, 61–69.

Thompson, A.R., Nisbet, R.M., and Schmitt, R.J. (2006). Dynamics of mutualist populations that are demographically open. *Journal of Animal Ecology*, **75**, 1239–1251.

Turchin, P. (2003). *Complex Population Dynamics*. New Jersey: Princeton University Press.

Vandermeer, J.H., and Boucher, D.H. (1978). Varieties of mutualistic interaction in population models. *Journal of Theoretical Biology*, **74**, 549–558.

Vannette, R.L., and Hunter, M.D. (2011). Plant defence theory re-examined: nonlinear expectations based on the costs and benefits of resource mutualisms. *Journal of Ecology*, **99**, 66–76.

Vazquez, D.P., Lomascolo, S.B. Maldonado, M.B., Chacoff, N.P., Dorado, J., Stevani, E.L., and Vitale, N.L. (2012).

The strength of plant-pollinator interactions. *Ecology*, **93**, 719–725.

Vazquez, D.P., Morris, W.F., and Jordano, P. (2005). Interaction frequency as a surrogate for the total effect of animal mutualists on plants. *Ecology Letters*, **8**, 1088–1094.

Wolin, C.L., and Lawlor, L.R. (1984). Models of facultative mutualism: density effects. *American Naturalist*, **124**, 843–862.

Wright, D.H. (1989). A simple, stable model of mutualism incorporating handling time. *American Naturalist*, **134**, 664–667.

Yoo, H.J.S., and Holway, D.A. (2011). Context-dependence in an ant-aphid mutualism: direct effects of tending intensity on aphid performance. *Ecological Entomology*, **36**, 450–458.

CHAPTER 9

Mutualism in a community context

Todd M. Palmer, Elizabeth G. Pringle, Adrian Stier, and Robert D. Holt

9.1 Introduction

In the highlands of East Africa, a giraffe ambles through a savanna dotted with *Acacia drepanolobium* trees, which stand half the giraffe's height and bristle with spines and bulbous hollow structures known colloquially as whistling thorns. The trees are myrmecophytes, or ant-plants, members of an iconic class of protection mutualisms first elucidated by Janzen (1966) nearly fifty years ago. As the giraffe lowers its head to feed from a tree, ants stream from the hollow domatia and attack its lips and nostrils. Surprisingly, the giraffe is unperturbed: it continues to browse, using its long prehensile tongue to wipe the ants away. Closer inspection reveals another surprise: the *Crematogaster nigriceps* ants that occupy this plant have methodically pruned it, sterilizing the tree by destroying the buds that would have developed into flowers and fruits. This is all occurring even as the tree pays heavily to support *C. nigriceps*, channeling energy into the production of extrafloral nectar that feeds the ants and domatia that house them. Indeed, the metabolic cost of these investments is sufficient to reduce the plant's growth rate. From the classic pairwise perspective of species interactions, therefore, *C. nigriceps* appears to be a parasite of its host plant, imposing multiple costs while providing no obvious benefit. Yet we now know that this relationship is in fact mutualistic, providing both growth and reproductive benefits to host trees (Palmer et al. 2010). How can this be the case?

The answer lies in the community context of the interaction. To fully understand the ecology and evolution of mutualism, we must consider other partners in the mutualism, and also the many non-mutualistic interactors in the community. And so in turn we must broaden both the spatial and temporal scopes of mutualism studies: for *A. drepanolobium*, the balance sheet of costs and benefits may be kept for 150 years or longer, and lifetime fitness is the only currency that matters in the end.

By taking a community-based, long-term perspective, we can unpack the apparent paradox of this *A. drepanolobium*–*C. nigriceps* interaction. In terms of defense, *C. nigriceps* is not very effective at defending host plants from chronic giraffe herbivory (Stanton and Palmer 2011), but it is highly effective against rare but potentially lethal elephant herbivory (Goheen and Palmer 2010). Indeed, long-term studies of the interaction have shown that elephants appear to be a driving force in the evolution of this association (Stanton and Palmer 2011). In terms of sterilization, *C. nigriceps* is but one of four ant species that compete intensely for exclusive occupancy of the tree, and it is the only ant that sterilizes the tree. As a subordinate ant in the competitive hierarchy, *C. nigriceps* persists in part by being a strong colonizer of new host plants. It therefore tends to occupy plants at early life stages, when reproductive potential is low. For young plants, protection from elephants is more important for lifetime fitness than squeezing out a few seeds: it is better for the individual to survive to reproduce at a larger, less vulnerable, and more fecund stage. As plants mature, competitively superior, non-sterilizing ant species supplant *C. nigriceps*, releasing the plant from reproductive suppression. However, these other ants do not defend the tree as effectively against elephants. In the context of

Mutualism. Edited by Judith L. Bronstein.

long-term community dynamics, therefore, *C. nigriceps* enhances the tree's lifetime fitness by conferring increased survival at the vulnerable early stage. Moreover, the cost to the trees occupied by *C. nigriceps* in terms of current reproduction is outweighed by the sum of future offspring (Palmer et al. 2010).

Although the *Acacia drepanolobium*–ant mutualism is a highly specialized and geographically restricted interaction, it shares common features with many of the world's mutualisms. Mutualisms nearly always include a diversity of interacting species (Box 9.1, by Michael Stat, highlights the diversity within coral-algal symbioses), variation in the costs imposed and benefits provided by different mutualist partners, and interactions between mutualists and non-mutualist members of the broader community. In this chapter, we elaborate on the themes underpinning the ant–acacia example, arguing that much recent progress in our understanding of mutualism has been made by carefully considering spatiotemporal heterogeneity of the broader communities in which mutualisms are embedded. We begin by discussing the complexity inherent in mutualisms, which typically consist of multiple species on one or both sides of the interaction. We then move on to consider the role of the broader community in shaping mutualisms, briefly discussing how mutualisms arise from and are changed by the interactions with the organisms that surround them. Finally, we consider how mutualisms themselves feedback to modify the structure and stability of the communities in which they occur.

Box 9.1 The community ecology of coral-algal mutualism

Michael Stat

Many corals form a mutually beneficial interaction with single-celled algae belonging to the dinoflagellate genus *Symbiodinium*. The association is primarily defined by the exchange of nutritional compounds that increases the metabolic capacity of the partners in the otherwise nutrient-poor oligotrophic ocean waters where corals are found (Muscatine and Porter 1977). The animal host receives organic carbon fixed through algal photosynthesis as well as amino acids, while the dinoflagellate symbiont receives inorganic compounds as a byproduct of host respiration. This highly efficient cycling of nutrients between the partners is made possible by the spatially intimate localization of the symbionts that are housed within the gastrodermal tissue of the animal, and ultimately results in the growth of corals and formation of reef ecosystems (Box 9.1 Figure 1).

In coral-algal mutualism, it has become evident that a dynamic interaction between the host and a diverse community of dinoflagellate symbionts occurs. This discovery was made possible with the application of molecular tools to identify different populations of *Symbiodinium*. In contrast to the limited morphological variation among *Symbiodinium* algae that contributed to the earlier perception of a single species within the genus inhabiting all corals (and other animal hosts), molecular techniques have revealed a relatively large amount of genetic diversity within this taxonomic group as compared to other dinoflagellate genera. To date, nine divergent evolutionary lineages that have been called clades A–I have been identified in the genus, with each lineage containing varying numbers of strains or subclade symbiont types (Pochon and Gates 2010). The importance of the genetic diversity among members of *Symbiodinium* is recognized by the functional diversity displayed by different symbiont types and their effect on host fitness. For example, differences among *Symbiodinium* strains in photophysiology, thermal tolerance, and carbon fixation and translocation to the host, and their subsequent effect on coral growth rates have been observed (Little et al. 2004, Rowan 2004, Cantin et al. 2009).

The majority of corals associate with a single dominant strain of *Symbiodinium*. However, recent research has now shown that most, if not all, corals also harbor a diversity of cryptic symbionts or *Symbiodinium* strains in very low numbers (Silverstein et al. 2012). While there are biogeographic and host specificity patterns in the distribution of the dominant symbionts of corals, providing evidence for an evolutionary influence on host–symbiont assemblages (LaJeunesse 2005), the functional role and interaction of cryptic symbionts with corals and their contribution to the stability and overall persistence of the symbiosis is unclear. Interestingly, changes in the community composition of *Symbiodinium* in corals away from the dominant homologous symbiont have

Box 9.1 *Continued*

been observed during ontogeny of the coral and in response to local environmental changes.

Coral-algal symbioses are very sensitive to variation and changes in the environment. Elevated seawater temperature (and a variety of other abiotic factors) can cause a breakdown in the symbiosis between corals and their dinoflagellate algae, leading to the expulsion of the algal complement from the host tissue, and resulting in the phenotype described as coral bleaching (Hoegh-Guldberg and Smith 1989). Prolonged exposure to temperature increases as little as 1 °C above the local average maximum can potentially lead to coral bleaching and mortality. During thermal stress (and other factors that affect the symbiosis), however, a disruption in the symbiotic interaction between the partners provides the opportunity for different *Symbiodinium* strains to proliferate and occupy the coral host environment (i.e. opportunistic symbionts). For some corals, shifts toward a more thermally tolerant symbiont population during ocean warming have been shown to occur, buffering the coral against the stress event and increasing their resistance to bleaching (Jones et al. 2008).

While changes in the symbiont community to thermally tolerant *Symbiodinium* may provide a rapid mechanism of acclimatization for corals, there is a trade-off to such

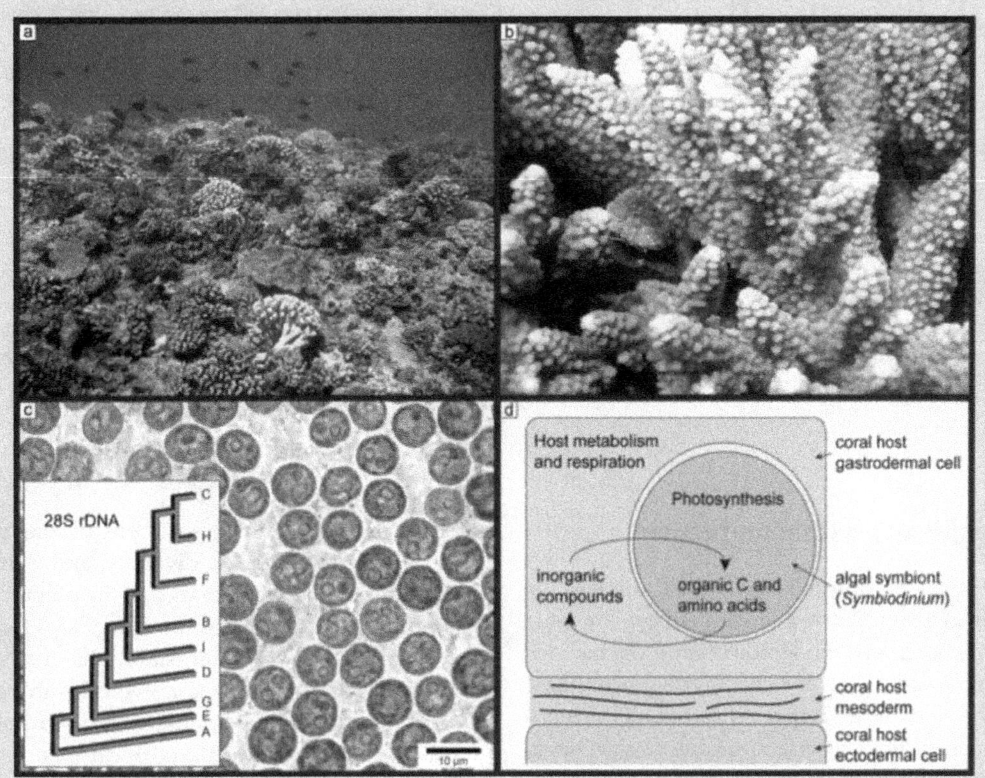

Figure 1 (a) The corals on the forereef of Moorea, French Polynesia, that grow and thrive as a result of the mutualistic association between corals and algae belonging to the genus *Symbiodinium*. (b) A close-up of a coral colony of a branching species of *Acropora*. (c) A bright-field microscope view of dinoflagellate algae in the genus *Symbiodinium* that form a mutualistic symbiosis with corals (image courtesy of Peta Clode, CMCA, The University of Western Australia). The inset shows a cartoon representation of the phylogeny of *Symbiodinium* and the evolutionary relationship of the nine lineages (clades A–I) within the genus based on the sequence of partial nuclear ribosomal DNA. (d) The localization of the algal symbionts that are housed within the gastrodermal cells of the coral host and the cycling of nutrients between the partners. See Plate 6 for color version.

continued

Box 9.1 *Continued*

symbiont transitions. Community shifts to thermally tolerant algae result in shifts away from the homologous symbionts of corals, and a breakdown in the host-specific patterns that have been observed. These altered coral-algal assemblages are not likely to be as productive as those developed over evolutionary time scales. Indeed, recent evidence shows thermally tolerant symbionts produce and translocate less carbon to their coral hosts as compared to heat-sensitive symbionts, resulting in slower coral growth rates (Little et al. 2004, Cantin et al. 2009, Jones and Berkelmans 2010). How these altered associations also affect other aspects of coral biology such as reproductive output is unknown. The role of cryptic, opportunistic, and thermally tolerant symbionts to the perpetuation and resilience of coral-algal symbioses is an area of current research, and may shed light on how this mutualism has persisted for millions of years.

References

Cantin, N.E., van Oppen, M.J.H., Willis, B.L., Mieog, J.C., and Negri, A.P. (2009). Juvenile corals can acquire more carbon from high-performance algal symbionts. *Coral Reefs*, **28**, 405–414.

Hoeghguldberg, O., and Smith, G.J. (1989). The effect of sudden changes in temperature, light and salinity on the population-density and export of zooxanthellae from the reef corals *Stylophora pistillata* Esper and *Seriatopora hystrix* Dana. *Journal of Experimental Marine Biology and Ecology*, **129**, 279–303.

Jones, A., and Berkelmans, R. (2010). Potential costs of acclimatization to a warmer climate: growth of a reef coral with heat tolerant vs. sensitive symbiont types. *Plos One*, **5**.

Jones, A.M., Berkelmans, R., van Oppen, M.J.H., Mieog, J.C., and Sinclair, W. (2008). A community change in the algal endosymbionts of a scleractinian coral following a natural bleaching event: field evidence of acclimatization. *Proceedings of the Royal Society B-Biological Sciences*, **275**, 1359–1365.

LaJeunesse, T.C. (2005). "Species" radiations of symbiotic dinoflagellates in the Atlantic and Indo-Pacific since the miocene-pliocene transition. *Molecular Biology and Evolution*, **22**, 570–581.

Little, A.F., van Oppen, M.J.H., and Willis, B.L. (2004). Flexibility in algal endosymbioses shapes growth in reef corals. *Science*, **304**, 1492–1494.

Muscatine, L., and Porter, J.W. (1977). Reef corals—mutualistic symbioses adapted to nutrient-poor environments. *Bioscience*, **27**, 454–460.

Pochon, X., and Gates, R.D. (2010). A new *Symbiodinium* clade (*Dinophyceae*) from soritid foraminifera in Hawai'i. *Molecular Phylogenetics and Evolution*, **56**, 492–497.

Rowan, R. (2004). Coral bleaching—thermal adaptation in reef coral symbionts. *Nature*, **430**, 742–742.

Silverstein, R.N., Correa, A.M.S., and Baker, A.C. (2012). Specificity is rarely absolute in coral-algal symbiosis: implications for coral response to climate change. *Proceedings of the Royal Society B-Biological Sciences*, **279**, 2609–2618.

9.2 Complexity within mutualism

Early studies of mutualism focused on species pairs, but we now know that mutualisms are usually interactions between guilds of multiple partner species (Stanton 2003). We consider such mutualisms to be "complex." Importantly, not all partners within such mutualisms are equal. Per capita effects of a given partner on its mutualist can range from strongly positive to mostly negative (in which case the partner is an exploiter rather than a mutualist; see Chapter 6). In addition, even partners that have similar per capita effects on a mutualist can vary in abundance, and hence in the magnitude of their population-level effects. There may also be interactions among the co-occurring species within a partner guild. Such interactions can dramatically modify the community structure of the partner species (i.e., the identity, diversity, and density of partners) and the performance of the other mutualist. Going back to our earlier *C. nigriceps* example, it was necessary to understand the interactions among the different acacia ant species in order to explain spatiotemporal patterns in both ant community structure and host tree performance.

In this section, we focus on three important questions that arise from our recognition of the ubiquitous complexity in mutualism. First, what maintains the diversity of potential mutualistic partners? Second, how does such complexity influence the fitness and performance outcomes of these interactions for the mutualists? Finally, can we understand multispecies mutualisms by studying their constituent pairwise associations, or do the collective interactions lead to emergent properties that require us to study all partners at once? Below we address each of these questions in turn.

9.2.1 Maintenance of mutualist diversity

Addressing what generates and maintains mutualist species diversity first requires us to define our terms. Mutualism is a very broad class of interactions (involving benefits such as seed dispersal, pollination, nutrition, and protection; see Chapter 1), and the mechanisms that mediate diversity in each of these interactions could be very different. Drawing generalities about how species diversity is maintained in diffuse, free-living mutualisms will probably be particularly challenging because the partners in such mutualisms are often taxonomically disparate. For example, in dispersal mutualisms, the seeds of a particular tree species may be primarily or secondarily dispersed by birds, primates, insects, and rodents, and species diversity in each guild may be maintained by different mechanisms. However, the theories of island biogeography, metapopulation dynamics, and metacommunities seem to offer useful conceptual frameworks for explaining diversity in interactions where a mutualist serves as a discrete island (host) for one or more partner species (e.g., coral-algal, plant–ant, plant–endophyte, and other mutualistic symbioses).

Drawing from these theories, diversity in antagonistic interactions, such as host–pathogen relationships, is thought to arise from variability within both individuals and communities, among populations, and through time (Holt and Dobson 2006). Individual hosts act as islands, which first serve as venues for pathogen colonization and later for local extinction. The success of colonization, and the likelihood of extinction, can be determined by competitive interactions among pathogens that share the same host(s). Similarly, when mutualists serve as hosts for multiple symbionts, processes of colonization and extinction, as well as competition among symbionts, should play key roles in mediating partner diversity and abundance. For example, 20 cm-diameter corals in the South Pacific can be inhabited simultaneously by assemblages of more than ten species of crabs that collectively defend the coral from predatory seastars. Adult crabs do not tolerate other adults of the same species on a single coral head, and this negative intraspecific density dependence creates opportunities for stable coexistence of multiple partner crabs on a single host coral (Stier et al. 2012). In another marine example, gutless polychaete *Osedax* worms that colonize "whalefall" (carcasses that have fallen to the ocean floor) host nutritional endosymbiotic bacteria whose diversity appears to be determined by the interplay of ocean depth (and associated abundance of the horizontally transmitted bacteria in the environment), host individual, host ontogenetic stage, and competitive interactions among the bacteria (Verna et al. 2010).

Do certain conditions generally favor low- versus high-diversity mutualist guilds? In the simplest case, strong partner choice by a mutualist can effectively screen out other potential partners, leading to high specificity within the interaction. But in other cases, the ecological and evolutionary factors that generate the observed continuum from monospecific dominance in some interactions to diverse guilds in others are not as obvious. One possibility is that the strength of competition among mutualist partners controls the diversity of the partner community. For example, intraspecific competition within guilds of partner species should increase diversity, as in the coral–crab mutualism described above, and interspecific competition should decrease diversity, as has been reported for plant-mycorrhizal mutualisms (Jansa et al. 2008). Alternatively, when different partners provide complementary benefits to mutualists, those partners may foster diversity by becoming indirect mutualists of one another. For example, the gastropods *Anachis lafresnayi* and *Mitrella lunata* remove different types of fouling epibionts from the red algae *Chondrus crispus*, and both gastropod species are required for the algae to grow, which protects the gastropods from crab predation (Stachowicz and Whitlatch 2005). When selection on competing partner species to reduce niche overlap leads to functional complementarity in the benefits they provide to their mutualist, positive relationships between partner diversity and mutualism stability can arise and affect entire networks (Bastolla et al. 2009). A key remaining question is whether particular types of mutualisms are more conducive than others to spatial and/or temporal niche partitioning among their partners. Whatever the ultimate causes of mutualist partner diversity, it is clear that this complexity has pervasive consequences for the performance and fitness outcomes of mutualistic interactions.

9.2.2 Consequences of diversity for mutualistic outcomes

When mutualists interact with many partner species, either serially or simultaneously, differences in partner identity and/or composition can substantially alter the performance of the other mutualist in at least two ways. First, when the total number of partners is constrained by a limiting resource supplied by the other mutualist (e.g., plant carbohydrates for mycorrhizal associates, Kiers et al. 2011), the resource-supplying mutualist is harmed by interactions with less-beneficial partners because they deplete the resource and reduce the potential for future interactions with better partners. In such cases, changes in the community of partner species, or more specifically in the proportion of partners of different quality, should alter the performance of the other mutualist. For example, the loss of large herbivores in an African savanna has caused *A. drepanolobium* trees to reduce the overall quantity of rewards for ants (Figure 9.1). This led to a shift in the community of ant associates of *A. drepanolobium* from more- to less-beneficial species, which, in turn, reduced rates of plant growth and increased plant mortality (Palmer et al. 2008). Second, changes in the partner community can shift the balance between cooperative and antagonistic interactions among partners, with a range of indirect consequences for the other mutualist. For example, van der Heijden et al. (1998) found a positive correlation between mycorrhizal fungal diversity and plant productivity in an old-field community. This positive correlation could be driven either by complementarity, if different mycorrhizal species perform non-redundant functions, or, possibly, by cooperative interactions among the mycorrhizal species themselves (Jansa et al. 2008). Quantifying the mechanisms underlying the effects of changes in partner diversity is an important future step in mutualism ecology. Importantly, understanding these mechanisms could help elucidate how anthropogenic disturbance to ecological communities will affect ecosystem function.

Despite the recognition that partner diversity can strongly affect the performance of mutualists (In Box 9.2, Daniel Doak shows how guilds of poor mutualists may benefit their partner as a group), however, we do not yet know whether there is any general pattern in the direction of outcomes

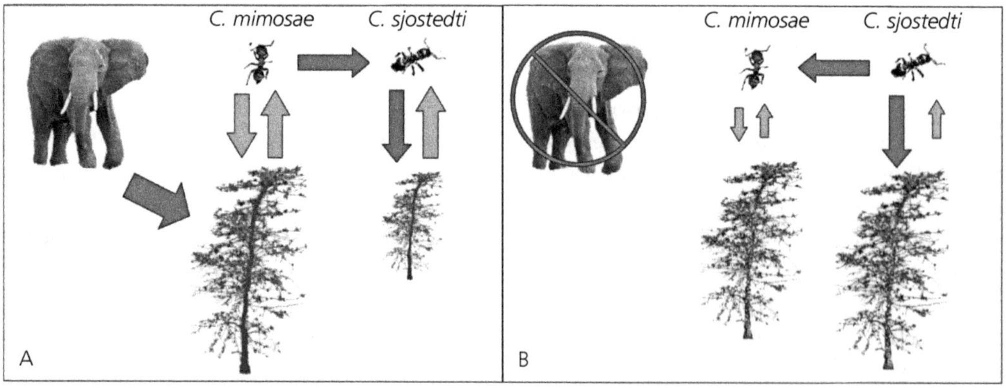

Figure 9.1 Effects of large mammal extinction on an ant–*Acacia* mutualism in East Africa. Browsing by large mammals (Panel A) induces high levels of nectar production in *Acacia drepanolobium*, which favors the competitive dominance of the strongly mutualistic and nectar-dependent *Crematogaster mimosae* over its competitor *C. sjostedti*, an antagonist of the host plant which does not depend on host plant nectar. Following the extinction of large mammals (Panel B), host plants reduce their production of extrafloral nectar, shifting the balance of competition among ant species towards dominance by the antagonistic *C. sjostedti*. Increased occupancy of host plants by *C. sjostedti* results in slower rates of tree growth and dramatically increased host plant mortality (Palmer et al. 2008).

produced by higher partner diversity (Morris et al. 2007). Expanding the temporal scales of mutualism studies seems particularly critical to assessing this question. For one, the magnitude of benefits derived from different partner species may vary with environmental conditions (Kersch and Fonseca 2005). For another, variation in the year-to-year reliability of a particular partner species may modify the "value" of alternate partners to the mutualist (Palmer et al. 2003). For example, a highly beneficial partner whose populations are highly variable will not confer many benefits to its mutualist in years when its densities are low. In general, the impact of partner diversity on host performance is probably highly context dependent (see Chapter 10), and a significant challenge is to identify the mechanisms underlying this variability. Moreover, we must consider to what extent we can actually quantify the overall outcomes of complex mutualistic webs by studying the individual, experimentally amenable components of a given mutualism.

Box 9.2 How imperfect relationships with multiple, exploiting partners can lead to mutualistic success

Daniel F. Doak

For a relationship between individuals of two species to be a mutualism, each partner must gain more benefits than it loses. If one partner is, on balance, hurt by the interaction and the interacting species is also supplanting the role of what we would otherwise consider a mutualist, we would generally classify it as an "exploiter" (see Chapter 6). While using the criterion that a mutualism is based on net positive effects for both partners is simple in theory, in many mutualisms the time span of the relationship and the longevity of at least one of the partners can conspire to make the net effects of the interaction quite difficult to unravel. In particular, these complications can allow guilds of poor mutualists to benefit their partner as a group, even when each alone would clearly be considered an exploiter.

Principles of stochastic demography (Tuljapurkar 1990, Doak et al. 2005) suggest that in the most simple circumstances three conditions must be met for an individual of species A to be sustained and even benefited by multiple, imperfect partners of a mutualist guild B (partners B1, B2, B3, etc.). Namely: (1) each alternate B partner must increase some aspects of A's fitness, even if diminishing fitness in other ways; (2) the different B partners must contribute contrasting costs and benefits to A; and (3) an individual of A must have a good chance of experiencing different B partners over its lifetime, either randomly or in a more ordered sequence. In essence, from A's perspective, association with different Bs creates different environments that have negatively correlated effects on different aspects of fitness—if these negative correlations are strong enough, association with multiple Bs can in some cases dramatically elevate A's fitness. Associating with diverse guilds of exploiters is most likely to produce benefits when interactions are relatively prolonged because this creates the opportunity for guild B to influence multiple aspects of species A's fitness. A good example is the relationship described at the outset of this chapter between myrmecophytic trees and ants, where trees are inhabited by many ant species over their long lifetimes (Palmer et al. 2010), and ants can influence the survival, reproduction, and growth of their hosts.

It is easiest to understand these effects with a simple model. Consider a myrmecophyte A with three life stages (seeds, juveniles, and adult plants) that associates with one of two ant species, B1 or B2. B1 is a highly effective defender of its plants, which leads to high survival rates of both juveniles and adults, but B1 also feeds on flower buds, which reduces reproductive success. In contrast, B2 is a poor defender of its plants, which leads to low survival, but it increases fruiting success, perhaps by pollinating or by fertilizing the plant with discarded insect prey. We can make two stage-structured demographic models, one for the demography of A individuals associated with B1, and one for A individuals associated with B2. These demographic transition models are in the standard form, showing transitions between the three stages of the plant's life cycle:

$$A_{B1} = \begin{pmatrix} 0 & 0.06 & 0.45 \\ 0.2 & 0 & 0 \\ 0 & 0.6 & 0.9 \end{pmatrix}, A_{B2} = \begin{pmatrix} 0 & 0.6 & 6 \\ 0.2 & 0 & 0 \\ 0 & 0.2 & 0.6 \end{pmatrix}$$

For our example, the annual growth rates of plants associated with B1 and B2 are $\lambda = 0.96$ and $\lambda = 0.93$,

continued

Box 9.2 *Continued*

respectively. In other words, in both cases a lineage of plants associated with only one ant species will rapidly decline. However, if each year a plant has an equal probability of associating with one or the other of the two ant species, the annual population growth rate (which is also a measure of individual fitness) is a little over 1.01—a growing lineage. Each ant in isolation is harmful, but in combination, they can have strong benefits for their common partner.

While this ability to benefit from two poor mutualists is striking, the scenario we have just described is not the optimal one for the plant. The reproductive costs of the B1 ant are small for the always low-reproducing juvenile plants, whereas the advantage of surviving to be an adult is high; for adults, the relative importance of reproduction is far higher. Thus, selection should favor the plant's ability to manipulate which ant is most likely to occupy it at each of these two life stages. At the extreme, if the plant can manipulate its associations such that juveniles mostly have B1 ants, and adults mostly have B2s, the plant achieves dramatically higher fitness (see Box 9.2 Figure 1). In one real study system we have seen signs of just such advantageous differential occupancy by alternate mutualists at different life stages (Palmer et al. 2010), and we expect similar patterns can be found in other systems.

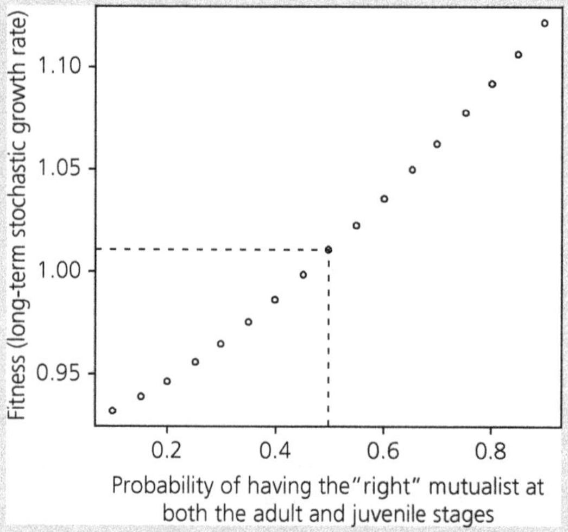

Figure 1 Control of the frequency of association with different mutualists at different life stages can yield strong benefits. In the text, we discuss how randomly alternating association with two partner species can yield benefits for a plant. Here we show results when juvenile plants have increasing probabilities of associating with B1, and adults have the same increasing probabilities of associating with B2. Compared to the random probability that juvenile or adult will have a 50% chance of hosting B1 or B2, there are enormous gains in biasing associations for each life stage of the plant.

Here we've focused on the advantages of interactions with different partners in the case where alternative partners can be considered exploiters. It is worth noting that the same general advantage of association with multiple partners should also in theory occur when alternative partners do not have net negative effects in isolation, but do still have contrasting effects on different aspects of fitness (e.g., a plant associating with nitrogen fixers and with protective ants, or lichen fungi that maintain both nitrogen-fixing bacteria and more photosynthetically efficient green algae).

As our simple analyses show, the line between good mutualists, poor mutualists, and exploiters can be a gray one, and depends on the patterns of association between each species and the entire suite of alternate partners. These examples also show that we should be clear that short-term "exploitation" may not accurately reflect the lifetime costs and benefits of associating with any one partner, when multiple partners interact with single individuals over their lives. Finally, these results point to the interesting hypothesis that mutualists can achieve strong fitness advantages by manipulating the sequence of partners they associate with during different life stages. This is a particularly fascinating question for systems more complex than the one we modeled, where both partners have some control over the interaction.

References

Doak, D.F., Morris, W.F., Pfister, C., Kendall, B.E., and Bruna, E.M. (2005). Correctly estimating how environmental stochasticity influences fitness and population growth. *American Naturalist*, **166**, E14–E21.

Palmer, T.M., Doak, D.F., Stanton, M.L., et al. (2010). Synergy of multiple partners, including freeloaders, increases host fitness in a multispecies mutualism. *Proceedings of the National Academy of Sciences of the United States of America*, **107**, 17234–17239.

Tuljapurkar, S. (1990). Population dynamics in variable environments. *Lecture Notes in Biomathematics*, **85**, 1–154.

9.2.3 Approaches to studying complex mutualisms

Speciose mutualist networks, like all diverse interaction webs, are inherently complex. There are two main approaches to the study of such complexity. We can either focus on a module of interactions among a small number of key players and build up to larger communities, or we can characterize community-wide interaction patterns holistically with network metrics (see Chapter 11). In recent decades, three- to four-species community modules (McCann 2012) have been the staple units for theoretical and experimental investigation of indirect and higher-order interactions (e.g., the non-additive effects of multiple predators or mutualists, apparent competition, intraguild predation, and competitive networks). Two factors motivate the use of community modules. First, modules are empirically manageable and analytically tractable structures within diverse interaction networks. Second, many networks are compartmentalized (Chapter 11), which may make it possible to quantify interactions at the module level and then scale up to entire networks. Although this latter rationale sounds good in theory, the logistical constraints of factorially manipulating more than a few species at a time have so far limited our ability to combine module-level studies in order to elucidate interactions within diverse networks.

When experiments become too onerous, network analyses provide an appealing alternative approach to studying complex webs of mutualistic interactions (Chapter 11). Recently, theoretical ecologists have begun to integrate classic food web theory with mutualistic network theory (Fontaine et al. 2011). Despite the appeal of network analyses, however, community-module studies suggest that the abundance of higher-order interactions in real communities will complicate our efforts to understand them using networks alone (Cohen et al. 2009). Higher-order interactions, in which the properties of a system emerge non-additively, make it difficult to recapitulate the dynamics, structure, and stability of complex systems by analyzing their parts. For example, experimental studies of predator–prey interactions have found that the combined effects of two or more predator species frequently differ from those predicted for a single predator (Vance-Chalcraft et al. 2007). These studies evaluate if effects are independent (i.e., the predator species do not interact), synergistic (i.e., effect on prey is greater than additive), or antagonistic (i.e., effect on prey is less than additive). In many cases, multiple predator effects are sub-additive, as a result of interference among different predator species competing for the same prey. A comparable framework could be employed to study the effects of multiple mutualists (Stier et al. 2012), using experiments to evaluate how the presence of multiple partners influences the net benefits of mutualism, relative to the benefits that result from pairwise interactions alone. If higher-order interactions are indeed common in mutualistic networks, then determining when these interactions are antagonistic and when they are synergistic, and by what mechanisms, will provide insight into how mutualists will respond to shifts in the density and diversity of their partners. For example, it has been suggested that antagonism among pollinator species reduces the pollination of plants by competitively subordinate pollinators (Palmer et al. 2003). If these subordinate competitors provide superior pollination service, interference between pollinators has the potential to influence plant fitness sub-additively.

Recent studies have suggested that higher-order interactions may indeed be of general importance to the structure and stability of mutualistic networks. For example, "nurse" plant species that increase the survival of heterospecific seedlings (e.g., by ameliorating evaporative stress) promote a stable, nested network structure, wherein some nurse-plant species facilitate many more seedling species than others (Verdu and Valiente-Banuet 2008). Kefi et al. (2012) demonstrated that such higher-order, non-trophic interactions among plants can both stabilize competitive interactions and increase network diversity. Despite these findings, many important questions remain unanswered. We do not know, for example, how higher-order interactions modify mutualistic coevolution, nor how variation in the frequency and strength of higher-order interactions affect mutualistic networks. Such questions deserve further study.

Approaches designed to understand complexity within mutualisms may be similarly useful in beginning to expand our view outwards, toward the broader community. Mutualisms, of course, are not interactions occurring in a vacuum; they occur within diverse, interacting ecological communities, which both affect and are affected by the mutualisms themselves. Next, we consider the impacts of the broader community on mutualisms.

9.3 How communities influence mutualisms

Comparable to the impact of multiple mutualists (Stanton 2003), species in the community that are external to the focal mutualism, i.e., third-party species, can drive context-dependent outcomes of mutualisms across time and space (Bronstein and Barbosa 2002). Many good examples of the importance of third-party species come from studies of pollination. Strauss (1997) argued that leaf-eating herbivores could alter floral traits, with consequences for pollinator attraction. We now know that herbivores can alter not only floral number and morphology, but also chemical traits and phenology, with strong indirect effects on pollinators (Gómez et al. 2009, Kessler et al. 2011). Additionally, there is evidence that plants benefit from pollination only in the absence of herbivores when herbivores preferentially attack plants with more flowers (Herrera et al. 2002), that induced defense traits reduce pollination more in the presence of other nectar-providing plant species (Gegear et al. 2007), and that such interactions can alter macroevolutionary patterns in flowering plants (Armbruster et al. 2009). Competitive interactions can also affect pollination mutualisms. For instance, in herbaceous plants, light competition from the woody overstory can indirectly reduce flowering success by lowering the average temperature, which decreases pollinator activity (Kilkenny and Galloway 2008). At an evolutionary scale, this may explain why herbaceous species in deciduous forests tend to flower in the spring and fall, when light penetrates well to the forest floor, and not in the summer (Schemske et al. 1978).

Recognition of mutualisms as integral components of larger interaction webs was a vital step forward for community ecology (Fontaine et al. 2011), and recent studies have begun to integrate trophic and mutualistic higher-order interactions into network theory (Stouffer 2010). Nevertheless, studying mutualistic interactions, themselves quite complex, in the context of all of the other potential interactions in a given community raises the spectre of overwhelming contingency (see Chapter 11). With this challenge in mind, we suggest a systematic approach to studying how third-party species affect mutualisms. Specifically, we must: (1) establish which third-party species exert the strongest effects on a given mutualist complex; (2) evaluate the importance of variation—both spatial and temporal—in the effects of third-party species on mutualists; and (3) incorporate metapopulation and metacommunity theory, in combination with molecular tools, to test whether there are predictable patterns of (co)evolution in such interactions.

Below, we first consider how third-party species can exert effects on mutualism. We then discuss potentially fruitful approaches to studying the importance of third-party species in the outcomes of a given mutualism, revisiting the three steps above.

9.3.1 Mechanisms of third-party effects

All mutualisms are affected by third-party species. A substantial subset of mutualisms—those that involve protection from competitors or natural enemies—are mutualistic precisely *because* of the presence of third-party antagonists. The influence of third parties on the dynamics of other mutualisms can be subtle. For example, dispersal mutualists may transport propagules to sites where competition and/or seed predation is reduced, but the benefits of this service will depend on the abundance of each community's competitors and consumers. As another example, the benefits of pollination mutualisms depend on the relative importance of sexual versus vegetative reproduction, which itself may depend on the intensity of the coevolutionary arms races between plants and their pathogens (Seger 1988).

Third-party species shape the outcomes of mutualisms via interactions, both direct and indirect, with the mutualists. Interactions with a third party can alter mutualistic outcomes by changing: (1) the density of mutualists (e.g., Knight et al. 2005); (2) the rewards offered by one mutualist to another (Strauss 1997, Palmer et al. 2008); and/or (3) the costs of reward production or receipt (Strauss 1997). For example, ant visitation to flowers can decrease bumblebee pollination, presumably because bees both reduce flower visits (lower density) and spend less time on flowers (reduced perceived rewards) when subject to the possible costs of ant harrassment or attack (Cembrowski et al. 2014). The substantial interest in the balance of trade between mutualistic partners (i.e., the continuum of outcomes between net benefits and net costs) has resulted in more research on how traits influence mutualism, and thus on how third-party species can alter such traits, than on how third-party species affect the population densities of mutualists (Bronstein 1998). Although empirical studies of how third parties alter mutualist densities are necessary, they will be challenging because such effects play out over multiple generations (Knight et al. 2006). In addition, we note that there is still very little specific information on the costs of mutualism, which limits our ability to infer how third-party species might affect such costs. For example, although herbivory leads to increased plant toxins in both flower nectar (Strauss 1997) and fruit pulp (Whitehead and Bowers 2013), the fitness effects of toxin allocation are rarely measured in plants and are even less understood for pollinators and seed dispersers.

Bronstein and Barbosa (2002) summarize several specific pathways by which a third species can affect mutualistic outcomes, including effects caused by antagonists, other mutualists, and exploiters of the mutualism. As in most interaction networks, however, the relative importance of direct versus indirect effects of third-party species on mutualists remains unclear (but see Menge 1995 for efforts to understand the balance of these effects in food webs). In one of the rare examples of indirect effects of third-party species on mutualism, Knight et al. (2005) showed that fish increase the benefits of a pollination mutualism to a plant by reducing densities of dragonflies, which eat the plant's pollinators. Examples of direct interactions between mutualists and third-party species are more common, and these can be comparable in strength to the interactions between the mutualists themselves, even causing the net benefits of the original mutualism to disappear (Palmer et al. 2008). It seems unlikely that a single indirect interaction could have an equally strong effect, but again, such studies will require a better understanding of how third-party species affect the population dynamics of mutualist species.

Finally, third-party species also affect mutualisms by modifying the environment. For example, falling tree branches that damage myrmecophytic *Leonardoxa* plants create the disturbance regime necessary for the coexistence of two symbiotic ant species, because one ant species is a strong colonist of damaged plants, whereas the other forms larger, more competitive colonies (Debout et al. 2009). Other examples include subterranean termites that shift the competitive balance among defensive ant species in acacia trees by increasing local primary and secondary productivity (Palmer 2003), and soil microbial communities that alter phosphorus availability, with cascading effects on plant–mycorrhizal mutualisms (Johnson et al. 2010). Such environmental effects by third-party species are doubtless pervasive, but deserve more explicit study. Spatiotemporal variation in such effects will also generate substantial variation in the dynamics of mutualism across space and time.

9.3.2 How do we study the effects of third-party species?

To understand the cumulative effects of third-party species on mutualisms, we return to our three goals from above. To establish which third-party species have the strongest effects, we must determine how much of the community must be considered explicitly to understand the outcomes of a particular mutualism. There is reason to hope that it will suffice to consider relatively small numbers of associated species. Similar to the way in which omnivory shortens the length of the average interaction chain in food webs (Menge 1995), third parties can affect both mutualistic partners simultaneously. For example, a grazing ungulate can both eat plants and reduce mycorrhizae by compacting soil (Eom et al. 2001). In addition, some important non-mutualistic interactions may involve the same species that function as mutualists: seed dispersers can eat defensive ants (Janzen 1966), defensive ants can deter pollinators (Ness 2006), and pollinators can also eat the plants they pollinate (Adler and Bronstein 2004). Finally, there are usually keystone third-party species with disproportionate effects on any given mutualism (Gómez et al. 2009), limiting the fraction of the community that may need to be studied in detail to understand a given mutualism. Of course, one of the primary challenges for establishing generality is that many effects of third parties will be strongly context dependent (Chapter 10).

An important step toward understanding context-dependent outcomes will be to investigate the role of third-party species in shaping the community-scale diversity of potential mutualists (see Section 9.2). The theoretical framework of evolving metacommunities (Leibold et al. 2004, Urban and Skelly 2006) predicts that communities are shaped by local and regional processes, linked by the dispersal of interacting, evolving populations. The processes of community assembly predicted by metacommunity theory can profitably be applied to mutualisms and third-party interactors. Dispersal, colonization, and local extinction lead to heterogeneity in local mutualist communities, which may be driven by competition–colonization trade-offs ("patch dynamics"), the effects of heterogeneity in local environmental conditions on the outcomes of competition ("species sorting"), and the persistence of populations that would otherwise be unsuited to local conditions by means of persistent dispersal ("mass effects"; see Mihaljevic 2012 for a recent exploration of this theme). Importantly, these forms of spatial dynamics have been shown to stabilize models of obligate mutualisms (Hoopes et al. 2005) by ensuring species persistence at larger spatial scales (i.e., the scale of the metacommunity).

Conversely, mutualisms themselves play important and little-studied roles in metacommunity dynamics. Most fundamentally, pollination and dispersal mutualisms provide the essential services of transporting genes and individuals across space. Mutualisms can also alter the balance of competition within trophic levels (Clay and Holah 1999) or the strength of interactions across trophic levels, and increase spatial heterogeneity in communities (Mihaljevic 2012). This heterogeneity, in turn, can play an important role in species coexistence by shifting competitive balance and/or creating enemy-free space (Hoopes et al. 2005). Incorporating mutualisms into both theoretical and empirical analyses of spatiotemporal heterogeneity in interaction webs (Gómez et al. 2009) is in our view crucial for gauging the role of the community in shaping mutualisms, and in turn of mutualisms in altering the broader community, as we will discuss below.

Moving forward, the complementary use of metacommunity theory and molecular tools will help to elucidate spatiotemporal variability in community assembly and its cascading effects on the ecology and evolution of species interactions. For example, phylogeographic studies of mutualist species can be used to estimate gene flow among metapopulations and to compare historical demographies to investigate codispersal of interacting species (e.g., Smith et al. 2011). By identifying similar phylogeographic patterns of yucca plants and moth pollinators that indicated historical connections, Smith et al. (2011) were able to investigate and rule out the possibility that plants had experienced historical dispersal limitation due to the extinction of their megafaunal seed dispersers. As molecular tools continue to develop, it may soon be possible to move beyond studies of neutral loci to investigate loci under selection in geographic mosaics (Vermeer et al. 2011) and to associate these loci with phenotypic traits that are relevant to a given mutualism (Kloth et al. 2012). These studies will require in-depth knowledge of the selective pressures on traits that can only come from increased effort to investigate interactions experimentally and measure traits of interest in the field.

What are the traits that mediate interactions between mutualists and third parties? In the special case of protection mutualisms, the importance of third-party natural enemies may be quantified indirectly in reward production, a trait of the protected species. For example, lycaenid caterpillars secrete nearly double the amount of nectar for attending ants when threatened by simulated enemy attack (Leimar and Axen 1993). Conversely, when enemies or competitors are absent, we should expect the disappearance of rewards, either plastically or adaptively. For example, Palmer et al. (2008)

demonstrated that these circumstances can lead to the disappearance of mutualistic rewards offered by the protected species on ecological time scales. In many mutualisms, the traits of interest will be behavioral: pollination, dispersal, and protection mutualisms all involve an exchange of resources for mutualistic services, and services usually depend on behaviors. Behavioral traits have been less studied in the context of mutualistic coevolution than other phenotypic traits, perhaps because they are more difficult to quantify, but they are no less important. For example, studies have shown that predators strongly affect pollination mutualisms not only by directly consuming pollinators, but also by reducing pollinator visitation rates and the time spent on flowers (Romero et al. 2011). Gómez et al. (2009) provide an example of how structural equation modeling can be used to identify strongly interacting third-party species and their effects on the evolution of plant morphological traits in a diffuse pollination spatial mosaic. An important next step will be to look simultaneously at pollinator traits in this network, and to do so will require behavioral studies.

It is our hope that the steps above provide a tractable roadmap for deepening our understanding of how the broader ecological community affects mutualism outcomes. We now turn our attention to the converse problem: how do mutualisms affect their broader communities?

9.4 How mutualisms affect communities

Because virtually every species on earth is involved in at least one mutualistic partnership, and because so many species rely on mutualisms for reproduction and/or survival, mutualism clearly plays a central role in the assembly, structure, and persistence of ecological communities. However, our understanding of how mutualisms affect communities is still limited. Here we will explore a few of the many important questions that remain unanswered. First, what determines whether mutualisms modify community diversity, and can we predict a priori whether they will increase or decrease species richness? Then, what is the role of mutualism in delineating species ranges, which are important determinants of community assembly and membership? Finally, do mutualisms tend to stabilize or destabilize the dynamics of species' populations that comprise communities?

9.4.1 Effects of mutualisms on community diversity

Mutualisms can reduce species diversity by enhancing the performance, and therefore the competitive dominance, of their partner species. These effects may in turn reduce diversity at other trophic levels (e.g., Ringel et al. 1996). A notable example of this phenomenon is found in specialized plant–endophyte mutualisms (Rudgers and Clay 2008). Alkaloid-producing fungal endophytes that live in tall fescue grass (*Lolium arundinaceum*) enhance the grass's competitive dominance by reducing herbivory, which results in strong suppression of both plant and arthropod species richness in the broader community (Rudgers and Clay 2008, but see Faeth 2010). Similarly, the presence of arbuscular mycorrhizae can disproportionately benefit the invasive and more mycotrophic grass species spotted knapweed, reducing prairie grass diversity (Harner et al. 2010), and the aggressive acacia ants from our opening example suppress elephant herbivory, which contributes to the monodominance of *A. drepanolobium* trees at ecosystem scales (Goheen and Palmer 2010). Indeed, these processes can have dramatic effects on entire ecosystems. In Hawaii, invasion of lowland forests by *Falcataria moluccana* trees that engage with nitrogen-fixing bacterial symbionts has dramatically increased the availability of limiting nitrogen in the soil. This in turn has facilitated the growth and spread of several invasive plant species, transforming lowland forests that had formerly been dominated by native plants (Hughes and Denslow 2005).

The above examples come from highly specialized mutualisms, but of course mutualisms do not have to be specialized to reduce community diversity. In fact, it is the diffuse, generalized nature of many mutualisms that has frequently allowed invasive species to establish new mutualistic partnerships that allow them to spread in introduced areas (Richardson et al. 2000). Although variation in partner quality is the norm within speciose mutualisms, we know relatively little about how the availability of any given partner affects the abundance or diversity of other partners on the same side of the mutualism. In some instances, a particular partner species may provide such attractive rewards or services that it effectively "steals" mutualists away from other members of its guild. For example, plant species that provide large rewards to mutualistic pollinators can increase the degree of pollen limitation in other plants (Traveset and Richardson 2006), although we know little about the consequences of these effects at the population or community level. Again, invasion ecology can provide insights into the mechanisms behind such effects. In central Europe, for example, the invasive annual *Impatiens glandulifera* produces superabundant floral nectar, drawing pollinators away from native plants and reducing the native-plant seed set (Chittka and Schurkens 2001).

Mutualism can also clearly increase community diversity in both evolutionary and ecological time. Among the most striking examples of this are mutualisms that enable the persistence of foundational (e.g., coral–zooxanthellae mutualisms) or keystone species (e.g., figs and their pollinating wasps), which in turn provide critical habitat, food, or other resources for myriad species. Mutualisms can also foster species diversity simply by permitting rare species to persist at lower densities or in harsher environments than they would without mutualism. For example, mixed-species flocks of foraging birds may permit some species to persist in low-resource conditions (e.g., northern winters for insectivorous birds) where otherwise they might disappear (Morse 1970). In addition, flocking provides protection from predation, which should weaken top-down effects, e.g., of raptors and other specialized bird predators. As another example, animals can facilitate long-distance dispersal, which can be important for species persistence at the metacommunity scale when local populations suffer recurrent extinctions. Mutualism also promotes diversity through other, less direct pathways. In Box 9.3, Charlotte Lee, Brian Inouye and Tom Miller discuss how interactions between competing partners and their shared mutualist can promote coexistence of the competitors. Additionally, mutualism can exert "equalizing" (Chesson 2000) effects on species coexistence when inferior competitors are preferentially chosen by superior mutualists (Bever 2002), or when superior competitors facilitate inferior competitors more than they harm them (Schmitt and Holbrook 2003).

An important unanswered question is whether we can predict the strength and direction of the effects on community diversity of different types of mutualisms (e.g., pollination, dispersal, nutrition, protection) that are more or less specialized. For example, are generalized mutualisms more likely to increase community diversity over ecological or evolutionary time scales since many species can openly access the resources exchanged? Conversely, are highly specialized mutualisms more likely to decrease diversity by conferring advantages to only certain mutualist species (e.g., Rudgers and Clay 2008)? Or are highly specialized mutualisms more "compartmentalized," reducing the impact of these interactions on the broader community? In general, while numerous studies have documented the positive and negative effects of mutualism on guild- or community-level diversity, we still lack a mechanistic understanding of how these effects are wrought. In the future, eradication programs for invasives may provide excellent large-scale experimental templates for understanding how particular mutualists drive changes in community diversity. In addition, the widespread, anthropogenically driven loss of particular guilds of mutualists, such as the loss of large mammalian dispersers caused by overhunting in tropical forests, is unfortunately providing opportunities to test how mutualists influence the structure of entire communities (Terborgh et al. 2008). Changes in seed-disperser communities affect how far juvenile plants can disperse from their parents, and other types of mutualisms also play important roles in establishing range limits.

9.4.2 Mutualism and species range limits

We know much less about the role of mutualism in determining species range limits than we know about the roles of abiotic factors and other biotic interactions (e.g., competition, Sexton et al. 2009). Local communities are assembled from the regional species pool, which itself is comprised of species with overlapping ranges. The factors that determine range limits are therefore important in

Box 9.3 Mutualism alters competition: coexistence of mutualists and exploiters

Charlotte T. Lee, Brian D. Inouye, and Tom E. X. Miller

The existence of exploiters is a prominent problem in the community ecology of mutualism (e.g., Bronstein 2001, Morris et al. 2003; see Chapter 6). Competition plays a central role in this problem because an exploiter might be expected to enjoy a competitive advantage as a result of not paying the costs of providing mutualistic benefits to a partner species. Why don't exploiters always disrupt mutualisms by driving one of the partners extinct through competitive exclusion? One possible answer to this question is that mutualism can alter the dynamics of competition in ways that promote coexistence of the mutualistic and exploitative competitors.

Recent theoretical work shows that the inclusion of key demographic detail in competition models can enable coexistence between competitors that differ in their effects on their resource species (Lee and Inouye 2010, Lee et al. 2011). This includes the case when one competitor is a mutualist and the other an exploiter (Box 9.3 Figure 1); no restrictive assumptions about competitive asymmetries, trade-offs, or spatial or temporal heterogeneity are required to enable coexistence. When interactions between individuals are prolonged and exclusive, such as those between ant-adapted plants and the ant colonies that nest in them, competitors' differing influences over the vital rates of their resources can alone enable coexistence. When interactions are shorter in duration, competitors' effects on their resource can enable coexistence through interaction with age or stage structure in the resource species. We discuss examples of each case below. In both, the novel stabilizing effect of mutualism arises when each competitor influences the resource population in a way that benefits the other species, i.e. in a way harmful to itself.

Mutualisms between ants and ant-adapted plants (myrmecophytes) provide excellent examples of prolonged interactions. In ant–myrmecophyte mutualisms, multiple ant species may compete to establish colonies within specialized plant structures. If ant colonies cannot be displaced once established, competition follows the lottery model (Sale 1977). The key departure from basic lottery competition is that ant colonies may protect their host plant from herbivores and/or competitors, which can "feed back" to the space they occupy by increasing plant vital rates such as survival, growth, and reproduction (e.g., Vasconcelos 1991, Palmer et al. 2010). These benefits feed back to affect the number and type of plants available to ants, altering the dynamics of competition between ant species (Lee and Inouye 2010). Stable coexistence is most likely when one ant species has greater rates of colony reproduction but less effectively protects plant survival and fecundity (and thus is an exploiter), and the other species has higher persistence on plants and lower reproductive rates, while promoting stronger plant performance (and thus is a mutualist).

Many mutualists do not form long-term relationships, but instead collect food from their hosts on a short-term basis (Holland and DeAngelis 2010). Prominent examples include the many animals that collect nectar, food bodies, or pollen from plants while pollinating or protecting them. Plant survival, growth, and reproductive rates not only are influenced by these activities, but also typically vary between plant size classes or life history stages. Numerous studies document effects of animals on stage-specific plant vital rates that can be expected to alter the dynamics of the host population (e.g., Maron and Crone 2006, Underwood and Halpern 2012). As in the longer-lasting associations described above, competitors' effects on plant demographic rates can feed back to stabilize coexistence (Lee et al. 2011). The competitors must differ in their effects on the plant population, in

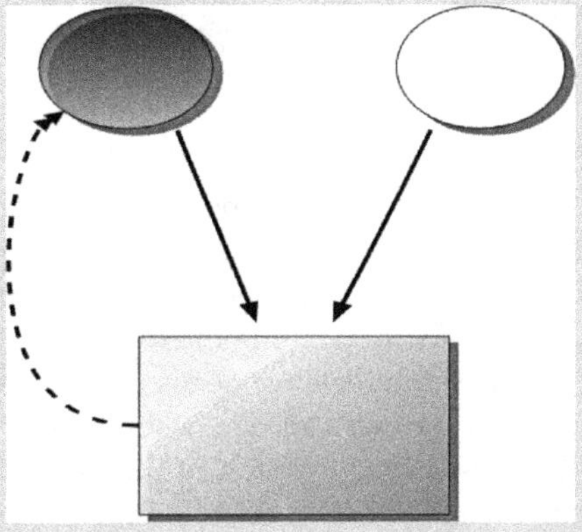

Figure 1 A mutualist (gray circle) and an exploiter (white circle) compete (black arrows) for resources supplied by a resource species (gray rectangle), but the resource species benefits from the mutualistic competitor (dashed arrow) at a cost to that competitor. This simple diagram highlights the importance of competition to the problem exploiters pose to mutualisms, but suppresses key demographic detail which can promote the coexistence of the two competitors.

continued

Box 9.3 *Continued*

the same way as competing mutualists and exploiters do above. Due to the shorter duration of interactions, however, it is also necessary for the resource population to exhibit some underlying age or stage structure, so that via their effects on resource demography, the competitors can shift the relative abundance of resource stages in ways that favor their competitors.

That mutualism could be ecologically vulnerable to invasion by an exploiter, yet not vulnerable to extirpation by that exploiter, emerges from two different models appropriate to mutualisms that involve prolonged symbiosis and short-term interaction, respectively. The key similarity in the models is that the underlying mechanisms of both mutualism and competition be expressed explicitly—by a careful accounting of competitors' direct effects on one another and indirect effects mediated through their impact on their shared resource's demography. When this is accomplished, mutualism and competition can interact in ways that can yield coexistence surprisingly easily.

References

Bronstein, J.L. (2001). The exploitation of mutualisms. *Ecology Letters*, **4**, 277–287.

Holland, J.N., and DeAngelis, D.L. (2010). A consumer-resource approach to the density-dependent population dynamics of mutualism. *Ecology*, **91**, 1286–1295.

Lee, C.T., and Inouye, B.D. (2010). Mutualism between consumers and their shared resource can promote competitive coexistence. *American Naturalist*, **175**, 277–288.

Lee, C.T., Miller, T.E.X., and Inouye, B.D. (2011). Consumer effects on the vital rates of their resource can determine the outcome of competition between consumers. *American Naturalist*, **178**, 452–463.

Maron, J.L., and Crone, E. (2006). Herbivory: effects on plant abundance, distribution and population growth. *Proceedings of the Royal Society B-Biological Sciences*, **273**, 2575–2584.

Morris, W.F., Bronstein, J.L., and Wilson, W.G. (2003). Three-way coexistence in obligate mutualist-exploiter interactions: the potential role of competition. *American Naturalist*, **161**, 860–875.

Palmer, T.M., Doak, D.F., Stanton, M.L., et al. (2010). Synergy of multiple partners, including freeloaders, increases host fitness in a multispecies mutualism. *Proceedings of the National Academy of Sciences of the United States of America*, **107**, 17234–17239.

Sale, P.F. (1977). Maintenance of high species diversity in coral reef fish communities. *American Naturalist*, **111**, 337–359.

Underwood, N., and Halpern, S.L. (2012). Insect herbivores, density dependence, and the performance of the perennial herb *Solanum carolinense*. *Ecology*, **93**, 1026–1035.

Vasconcelos, H.L. (1991). Mutualism between *Maieta guianensis Aubl.*, a myrmecophytic melastome, and one of its ant inhabitants: ant protection against insect herbivores. *Oecologia (Berlin)*, **87**, 295–298.

determining community membership. Mutualistic interactions probably help to delimit species' ranges because of their role in metacommunity dynamics, as discussed previously. Strong interdependence in more specialized interactions means that the range limits of either partner can be constrained by the availability/dispersal limitation of the other. Both colonization and extinction could thus be strongly influenced by the presence or absence of mutualist partners.

We are just beginning to characterize the mechanisms by which interdependence between mutualists determines species' ranges. For example, plant species with a strong degree of pollen limitation may be limited at range edges by the availability or performance of their dominant pollinators (Chalcoff et al. 2012). Nutritional mutualisms between plants and their belowground microbial associates may be similarly important to spatial distributions: although many of these interactions are generalized, there is evidence that a lack of sufficient mycorrhizal or rhizobial partners can limit both the spread of invasive species (Parker et al. 2006) and the distribution of native species (Stanton-Geddes and Anderson 2011). For example, pines were slow to invade the southern hemisphere, where there are no native ectomycorrhizal partners (Richardson et al. 2000), until introduced fungal networks (e.g., from pine plantations) accumulated sufficiently to insure inoculation. At a more local scale, Terwilliger and Pastor (1999) suggested that black spruce (*Picea mariana*) were slow to colonize meadows located in old beaver ponds because of the compounded effects of absent mutualisms. The tree requires ectomycorrhizal fungi to grow, and the fungi cannot tolerate submergence at the initial stages of beaver pond succession. In addition, the animal that disperses the fungus, the redbacked vole (*Clethrionomys gupperi*), is prevented from entering the meadows by the dominant rodent in meadows, the meadow vole (*Microtus pennsylvanicus*; Terwilliger and Pastor 1999). Thus, abiotic conditions and competitive interactions between dispersers together limit the spatial distribution of a host that requires specialized mutualisms.

Seed-dispersal mutualisms also play a pivotal role in determining species ranges (Svenning and Skov 2007), but we know very little about whether and how dispersal mutualisms are degraded or altered at range edges. These dynamics will be important to understand because they affect the plasticity of plant responses to future climate change. The establishment and persistence of plant populations depends in large part on the frequency and timing of dispersal events, which provide opportunities for range expansion and "rescue" from local extinction. Case et al. (2005) suggest that gradients in the abundance of generalist animal dispersal agents could lead to range limits in plant species; any factor that reduces colonization relative to extinction could generate a range limit. Long-distance dispersal events in particular can play a powerful role in shaping plant distributions, but the rarity of these events poses challenges to empirical assessments of how much dispersal mutualisms constrain species ranges. Seed addition experiments, in which seeds are planted into novel habitats beyond range edges, will be informative in assessing the relative importance of dispersal limitation (Clark et al. 2007) versus factors such as habitat availability. Such studies can be paired with more mechanistic studies to evaluate how disperser guilds and plant performance vary from range centers to boundaries.

9.4.3 Mutualism and community stability

Ecologists are beginning to recognize an important role for mutualism in determining community stability. Recent models suggest that diverse interaction types, including mutualism, competition, and predation, stabilize the population dynamics of species within communities (Mougi and Kondoh 2012). This of course suggests that positive interactions play a crucial role in maintaining community-wide stability, contrary to classic models in which the positive feedbacks inherent in mutualism were considered to be destabilizing (May 1981). Yet the biological mechanisms that underlie the stabilizing effects of mutualism remain unclear. Are mutualisms stabilizing because they introduce positive feedbacks between constituent species that are otherwise vulnerable to overexploitation by predators or extirpation by competitors? Does the diffuse nature of most mutualistic interaction networks cause weak interactions, which enforce stability, to dominate (Emmerson and Yearsley 2004)? Or does the disparate nature of the broad class of interactions that we identify as mutualisms, and the multiple mechanisms that therefore underlie them, preclude any meaningful generality?

One profitable research avenue will be to explore the effects of mutualism on the stability of community modules that lie somewhere between highly simplified two-species studies, and the (largely phenomenological) modeling studies that embrace entire ecological communities across multiple trophic levels. For example, Benadi et al. (2012) recently modeled the effects of pollinators on the stability of the guild of plants they pollinate. They found that whether pollinators stabilized plant guilds depended on both the extent to which plant resource use overlapped, and the degree of pollinator specialization. When plant species had overlapping resource niches, pollinators strongly destabilized plant guilds irrespective of the degree of pollinator specialization; in effect, rarer plants received lower levels of conspecific pollen transfer than more common species, decreasing their relative contribution to the seed bank in each successive year. However, when plant species had very low levels of niche overlap (contributing "stabilizing" effects to coexistence, Chesson 2000), pollinators stabilized plant guilds in all cases except for instances in which pollinators were completely generalized (Benadi et al. 2012). Importantly, this study considered the stability of the pollination network in a broader context, evaluating both variation in the pollination mutualism and the degree of resource competition among plant species. Because niche overlap among coexisting plant species is often high, the results from this study suggest that other mechanisms in addition to niche overlap may need to be invoked to explain coexistence of plant species that share pollinators. Nevertheless, this work provides an excellent example of how viewing mutualism within a broader community context can shed light on the extent to which mutualism, in concert with other species interactions, influences the stability of ecological communities.

Empirical studies of the effects of mutualism on community stability are scarce, but one emerging generality seems to be that the mechanisms by which the mutualists interact with the broader community determine whether these interactions stabilize or destabilize community dynamics. For example, nutritional mutualisms that fuel species' dominance are likely to stabilize communities where such dominants provide heterogeneous habitats (e.g., corals, Hay et al. 2004), but destabilize communities where these dominants exert strong competitive effects on other guild members (e.g., invasive ants, Shik and Silverman 2013). The effects of protection mutualisms on communities may also depend in part on the trophic level of the protected partner. If the protected species is an autotroph, mutualism can stabilize communities by buffering plants from consumption, for example where ants stabilize tree cover by protecting their acacia hosts from catastrophic elephant herbivory (Goheen and Palmer 2010). Contrastingly, in cases where the protected species is a consumer, mutualism may lead to strong cycles of exploitation and recovery. For example, Morales et al. (2008) showed that when aphids were protected by ant mutualists from predation, they exerted strong consumptive effects on plants, with destabilizing effects on the food web. More generally, where protection mutualisms simplify the trophic structure of communities, the likelihood of strong trophic interactions could potentially increase, with destabilizing effects on communities. These ideas provide testable propositions for empirical work.

9.5 Conclusions and future directions

Over the past thirty years, a groundswell of interest in interspecific cooperation has greatly increased our understanding of the ecology and evolution of these ubiquitous interactions. We have learned that mutualisms are themselves complex associations of interacting species, that third-party species greatly modify the outcomes of interactions between mutualist guilds, and that mutualisms transform communities by creating foundational habitat, extending species ranges, and linking populations across space, through time, and via webs of interactions. Yet in many ways, our understanding of mutualism is still in its infancy: relative to interactions such as competition and predation, there remains a great need for research—both theoretical and empirical—on the interplay between mutualisms and the broader communities in which they are embedded. We have highlighted recent research that illustrates some general phenomena about feedbacks between mutualisms and their community context, but many questions remain. Below, we highlight a number of unanswered questions (by no means exhaustive) that we think are important to consider as the field moves forward.

1. Because the benefits of mutualism vary with abiotic and biotic context, the ranking of the "best" partners for a particular mutualist may also vary from year to year, or from one place to another. In light of this, does partner diversity buffer mutualism from temporal and spatial environmental variation (Box 9.1)? Are there abiotic conditions which foster the efflorescence of mutualisms, or conversely hamper their formation or retention?
2. How does species diversity in mutualistic assemblages, whereby a given mutualist interacts with multiple partners either serially or simultaneously, integrate to influence long-term fitness of each participant (Box 9.2)? Because natural selection operates on lifetime fitness, this is a particularly critical area of study for understanding the maintenance of cooperation (and related traits and behaviors), particularly among partners differing in generation length and dispersal range.
3. When a mutualist interacts sequentially with a number of partners, how does the timing, duration, or sequence of partnerships influence fitness? Does the rank order of the quality of partners remain constant or change in the course of mutualist ontogeny?

4. How important are indirect effects provided by third parties to the structure and functioning of mutualistic interactions (e.g., Knight et al. 2005), and do third parties tend to have stronger effects on mutualisms by changing mutualist traits or population densities?
5. Are mutualisms, which often result in transport of genes and individuals, disproportionately important to metacommunity dynamics, relative to interactions such as competition or predation? How does the degree of specialization in mutualism influence the persistence of the participant species within metacommunities?
6. Especially for obligate mutualisms, to what degree do the costs of mutualism constrain the range limits of species?
7. To what degree can a quantitative, theoretical framework for the community ecology of mutualisms be devised, comparable to say niche partitioning or food-web models? To what extent does the broad natural history of mutualisms, from mycorrhizae to seed dispersal, preclude the development of generalized models?
8. How can an understanding of mutualism in community settings inform issues in applied ecology? For example, can we use trait-based approaches to predict the success of invasive mutualists in novel habitats (see also Chapter 13)? Can one make strong predictions about how communities and ecosystems will change when mutualisms dissipate or disappear? For example, the loss of specialized pollinators will lead to plant extinctions, but what other broad-scale community-wide or ecosystem-level consequences (discussed in Chapter 14) will there be?

We began this chapter by highlighting how the acacia ant *Crematogaster nigriceps*, an apparent sterilization "parasite" of the East African ant-plant *Acacia drepanolobium*, actually exerts strong positive effects on the lifetime fitness of these acacias. Yet this positive effect only emerged by considering the broader community in which this particular pairwise association is embedded. To illuminate the mutualistic role of this species, it was necessary to understand the dynamics of competition among the various ant partners, the impact of a broad suite of herbivores and other mortality factors on the plant, the costs and benefits of different ant–plant pairings across acacia ontogeny, and the long-term demography of the host plant. In essence, broadening the spatial and temporal scope of the study of this mutualism yielded insights that would not have been possible using more conventional pairwise, single-life-stage study approaches. In the same way, a recurrent theme throughout this chapter is the need to identify the mechanisms underlying the interactions between mutualists and their communities. Expanding our studies of mutualism to consider the surrounding community and longer temporal scales—and linking ecological and evolutionary processes over different spatial and temporal scales—will be crucial for identifying such mechanisms. As we expand the scope of our studies to include broader ecological contexts, mutualism will continue to claim an overdue and more prominent role in community ecology.

References

Adler, L.S., and Bronstein, J.L. (2004). Attracting antagonists: does floral nectar increase leaf herbivory? *Ecology*, **85**, 1519–1526.

Armbruster, W.S., Lee, J., and Baldwin, B.G. (2009). Macroevolutionary patterns of defense and pollination in *Dalechampia* vines: adaptation, exaptation, and evolutionary novelty. *Proceedings of the National Academy of Sciences of the United States of America*, **106**, 18085–18090.

Bastolla, U., Fortuna, M.A., Pascual-Garcia, A., Ferrera, A., Luque, B., and Bascompte, J. (2009). The architecture of mutualistic networks minimizes competition and increases biodiversity. *Nature*, **458**, 1018–1019.

Benadi, G., Bluthgen, N., Hovestadt, T., and Poethke, H.J. (2012). Population dynamics of plant and pollinator communities: stability reconsidered. *American Naturalist*, **179**, 157–168.

Bever, J.D. (2002). Negative feedback within a mutualism: host-specific growth of mycorrhizal fungi reduces plant benefit. *Proceedings of the Royal Society B-Biological Sciences*, **269**, 2595–2601.

Bronstein, J.L. (1998). The contribution of ant plant protection studies to our understanding of mutualism. *Biotropica*, **30**, 150–161.

Bronstein, J.L., and Barbosa, P. (2002). Multi-trophic/multi-species mutualistic interactions: the role of non-mutualists in shaping and mediating mutualisms. In B. Hawkins and T. Tscharntke (eds.), *Multitrophic Level Interactions*, pp. 44–65. Cambridge: Cambridge University Press.

Case, T.J., Holt, R.D., McPeek, M.A., and Keitt, T.H. (2005). The community context of species' borders: ecological and evolutionary perspectives. *Oikos*, **108**, 28–46.

Cembrowski, A.R., Tan, M.G., Thomson, J.D., and Frederickson, M.E. (2014). Ants and ant scent reduce bumblebee pollination of artificial flowers. *American Naturalist*, **183**, 133–139.

Chalcoff, V.R., Aizen, M.A., and Ezcurra, C. (2012). Erosion of a pollination mutualism along an environmental gradient in a south Andean treelet, *Embothrium coccineum* (Proteaceae). *Oikos*, **121**, 471–480.

Chesson, P. (2000). Mechanisms of maintenance of species diversity. *Annual Review of Ecology And Systematics*, **31**, 343–366.

Chittka, L., and Schurkens, S. (2001). Successful invasion of a floral market—an exotic Asian plant has moved in on Europe's river-banks by bribing pollinators. *Nature*, **411**, 653–653.

Clark, C.J., Poulsen, J.R., Levey, D.J., and Osenberg, C.W. (2007). Are plant populations seed limited? A critique and meta-analysis of seed addition experiments. *American Naturalist*, **170**, 128–142.

Clay, K., and Holah, J. (1999). Fungal endophyte symbiosis and plant diversity in successional fields. *Science* (Washington D.C.), **285**, 1742–1744.

Cohen, J.E., Schittler, D.N., Raffaelli, D.G., and Reuman, D.C. (2009). Food webs are more than the sum of their tritrophic parts. *Proceedings of the National Academy of Sciences of the United States of America*, **106**, 22335–22340.

Debout, G.D.G., Dalecky, A., Ngomi, A., and McKey, D.B. (2009). Dynamics of species coexistence: maintenance of a plant-ant competitive metacommunity. *Oikos*, **118**, 873–884.

Emmerson, M., and Yearsley, J.M.(2004). Weak interactions, omnivory and emergent food- web properties. *Proceedings of the Royal Society B-Biological Sciences*, **271**, 397–405.

Eom, A.H., Wilson, G.W.T., and Hartnett, D.C. (2001). Effects of ungulate grazers on arbuscular mycorrhizal symbiosis and fungal community structure in tallgrass prairie. *Mycologia*, **93**, 233–242.

Faeth, S.H. (2010). Occam's razor cuts both ways: endophytes, resource allocation, herbivory, and mutualism: a reply to Rudgers et al. *American Naturalist*, **176**, 104–110.

Fontaine, C., Guimaraes, P.R., Kefi, S., et al. (2011). The ecological and evolutionary implications of merging different types of networks. *Ecology Letters*, **14**, 1170–1181.

Gegear, R.J., Manson, J.S., and Thomson, J.D. (2007). Ecological context influences pollinator deterrence by alkaloids in floral nectar. *Ecology Letters*, **10**, 375–382.

Goheen, J.R., and Palmer, T.M. (2010). Defensive plant-ants stabilize megaherbivore-driven landscape change in an African savanna. *Current Biology*, **20**, 1768–1772.

Gómez, J.M., Perfectti, F., Bosch, J., and Camacho, J.P.M. (2009). A geographic selection mosaic in a generalized plant-pollinator-herbivore system. *Ecological Monographs*, **79**, 245–263.

Harner, M.J., Mummey, D.L., Stanford, J.A., and Rillig, M.C. (2010). Arbuscular mycorrhizal fungi enhance spotted knapweed growth across a riparian chronosequence. *Biological Invasions*, **12**, 1481–1490.

Hay, M.E., Parker, J.D., Burkepile, D.E., Caudill, C.C., Wilson, A.E., Hallinan, Z.P., and Chequer, A.D. (2004). Mutualisms and aquatic community structure: the enemy of my enemy is my friend. *Annual Review of Ecology Evolution and Systematics*, **35**, 175–197.

Herrera, C.M., Medrano, M., Rey, P.J., Sanchez-Lafuente, A.M., Garcia, M.B., Guitian, J., and Manzaneda, A.J. (2002). Interaction of pollinators and herbivores on plant fitness suggests a pathway for correlated evolution of mutualism- and antagonism-related traits. *Proceedings of the National Academy of Sciences of the United States of America*, **99**, 16823–16828.

Holt, R.D., and Dobson, A. (2006). Extending the principles of community ecology to address the epidemiology of host-pathogen systems. In S.K. Collinge and C. Ray (eds.), *Disease Ecology: Community Structure and Pathogen Dynamics*, Oxford: Oxford University Press.

Hoopes, M.F., Holt, R.D., and Holyoak, M. (2005). The effects of spatial processes on two species interactions. In M. Holyoak, M.A. Leibold, and R.D. Holt (eds.), *Metacommunities: Spatial Dynamics and Ecological Communities*, pp. 35–67. Chicago, IL: University of Chicago Press.

Hughes, R.F., and Denslow, J.S. (2005). Invasion by a N-2-fixing tree alters function and structure in wet lowland forests of Hawaii. *Ecological Applications*, **15**, 1615–1628.

Jansa, J., Smith, F.A., and Smith, S.E. (2008). Are there benefits of simultaneous root colonization by different arbuscular mycorrhizal fungi? *New Phytologist*, **177**, 779–789.

Janzen, D.H. (1966). Coevolution of mutualism between ants and acacias in Central America. *Evolution*, **20**, 249–275.

Johnson, N.C., Wilson, G.W.T., Bowker, M.A., Wilson, J.A., and Miller, R.M. (2010). Resource limitation is a driver of local adaptation in mycorrhizal symbioses. *Proceedings of the National Academy of Sciences of the United States of America*, **107**, 2093–2098.

Kefi, S., Berlow, E.L., Wieters, E.A., et al. (2012). More than a meal . . . integrating non-feeding interactions into food webs. *Ecology Letters*, **15**, 291–300.

Kersch, M.F., and Fonseca, C.R. (2005). Abiotic factors and the conditional outcome of an ant-plant mutualism. *Ecology*, **86**, 2117–2126.

Kessler, A., Halitschke, R., and Poveda, K. (2011). Herbivory-mediated pollinator limitation: negative impacts of induced volatiles on plant-pollinator interactions. *Ecology*, **92**, 1769–1780.

Kiers, E.T., Duhamel, M., Beesetty, Y., et al. (2011). Reciprocal rewards stabilize cooperation in the mycorrhizal symbiosis. *Science*, **333**, 880–882.

Kilkenny, F.F., and Galloway, L.F. (2008). Reproductive success in varying light environments: direct and indirect effects of light on plants and pollinators. *Oecologia*, **155**, 247–255.

Kloth, K.J., Thoen, M.P.M., Bouwmeester, H.J., Jongsma, M.A., and Dicke, M. (2012). Association mapping of plant resistance to insects. *Trends in Plant Science*, **17**, 311–319.

Knight, T.M., Chase, J.M., Hillebrand, H., and Holt, R.D. (2006). Predation on mutualists can reduce the strength of trophic cascades. *Ecology Letters*, **9**, 1173–1178.

Knight, T.M., McCoy, M.W., Chase, J.M., McCoy, K.A., and Holt, R.D. (2005). Trophic cascades across ecosystems. *Nature*, **437**, 880–883.

Leibold, M.A., Holyoak, M., Mouquet, N., et al. (2004). The metacommunity concept: a framework for multiscale community ecology. *Ecology Letters*, **7**, 601–613.

Leimar, O., and Axen, A.H. (1993). Strategic behavior in an interspecific mutualism: interactions between lycaenid larvae and ants. *Animal Behaviour*, **46**, 1177–1182.

May, R.M. (1981). Models for two interacting populations. In R.M. May (ed.), *Theoretical Ecology: Principles and Applications*, pp. 5–29. Sunderland, MA: Sinauer.

McCann, K.S. (2012). *Food Webs*. Princeton: Princeton University Press.

Menge, B.A. (1995). Indirect effects in marine rocky intertidal interaction webs—patterns and importance. *Ecological Monographs*, **65**, 21–74.

Mihaljevic, J.R. (2012). Linking metacommunity theory and symbiont evolutionary ecology. *Trends in Ecology and Evolution*, **27**, 323–329.

Morales, M.A., Morris, W.F., and Wilson, W.G. (2008). Allee dynamics generated by protection mutualisms can drive oscillations in trophic cascades. *Theoretical Ecology*, **1**, 77–88.

Morris, W.F., Hufbauer, R.A., Agrawal, A.A., et al. (2007). Direct and interactive effects of enemies and mutualists on plant performance: a meta-analysis. *Ecology*, **88**(4), 1021–1029.

Morse, D.H. (1970). Ecological aspects of some mixed-species foraging flocks of birds. *Ecological Monographs*, **40**(1), 119–168.

Mougi, A., and Kondoh, M. (2012). Diversity of interaction types and ecological community stability. *Science*, **337**, 349–351.

Ness, J.H. (2006). A mutualism's indirect costs: the most aggressive plant bodyguards also deter pollinators. *Oikos*, **113**, 506–514.

Palmer, T.M. (2003). Spatial habitat heterogeneity influences competition and coexistence in an African acacia ant guild. *Ecology*, **84**, 2843–2855.

Palmer, T.M., Doak, D.F., Stanton, M.L., et al. (2010). Synergy of multiple partners, including freeloaders, increases host fitness in a multispecies mutualism. *Proceedings of the National Academy of Sciences of the United States of America*, **107**, 17234–17239.

Palmer, T.M., Stanton, M.L., and Young, T.P. (2003). Competition and coexistence: exploring mechanisms that restrict and maintain diversity within mutualist guilds. *American Naturalist*, **162**, S63–S79.

Palmer, T.M., Stanton, M.L., Young, T.P., Goheen, J.R., Pringle, R.M., and Karban, R. (2008). Breakdown of an ant-plant mutualism follows the loss of large herbivores from an African Savanna. *Science*, **319**, 192–195.

Parker, M.A., Malek, W., and Parker, I.M. (2006). Growth of an invasive legume is symbiont limited in newly occupied habitats. *Diversity and Distributions*, **12**, 563–571.

Richardson, D.M., Allsopp, N., D'Antonio, C.M., Milton, S.J., and Rejmanek, M. (2000). Plant invasions—the role of mutualisms. *Biological Reviews*, **75**, 65–93.

Ringel, M.S., Hu, H.H., and Anderson, G. (1996). The stability and persistence of mutualisms embedded in community interactions. *Theoretical Population Biology*, **50**, 281–297.

Romero, G.Q., Antiqueira, P.A.P., and Koricheva, J. (2011). A meta-analysis of predation risk effects on pollinator behaviour. *Plos One*, **6**.

Rudgers, J.A., and Clay, K. (2008). An invasive plant-fungal mutualism reduces arthropod diversity. *Ecology Letters*, **11**, 831–840.

Schemske, D.W., Willson, M.F., Melampy, M.N., Miller, L.J., Verner, L., Schemske, K.M., and Best, L.B. (1978). Flowering ecology of some spring woodland herbs. *Ecology*, **59**, 351–366.

Schmitt, R.J., and Holbrook, S.J. (2003). Mutualism can mediate competition and promote coexistence. *Ecology Letters*, **6**, 898–902.

Seger, J., and Hamilton, W.D. (1988). Parasites and sex. In R.E. Michod and B.R. Levin (eds.), *The Evolution of Sex: An Examination of Current Ideas*, pp. 176–193. Sunderland, MA: Sinauer Press.

Sexton, J.P., McIntyre, P. J., Angert, A.L., and Rice, K.J. (2009). Evolution and ecology of species range limits. *Annual Review of Ecology Evolution and Systematics*, **40**, 415–436.

Shik, J.Z., and Silverman, J. (2013). Towards a nutritional ecology of invasive establishment: aphid mutualists provide better fuel for incipient Argentine ant colonies than insect prey. *Biological Invasions*, **15**, 829–836.

Smith, C.I., Tank, S., Godsoe, W., Levenick, J., Strand, E., Esque, T., and Pellmyr, O. (2011). Comparative phylogeography of a coevolved community: concerted population expansions in Joshua trees and four yucca moths. *PLoS ONE*, **6**, 1–18.

Stachowicz, J.J., and Whitlatch, R.B. (2005). Multiple mutualists provide complementary benefits to their seaweed host. *Ecology*, **86**, 2418–2427.

Stanton, M.L. (2003). Interacting guilds: moving beyond the pairwise perspective on mutualisms. *American Naturalist*, **162**, S10–S23.

Stanton, M.L., and Palmer, T.M. (2011). The high cost of mutualism: effects of four species of East African ant symbionts on their myrmecophyte host tree. *Ecology*, **92**, 1073–1082.

Stanton-Geddes, J., and Anderson, C.G. (2011). Does a facultative mutualism limit species range expansion? *Oecologia*, **167**, 149–155.

Stier, A.C., Gil, M.A., McKeon, C.S., Lemer, S., Leray, M., Mills, S.C., and Osenberg, C.W. (2012). Housekeeping mutualisms: do nore symbionts facilitate host performance? *Plos One*, **7**, e32079, 32071–32076.

Stouffer, D.B. (2010). Scaling from individuals to networks in food webs. *Functional Ecology*, **24**, 44–51.

Strauss, S.Y. (1997). Floral characters link herbivores, pollinators, and plant fitness. *Ecology*, **78**, 1640–1645.

Svenning, J.C., and Skov, F. (2007). Could the tree diversity pattern in Europe be generated by postglacial dispersal limitation? *Ecology Letters*, **10**, 453–460.

Terborgh, J., Nunez-Iturri, G., Pitman, N.C.A., et al. (2008). Tree recruitment in an empty forest. *Ecology*, **89**, 1757–1768.

Terwilliger, J., and Pastor, J. (1999). Small mammals, ectomycorrhizae, and conifer succession in beaver meadows. *Oikos*, **85**, 83–94.

Traveset, A., and Richardson, D.M. (2006). Biological invasions as disruptors of plant reproductive mutualisms. *Trends in Ecology and Evolution*, **21**, 208–216.

Urban, M.C., and Skelly, D.K. (2006). Evolving metacommunities: toward an evolutionary perspective on metacommunities. *Ecology*, **87**, 1616–1626.

van der Heijden, M.G.A., Klironomos, J.N., Ursic, M., et al. (1998). Mycorrhizal fungal diversity determines plant biodiversity, ecosystem variability and productivity. *Nature*, **396**, 69–72.

Vance-Chalcraft, H.D., Rosenheim, J.A., Vonesh, J.R., Osenberg, C.W., and Sih, A. (2007). The influence of intraguild predation on prey suppression and prey release: a meta-analysis. *Ecology*, **88**, 2689–2696.

Verdu, M., and Valiente-Banuet, A. (2008). The nested assembly of plant facilitation networks prevents species extinctions. *American Naturalist*, **172**, 751–760.

Vermeer, K., Dicke, M., and de Jong, P.W. (2011). The potential of a population genomics approach to analyse geographic mosaics of plant-insect coevolution. *Evolutionary Ecology*, **25**, 977–992.

Verna, C., Ramette, A., Wiklund, H., Dahlgren, T.G., Glover, A.G., Gaill, F., and Dubilier, N. (2010). High symbiont diversity in the bone-eating worm *Osedax mucofloris* from shallow whale-falls in the North Atlantic. *Environmental Microbiology*, **12**, 2355–2370.

Whitehead, S.R., and Bowers, M.D. (2013). Evidence for the adaptive significance of secondary compounds in vertebrate-dispersed fruits. *American Naturalist*, **182**, 563–577.

CHAPTER 10

Context-dependent outcomes of mutualistic interactions

Jason D. Hoeksema and Emilio M. Bruna

10.1 Introduction

The common conception of mutualisms is of interacting species that experience enhanced performance in the presence of each other. Most mutualisms do not always conform to this stereotype, however. The outcome of interactions can vary over time and space along a continuum from strong mutualism to commensalism, or even change sign from mutualism to parasitism (Bronstein 1994a, Herre et al. 1999, Sachs and Simms 2006, Thompson 1988). This plasticity is often driven by the abiotic and biotic context in which interactions take place, such as the abundance of key nutrients or the identity of species present in the community. It can also be driven by genetic context, i.e., genetic variation in one or both species. Although for some species interactions many studies currently exist addressing context dependency, for others we know next to nothing, and many critical questions remain to be answered across all major types of mutualisms. Here, we aim to articulate some of those questions, assess progress to date on answering them, and highlight areas in particular need of future research.

We define context dependency of a mutualism to be a change in interaction net outcome; i.e., a change in the magnitude of how one species influences the average fitness correlates of another species. This change is in response to variation in the abiotic, biotic, or genetic context in which those two species are interacting, assuming that both species benefit from the interaction under at least some context. This definition includes cases in which a strong mutualism, wherein both species benefit substantially from the interaction, shifts to a weak mutualism or even changes sign to a parasitism from the perspective of at least one of the two species. Both of these cases involve context-dependent changes in the magnitude of an interaction. Thus, a change in sign of the interaction (from mutualism to parasitism) is a special case of the more general definition of context dependency—any change of interaction magnitude in response to context. In a mutualism with multiple potential partner species on both sides of the interaction, we do not consider a change in performance of a focal species when paired with alternate partner species to constitute context dependency. Rather, context dependency in such interactions would be observed as a change in interaction outcome between a particular focal pair of species, when compared between different background communities (potentially including other species participating in the interaction). Our definition of context dependency is broadly consistent with that used in previous reviews. For example, in a meta-analysis of 247 published studies of context dependency across mutualism, predation, and competition, Chamberlain et al. (2014) quantified context dependency in two ways: the coefficient of variation (CV) of interaction magnitude among contexts, and a change in sign of the interaction among contexts.

Chamberlain et al. (2014) showed that context dependency is often substantial in mutualisms, albeit not consistently more than in predator/prey or competitive interactions. The implications of such

Mutualism. Edited by Judith L. Bronstein.

context dependency may be far-reaching (Agrawal 2001, Bronstein 1994a, Thompson 1988, 2005, Thrall et al. 2007). If outcomes of mutualisms are context dependent, then interaction strengths estimated in a particular context cannot be extrapolated over time or space. Instead, they may be complex functions of their environment. Moreover, plasticity of outcomes can potentially stabilize mutualisms by ameliorating conflicts of interest between partners and promoting partner fidelity (Agrawal, 2001). If outcomes of a mutualism depend on contextual factors that vary over space and/or time, and if variation in outcomes is genetically based, then natural selection by species on each other's traits may also vary over time and space, potentially generating polymorphisms in populations and driving interpopulation divergence. It is therefore essential to understand the genetic basis of distributions of outcomes within populations, as well as genetically based variation in outcomes of interactions among abiotic or biotic environmental contexts (i.e. interaction norms) to predict how interactions will evolve (Thompson 1988).

The distribution of published studies on context dependency is extremely uneven among the major classes of mutualisms. For example, thousands of field and laboratory experiments have investigated the effects of mycorrhizal fungi, which inhabit the roots of plants and potentially enhance plant nutrient uptake from soils, on plant performance. Many of those experiments have recorded and/or manipulated contextual factors that could influence the outcomes of the interactions. Several reviews and meta-analyses have synthesized the results of these experiments (e.g., Hoeksema et al. 2010, Johnson et al. 1997, Jones and Smith 2004, Karst et al. 2008). Similarly, there is a relatively long history in agricultural ecology of exploring the dependence of the legume–rhizobia symbiosis, in which bacteria inhabit the roots of plants and provide fixed nitrogen (N) in exchange for plant photosynthetic products, on a variety of abiotic contextual factors such as soil N or phosphorus (P) availability, soil salinity, or water availability (see reviews by Graham et al. 2003, Serraj et al. 1999, Streeter and Wong 1988, Zahran 1999).

Plant–pollinator interactions are also among the most studied of mutualisms with regard to context dependency, with much of the attention focusing on how the outcome of these interactions changes as a function of local abiotic and biotic environments. For instance, a large body of experimental and observational work has documented that pollinator effects on plant fruit or seed production can vary with resources, the diversity and structure of the local habitat, and the presence of other herbivores, predators, and other species with which pollinators and plants interact (reviewed in Knight et al. 2005, Garcia-Camacho and Totland 2009, Ashman et al. 2004). The magnitude of pollinators' beneficial effects can also be affected by traits such as floral shape and color, nectar production, volatile emission, and other such cues used to attract pollinators, all of which can also be influenced by the local environmental or genetic context (Bradshaw and Schemske 2003, Ashman et al. 2005, Temeles and Kress 2003).

In contrast, surprisingly few studies have explicitly investigated effects of contextual factors on mutualisms formed by grasses and fungal endophytes, plants and their protective ants, cleaner fish and the clients from which they remove parasites, the nutritional symbiosis between cnidarian animals and dinoflagellate algae, or protective mutualisms between insects and symbiotic bacteria. For example, in the grass–endophyte symbiosis, very few studies have investigated contextual factors besides water and nutrient availability, despite the obvious and important influence that light, fire and other factors could have on the outcome of the symbiosis (Cheplick and Faeth 2009). For ant–plant mutualisms, meta-analyses have found that plant growth form or life-history strategy (annual or perennial) can influence the outcome of interactions, as does the diversity of ants with which plants interact (Chamberlain and Holland 2009, Trager et al. 2010). However, the way in which the abiotic or genetic context influences their outcome remains largely unknown. For instance, few studies have experimentally manipulated soil nutrients to explore how changes in characteristics such as plant quality and reward production influence ant and herbivore behavior; not unexpectedly, those that have suggest local resource availability can have far-reaching effects (e.g., Trager and Bruna 2006, Kersch and Fonseca 2005, Heil et al. 2001). Microclimate and light, which can influence ant community composition, rewards to partners, and plant growth, also appear to be important (Alonso 1998, Kersch and

Fonseca 2005, Folgarait and Davidson 1995). For cleaner–client fish interactions, only recently have experimental studies begun to accumulate showing mutual benefits for both cleaners and hosts, much less elucidating abiotic or biotic contextual factors that may alter interaction outcomes (but see, e.g., Cheney and Cote 2005).

Even in the most heavily studied mutualisms, many questions on context dependency are only in the earliest stages of being answered. Our overall goal for the remainder of this chapter is to suggest a roadmap for fruitful areas of investigation on context dependency in mutualisms. We start by outlining the four major types of research tools that can be used in such investigations. We then identify eight big questions on context dependency in mutualisms, assess progress to date in the application of the four major research tools to answering them, and point out areas in particular need of future research.

10.2 Tools for studying context dependency of mutualisms

We recognize four primary tools that can be used to explore context dependency in mutualisms: models, observational field studies, experiments, and meta-analyses.

10.2.1 Mathematical, graphical, and conceptual models

Models explicitly focusing on mechanisms of context dependency can help formalize and explore testable hypotheses on context dependency, and also allow the modeling of long-term dynamics (see also Chapter 4). In Box 10.1, John Jaenike uses a mathematical model to predict how the benefits conferred by bacterial endosymbionts to their insect hosts may vary in response to contextual factors,

Box 10.1 Context dependency in defensive endosymbioses

John Jaenike

Many insects harbor bacterial endosymbionts that may defend their insect hosts against pathogens. Although little empirical work has confirmed how contextual factors may alter outcomes of these putative mutualisms, these systems may eventually provide substantial insights into factors driving variation in animal–microbe mutualisms. Outcomes of defensive symbioses are hypothesized to depend on how context-dependent factors affect two key variables controlling the impact of a parasite or pathogen on a host individual or population: the force of infection (the rate at which susceptible individuals become infected) and pathogen virulence (the extent to which host fitness is decreased by the infection). The review and mathematical model below illustrate how these two variables can be influenced by several biotic and abiotic variables, including community diversity and composition, epidemic fluctuations or outbreaks in pathogen populations, and ambient environmental temperatures.

In most models of pathogen dynamics, the force of infection increases with the number of infected hosts, which depends on host population size or density and the prevalence of infection (McCallum et al. 2001). For a single host species with age structure, the force of infection, λ, is $\beta\int_0^{\infty} y(a,t)\,da$, where $Y(a,t)$ is the number of infected individuals of age a at time t, and β is the transmission rate, which, in this model, is assumed to be independent of host age (Anderson and May 1991). Now consider a community in which a pathogen can be passed among several host species. In a community of n host species sharing a common pathogen, the above equation can be modified to say that the force of infection for focal host species i is $\lambda_i = \sum_{j=1}^{n} \beta_{ij} y_j$, where β_{ij} is the transmission rate of the pathogen from individuals of species j to species i, including transmission from species i to itself, and Y_j is the number or density of infected individuals of species j. (This formulation ignores age structure within the host populations.) Therefore, if the density of species i remains the same, then the force of infection is greater in a community in which more species share a given pathogen and in which there are more opportunities for interspecific transmission. This would suggest that, in the case of protection against pathogens with a broad host range, there will be greater selection for protection by a defensive endosymbiont in species-rich communities, such as those in tropical regions. In addition, because defensive symbionts are acquired via lateral transfer from other host species, tropical communities, being more diverse, probably have more symbiont-infected host species that can serve as sources of

continued

Box 10.1 *Continued*

defensive symbionts for other species (Jaenike 2012). Moreover, the net fitness effect on an insect host of carrying a particular symbiont can depend on the host plant species on which the insect is feeding (e.g., Tsuchida et al. 2011). Thus, the species diversity and composition of a community in which a host species resides is likely to influence both selection for and opportunities to utilize defensive symbionts.

The force of infection by many pathogens is temporally variable. Many parasites, especially microparasites of insects, undergo wide fluctuations in prevalence, characterized by periodic outbreaks interspersed with longer periods of low infection prevalence (Briggs et al. 1995). Because these epidemics are associated with large changes in insect population size and the prevalence of parasitism, there is corresponding variation in the force of infection and thus the strength of selection for a defensive symbiont. One might think that the incidence of mutualism with a defensive symbiont would depend on the average prevalence of pathogen infection, but this is not the case. Consider a host-sterilizing parasite that either infects a constant proportion (p) of the host population every generation, or that undergoes periodic epidemics, infecting np of the hosts once every n years. If the host species carries a heritable defensive symbiont, then, for a constant rate of parasitism, the prevalence of symbiont inheritance will reach a stable equilibrium frequency, where the advantage due to selection is balanced by losses due to imperfect maternal transmission (Jaenike 2012). A numerical model shows that pathogen epidemics result in fluctuations in the prevalence of symbiosis with a defensive symbiont (Box 10.1 Figure 1 A), and, perhaps more importantly, a greater mean frequency of mutualism establishment by the symbiont, even though the mean prevalence of pathogen infection is the same in the different cases (Box 10.1 Figure 1 B). In essence, the epidemic ratchets up the frequency of symbiosis occurrence, which then gradually declines during disease-free periods due to imperfect maternal transmission or the costs of symbiosis in unparasitized hosts, as has been found for pea aphids harboring symbiotic *Hamiltonella defensa* (Oliver et al. 2008). From a practical standpoint, the persistence of a defensive symbiont through long disease-free periods could make it difficult to identify the selective agent responsible for the persistence of a defensive symbiont in natural populations.

The prevalence of a defensive symbiont, and therefore the expected virulence of pathogen infections, can vary geographically due to environmental tolerance limits of a defensive symbiont. For example, parasites and/or defensive symbionts can have narrower temperature ranges than their hosts. The *Drosophila*-parasitic nematode *Howardula aoronymphium* is less tolerant of high temperatures than several of its host species (Jaenike 1995). *Drosophila hydei* carries a strain of *Spiroplasma* that confers protection from wasp parasitism (Xie et al. 2010) and whose transmission fidelity drops from 0.99 at 25 °C to 0 at 15 °C (Osaka et al. 2008). Thus, the level of protection that a defensive symbiont can provide may vary geographically and seasonally, and this could lead to spatial or temporal variation in the prevalence of parasitism and of symbiont infection (Box 10.1 Figure 2).

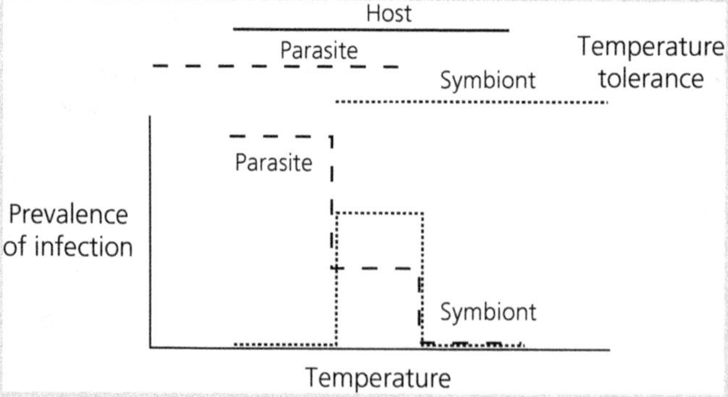

Figure 1 Variation among interacting hosts, parasites, and defensive symbionts in temperature tolerance could lead to geographic variation in the prevalence of parasitism and symbiont infection. It is assumed that, in the absence of parasites, the symbiont will be lost due to imperfect maternal transmission or the cost of infection.

Box 10.1 *Continued*

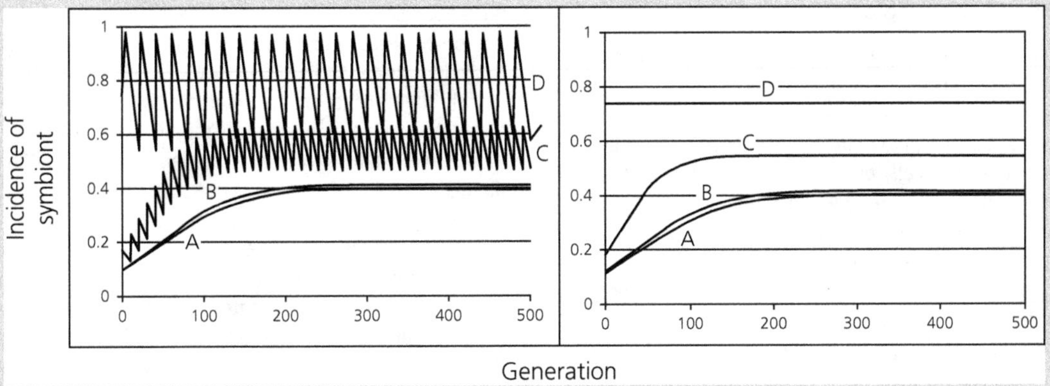

Figure 2 Incidence of a defensive endosymbiont that protects against a pathogen that either kills or sterilizes its host. The model assumes that the fitness of unparasitized hosts or parasitized hosts carrying the symbiont is 1, while that of parasitized hosts lacking the symbiont is 0. Maternal transmission is set at 0.97, which corresponds to what is often seen in nature. The prevalence of parasitism = 0.05 every generation (A), 0.1 every other generation (B), 0.5 once every 10 generations (C), or 100% once every 20 generations (D). The generation to generation fluctuations in the prevalence of symbiont infection and the 20-generation moving average are shown on the left and right, respectively.

References

Anderson, R.M., and May, R.M. (1991). *Infectious Diseases of Humans: Dynamics and Control.* Oxford: Oxford University Press.

Briggs, C.J., Hails, R.S., Barlow, N.D., and Godfray, H.C.J. (1995). The dynamics of insect-pathogen interactions. In B.T. Grenfell and A.P. Dobson (eds.), *Ecology of Infectious Diseases in Natural Populations*, pp. 295–326. Cambridge: Cambridge University Press.

Jaenike, J. 1995. Interactions between mycophagous *Drosophila* and the nematode parasites: from physiological to community ecology. *Oikos*, **72**, 235–244.

Jaenike, J. 2012. Population genetics of beneficial heritable symbionts. *TREE*, **27**, 226–232.

McCallum, H., Barlow, N., and Hone, J. (2001). How should pathogen transmission be modeled? *TREE*, **16**, 295–300.

Oliver, K.M., Campos, J., Moran, N.A., and Hunter, M.S. (2008). Population dynamics of defensive symbionts in aphids. *Proc. Roy. Soc. Lond. B*, **275**, 293–299.

Osaka. R, Nomura, M., Watada, M., and Kageyama, D. (2008). Negative effects of low temperatures on the vertical transmission and infection density of a *Spiroplasma* endosymbiont in *Drosophila hydei. Curr. Microbiol.*, **57**, 335–339.

Tsuchida, T., Koga, R., Matsumoto, S., and Fukatsu, T. (2011). Interspecific symbiont transfection confers a novel ecological trait to the recipient insect. Biology Letters, **7**, 245–248.

Xie, J.L., Vilchez, I., and Mateos, M. (2010). Spiroplasma bacteria enhance survival of *Drosophila hydei* attacked by the parasitic wasp *Leptopilina heterotoma. PLoS ONE*, **5**, e12149.

including the species diversity and composition of the surrounding community, fluctuations in pathogen populations, and abiotic environmental factors. Such theoretical studies have the advantage of being able to explore a range of potential scenarios, which is especially helpful for interactions that have received limited empirical attention. However, it must be kept in mind that simple mathematical models will often make necessarily unrealistic assumptions.

10.2.2 Observational field studies

Detailed observations can measure the outcomes of mutualisms along biotic and abiotic gradients. For example, in the potential mutualism between coral reef fish and cleaner goby fish (Gobiidae), observational studies documented that environments with higher parasite loads were associated with lower rates of cheating (i.e., consumption of material such as mucus and scales) by the cleaning gobies (Cheney and Cote 2005). While such studies have the advantage of showing how mutualistic outcomes vary in a realistic context, a key limitation is that relationships of outcomes to contextual factors are correlative. Such correlations can suggest hypotheses for further testing, but can also be driven by other, unmeasured variables. In Box 10.2, Silvia Lomáscolo

Box 10.2 Elucidating context dependency in seed dispersal interactions

Silvia B. Lomáscolo

The outcome of seed dispersal interactions is difficult to assess because seeds do not have a specific "target" as pollen does in plant–pollinator interactions. Whether a seed lands in a site that is appropriate for germination and establishment depends greatly on the general characteristics of the surrounding environment, which often vary among microsites and with climate stochasticity. Quantity and quality of dispersal (Schupp et al. 2010) will also determine the fate of seeds. Quantity of dispersal is the number of seeds removed by a specific seed disperser agent, which may vary with fruit availability (Perea et al. 2013). Quality includes the treatment of seeds in frugivores' guts and the behavior of the seed disperser, which ultimately determine where the seed is deposited. Seed treatment is probably the least variable of all the factors considered above, while disperser behavior and concomitant seed deposition may greatly vary according, among other things, to overall characteristics of the environment. Variation in one or more of these factors, i.e., the context where the interaction takes place, will determine its outcome. The resulting complexity in quantifying the effect of a seed disperser on plant reproduction has led to the commonly cited statement that "the implied marriage of animal foraging with plant demography is rarely consummated" (Howe 1989).

The quantitative component of seed dispersal, i.e., the number of seeds removed by a specific frugivore, has been shown to vary among populations of a specific plant species (Perea et al. 2013). Seed dispersal effectiveness (SDE) of a frugivore, defined as the number of seedlings resulting from dispersal by a specific seed disperser relative to other seed dispersers (or relative to no dispersal; Schupp 1993, Schupp et al. 2010), includes the quantitative and the qualitative aspects together, and it has also been shown to vary according to the environment in which the population is located, and with overall fruit availability (Calviño-Cancela and Martín-Herrero 2009). As an example, a study done on the fleshy-fruited shrub, *Corema album*, in northwest Spain (Calviño-Cancela and Martín-Herrero 2009) shows that in two out of three populations of this plant, gulls are the most efficient dispersers as they deposit seeds in open habitat, compared to blackbirds, which perch on shrubs under which defecated seeds are not highly successful. In a third population, however, blackbirds were just as efficient dispersers as gulls were. The suggested mechanism is that very few shrubs grow where this third population is located, and hence blackbirds take seeds to an open habitat, just as gulls do. Also, competition for fruit may decrease the proportion of seeds dispersed by the most efficient disperser (Calviño-Cancela and Martín-Herrero 2009). Therefore, the overall effectiveness of the complete assemblage of seed dispersers, which is included in the SDE concept (Schupp et al. 2010), may vary if fruit availability varies between populations.

Although the above-mentioned examples are first attempts to identify context dependency and the environmental variables associated with it, we are still far from understanding how much context dependency really exists in fruit–frugivore systems (Schupp et al. 2010), and whether the environmental variables found to be associated with context dependency in our examples are really the cause of the variation observed in SDE. Does SDE generally vary with microhabitat availability? Or with fruit limitation? Or with variability in roosting/perching/nesting site of frugivores? To answer these types of questions, we need to identify populations of plants located in habitats that differ in the variable of interest (e.g. microhabitat availability, degree of fruit limitation, roosting/perching/nesting site of frugivores), and measure SDE for those populations, either for a specific seed disperser or for the complete assemblage of dispersers. But populations may share a genetic background that will determine the requirements for germination and establishment. Hence, they may not be treated as independent data points to be compared. This is similar to the problem faced when trying to compare species (Felsenstein 1985), which violates the assumptions of most statistical tests and often leads to inappropriate interpretation of the patterns in the data. Phylogenetic comparative methods (PCMs) have been designed to ensure that taxa to be compared are independent (Harvey and Pagel 1991), and they can help in the identification of the causes of context dependency in the questions proposed for fruit–frugivore systems. Using the example above, say that we are interested in identifying whether perching site availability is causing the variation observed in SDE by blackbirds. We could take the experimental route, which may prove efficient for addressing some questions, and quantify SDE for blackbirds while controlling the number of perches in the habitat. Alternatively, we could identify several populations of *Corema album* where SDE by blackbirds varies and measure perching site availability in the habitat. These two variables can be plotted on the phylogeny of the plant populations, and their correlation can be quantified using different PCMs, depending on whether the variables are quantitative or categorical (Felsenstein 1985,

Box 10.2 *Continued*

Martins and Hansen 1997, Giannini 2003). PCMs may help to identify general causes of context dependency in plant–seed disperser systems, a true challenge for future research on context dependency in interspecific interactions.

References

Calviño-Cancela, M., and Martín-Herrero, J. (2009). Effectiveness of a varied assemblage of seed dispersers of a fleshy-fruited plant. *Ecology*, **90**, 3503–3515.

Felsenstein, J. (1985). Phylogenies and the comparative method. *The American Naturalist*, **125**, 1–15.

Giannini, N.P. (2003). Canonical phylogenetic ordination. *Systematic Biology*, **52**, 684–695.

Harvey, P.H., and Pagel, M.D. (1991). *The Comparative Method in Evolutionary Biology*. Oxford: Oxford University Press.

Howe, H.F. (1989). Scatter- and clump-dispersal and seedling demography: hypotheses and implications. *Oecologia*, **79**, 417–426.

Martins, E.P., and Hansen, T.F. (1997). Phylogenies and the comparative method: a general approach to incorporating phylogenetic information into the analysis of interspecific data. *The American Naturalist*, **149**, 646–667.

Perea, R., Delibes, M., Polko, M., Suárez-Esteban, A., and Fedriani, J.M. (2013). Context-dependent fruit–frugivore interactions: partner identities and spatio-temporal variations. *Oikos*, **122**, 943–951.

Schupp, E.W. (1993). Quantity, quality and the effectiveness of seed dispersal by animals. *Vegetatio*, **107/108**, 15–29.

Schupp, E.W., Jordano, P., and Gómez, J.M. (2010). Seed dispersal effectiveness revisited: a conceptual review. *New Phytologist*, **188**, 333–353.

describes how observational field studies can be combined with phylogenetic comparative methods to advance our understanding of context dependency in plant–frugivore seed-dispersal mutualisms.

10.2.3 Manipulative field and laboratory experiments

Manipulating the biotic, abiotic, and genetic context can isolate the relative importance of particular factors. For example, in a study of a potential mutualism between leguminous plants and their root-symbiotic rhizobia bacteria that fix atmospheric nitrogen, Sprent (1976) found that *Phaseolus vulgaris* plants subjected to water stress exhibited a 70% reduction in the number of bacterial root nodules and a 90% reduction in nitrogen fixation rate per gram of nodule when compared to well-watered plants. Because contextual factors are manipulated by the experimenter and randomly assigned to experimental units independent of other potential causal factors, and outcomes of mutualisms can be compared among treatments and appropriate controls, such experimental studies have the advantage of stronger inference with respect to causation, compared to field observational studies. On the other hand, although contextual factors can often be most unambiguously manipulated in laboratory or greenhouse studies, the latter can be limited in realism and in the diversity of contextual factors that are explored.

One novel and potentially powerful experimental technique is the manipulation of genomic regions responsible for key traits governing the establishment or function of a mutualism. Mutant and wild type genomes can then be compared to assess how genetic context influences the outcome of interactions. For example, we now have the ability to manipulate regions of the genome responsible for traits attracting pollinators to plants (Bradshaw and Schemske 2003, Kessler et al. 2008; see also Chapter 5). Using such tools allows a comparison of the magnitude of benefits accrued by individual plant genotypes under different ecological conditions. In so doing, we can begin to evaluate at which level of trait expression these interactions shift from being commensal to mutualistic.

10.2.4 Meta-analyses

Synthesizing variation in the outcomes of mutualisms across accumulated empirical studies using formal meta-analysis can allow for more general conclusions than are possible from any particular empirical study. Typically, ecological meta-analyses are conducted by accumulating measures of effect size from each of many studies, where effect size is a metric for an outcome of interest. There are two distinct ways that effect sizes can be calculated for meta-analysis on context dependency of mutualism. In the between-studies approach, effect size simply quantifies the outcome of a particular mutualism in each study, for example, the log ratio of mean plant performance with versus without an ant mutualist (e.g., Chamberlain and Holland 2009). In this case, context dependency is explored by modeling variation in effect size among studies as a function of contextual factors that vary among studies. These contextual factors are treated as explanatory variables in separate univariate meta-analysis models, or used together as predictors in a multiple regression meta-analysis model. A benefit of this between-studies approach is that it takes advantage of variation among studies in contextual factors, so that individual studies can be used in the meta-analysis even when they did not manipulate contextual factors. However, a caveat is that in such an analysis, associations of effect sizes with variation in contextual factors among studies are correlative, since this variation in context is observed, not experimentally manipulated.

A key concept in the interpretation of such between-studies meta-analyses is that although they may quantify the overall average outcome of a mutualism across many studies, and this average effect size may be useful for some questions, its average magnitude does not tell us a great deal about the degree of context dependency. Rather, to understand context dependency we must focus on the variation around this average effect size, and ask which contextual factors best explain this variation. For example, Chamberlain and Holland (2009) found that in a meta-analysis across 76 studies, ants on average had significantly positive effects on their host plants, in terms of plant performance or reduced herbivory. In addition, factors such as ant diversity—a form of local biotic environmental context—helped explain significant variation in benefits of ants to plants.

However, meta-analyses often also consider explanatory variables that vary among the species used in the studies being reviewed, such as growth habit and other life-history traits. Because they are fixed within species, variables such as these may not help advance our understanding of context dependency in a particular mutualistic interaction between two species. An alternative approach, the within-studies meta-analysis, utilizes only those studies in which a contextual factor of interest was experimentally manipulated or otherwise varied within the study, and an effect size is calculated for each study that quantifies not only the outcome of the mutualism, but how it was altered by the contextual factor in each study (e.g., Chamberlain et al. 2014). This approach may allow for stronger inference on any particular contextual factor, but fewer studies exist that can be utilized in this kind of meta-analysis.

Below, as we review progress in answering eight major questions about context dependency in mutualism, we highlight how each of the four main tools—models, observational field studies, experiments, and meta-analysis—have so far been applied, or could be applied, to make progress on these questions.

10.3 Eight big questions on context dependency in mutualisms

10.3.1 Question 1: What is the relative importance of multiple contextual factors for outcomes of mutualisms, and when do they interact to alter these outcomes?

Organisms engaged in mutualisms are subject to the influence of numerous contextual factors simultaneously. Some of these factors are likely to be significantly more important than others in influencing outcomes of mutualisms, and interactions among these contextual factors are likely common. Yet, relatively few studies in any mutualism system have explicitly examined both the main and interactive effects of multiple contextual factors.

Models of resource-exchange mutualisms between plants and microbial root symbionts (mycorrhizal fungi and rhizobia bacteria) in which plants receive nutrients such as N or P from the microbes in exchange for fixed carbon (C), have focused on abiotic factors such as ambient light and soil nutrient availability as key contextual factors that may influence ecological outcomes (e.g., Akcay and Simms 2011; see also Chapter 4). Typically, species are predicted to benefit most from the interactions when the resources they are trading are most limiting to their performance, and when their partner has a relative surplus of that limiting resource. For example, models of legume–rhizobia mutualisms predict that the symbiosis will become less beneficial (or parasitic) for the plant host as soil N availability increases, since it may be less beneficial for the plant to trade C for N when it can obtain N directly from the soil. Furthermore, rhizobia may evolve to fix less atmospheric N under these conditions (West et al. 2002). If the plant lacks control over the interaction, which is often the case, then increased soil N availability can lead to parasitism, wherein the costs of the interaction outweigh the benefits for the plant. Under plant control, soil N fertilization should lead to reduced nodulation. Models also predict that when light is limiting to plant photosynthetic rates, the symbiosis should be less beneficial to both partners, as C becomes a more valuable commodity for the plant (see also Chapter 4). Most experiments on context dependency in plant–microbe mutualism have focused on single abiotic factors, finding general support for many of these predictions, although results vary greatly depending on the plant and microbial taxa involved (e.g., Johnson et al. 1997).

An important subset of conceptual models, however, has highlighted the likely importance of interactive effects of multiple abiotic resources, suggesting situations in which contingency of the mutualism on one resource depends on availability of a second resource. For example, Johnson (2010) developed the stoichiometric trade-balance model, which highlights the fact that N is incorporated in large quantities into the photosynthetic apparatus of plants, and also limits fungal assimilation of C received from hosts (see Figure 1 in Johnson 2010). This model predicts that although the arbuscular mycorrhizal (AM) symbiosis will be somewhat beneficial to both partners when overall soil fertility is low, i.e. when N and P are both relatively scarce, the mutualism will be strongest when soil N availability is high and soil P availability is low. Under these conditions, plant photosynthesis will not be N-limited, and C-for-P trade can be maximized. Experimental results from AM systems tend to support the trade-balance model (Johnson 2010), wherein N fertilization in P-rich soil results in depressed AM fungal biomass and negative effects of AM fungi on host plants, but N fertilization in P-poor soils increases fungal biomass and enhances positive effects of AM fungi on host plants (reviewed by Johnson 2010). Wallander (1995) developed a related hypothesis suggesting that the degree of benefit ectomycorrhizal (EM) fungi derive from mycorrhizal symbiosis depends on the relative availability of N and P in the soil.

Lau et al. (2012) argued that it is essential to understand interactive effects of N and light on legume-rhizobial symbiosis, since N availability and plant N status may influence photosynthate production by the host (also see Son and Smith 1988). To test their idea, they manipulated light levels and soil fertility in the symbiosis between soybean (*Glycine max*) and *Bradyrhizobium japonicum*. Although they found no evidence for interactive effects of the two resources on plant benefits, they did find an interaction between light and soil nutrients for the number of nodules produced, suggesting the potential for non-additive effects of light and soil nutrients on rhizobial fitness benefits. Kersch and Fonseca (2005) performed a similar experiment on the ant–plant mutualism between the legume *Inga vera* and its ants, and found no evidence for interactions between ambient light levels and available soil nutrients. Clearly, interactions between light and soil nutrients are not universal in their influence on how plants benefit from mutualisms. Theoretical explorations are needed of how interactive effects of abiotic and biotic factors could impact particular mutualisms, based on the natural history of these systems.

The recent meta-analysis by Chamberlain et al. (2014) compared context dependency of mutualisms among different types of contextual gradients, including abiotic gradients versus biotic gradients (what they called "third party" gradients, which were usually the presence of a third species). One measure of context dependency in mutualisms, the CV of interaction magnitude among contexts, did not vary substantially between abiotic and biotic gradients. However, their second measure of context dependency, measured as a change in sign of the interaction outcome, was much larger for abiotic than biotic gradients. This result supports the general hypothesis that abiotic gradients may have more dramatic effects on outcomes of mutualisms than biotic gradients, often causing a change from mutualism to parasitism or vice versa.

Hoeksema et al. (2010) used a between-studies meta-analysis to estimate the relative importance of different abiotic and biotic contextual factors for plant responses to inoculation with mycorrhizal fungi. In one analysis across 306 laboratory studies of plant responses to AM fungi, they estimated the relative importance of seven contextual factors, and found that plant benefits were more than twice as large when non-mycorrhizal microbes were present (compared to when the background soil was sterile) and that plants responded much more positively when the fungal inoculum contained more than one species (Hoeksema et al. 2010). Those factors were significantly more important than whether or not fertilizer was added to the background soil in which the experiment was conducted, or whether that soil was sterilized at the outset of the experiment. In another analysis across 130 studies of plant responses to AM fungi, they found support for the trade balance model, as plant growth responses were more positive when plants were relatively more limited by soil P than by soil N (Hoeksema et al. 2010). In theory, this approach could be applied to any mutualism, once sufficient studies have accumulated in which the benefits to one or both partners in the mutualism have been measured. An important caveat to this approach, however, is that associations detected between effect size and study-level characteristics are correlational, since variation in predictors among studies was not manipulated experimentally. Thus, it will continue to be important to use experimental approaches in which multiple abiotic and biotic factors are manipulated to test specific hypotheses from theory to application to particular systems (Chamberlain and Holland 2009).

10.3.2 Question 2: Which aspects of biotic context have the most substantial effects on outcomes of mutualisms?

Since John Thompson (1988) wrote that "There are no analyses showing how the distribution of outcomes in a two-species interaction is affected by interactions with other species," significant progress has been made on this question. For all of the major mutualisms we surveyed here, we found at least a few examples in which mutualism outcomes were found to vary among different biotic contexts (see also Chapter 9). For example, well-studied mutualisms involving a plant partner (mycorrhizae, legume–rhizobia, plant–pollinator) seem to be commonly influenced by the presence or density of herbivores (Bethlenfalvay and Dakessian 1984, Heath and Lau 2011, Strauss et al. 1996), and those involving an animal partner (coral-algal, cleaner–client, and insect–bacteria) are influenced by the presence or density of pathogens or parasites (Cheney and Cote 2005, Douglas 2003; Box 10.1). Mutualisms involving microbial symbionts (mycorrhizae, legume–rhizobia, plant–endophyte) have been shown to be influenced by the presence or composition of non-mutualistic microbes (Larimer et al. 2010). For most mutualisms, however, there are still relatively few studies explicitly investigating the effects of local biotic context, much less comparing the magnitude of effects from different components of the biotic environment. For example, in ant–plant mutualisms, only a few studies have explicitly tested how the composition of the local ant community influences plant benefits from particular ants (Bruna et al. 2004, Palmer et al. 2010; see also Chapter 9). While many ant species can occupy or feed from different plant species, the effects of local plant community composition on ant fitness are largely unknown. In coral-algal mutualisms, explicit studies of the influence of biotic context have been even more limited; obvious needs are to understand how the outcomes of interactions between particular coral and algal species are affected by the presence or density of additional coral or algal species, or by the presence of particular coral pathogens or consumers.

Despite the relative paucity of studies on biotic context in some major mutualisms, one meta-analysis has already been published utilizing studies from the most heavily studied mutualisms, and it suggests some general conclusions relevant to Question 2. Morris et al. (2007) compiled 160 studies of plant performance in experiments in which the presence of putative mutualists and other species (either "enemy" species or other mutualists) were crossed factorially. This set of studies allowed a meta-analysis of whether the effects of particular mutualists on plants differed depending on the type of mutualist, the presence of additional mutualist species, or the presence of enemy species. One key conclusion was that the effects of mutualists on plants were not altered on average by the presence of additional mutualists, but plant response to mutualists was enhanced by the presence of enemy species. This pattern apparently emerged because belowground microbial mutualists (rhizobia and mycorrhizal fungi) commonly protect plants from pathogens. Larimer et al. (2010) performed a similar meta-analysis across 31 studies of plant responses to factorial combinations of at least two different microbial symbionts. Similar to Morris et al. (2007), they found a synergistic effect on plants coinfected by mutualistic mycorrhizal fungi and pathogenic fungal endophytes, wherein plant benefits from mycorrhizal fungi were enhanced by the presence of the endophyte. In contrast, they found that the effects of mycorrhizal fungi, beneficial endophytes, and rhizobia on plant growth were not synergistic with each other. A broad implication of the results from these two meta-analyses is that to accurately assess the importance of a particular mutualism, it should ideally be measured in the presence of a relatively realistic community context, and that omitting natural enemies or pathogens of the host may likely underestimate the host benefit from the mutualism.

10.3.3 Question 3: How do mutualisms shift along gradients of environmental productivity?

Thrall et al. (2007) developed a set of general hypotheses for how symbiotic mutualisms and parasitisms should be expected to shift along gradients of environmental productivity. They predicted that increasing productivity—for which availability of resources such as nutrients, light, and water are typically used as surrogates—should favor antagonism over mutualism in symbioses, and therefore parasitic versus mutualistic individuals, genotypes, or species. Some theoretical and empirical studies on the major symbiotic mutualisms align well with this prediction. In mycorrhizal and legume–rhizobia interactions, for example, manipulative experiments have shown that weak mutualism or even parasitism by symbionts on their host plants is often favored when overall soil fertility is high (Jones and Smith 2004, Zahran 1999) and when plant or fungal density is maximized (Bever and Schultz 2005), and observational field studies have shown that reef-building corals are especially common in marine waters of low productivity (Thompson 1988).

On the other hand, extreme nutrient limitation (arguably the opposite of high productivity) can lead to the breakdown of the mycorrhizal mutualism (Treseder and Allen 2002). Moreover, more beneficial mycorrhizal mutualisms are favored when plant photosynthesis is enhanced under high light and/or atmospheric CO_2 concentrations (Johnson 2010, Johnson et al. 1997, Jones and Smith 2004, Treseder 2004), which could be considered higher productivity environments. Similarly, water and salinity stress (again, opposite of high productivity) tend to reduce benefits conferred in the legume–rhizobia mutualism (Serraj et al. 1999, Zahran 1999). Other symbiotic mutualisms, including corals (Douglas 2003) and grass–endophyte symbioses (Cheplick and Faeth 2009, Saikkonen et al. 2006), exhibit inconsistent changes in outcomes along manipulated gradients in productivity. Among non-symbiotic mutualisms, plant–pollinator interactions are inconsistent in their responses to gradients of productivity (Burkle and Irwin 2009). Others, e.g., cleaner–client fish interactions and ant–plant mutualisms, have received very little empirical attention on this question (but see Trager and Bruna 2006).

Thus, to date, it does not appear that patterns of how symbiotic (or other) mutualisms respond to variation in abiotic context support the general prediction that increasing environmental quality should favor antagonism over mutualism. However, this question might be best approached using within-study meta-analysis, with multiple studies in which mutualism outcomes were measured along productivity gradients, or between-study meta-analysis, with multiple studies varying in the productivity of the ambient conditions in which mutualism outcomes were measured.

10.3.4 Question 4: How do mutualisms shift along gradients of diversity?

Thrall et al. (2007) also hypothesized that more diverse community contexts may present more opportunities for conflict, and predicted that the average effectiveness of particular symbiotic mutualisms might generally decline along gradients of increasing diversity. As far as we are aware, this question has not been explicitly tested empirically, although studies of insect–bacteria defensive endosymbiosis seem to support the idea (Box 10.1). We agree with Thrall et al. (2007) that the plant resource-exchange mutualisms (e.g., mycorrhizae, legume–rhizobia) might provide ideal opportunities to test such predictions, since they range from mutualism to antagonism and occur across gradients of host plant and symbiont diversity. In one experimental study, van der Heijden et al. (1998) found that plant diversity and biomass, plant tissue P content, and fungal hyphal length all increased along an experimental gradient of increasing AM fungal species richness. The meta-analysis by Hoeksema et al. (2010) across laboratory studies of plant responses to AM fungi found that plant responses to inoculation with AM fungi were substantially more positive when multiple fungal species were present in the inoculum, compared to single-species inoculations. In support of the prediction by Thrall et al. (2007), a meta-analysis found that plants tended to benefit less from ant mutualists, in terms of decreased herbivory, when the ant mutualist community was more diverse (Chamberlain and Holland 2009; see also Chapter 9). However, none of these studies has explicitly tested whether or how the benefits exchanged between two particular interacting mutualists were altered by the diversity of the community. More experimental studies are needed measuring mutualism outcomes for particular pairs of interacting species, along explicitly manipulated gradients of diversity.

10.3.5 Question 5: Are mutualism outcomes density dependent, and if so, in what manner?

If mutualisms generally decline in strength with increasing density, this form of context dependency could act as a general mechanism by which populations are regulated through negative feedback. If mutualism outcomes also decline in strength at very low densities, they could contribute to Allee effects, whereby rare species decline precipitously. Holland and DeAngelis (2001) developed analytical and simulation models of density-dependent outcomes for mutualisms between plants and their pollinating seed predators, and found a broad range of conditions under which populations of one mutualist could be regulated by density-dependent outcomes of its interactions with the other species. Holland et al. (2004) found empirical support for these ideas in experimental and observational field studies of the interaction between the senita cactus (*Pachycereus schottii*) and the senita moth (*Upiga virescens*), whose populations are regulated by density-dependent responses of the cactus. Holland and DeAngelis (2010) developed a general consumer–resource model that shows the generality of density-dependent population regulation in mutualisms (see also Chapter 8). Such phenomena are worthy of study across additional mutualisms, as they could prove to be universal principles of population regulation in mutualisms (Dean 1983).

Neighbor density significantly influences outcomes of mycorrhizal interactions, and results have been fairly consistent among studies. Typically, plant benefits from mycorrhizal fungi decrease as plant density increases in manipulative experiments (reviewed by Koide and Dickie 2002). One explanation for this observation is that at high root densities, overlap of nutrient depletion zones around roots is extensive enough to render mycorrhizal hyphae superfluous to plant nutrient absorption. Variation in fungal densities may similarly affect the benefits derived by plants from the mutualism (Bever and Schultz 2005). Such phenomena could provide density-dependent regulation of populations in mycorrhizal interactions.

Perhaps some of the most comprehensive work on the way in which the outcome of mutualisms is density dependent has been conducted on plant–pollinator interactions. The visitation rates to plants can increase with conspecific density, though this can depend on local diversity and pollinator preference in complex ways (Kunin 1993). All of these factors will interact with local landscape structure to influence the rates of fruit production and gene flow (Côrtes et al. 2013). Neverthless, much remains to be learned about the way in which pollinator benefts vary with local density. As with other mutualisms, most attention has been focused on only one partner, in this case the sessile plants in which it is easier to assess surrogates of fitness and local recruitment.

Meta-analysis could potentially be used to analyze how outcomes of multiple types of mutualisms vary with densities of the partner species. However, a recent literature search (Chamberlain et al. 2014) found insufficient numbers of studies that manipulated the density of one or both mutualists, so currently meta-analysis on this question would require using a between-studies approach, in which outcomes of particular mutualisms are compared across studies differing in the density at which experimental individuals were grown.

10.3.6 Question 6: Are patterns and consequences of context dependency typically similar for both partner species in mutualisms?

Answering this question is especially essential if we wish to understand the consequences of context dependency in mutualisms for coevolutionary dynamics. Friesen (2012) found sufficient studies to conduct a meta-analysis across experiments in which variation in both host plant and rhizobium fitness components were estimated. Across several methods of analysis, results from the meta-analysis point toward positive fitness correlations between hosts and symbionts, suggesting that antagonistic genotypes are rarely observed and that evolutionary mechanisms that align host and symbiont fitness may be prevalent, leading to mutualistic coevolution (Friesen 2012). Similar meta-analyses may be possible for experiments on mycorrhizal mutualisms in which it is relatively common to quantify some aspect of both fungal and plant performance. See Chapter 4 for a discussion of models of evolutionary processes that could lead to alignment of host and symbiont interests.

Consistent with Bronstein's (1994b) observation that studies of mutualisms are rarely reciprocal, two decades later, few studies quantify consequences of mutualisms for both partners. Moreover, studies are usually short term, which limits the degree to which benefits can be assessed for the focal taxon. As a consequence, Question 6 remains largely unanswered for several major classes of mutualisms. For example, while the costs and benefits for plants of participating in ant–plant mutualisms are well documented, virtually nothing is known about the costs and benefits for their ant partners (but see, e.g., Frederickson and Gordon 2009, Nery and Vasconcelos 2003). While the conclusion that participating in these interactions is generally beneficial will almost certainly hold true for some ant partners—for instance, ants that obligately reside in plant domatia (plant structures apparently evolved for housing ants)—for facultative participants such as those feeding on extrafloral nectaries the benefits derived may be incidental.

One-sided studies also dominate the literature on plant–pollinator interactions. In much the same way we know little about the benefits of participation in plant–ant interactions for ants, almost nothing is known about the fitness benefits of participation in pollination for animal pollinators, or how these benefits vary as a function of local abiotic and biotic conditions (but see Crone (2013) for a rare example of a demographic response observed in bees in response to changes in floral resources). This is hardly surprising. The tracking of individual pollinators and assessing their fitness is logistically difficult, especially in a field setting. However, it is nevertheless surprising to see that with the exception of highly specialized pollinating seed consumers such as fig wasps or yucca moths, where assessing fitness of pollinators is straightforward, there are almost no attempts to measure how pollinator fitness varies as a function of local environmental conditions or partner identity. While challenging, doing so is an essential component in modeling the evolution of pollination, in that fitness is the currency that ultimately drives the maintenance of mutualisms in ways that more commonly recorded data on pollinator behavior or energetic gains does not.

In the coral reef realm, studies on the reciprocal benefits to the mutualists exist for coral reef fish that obligatively inhabit either corals or sea anemones. Given the obligatory nature of the mutualism for the fish, the benefits to the fish from the symbiosis are relatively straightforward. However, since the host is a sedentary organism that potentially cannot get rid of its inhabitants, it is unclear whether its inhabitants affect the fitness of the host. Liberman et al. (1995), both in a seven-month experimental fish removal study and in following naturally inhabited and uninhabited corals, demonstrated that the coral *Stylophora pistillata* grew more in the presence of the coral reef fish *Dascyllus marginatus* than corals without fish. In addition, since coral reproduction is a function of the coral surface area, the coral's reproductive output also increased in fish-inhabited corals. Similarly, the sea anemone *Entacmaea quadricolor*'s growth was detrimentally affected by the removal of the anemonefish *Amphiprion bicinctus*, due to the host contracting in the absence of the symbionts and to increased predation on the host by butterflyfish (Porat and Chadwick-Furman 2004).

Symbioses between *Symbiodinium* algae and cnidarian animals form the foundation for coral reefs worldwide, and these ecosystems are under threat from the breakdown of this symbiosis via the phenomenon known as coral bleaching. In Box 10.3,

Box 10.3 The importance of the combined host–symbiont (holobiont) perspective in studies of cnidarian–*Symbiodinium* mutualisms

Tamar L. Goulet

The obligate mutualisms between many tropical cnidarians (e.g. corals, octocorals, sea anemones) and intracellular dinoflagellates (*Symbiodinium* spp.) are the centerpiece of coral reefs. These interactions can be highly context dependent, as changes in environmental conditions such as elevated or reduced seawater temperatures, higher or lower irradiances, ocean acidification, sedimentation, and even pathogenic organisms can lead to a reduction in *Symbiodinium* density and/or photosynthetic pigments (reviewed in Douglas 2003). This mutualism breakdown, commonly called coral bleaching, has been increasing in frequency and intensity on coral reefs worldwide (Hoegh-Guldberg 1999). The extent of coral bleaching, and whether it eventually leads to host death, differs between cnidarian species and even within a species (Goulet et al. 2008, Baird et al. 2009). With the continuation of global climate change, the projections for coral reef ecosystems are dire, although recently there has been an increased awareness that coral reef response is context dependent (Pandolfi et al. 2011). In an attempt to understand, predict, and potentially ameliorate coral bleaching, comprehending the context dependence of perturbations on the host, the endosymbiont, and the host–symbiont combination (holobiont) is imperative.

A hindrance to this goal is that much of the cnidarian–*Symbiodinium* research approaches the symbioses from the host or symbiont point of view, rarely assessing both partners' contribution to the combined entity of the holobiont. Some studies attribute the demise of the symbiosis to the symbiont present (Strychar and Sammarco 2009). Conversely, coral hosts themselves vary in their ability to withstand environmental stressors such as elevated temperatures (Baird et al. 2009), and hosts exposed to a prior environmental perturbation seem to be hardier facing a new perturbation (Pandolfi et al. 2011). But, since the symbiont resides within the host, the potential of the holobiont is a product of a given host–symbiont genotypic combination. Although the holobiont's physiological properties are based on the symbiotic partners involved, the host–symbiont combination yields a physiology unique to that holobiont. The sea anemone *Aiptasia pallida*, for example, hosts either clade A or B *Symbiodinium* in the Florida Keys and clade B *Symbiodinium* in Bermuda and other geographic locations (Goulet et al. 2005). At ambient temperatures, artificial holobionts, composed of *A. pallida* from Bermuda hosting clade A symbionts, produced oxygen fluxes that differed from either natural host–symbiont combinations of Bermuda host–clade B symbionts or Florida host–clade A symbionts. The artificial holobiont's oxygen production at elevated temperatures was in fact higher than either one of the natural holobionts (Goulet et al. 2005). Infecting the coral *Acropora millepora* with different symbionts, including those not naturally found in the symbiosis, yielded differences in thermal tolerance amongst the holobionts (Mieog et al. 2009). Therefore, reaching conclusions about a given holobiont based on data about one of the partners may not represent the holobiont's capabilities.

The holobiont is a composite of the forces working on the symbiotic partners. External forces may vary in their positive or negative effects on the partners, making the holobiont a fluid entity that may be robust or fragile in any given environment (Box 10.3 Figure 1). Consequently, trade-offs may drive the persistence of a symbiosis in a given environment. For example, *A. millepora* coral colonies hosting clade D *Symbiodinium* are more tolerant of elevated seawater temperatures compared to colonies predominantly hosting clade C (Berkelmans and van Oppen 2006). On the other hand, *A. millepora* colonies hosting clade D grow significantly less compared to colonies hosting clade C *Symbiodinium* (Jones and Berkelmans 2010), but this effect is specific to the coral reef site (Mieog et al. 2009). Therefore, a coral holobiont's survival is context dependent, combining the effects of current environmental conditions with prior perturbation history, host and symbiont physiologies, and the unique physiology of the holobiont. Consequently, when assessing the ability of a holobiont to withstand environmental change, multiple parameters need to be concurrently measured in the host, symbiont, and the resulting holobiont since complex trade-offs may occur within the symbiosis in a given environmental context.

Box 10.3 *Continued*

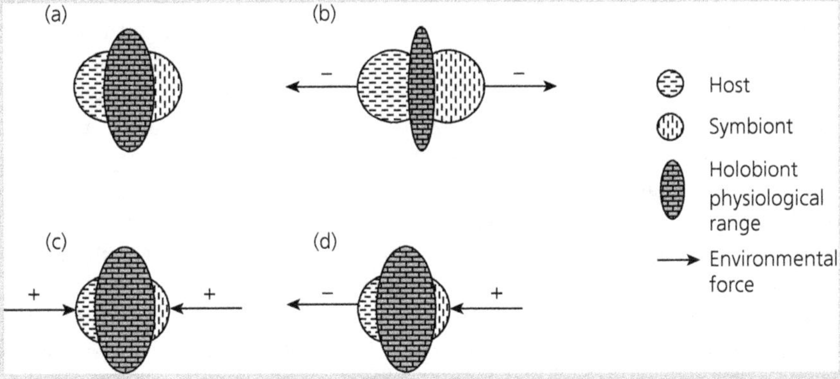

Figure 1 The holobiont is a unique physiological entity affected by environmental forces acting on the host and symbiont. (a) A host–symbiont combination produces a holobiont with its own unique physiological range. (b) When environmental forces detrimentally (−) affect both the host and symbiont, the holobiont's physiological range may narrow and the demise of the holobiont may ensue. (c) When environmental forces positively (+) affect both host and symbiont, the resulting holobiont may be robust, surviving, and even flourishing, in that environment. (d) When environmental forces negatively affect the host (−) while positively (+) affecting the symbiont, the resulting holobiont may be robust to environmental change. The opposite scenario, a positive effect on the host and a negative effect on the symbiont, may also produce a robust holobiont.

References

Baird, A.H., Bhagooli, R., Ralph, P.J., and Takahashi, S. (2009). Coral bleaching: the role of the host. *Trends in Ecology & Evolution*, **24**, 16–20.

Berkelmans, R., and van Oppen, M.J.H. (2006). The role of zooxanthellae in the thermal tolerance of corals: a "nugget of hope" for coral reefs in an era of climate change. *Proceedings of the Royal Society B-Biological Sciences*, **273**, 2305–2312.

Douglas, A.E. (2003). Coral bleaching—how and why? *Marine Pollution Bulletin*, **46**, 385–392.

Goulet, T.L., Cook, C.B., and Goulet, D. (2005). Effect of short-term exposure to elevated temperatures and light levels on photosynthesis of different host-symbiont combinations in the *Aiptasia pallida/Symbiodinium* symbiosis. *Limnology and Oceanography*, **50**, 1490–1498.

Goulet, T.L., LaJeunesse, T.C., and Fabricius, K.E. (2008). Symbiont specificity and bleaching susceptibility among soft corals in the 1998 Great Barrier Reef mass coral bleaching event. *Marine Biology*, **154**, 795–804.

Hoegh-Guldberg, O. (1999). Climate change, coral bleaching and the future of the world's coral reefs. *Marine and Freshwater Research*, **50**, 839–866.

Jones, A., and Berkelmans, R. (2010). Potential costs of acclimatization to a warmer climate: growth of a reef coral with heat tolerant vs. sensitive symbiont types. *PLoS ONE*, **5**.

Mieog, J.C., Olsen, J.L., Berkelmans, R., Bleuler-Martinez, S.A., Willis, B.L., and van Oppen, M.J.H. (2009). The roles and interactions of symbiont, host and environment in defining coral fitness. *PLoS ONE*, **4**.

Pandolfi, J.M., Connolly, S.R., Marshall, D.J., and Cohen, A.L. (2011). Projecting coral reef futures under global warming and ocean acidification. *Science*, **333**, 418–422.

Strychar, K.B., and Sammarco, P.W. (2009). Exaptation in corals to high seawater temperatures: low concentrations of apoptotic and necrotic cells in host coral tissue under bleaching conditions. *Journal of Experimental Marine Biology and Ecology*, **369**, 31–42.

Tamar Goulet explores how analyzing both sides of the interaction, i.e. understanding context dependency for both the cnidarian and the algae through a "holobiont" approach, is essential for predicting and potentially mitigating the effects of coral bleaching on coral reef ecosystems.

The literature on ant–plant mutualisms illustrates the importance of long-duration studies for answering the question of how species on both sides of an interaction benefit from the mutualism. Most studies of ant–plant mutualisms are of relatively short duration, and as the landscape in which species are interacting changes, the costs and benefits provided by partners may change as well. Herbivore pressure, rewards offered to partners, the dispersal capability of partner species, and the diversity and abundance of partners can all vary temporally (Diaz-Castelazo et al. 2010, Bruna et al. 2005, 2011, Moraes and Vasconcelos 2009) or as a result of anthropogenic activity (Bruna et al. 2005, Moraes and Vasconcelos 2009). This may be why even well-studied ant–plant systems can show very different dynamics across their range or in different years (e.g., Rudgers and Strauss 2004). While perhaps not as dramatic as the context dependency observed in other systems, the extent to which ant–plant systems exhibit context dependency requires studies evaluating the costs and benefits for both partner species in light of changes to the habitat in which they are found.

10.3.7 Question 7: How important are "cryptic" benefits for outcomes of context-dependent mutualisms?

Studies documenting context dependency in mutualisms necessarily quantify the performance of species on at least one side of the interaction, but do not necessarily measure the actual benefits being exchanged between putative mutualist species. For example, many studies of the mycorrhizal mutualism quantify plant growth with and without mycorrhizal fungi, but do not measure the soil nutrients passing from fungus to plant or the carbohydrates passing from plant to fungus. Despite the lack of quantification, these commodities are often assumed to be the primary mechanisms by which partners benefit from mutualisms. Conversely, cryptic or alternative benefits, such as pathogen protection of host plants by mycorrhizal fungi (Marx 1969, Wehner et al. 2010), are increasingly being discovered in diverse mutualisms. For example, tactile stimulation ("tickling") of client host fish by cleaners may reduce stress in hosts (Bshary et al. 2007, Soares et al. 2011), conveying a cryptic benefit that does not depend on parasite removal. Infection of cleaners by the parasites they have removed from hosts may be a cryptic cost to cleaners; cleaner wrasses have been found to ingest, harbor, and transmit infective stages of bucephalid trematodes (Jones et al. 2004).

If cryptic benefits are as important as the benefits typically measured, then they have important implications for the design of experiments meant to measure context dependence in mutualisms. Indeed, hypotheses on key contextual factors affecting mutualistic outcomes are usually explicitly grounded in some assumption about the benefits being exchanged. For example, if tactile stimulation is more important than parasite removal as a benefit in cleaner–client mutualisms, then ambient densities of parasites would not be expected to be as significant as a contextual factor affecting the outcome of the mutualism. In studies of the mycorrhizal mutualism, if protection from pathogens is an important cryptic benefit that fungi convey to plant hosts, then pathogen presence is a key contextual factor that should be manipulated or explored. A corollary of this point is that experiments performed in environments from which pathogens have been removed or excluded would be expected to observe smaller benefits to plants from mycorrhizal fungi. This is one possible explanation for why the meta-analysis by Hoeksema et al. (2010) found that plant responses to inoculation with AM fungi were substantially more positive when the background soil for the experiment was not sterile but rather contained a diverse community of non-mycorrhizal microbes (such as pathogens). We suggest that careful quantification and consideration of not only standard but also cryptic benefits in mutualisms will lead to studies that provide a much clearer understanding of the reasons for variation in mutualistic outcomes.

10.3.8 Question 8: How often do outcomes of mutualisms depend on G × E, G × G, and G × G × E interactions?

Outcomes of mycorrhizal interactions (reviewed by Hoeksema 2010, Johnson et al. 1997), legume–rhizobia interactions (e.g., Miller and Sirois 1982), and grass–endophyte symbiosis (Tintjer and Rudgers 2006) can vary substantially across different genotypes of the same microbial or host plant species. The magnitude of such genetic effects is often as large or larger than variation among species

(Johnson 2010), and can span the range from mutualism to parasitism. However, it is the context dependency of this genetic variation that determines the potential for direct selection on mutualism traits from abiotic or biotic factors, coevolution, and geographic mosaics of coevolution.

Within-population genetic variation for how mutualisms respond to abiotic and biotic contextual factors provides the raw material for context-dependent selection on the traits of mutualists. Conversely, context-dependent selection can play a key role in maintaining genetic variation within mutualist populations. Specifically, genotype-by-environment (G × E) interactions for fitness of one or both partners in a mutualism enable spatially or temporally variable selection on a mutualist species, potentially leading to local adaptation, because genotypes have different relative fitnesses in different environments. Similarly, genotype-by-genotype (G × G) interactions for fitness of mutualist partners are a prerequisite for coevolutionary selection, wherein selection on one species differs among genotypes of the other (Heath 2010, Thompson 1988). Finally, genotype-by-genotype-by environment (G × G × E) interactions allow geographic selection mosaics (Thompson 2005) whereby coevolutionary selection varies spatially among different environmental contexts (see also Chapter 7). Thus, quantifying these interactions is an essential prerequisite to understanding whether context-dependent natural selection may generate and maintain trait diversity in mutualisms.

Among all the major mutualisms, mycorrhizal and legume–rhizobia interactions have been the most frequent subjects of studies explicitly manipulating the genotypic composition of both partners. As a consequence, we now have convincing experimental evidence for G × E, G × G, and G × G × E interactions in both of those mutualisms. In some mycorrhizal mutualisms, for example, significant interactions exist between host plant or fungal genotypes and the presence of heavy metal contamination in soils (reviewed by Meharg and Cairney 2000), showing that plant and mycorrhizal fungal populations can adapt rapidly in response to that direct abiotic selection (see Stahl and Smith 1984 for evidence of a similar G × E interaction with respect to drought). In studies of legume–rhizobia mutualisms, salt-tolerant rhizobia strains improve legume host plant performance in the presence of high salinity (e.g., Zou et al. 1995), implying that such G × E interactions would drive local adaptation of the symbiosis in response to salinity gradients. In contrast, other studies have found that salt-tolerant rhizobial strains do not form functional nodules (reviewed by Zahran 1999) or do not improve host performance in saline soils (e.g., Lal and Khanna 1994), or that salt-tolerant host species are less responsive to rhizobial inoculation (Thrall et al. 2008). The latter results imply that salinity gradients could drive evolutionary divergence in dependency on the symbiosis, due to genetic trade-offs in rhizobia or hosts between tolerance of salt stress and symbiotic compatibility.

An experimental study of the interaction between the leguminous plant barrel medic (*Medicago truncatula*) and its *Sinorhizobium* symbionts (Heath 2010) found that both plant and bacterial fitness were substantially altered by an interaction between host and symbiont genotype, i.e. G × G (see also Heath and Tiffin 2007, Parker 1995). If such variation is common, it implies that the outcome of a mutualism involving a particular host or symbiont genotype depends on which partner genotypes it interacts with, and suggests potential for ongoing coevolutionary dynamics in legume–rhizobia interactions. In mycorrhizal interactions, recent studies have tested for three-way interactions between host plant genotype, mycorrhizal fungal genotypes or species, and some aspect of the abiotic or biotic context (Johnson et al. 2010, Piculell et al. 2008), i.e. G × G × E interactions. In an experimental study of interactions between the grass *Andropogon gerardii* and its symbiotic AM fungi, Johnson et al. (2010) estimated plant and fungal performance in all reciprocal combinations of plant populations (G), whole AM fungal guilds of species (G), and sterile soils (E) from three geographic sites differing in climatic and soil characteristics. The authors found a significant G × G × E interaction for the formation of arbuscules, the fungal structures formed inside the plant root through which nutrient exchange occurs. Arbuscule formation was highest in local combinations of plants, fungi, and soils, suggesting the potential for natural selection to drive three-way local adaptation among plants, fungi, and soils. Similarly, experiments have shown that soil N availability can alter the nature of particular G × G interactions between legume and rhizobia genotypes. In a series of greenhouse experiments on the host plant barrel medic and bacterial symbionts in the genus *Sinorhizobium*, Heath and Tiffin (2007) and Heath et al. (2010) found that nodulation depended on a three-way interaction (G × G × E) between rhizobial strain, host genotype, and soil nitrogen availability, with variation in nodulation across host–symbiont combinations depending on nitrogen environment (see Figure 2 in Heath et al. 2010).

Clearly both mycorrhizal and legume-rhizobial mutualisms harbor enough genetic variation in context dependency that there is strong potential for ongoing direct abiotic and biotic selection, local adaptation, coevolution, and geographic mosaics of coevolution in these mutualisms. We still, however, understand little about the relative importance of abiotic and biotic contextual factors in driving evolution of mutualism traits. Moreover, in most cases the demonstrations of genetic variation affecting fitness outcomes are not tied directly to specific traits, so we do not know what traits are under selection, nor do we know the form of coevolutionary selection (e.g., positive versus negative frequency dependence). For contextual variation in abiotic resources that are traded in these resource-exchange interactions, variation in evolutionary trajectories will likely be driven by how environmental variation in a resource alters the costs and benefits of the trading dynamics for different genotypes. Studies are needed that not only test for interactions between genetic variation and these environmental gradients, but also estimate selection on specific traits from specific sources (Ridenhour 2005).

Among the other major mutualisms (besides mycorrhizae and legume–rhizobia), quantification of G × E, G × G, and G × G × E interactions for fitness of one or both partners has been more limited. In the grass–endophyte symbiosis, for example, a reasonable number of studies provide evidence for significant host genotype by endophyte infection (+/−) interactions on components of host fitness, or interactions among host genotype, endophyte infection, and abiotic factors. For example, Krauss et al. (2007) found that biomass response of *Lolium perenne* host plants to infection with *Neotyphodium lolii* fungal endophytes was significantly influenced by an interaction between host cultivar and fertilization. Our ability to understand the potential for ongoing *coevolutionary* selection in this symbiosis is severely limited by the paucity of studies controlling both host and endophyte genotype and thus exploring G × G or G × G × E interactions. In the coral-algal symbiosis some evidence for G × G interactions exists from studies showing that properties of the holobiont (the host–symbiont combination) are unique to particular combinations of host and algae (Box 10.3). As in mycorrhizal and legume–rhizobia interactions, studies in these systems have not yet taken the step of showing which traits might be under natural selection, and from which abiotic and biotic sources. In bacterial defensive endosymbioses of insects, preliminary evidence demonstrates the potential of G × G × E interactions, whereby variation among environments in pathogen prevalence may cause variation in selection on infection status, which is a trait that may be controlled by the genotypes of both the host and symbiont (Box 10.1; J. Jaenike pers. comm.). Rudgers and Strauss (2004) showed that natural selection on wild cotton plants to allocate resources toward extrafloral nectar for attracting ant protection mutualists varied among wild cotton populations, providing a rare example in which context-dependent outcomes in a mutualism have been shown to lead to altered natural selection on the interaction. Until more studies of diverse mutualisms accumulate, in which genotypes and associated traits of one or both partners are varied experimentally, and their fitness measured across key abiotic and biotic contextual gradients, and the targets and sources of natural selection are estimated, we will have only a rudimentary understanding of how context dependence may be driving trait evolution in mutualisms.

10.4 Concluding thoughts

Most major mutualisms have the potential to dissolve or shift to parasitism under particular conditions. Even interactions with mean outcomes that are consistently mutualistic exhibit substantial variation among contexts (e.g., ant–plant mutualisms; Chamberlain and Holland 2009), much of which variation remains unexplained. Explaining this variation can aid in answering a wide variety of questions in ecology and evolution, including those outlined here (see also Chapters 4 and 9).

The questions we have highlighted vary in the degree to which they can or have been answered with studies to date. In several cases (Questions 1, 3, 5, 6, and 8), most progress to date has come from studies of mutualisms involving plants. We hope that this review will stimulate further observational and experimental work in other mutualisms. In a few cases (Questions 1, 2, 4, and 6), sufficient studies have accumulated to support relevant meta-analyses, but these analyses are only as informative as the set of studies available. Many more studies could be utilized in new or repeated meta-analyses, especially within-study meta-analyses using studies that have explicitly manipulated contextual factors. In other cases (especially Questions 3, 5, and 6), meta-analyses seem possible with the existing literature, but have not yet been conducted. Overall, however, despite the progress highlighted here, all of the questions we have raised are in need of not only further exploration in the most heavily studied mutualisms, but especially groundbreaking studies in the more empirically challenging systems. Our hope is that by raising these questions, which are applicable across diverse systems, readers will be inspired to forge new ground in search of general principles of context dependency in mutualisms.

Acknowledgments

J.D.H. was supported by National Science Foundation DEB Award #1119865. E.M.B. was supported by National Science Foundation DEB Awards 0614149 and 0453631.

References

Agrawal, A.A. (2001). Phenotypic plasticity in the interactions and evolution of species. *Science*, **294**, 321–326.

Akcay, E., and Simms, E.L. (2011). Negotiation, sanctions, and context dependency in the legume-rhizobium mutualism. *The American Naturalist*, **178**, 1–14.

Alonso, L.E. (1998). Spatial and temporal variation in the ant occupants of a facultative ant-plant. *Biotropica*, **30**, 201–213.

Ashman, T.L., Bradburn, M., Cole, D.H., Blaney, B.H., and Raguso, R.A. (2005). The scent of a male: the role of floral volatiles in pollination of a gender dimorphic plant. *Ecology*, **86**, 2099–2105.

Ashman, T.L., Knight, T.M., Steets, J.A., Amarasekare, P., Burd, M., Campbell, D.R., Dudash, M.R., Johnston, M.O., Mazer, S.J., Mitchell, R.J., Morgan, M.T., and Wilson, W.G. (2004). Pollen limitation of plant reproduction: ecological and evolutionary causes and consequences. *Ecology*, **85**, 2408–2421.

Bethlenfalvay, G.J., and Dakessian, S. (1984). Grazing effects on mycorrhizal colonization and floristic composition of the vegetation on a semiarid range in northern Nevada. *Journal of Range Management*, **37**, 312–316.

Bever, J.D., and Schultz, P.A. (2005). Mechanisms of arbuscular mycorrhizal mediation of plant-plant interactions. In Dighton, J., White, J., and Oudemans, P. (eds.) *The Fungal Community: Its Organization and Role in the Ecosystem*. New York: CRC Press.

Bradshaw, H.D., and Schemske, D.W. (2003). Allele substitution at a flower colour locus produces a pollinator shift in monkeyflowers. *Nature*, **426**, 176–178.

Bronstein, J.L. (1994a). Conditional outcomes in mutualistic interactions. *Trends in Ecology & Evolution*, **9**, 214–217.

Bronstein, J.L. (1994b). Our current understanding of mutualism. *Quarterly Review of Biology*, **69**, 31–51.

Bruna, E.M., Izzo, T.J., Inouye, B.D., Uriarte, M., and Vasconcelos, H.L. (2011). Asymmetric dispersal and colonization success of Amazonian plant-ants queens. *PLoS One*, **6**.

Bruna, E.M., Lapola, D.M., and Vasconcelos, H.L. (2004). Interspecific variation in the defensive responses of obligate plant-ants: experimental tests and consequences for herbivory. *Oecologia*, **138**, 558–565.

Bruna, E.M., Vasconcelos, H.L., and Heredia, S. (2005). The effect of habitat fragmentation on communities of mutualists: Amazonian ants and their host plants. *Biological Conservation*, **124**, 209–216.

Bshary, R., Oliveira, R.F., Oliveira, T.S.F., and Canã¡Rio, A.V.M. (2007). Do cleaning organisms reduce the stress response of client reef fish? *Frontiers in Zoology*, **4**, 21.

Burkle, L.A., and Irwin, R.E. (2009). The effects of nutrient addition on floral characters and pollination in two subalpine plants, *Ipomopsis aggregata* and *Linum lewisii*. *Plant Ecology*, **203**, 83–98.

Chamberlain, S.A., Bronstein, J.L., and Rudgers, J.A. (2014). How context dependent are species interactions? *Ecology Letters*, **17**, 881–890.

Chamberlain, S.A., and Holland, J.N. (2009). Quantitative synthesis of context dependency in ant-plant protection mutualisms. *Ecology*, **90**, 2384–2392.

Cheney, K.L., and Cote, I.M. (2005). Mutualism or parasitism? The variable outcome of cleaning symbioses. *Biology Letters*, **1**, 162–165.

Cheplick, G.P., and Faeth, S.H. (2009). *Ecology and Evolution of the Grass–Endophyte Symbiosis*. Oxford: Oxford University Press.

Côrtes, M., Uriarte, M., Lemes, M., Gribel, R., Kress, W.J., Smouse, P., and Bruna, E.M. (2013). Low plant density enhances gene dispersal in the Amazonian understory herb *Heliconia acuminata*. *Molecular Ecology*, **22**, 5716–5729.

Crone, E.E. (2013). Responses of social and solitary bees to pulsed floral resources. *The American Naturalist*, **182**, 465–473.

Dean, A.M. (1983). A simple model of mutualism. *American Naturalist*, **121**, 409–417.

Diaz-Castelazo, C., Guimaraes, P.R., Jordano, P., Thompson, J.N., Marquis, R.J., and Rico-Gray, V. (2010). Changes of a mutualistic network over time: reanalysis over a 10-year period. *Ecology*, **91**, 793–801.

Douglas, A.E. (2003). Coral bleaching—how and why? *Marine Pollution Bulletin*, **46**, 385–392.

Folgarait, P.J., and Davidson, D.W. (1995). Myrmecophytic *Cecropia*: antiherbivore defenses under different nutrient treatments. *Oecologia (Berlin)*, **104**, 189–206.

Frederickson, M.E., and Gordon, D.M. (2009). The intertwined population biology of two Amazonian myrmecophytes and their symbiotic ants. *Ecology*, **90**, 1595–1607.

Friesen, M.L. (2012). Widespread fitness alignment in the legume-rhizobium symbiosis. *New Phytologist*, **194**, 1096–1111.

Garcia-Camacho, R., and Totland, O. (2009). Pollen limitation in the Alpine: a meta-analysis. *Arctic Antarctic and Alpine Research*, **41**, 103–111.

Graham, P.H., Rosas, J.C., Estevez De Jensen, C., Peralta, E., Tlusty, B., Acosta-Gallegos, J., and Arraes Pereira, P.A. (2003). Addressing edaphic constraints to bean production: the bean/cowpea CRSP project in perspective. *Field Crops Research*, **82**, 179–192.

Heath, K.D. (2010). Intergenomic epistasis and coevolutionary constraint in plants and rhizobia. *Evolution*, **64**, 1446–1458.

Heath, K.D., and Lau, J.A. (2011). Herbivores alter the fitness benefits of a plant–rhizobium mutualism. *Acta Oecologica*, **37**, 87–92.

Heath, K.D., Stock, A.J, and Stinchcombe, J.R. (2010). Mutualism variation in the nodulation response to nitrate. *Journal of Evolutionary Biology*, **23**, 2494–2500.

Heath, K.D., and Tiffin, P. (2007). Context dependence in the coevolution of plant and rhizobial mutualists. *Proceedings of the Royal Society B-Biological Sciences*, **274**, 1905–1912.

Heil, M., Hilpert, A., Fiala, B., and Linsenmair, K.E. (2001). Nutrient availability and indirect (biotic) defence in a Malaysian ant-plant. *Oecologia*, **126**, 404–408.

Herre, E.A., Knowlton, N., Mueller, U.G., and Rehner, S.A. (1999). The evolution of mutualisms: exploring the paths between conflict and cooperation. *Trends in Ecology and Evolution*, **14**, 49–53.

Hoeksema, J.D. (2010). Tansley review. Ongoing coevolution in mycorrhizal interactions. *New Phytologist*, **187**, 286–300.

Hoeksema, J.D., Chaudhary, V.B., Gehring, C. A., et al. (2010). A meta-analysis of context-dependency in plant response to inoculation with mycorrhizal fungi. *Ecology Letters*, **13**, 394–407.

Holland, J.N., Bronstein, J.L., and Deangelis, D.L. (2004). Testing hypotheses for excess flower production and low fruit-to-flower ratios in a pollinating seed-consuming mutualism. *Oikos*, **105**, 633–640.

Holland, J.N., and Deangelis, D.L. (2001). Population dynamics and the ecological stability of obligate pollination mutualisms. *Oecologia*, **126**, 575–586.

Holland, J.N., and Deangelis, D.L. (2010). A consumer-resource approach to the density-dependent population dynamics of mutualism. *Ecology*, **91**, 1286–1295.

Johnson, N.C. (2010). Tansley review: resource stoichiometry elucidates the structure and function of arbuscular mycorrhizas across scales. *New Phytologist*, **185**, 631–647.

Johnson, N.C., Graham, J.H., and Smith, F.A. (1997). Functioning of mycorrhizas along the mutualism-parasitism continuum. *New Phytologist*, **135**, 1–12.

Johnson, N.C., Wilson, G.W.T., Bowker, M.A., Wilson, J., and Miller, R.M. (2010). Resource limitation is a driver of local adaptation in mycorrhizal symbioses. *Proceedings of the National Academy of Sciences of the United States of America*, **107**, 2093–2098.

Jones, C.M., Grutter, A.S., and Cribb, T.H. (2004). Cleaner fish become hosts: a novel form of parasite transmission. *Coral Reefs*, **23**, 521–529.

Jones, M.D., and Smith, S.E. (2004). Exploring functional definitions of mycorrhizas: are mycorrhizas always mutualisms? *Canadian Journal of Botany*, **82**, 1089–1109.

Karst, J., Marczak, L., Jones, M.D., and Turkington, R. (2008). The mutualism-parasitism continuum in ectomycorrhizas: a quantitative assessment using meta-analysis. *Ecology*, **89**, 1032–1042.

Kersch, M.F., and Fonseca, C.R. (2005). Abiotic factors and the conditional outcome of an ant-plant mutualism. *Ecology*, **86**, 2117–2126.

Kessler, D., Gase, K., and Baldwin, I.T. (2008). Field experiments with transformed plants reveal the sense of floral scents. *Science*, **321**, 1200–1202.

Knight, T.M., Steets, J.A., Vamosi, J.C., Mazer, S.J., Burd, M., Campbell, D.R., Dudash, M.R., Johnston, M.O., Mitchell, R.J., and Ashman, T.L. (2005). Pollen limitation of plant reproduction: pattern and process. *Annual Review of Ecology Evolution and Systematics*, **36**, 467–497.

Koide, R.T., and Dickie, I.A. (2002). Effects of mycorrhizal fungi on plant populations. *Plant and Soil*, **244**, 307–317.

Krauss, J., Härri, S.A., Bush, L., Husi, R., Bigler, L., Power, S.A., and Muller, C.B. (2007). Effects of fertilizer, fungal endophytes and plant cultivar on the performance of insect herbivores and their natural enemies. *Functional Ecology*, **21**, 107–116.

Kunin, W.E. (1993). Sex and the single mustard: population density and pollinator behavior effects on seed set. *Ecology*, **74**, 2145–2160.

Lal, B., and Khanna, S. (1994). Selection of salt-tolerant *Rhizobium* isolates of *Acacia nilotica*. *World Journal of Microbiology and Biotechnology*, **10**, 637–639.

Larimer, A.L., Bever, J.D., and Clay, K. (2010). The interactive effects of plant microbial symbionts: a review and meta-analysis. *Symbiosis*, **51**, 139–148.

Lau, J. A., Bowling, E.J., Gentry, L.E., Glasser, P.A., Monarch, E.A., Olesen, W.M., Waxmonsky, J., and Young, R.T. (2012). Direct and interactive effects of light and nutrients on the legume-rhizobia mutualism. *Acta Oecologica*, **39**, 80–86.

Liberman, T., Genin, A., and Loya, Y. (1995). Effects on growth and reproduction of the coral *Stylophora pistillata* by the mutualistic damselfish *Dascyllus marginatus*. *Marine Biology*, **121**, 741–746.

Marx, D.H. (1969). The influence of ectotrophic mycorrhizal fungi on the resistance of pine roots to pathogenic infections. I. Antagonism of mycorrhizal fungi to root pathogenic fungi and soil bacteria. *Phytopathology*, **59**, 153–163.

Meharg, A.A., and Cairney, J.W.G. (2000). Co-evolution of mycorrhizal symbionts and their hosts to metal-contaminated environments. *Advances in Ecological Research*, **30**, 69–112.

Miller, R.W., and Sirois, J.C. (1982). Relative efficiency of different alfalfa cultivar-*Rhizobium meliloti* strain combinations for symbiotic nitrogen fixation. *Applied and Environmental Microbiology*, **43**, 764–768.

Moraes, S.C., and Vasconcelos, H.L. (2009). Long-term persistence of a Neotropical ant-plant population in the absence of obligate plant-ants. *Ecology*, **90**, 2375–2383.

Morris, W.F., Hufbauer, R.A., Agrawal, A.A., et al. (2007). Direct and interactive effects of enemies and mutualists on plant performance: a meta-analysis. *Ecology*, **88**, 1021–1029.

Nery, A.S., and Vasconcelos, H.L. (2003). Growth and survival of incipient ant-colonies in two Amazonian ant-plants: effects of host plant, habitat, and mode of colony founding (Hymneoptera: Formicidae). *Sociobiology*, **42**, 151–162.

Palmer, T.M., Doak, D.F., Stanton, M.L., Bronstein, J.L., Kiers, E.T., Young, T.P., Goheen, J.R., and Pringle, R.M. (2010). Synergy of multiple partners, including free-loaders, increases host fitness in a multispecies mutualism. *Proceedings of the National Academy of Sciences of the United States of America*, **107**, 17234–17239.

Parker, M.A. (1995). Plant fitness variation caused by different mutualist genotypes. *Ecology*, **76**, 1525–1535.

Piculell, B.J., Hoeksema, J.D., and Thompson, J.N. (2008). Interactions of biotic and abiotic environmental factors in an ectomycorrhizal symbiosis, and the potential for selection mosaics. *BMC Biology*, **6**, 23–34.

Porat, D., and Chadwick-Furman, N.E. (2004). Effects of anemonefish on giant sea anemones: expansion behavior, growth, and survival. *Hydrobiologia*, **530**, 513–520.

Ridenhour, B.J. (2005). Identification of selective sources: partitioning selection based on interactions. *American Naturalist*, **166**, 12–25.

Rudgers, J.A., and Strauss, S.Y. (2004). A selection mosaic in the facultative mutualism between ants and wild cotton. *Proceedings of the Royal Society of London Series B-Biological Sciences*, **271**, 2481–2488.

Sachs, J.L., and Simms, E.L. (2006). Pathways to mutualism breakdown. *Trends in Ecology & Evolution*, **21**, 585–592.

Saikkonen, K., Lehtonen, P.I., Helander, M., Koricheva, J., and Faeth, S.H. (2006). Model systems in ecology: dissecting the endophyte–grass literature. *Trends in Plant Science*, **11**, 428–433.

Serraj, R., Sinclair, T.R., and Purcell, L.C. (1999). Symbiotic N2 fixation response to drought. *Journal of Experimental Botany*, **50**, 143–155.

Soares, M.C., Oliveira, R.F., Ros, A.F.H., Grutter, A.S., and Bshary, R. (2011). Tactile stimulation lowers stress in fish. *Nature Communications*, **2**, 534.

Son, C.L., and Smith, S.E. (1988). Mycorrhizal growth responses: interactions between photon irradiance and phosphorus nutrition. *New Phytologist*, **108**, 305–314.

Sprent, J.I. (1976). Nitrogen fixation by legumes subjected to water and light stresses. In Nutman, P.S. (ed.) *Symbiotic Nitrogen Fixation in Plants*. Cambridge, UK: Cambridge University Press.

Stahl, P.D., and Smith, W.K. (1984). Effects of different geographic isolates of *Glomus* on the water relations of *Agropyron smithii*. *Mycologia*, **76**, 261–267.

Strauss, S.Y., Conner, J.K., and Rush, S.L. (1996). Foliar herbivory affects floral characters and plant attractiveness to pollinators: implications for male and female plant fitness. *American Naturalist*, **147**, 1098–1107.

Streeter, J., and Wong, P.P. (1988). Inhibition of legume nodule formation and N_2 fixation by nitrate. *Critical Reviews in Plant Sciences*, **7**, 1–23.

Temeles, E.J., and Kress, W.J. (2003). Adaptation in a plant-hummingbird association. *Science*, **300**, 630–633.

Thompson, J.N. (1988). Variation in interspecific interactions. *Annual Review of Ecology and Systematics*, **19**, 65–87.

Thompson, J.N. (2005). *The Geographic Mosaic of Coevolution*. Chicago: University of Chicago Press.

Thrall, P.H., Bever, J.D., and Slattery, J.F. (2008). Rhizobial mediation of *Acacia* adaptation to soil salinity: evidence of underlying trade-offs and tests of expected patterns. *Journal of Ecology*, **96**, 746–755.

Thrall, P.H., Hochberg, M.E., Burdon, J.J., and Bever, J.D. (2007). Coevolution of symbiotic mutualists and parasites in a community context. *Trends in Ecology & Evolution*, **22**, 120–126.

Tintjer, T., and Rudgers, J.A. (2006). Grass-herbivore interactions altered by strains of a native endophyte. *New Phytologist*, **170**, 513–521.

Trager, M.D., Bhotika, S., Hostetler, J.A., Andrade, G.V., Rodriguez-Cabal, M.A., Mckeon, C.S., Osenberg, C.W., and Bolker, B.M. (2010). Benefits for plants in ant-plant protective mutualisms: a meta-analysis. *PLoS ONE*, **5**, e14308.

Trager, M.D., and Bruna, E.M. (2006). Effects of plant age, experimental nutrient addition and ant occupancy on herbivory in a neotropical myrmecophyte. *Journal of Ecology*, **94**, 1156–1163.

Treseder, K.K. (2004). A meta-analysis of mycorrhizal responses to nitrogen, phosphorus, and atmospheric CO_2 in field studies. *New Phytologist*, **164**, 347–355.

Treseder, K.K., and Allen, M.F. (2002). Direct nitrogen and phosphorus limitation of arbuscular mycorrhizal fungi: a model and field test. *New Phytologist*, **155**, 507–515.

Van Der Heijden, M.G.A., Klironomos, J.N., Ursic, M., Moutoglis, P., Streitwolf-Engel, R., Boller, T., Wiemken, A., and Sanders, I.R. (1998). Mycorrhizal fungal diversity determines plant biodiversity, ecosystem variability and productivity. *Nature*, **396**, 69–72.

Wallander, H. (1995). A new hypothesis to explain allocation of dry matter between mycorrhizal fungi and pine seedlings in relation to nutrient supply. *Plant and Soil*, **168–169**, 243–248.

Wehner, J., Antunes, P.M., Powell, J.R., Mazukatow, J., and Rillig, M.C. (2010). Plant pathogen protection by arbuscular mycorrhizas: a role for fungal diversity? *Pedobiologia*, **53**, 197–201.

West, S.A., Kiers, E.T., Simms, E.L., and Denison, R.F. (2002). Sanctions and mutualism stability: why do rhizobia fix nitrogen? *Proceedings of the Royal Society of London. Series B: Biological Sciences*, **269**, 685–694.

Zahran, H.H. (1999). Rhizobium-legume symbiosis and nitrogen fixation under severe conditions and in an arid climate. *Microbiology and Molecular Biology Reviews*, **63**, 968–989.

Zou, N., Dart, P.J., and Marcar, N.E. (1995). Interaction of salinity and rhizobial strain on growth and N_2-fixation by *Acacia ampliceps*. *Soil Biology and Biochemistry*, **27**, 409–413.

CHAPTER 11

Mutualistic networks

Jordi Bascompte and Jens M. Olesen

11.1 A historical perspective on network thinking in mutualism

Plant–animal mutualisms such as those between flowering plants and their pollinators or seed dispersers have played a major role in the generation of biodiversity on earth. Beyond early notes and some predecessors of ideas, it is fair to say that studies of mutualistic interactions among free-living species began with the work of J.G. Kölreuter (1733–1806; Waser 2006), later followed by C.K. Sprengel (1793), who become a direct inspiration of Charles Darwin.

Darwin himself is often cited as one of the founders of research on mutualisms. His book *On the Various Contrivances by Which British and Foreign Orchids are Fertilised by Insects, and on the Good Effects of Intercrossing* (Darwin 1862) was very important in advancing the idea that one cannot understand the evolution of floral morphology without taking into account the evolution of insect morphology. That is, the evolution of plants and animals are not independent of each other.

With the development of the modern evolutionary synthesis in the 1930s, mutualisms began to be viewed from a population and evolutionary perspective. This involved formulation of systems of trait "syndromes" and studies of pairwise interactions (Baker 1983). Indeed, most early studies of plant–animal mutualisms focused on single pairs of species. This set up a research agenda organized around highly specific interactions. Paradigmatic examples of these specific interactions are fig wasps and figs (Cook and Rasplus 2003) and yucca moths and yuccas (Pellmyr 2003).

At the same time, however, there are early predecessors of community studies, some of them resulting in the compilation of large datasets of interactions later used in network research. One classic subject of discussion in these community-wide mutualistic studies was whether the majority of interactions are specialized or generalized. In this context, specialist species are those interacting with only one or a few other species. An important contribution here is the classic paper by Waser et al. (1996). This paper made the point that previous studies had overemphasized the prevalence of specific, one-on-one interactions, and that when viewed at a community level, most species interact with a large set of other species.

In another significant contribution, Daniel Janzen wrote a highly influential paper that significantly contrasted with the above dominant focus on the one-species-to-one-species approach (Janzen 1980). He introduced the idea that mutualisms occur in species-rich communities. That is, the interaction is not between one plant species and one animal species, but between one set of plants and one set of animals. The catch-all term "diffuse coevolution" was coined to refer to this situation. Several papers subsequently elaborated upon this concept, emphasizing the high variability of mutualistic interactions (Herrera, 1982). One drawback of this approximation, however, is that it seems to preclude any analytical tractability to community-wide mutualism (Thompson 2005). As a consequence, the theoretical framework in mutualistic studies seems to predict that these interactions have to lead to either highly specific or highly diffuse assemblages.

Mutualism. Edited by Judith L. Bronstein.

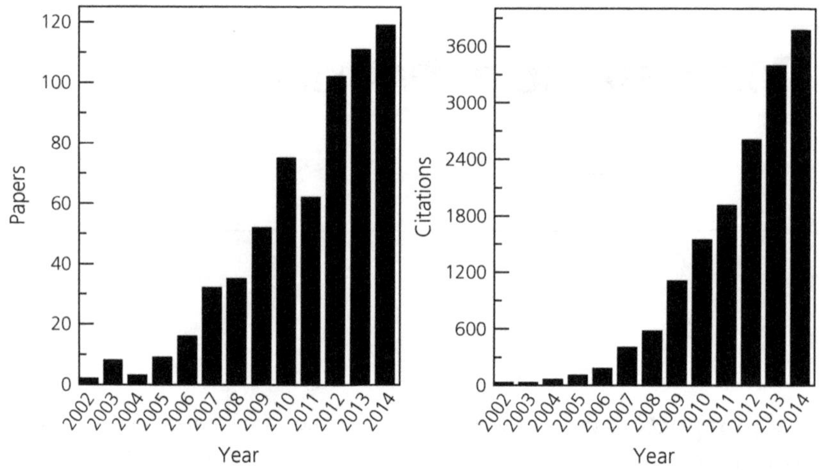

Figure 11.1 Temporal increase in the number of papers (left) and citations to papers (right) on mutualistic networks. Search performed on the Web of Science on 29 January 2015 (topic: *mutualistic networks*).

Several exceptions that brought tractability to the understanding of multispecific mutualism (in terms of allowing the search of repeated patterns) are worth mentioning. For example, the geographic mosaic theory of coevolution (Thompson 2005) described predictable patterns in the geographic distribution of mutualistic interactions involving small groups of species. Thus, the sign and strength of a mutualistic interaction may vary across the landscape. In some patches, the interaction is mutually beneficial, while it becomes antagonistic in other patches. Whether the net effect is positive or negative depends on the presence of other mutualists and antagonists. The geographic mosaic theory has represented a major advance in our understanding of how coevolution proceeds in complex natural settings (see Chapter 7).

The second major approach to community-wide mutualism has been brought by network analysis. The first papers introducing a network approach to mutualism mainly used concepts from food-web theory and looked at global properties such as connectance (Jordano 1987, Memmott 1999, Elberling and Olesen 1999). Several years later, there was a resurgence of interest in mutualistic networks (Bascompte and Jordano 2007, Bascompte and Jordano 2014). These foundational papers applied a new generation of tools and models to large and detailed datasets, facilitating a search for invariant principles or regularities across mutualistic networks despite obvious differences in latitude, habitat, species richness, and so on. The number of papers on mutualistic networks, and the citations these papers have gathered, have risen exponentially in the last few years (Figure 11.1).

The first studies on mutualistic networks focused on plant–pollinator and plant–seed disperser mutualisms, due to the main interests of the initial core of authors who introduced this approach. Thus, this reflects historical contingency rather than implying that other types of mutualisms cannot be seen as networks. Indeed, in subsequent years, network analysis has been applied to the study of plant–ant mutualisms, cleaning mutualisms, and several other systems, including cooperative interactions in socioeconomic systems (see Bascompte and Jordano 2014 for a review). Network theory has provided a framework that can be used to characterize all mutualisms. Thorough this chapter, however, we will mainly focus on pollination and seed dispersal mutualisms for illustrative purposes.

11.2 Rationale for a network approach to mutualism

Despite several recent claims and the early predecessors noted above, many coevolutionary researchers still think in a pairwise way. The reason is rooted in the reductionistic tradition of breaking down complex systems into their basic elements in the hope that their study can be scaled up to understand entire communities. Indeed, we have lacked a conceptual framework to address multispecies interactions. Network theory provides this conceptual framework. This approach treats species as nodes, and mutualistic interactions as links between two such nodes. Technically, the type of network that fits these systems is called a bipartite network, that is, a network formed by two sets of nodes (here, plants and animals) with interactions between, but not within, sets.

Network studies can bring insight in three important ways. First, network theory can unveil complex patterns; that is, it can help us to visualize the structure of highly diverse mutualisms. Second, network tools help us to analyze this complexity and to test hypotheses about its origins. Third, network theory emphasizes the relationship between network structure and dynamics, which may help us evaluate the community-wide effects of species extinctions and other drivers of global change.

Network theory has rapidly become a popular approach in community ecology in general, and in the study of mutualism in particular. As with any new paradigm, there is the risk that it can become a fancy new way of stating what we already knew, or a quantitative exercise poorly rooted in the ecological and evolutionary realities of the system being studied. Another potential problem with the application of network theory is that—as with any other approach—there are trade-offs between simplicity and realism. Treating all species as nodes necessarily masks plenty of very interesting natural history details. The underlying logic of the network approach is not to assume that these details are irrelevant, but rather to consider that for certain questions at certain scales, one can obtain a good global view of these systems without them.

With the above caveats in mind, one can ask whether studies of mutualistic networks have led to significant and novel understanding of the ecology and evolutionary biology of mutualisms. Our goal in this chapter is to focus on that question. We do not aim to provide another review of research on mutualistic networks (Bascompte and Jordano 2007). Rather, we discuss the ecological and evolutionary implications of network studies. We begin by describing an example of the potential of network research at the level of the entire network. This relates to classic work on specialization versus generalization in mutualistic communities. We devote the rest of this chapter to the study of a particular organization of mutualistic networks that allows linking the scale of the entire network to that of a small subgroup of species.

11.2.1 The example of asymmetric specialization

The network approach has introduced two advances in our understanding of the level of generalization of mutualistic interactions. First, it describes the specific form of the continuum between the frequency of specialists and that of generalists within a community. Second, it has made the point that specialists do not typically interact with specialists, as once believed.

Regarding the first point, network theory has established the concept of connectivity distributions, which describe the cumulative probability of a species interacting with one, two, . . ., k other species. Mutualistic networks have been found to be heterogeneous: while the majority of species have one or a few interactions with other species, a few species are super-generalists, acting like hubs in technological networks. Technically, the connectivity distribution in these networks is best described by a broad scale distribution (Jordano et al. 2003). This naturally leads to a situation in which "specialists" or "generalists" are not discrete entities, but extremes of a continuum. Network theory describes the mathematical function of this continuum. An analogy would be the scale-free distribution describing the frequency of earthquakes releasing a certain amount of energy. One does not need a different theory for small earthquakes than for large ones; these are just extremes of a continuum.

Second, network theory has addressed the novel question of whether specialization is symmetric or asymmetric, that is, whether specialist species interact with specialist species or with generalist ones. Note that we cannot address this sort of question by considering species and their interactions isolated from the community. Now we are looking at how these species and their interactions are embedded within the community; that is, we are now looking at a network level. This result was shown simultaneously by Bascompte et al. (2003) and Vázquez and Aizen (2004). Both sets of authors described a situation in which specialists tend to interact with generalist species, a pattern that has been proven quite robust when re-analyzed with more conservative null models (Joppa et al. 2009).

Does this network-like description of interactions bring new insight into mutualistic systems? We think so. All else being equal, a specialist interacting with a generalist has more chances to persist in variable environments, since it relies on a more abundant, relatively less fluctuating species. This concept was first advanced verbally by Bascompte et al. (2003), and later confirmed empirically by Ashworth et al. (2004) and analytically by Bastolla et al. (2009) and Rohr et al. (2014). This is the sort of insight we think is useful in the context of mutualism conservation, an idea developed by Jason Tylianakis in Box 11.1 (see also Chapter 14).

Box 11.1 Conservation and restoration implications of network studies

Jason Tylianakis

There has been growing recognition that the study of ecological networks has applied relevance, by linking persistence of one group of organisms (e.g., birds) with other species (e.g., plants), and by combining biodiversity with ecosystem processes (Tylianakis et al. 2010). Therefore, network studies clearly overlap with the interests of conservation, and their relevance in this regard is only just beginning to be understood (Tylianakis et al. 2010).

Understanding how environmental changes alter networks

Mutualistic networks can inform conservation by providing a context within which to understand how environmental changes affect the stability and functioning of entire communities. Networks can be altered via additions and/or deletions of species (e.g., via invasions, range shifts, or extinctions), or via changes in the strength of interactions (e.g., due to spatial or phenological mismatching or altered search efficiency; Tylianakis et al. 2008). Recent work has begun to illuminate both the community-wide impacts of these changes and some generalities that may inform conservation efforts.

One area that has received particular attention is the impact of invasive species on mutualist networks. Invaders are often "super-generalists," interacting with numerous other species throughout the network (Vilà et al. 2009). This increases overall network connectivity, and it may alter topological aspects of the network. For example, by interacting preferentially with the most generalist species, alien pollinators were found to enhance the nestedness of pollination networks on the Galapagos Islands (Traveset et al. 2013). Combined with previous findings that alien species may engage in highly asymmetric interactions (reviewed in Tylianakis et al. 2010), these results suggest that invaded communities may be more robust to species losses. However, Traveset et al. (2013) also found that these highly connected invaders tended to act as connectors, both within their own subnetworks and across the different islands that comprised distinct modules within the archipelago network. This may make invaded networks more susceptible to perturbations (e.g., diseases, contaminants) that may spread via interactions.

Mutualist networks have also provided significant insights into the effects of habitat fragmentation. For example, network structure is a key determinant of how pollination and seed-dispersal communities respond to fragmentation, with the nestedness and heterogeneous degree distribution of real networks causing them to decay sooner than random communities, but persist longer as habitat fragmentation continues (Fortuna and Bascompte 2006). In fact, the loss of particular interactions with decreasing fragment size is non-random, and can be predicted based on characteristics such as the rarity of the interaction or the connectedness of the species involved (Aizen et al. 2012).

Translating understanding into restoration practice

Given these and other findings, we can begin to derive recommendations for conservation practices such as restoration of habitat or removal of invaders. For instance, particular species could be targeted for reinstatement (e.g., by replanting) or removal to achieve a certain goal. Generalist species tend to play a central role in structuring the network and connecting modules, so these species could be targeted to generate the most rapid changes in network structure. However, several studies from the group of Jane Memmott have demonstrated potentially harmful indirect effects of removing invasive plants. For example, removal of a flowering invader may reduce the overall attractiveness of the plant community to pollinators, thereby reducing pollination of rare native plants that depend on a diverse, abundant pollinator community

Box 11.1 *Continued*

(Carvalheiro et al. 2008). Furthermore, by coupling separate modules (e.g., the networks of different habitats), invaders may play an important role in maintaining spatial stability in ecosystem functioning, again suggesting that knowledge of their role in networks could inform the costs or benefits of their removal. Even in the absence of local knowledge regarding the role of each species in the network, the general enhancement of plant diversity can generate more connected pollination networks that are more temporally stable in their pollination services (Ebeling et al. 2011).

Knowledge of network structure may also inform restoration planning. For example, knowing which components of a network occur in which type of habitat allows managers to determine how the loss of certain habitats from the landscape will affect different components of the local interaction network (Evans et al. 2013). Similarly, it can allow computer simulations of plantings of certain species to compare their effects on network structure (Devoto et al. 2012). This can be aimed at generating networks with emergent properties, such as pollination networks with high functional complementarity (to maximize pollination rates) or redundancy (to maximize resilience of the network in the face of extinctions; Devoto et al. 2012). Finally, by incorporating so much information on biodiversity and species interactions, mutualist networks can be a useful yardstick for measuring the success of previous restoration actions (Forup et al. 2008).

References

Aizen, M.A., Sabatino, M., and Tylianakis, J.M. (2012). Specialization and rarity predict nonrandom loss of interactions from mutualist networks. *Science*, **335**, 1486–1489.

Carvalheiro, L.G., Buckley, Y.M., Ventim, R., Fowler, S.V., and Memmott, J. (2008). Apparent competition can compromise the safety of highly specific biocontrol agents. *Ecology Letters*, **11**, 690–700.

Devoto, M., Bailey, S., Craze, P., and Memmott, J. (2012). Understanding and planning ecological restoration of plant-pollinator networks. *Ecology Letters*, **15**, 319–328.

Ebeling, A., Klein, A. M., and Tscharntke, T. (2011). Plant-flower visitor interaction webs: temporal stability and pollinator specialization increases along an experimental plant diversity gradient. *Basic and Applied Ecology*, **12**, 300–309.

Evans, D.M., Pocock, M.J.O., and Memmott, J. (2013). The robustness of a network of ecological networks to habitat loss. *Ecology Letters*, **16**, 844–852.

Fortuna, M.A., and Bascompte, J. (2006). Habitat loss and the structure of plant-animal mutualistic networks. *Ecology Letters*, **9**, 281–286.

Forup, M.L., Henson, K.S., Craze, P.G., and Memmott, J. (2008). The restoration of ecological interactions: plant-pollinator networks on ancient and restored heathlands. *Journal of Applied Ecology*, **45**, 742–752.

Traveset, A., Heleno, R., Chamorro, S., Vargas, P., McMullen, C.K., Castro-Urgal, R., Nogales, M., Herrera, H.W., and Olesen, J.M. (2013). Invaders of pollination networks in the Galapagos Islands: emergence of novel communities. *Proceedings of the Royal Society B: Biological Sciences*, **280**, 20123040.

Tylianakis, J.M., Didham, R.K., Bascompte, J., and Wardle, D.A. (2008). Global change and species interactions in terrestrial ecosystems. *Ecology Letters*, **11**, 1351–1363.

Tylianakis, J.M., Laliberté, E., Nielsen, A., and Bascompte, J. (2010). Conservation of species interaction networks. *Biological Conservation*, **143**, 2270–2279.

Vilà, M., Bartomeus, I., Dietzsch, A.C., Petanidou, T., Steffan-Dewenter, I., Stout, J.C., and Tscheulin, T. (2009). Invasive plant integration into native plant-pollinator networks across Europe. *Proceedings of the Royal Society B: Biological Sciences*, **276**, 3887–3893.

Having introduced the sort of insight that a network approach can provide for our understanding of mutualistic communities, we now expand on one particular way in which a property of a mutualistic network can be further tied to the natural history of its interactions. Specifically, we consider small groups of strongly interacting species that occupy an intermediate scale between a pairwise interaction and the entire network.

11.3 Syndromes, modularity, and coevolutionary units

While the scope of this chapter—like that of this entire volume—addresses mutualism in general, in this section we use pollination mutualisms for illustrative purposes. Similar historical accounts could be drawn for other types of mutualisms.

The first serious attempts to organize lists of visitors to the flowers of particular plant species were made by the Italian botanist F. Delpino in writings from 1868 to 1875 (Waser 2006). Delpino's classification of flower types was modified by several later pollination biologists, such as P. Knuth in his monumental *Handbook of Flower Biology*, published from 1898 on. These early schemes attempted to sort flowering plants into groups according to floral morphology and rewards in relation to their pollinators.

The concept of pollination syndromes was introduced in the 1960s by L. van der Pijl, K. Faegri, and H.G. Baker (see Faegri and van der Pijl 1971, Vogel 1996, Waser 2006). Syndromes are defined as co-occurring sets of flower (or fruit) traits matched to the morphology of different pollinators or seed dispersers (van der Pijl 1969, Jordano 1995, Ollerton et al. 2009, Lomáscolo et al. 2010). These groups of plants and their animal mutualists exhibit a complex of shared traits and are adduced to be the consequence of coevolutionary convergence. Throughout this chapter, we refer to these strongly interacting groups of species as coevolutionary units.

One problem with syndromes is that they are top-down creations. That is, it is the scientist who defines them on the basis of a somewhat subjective classification. Although syndromes have had a long tradition in shaping coevolutionary thinking about mutualisms, demonstrating their existence has been more elusive. Some notable exceptions are Lomáscolo et al. (2010), who found strong evidence of syndromes in a detailed study that explicitly considers the phylogenetic non-independence of species.

The identification of the groups of plant and animal species showing the complex of shared traits involved in syndromes is very much related to the problem of scale, which is at the core of ecological and evolutionary thinking (Levin 1992, Chave 2013). One might think that the relevant scale in a network is that of a node. It may also be that the relevant scale is that of the entire network. Finally, it is also possible that the relevant scale lies between the node and the entire network. Modularity analysis generates the best partition of a mutualistic network into modules, resulting in a bottom-up classification based on the patterning of interactions (Box 11.2; Figure 11.2). These modules are groups of strongly interacting nodes that can be a surrogate of the coevolutionary units arising from syndromes. As an example, Figure 11.3 illustrates five distinct modules, such as the one formed by plants whose seeds are dispersed by fishes. In the context of mutualism, modularity can help identifying relevant highly interacting units. Whether or not coevolution plays a role in the structure of these modules is a question that cannot be answered with static patterns. The point, however, is that should coevolution play a role at all in these multispecific systems, the process would take place under the template of these modules.

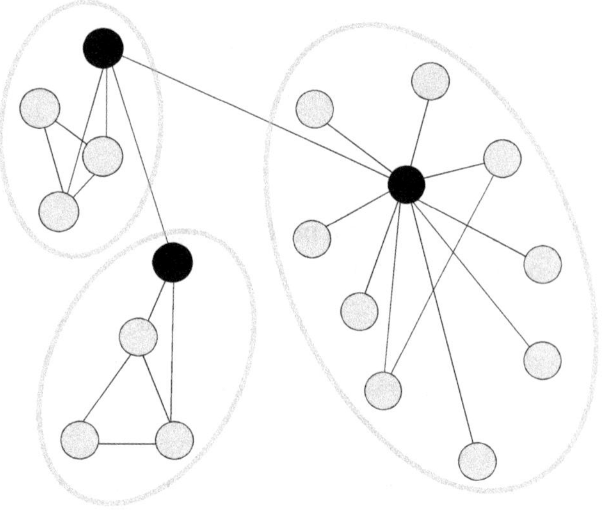

Figure 11.2 A module is a subset of nodes more densely connected among themselves than they are with nodes from other modules. The figure illustrates three modules highlighted by the ellipses. Within a modular network, different nodes have different roles. For example, the three black nodes serve as connectors of different modules. Modified from Bascompte and Jordano (2014).

Box 11.2 Definition and measure of modularity

Roger Guimerà

In nature, complex networks are modular. This means that, rather than being uniform objects whose nodes are all equally likely to connect to each other, real networks have groups, or modules, with high intra-group and low inter-group connectivity (Girvan and Newman 2002, Guimerà et al. 2007, Fortunato 2010, Newman 2011).

Modularity is important for at least two reasons. First, modules define an intermediate scale between the micro-scale of single nodes and the macro-scale of the whole network. This intermediate scale is the most relevant one for dynamical processes in the network because, in general, time scales for processes within modules are much shorter than the time scales for processes between modules (Arenas et al. 2006). In other words, when we consider a dynamical process within a network, we see modules evolving as units with only rapid fluctuations within them. Second, modules provide a convenient way to describe networks, and nodes in a module usually also share other properties (Guimerà et al. 2007). In the context of the subject of this book, modules define groups of species strongly interacting among themselves. If (co)evolution happens in these networks, then a module would be a good candidate for a (co)evolutionary unit.

Unfortunately, identifying modules within complex networks is challenging. The first challenge is how to precisely define what a module is. Indeed, the broad idea of "groups of nodes with higher within-group connectivity than between-group connectivity" is too general to be operational. Is a group of four fully connected nodes a module? Is the whole network a single module, given that if we have a single module all links are then within-module? The second challenge is that, even with a precise definition, there is a combinatorially large number of ways in which one can partition a network into modules. For example, a network with only 20 nodes can be partitioned in more than 5×10^{13} different ways.

A widespread approach to solving the first challenge is to use a modularity function M that quantifies the deviation of a partition from the null expectation of uniform connectivity between modules (Girvan and Newman 2002). Specifically, in the absence of connection bias, the expected number of links between two modules is $d_i d_j/2L$, where d_i and d_j are the sum of the degrees of the nodes in modules i and j, respectively, and L is the total number of links in the network. Therefore, if we denote by l_i the actual number of links between nodes in module i, the modularity function is

$$M=\sum_i\left[\frac{l_i}{L}-\left(\frac{d_i}{2L}\right)^2\right], \tag{1}$$

where l_i/L is the fraction of all links inside module i, and $(d_i/2L)^2$ is the random expectation for the fraction of links within i. Defined in this way, M is close to one when intra-module connectivity is much higher than inter-module connectivity, and close to zero when the modules are random, when all nodes are placed in a single module, or when each node is placed in a separate module (Girvan and Newman 2002).

One then needs to address the second challenge—finding, among all possible partitions, the one with maximum modularity. This is a combinatorial optimization problem for which it is not possible to find the exact solution in "reasonable" time. Therefore, it is necessary to resort to heuristic algorithms (Fortunato 2010). The simplest one would be to start from an arbitrary partition of the network into modules, and then repeatedly choose a random node and move it to a random new module if that increases the modularity of the partition. However, this approach leads to partitions that, while being locally optimal (in the sense that no single node move can increase the modularity), are not even close to the most modular partition.

One way to address this problem is to use simulated annealing (Kirkpatrick et al. 1983, Guimerà and Amaral 2005), a technique designed precisely to avoid local optima in combinatorial optimization problems. In simulated annealing, one attempts random moves of nodes and accepts those that increase modularity, just as in the simple approach above. Unlike there, moves that decrease modularity are also accepted with some small (and decreasing) probability.

This approach (modularity function optimization using simulated annealing) has the advantage that it is accurate, and that it can be generalized to weighted and bipartite networks (Guimerà et al. 2010). It also has limitations, such as the inability to detect certain small groups (Fortunato and Barthélemy 2007) or hierarchically nested modular structures (Sales-Pardo et al. 2007). Many other algorithms have also been developed in the last ten years, some of which address these problems, although often at the price of being less accurate in other situations (Fortunato 2010).

continued

Box 11.2 *Continued*

References

Arenas, A., Díaz-Guilera, A., and Pérez-Vicente, C.J. (2006). Synchronization reveals topological scales in complex networks. *Physic Review Letters*, **96**, 114102.

Fortunato, S., and Barthélemy, M. (2007). Resolution limit in community detection. *Proceedings of the National Academy of Sciences of the USA*, **104**, 36–41.

Fortunato, S. (2010). Community detection in graphs. *Physics Reports*, **486**, 75–174.

Girvan, M., and Newman, M.E.J. (2002). Community structure in social and biological networks. *Proceedings of the National Academy of Sciences of the USA*, **99**, 7821–7826.

Guimerà, R., and Amaral, L.A.N. (2005). Functional cartography of complex metabolic networks. *Nature*, **433**, 895–900.

Guimerà, R., Sales-Pardo, M., and Amaral, L.A.N. (2007). Classes of complex networks defined by role-to-role connectivity profiles. *Nature Physics*, **3**, 63–69.

Guimerà, R., Stouffer, D.B., Sales-Pardo, M., Leicht, E., Newman, M.E.J., and Amaral, L.A.N. (2010). Origin of compartmentalization in food webs. *Ecology*, **91**, 2941–2951.

Kirkpatrick, S., Gelatt, C.D., and Vecchi, M.P. (1983). Optimization by simulated annealing. *Science*, **220**, 671–680.

Newman, M.E.J. (2011). Communities, modules and large-scale structure in networks. *Nature Physics*, **8**, 25–31.

Sales-Pardo, M., Guimerà, R., Moreira, A.A., and Amaral, L.A.N. (2007). Extracting the hierarchical organization of complex systems. *Proceedings of the National Academy of Sciences of the USA*, **104**, 15224–15229.

Figure 11.3 Modules in a seed-disperser network in the Pantanal, Brazil. Circles and squares indicate animal and plant species, respectively. The size of the nodes represents animal body mass and fruit diameter, respectively. This mutualistic network is organized in five distinct modules highlighted by the different shades of gray. Specifically, there are two bird modules (top), each with 22 species; two mammal-dominated modules (bottom) with 25 and 18 species, respectively; and a fish module with 4 species. Based on Donatti et al. (2011).

In ecology, the concept of modularity has been at the forefront of important research on food webs. The classic paper by Robert May on stability and complexity in model ecosystems not only made the claim that food-web structure is related to stability, but actually suggested that an organization of trophic interactions in modules can increase the persistence of the food web (May 1972). This spurred a rich search for this relationship between structure and stability using other models and real food webs (May 1972, Pimm 1979, Raffaelli and Hall 1992, Melián and Bascompte 2004, Olesen et al. 2007, Guimerà et al. 2010). A recent paper, for example, has proven that the tendency of a food web to be organized in compartments increases its persistence in the sense of buffering the transmission of perturbations across the entire food web (Stouffer and Bascompte 2011).

Olesen et al. (2007) first introduced the concept of modularity in the context of mutualistic networks. These authors analyzed a large dataset of 51 pollination networks including almost 10,000 species and 20,000 interactions. Networks larger than a certain threshold were found to be significantly modular. As described in Olesen et al. (2007), the majority of modules involved a few species with convergent traits. In a pollination network from the Andes (Arroyo et al. 1982), for example, distinct modules include various Diptera species and many plant species with white flowers or large flies and small, yellow umbellifer flowers (see Figure 11.4).

The pollination modules described by Olesen et al. (2007) consist mainly of two broad types. The first one is formed by one or a few very abundant plant species with open flowers and easily accessible rewards, visited by many animal species, either simultaneously or in a temporal sequence. The second broad type is made up of tightly linked subsets of plants and animals, with their links organized in a nested fashion. Seed-dispersal networks, in turn, tend to consist of one or a few birds, bats, or reptiles (lizards, tortoises) visiting a suite of fleshy-fruited plants, with fruits being ripe simultaneously or in a temporal sequence.

Overall, analyzing the 29 significantly modular pollination networks, and the resulting 254 different modules, Olesen et al. (2007) found that the average number of modules per network was close to nine, with a minimum of five in a temperate forest meadow and a maximum of 19 in a network in the Amami Islands. The specific number of modules

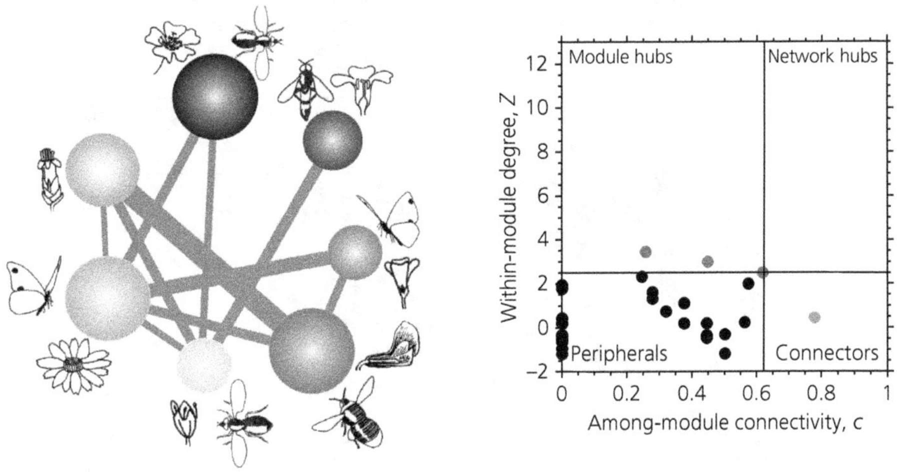

Figure 11.4 The figure at the left illustrates the different modules found in a plant–pollinator network in the Andes (Arroyo et al. 1982). The links between modules are weighted by both number of species links between modules and number of species in modules. The vignettes show dominant pollinator and flower types. From top and clockwise, the first module consists of a Diptera sp. and mainly white flowers; second module: small–medium-sized beetles, flies, an ant, and small, white/yellow flowers; third module: butterflies and one plant species, *Oxalis* sp.; fourth module: bees, birds, and large, mainly yellow flowers with a closed morphology; fifth module: large flies and small, yellow umbellifer flowers; sixth module: butterflies, a large fly, and white/yellow/pink/purple flowers; and seventh module: large flies and mainly small, white flowers. Right: classification of species roles both within their module and between modules, with three module hubs and one connector, but no network hub. From Olesen et al. (2007).

of a network was related to its number of species. As plot size grows or as species density (number of species per unit area) increases, the number of modules increases as well. In Olesen et al. (2007), small networks were shown to be non-modular, i.e., such networks could, in fact, be understood as one module.

Each one of these modules can be thought of as a potential coevolutionary unit. They represent groups of plants and animals that interact more frequently among themselves than they do with other species in the network. Therefore, if there is coevolution (Chapter 7), there are more chances that this occurs within these modules than between modules.

Next, we will consider how these modules combine to form a complex network of mutualistic interactions, and how this constrains other patterns at the network scale.

11.4 Modularity and the basic building blocks of mutualistic networks

The idea of modularity also helps bridge small groups of mutualisms to entire mutualistic networks. One could see a module as a basic building block of these networks. The way different modules are linked will determine other network properties such as nestedness. In a nested network, specialists interact with species that form perfect subsets of the species with which generalists interact (Figure 11.5). This gives rise to an organization as in Russian dolls, with smaller dolls within larger ones. Olesen et al. (2007) reported a positive correlation between

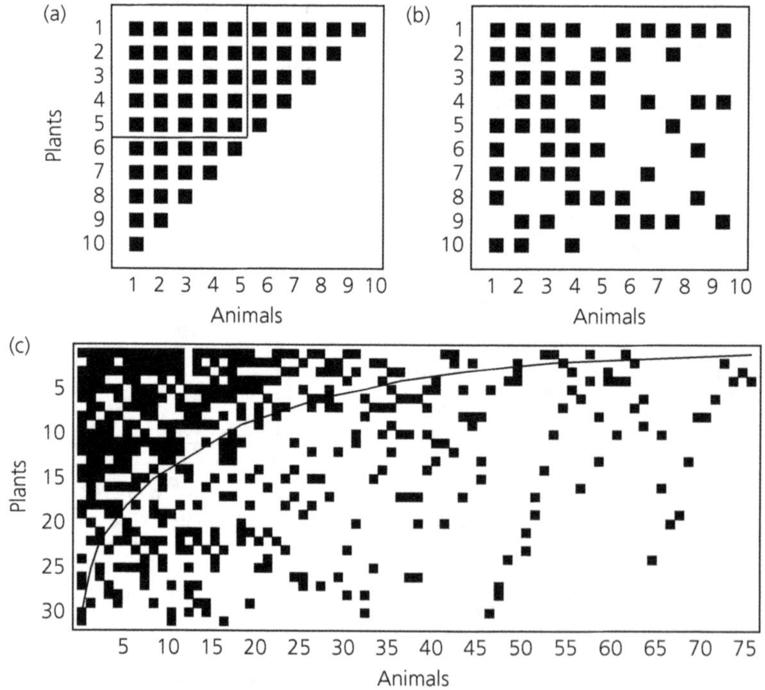

Figure 11.5 Plant–animal mutualistic networks show a nested structure when represented as a matrix. Here, each row and column represents a plant and animal species, respectively, and a square indicates an observed interaction between such a plant and animal. In a nested arrangement (a), specialists interact with species that form well-defined subsets of the species interacting with more generalist ones. (b) Represents a different matrix organization for the same network, which has been obtained by randomizing the distribution of interactions from the previous matrix. (c) Represents the actual arrangement of one plant–pollinator network. From Bascompte et al. (2003).

modularity and nestedness in pollination networks. This overall pattern was explored by Fortuna et al. (2010) in a wider range of ecological networks, using randomization schemes and measures that permit direct comparisons between these two network properties. They essentially confirmed the Olesen et al. (2007) results for low-connectance networks. At low connectivities, significantly nested matrices tend also to be significantly modular (Figure 11.6); as connectivity increases, the correlation between these two network patterns decreases in magnitude and sign. Highly connected matrices are either modular or nested. Put another way, at low connectivity values the modules are glued together in a way that creates a nested structure (Figure 11.6), but as connectivity is increased, the number of degrees of freedom in combining these basic blocks is very much reduced. One can no longer combine modules into a nested network.

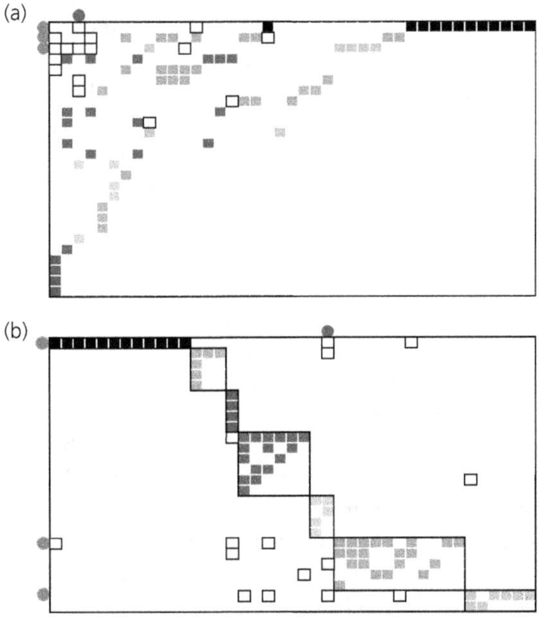

Figure 11.6 Two matrix representations of the pollination network displayed in Figure 11.4. (a) represents the matrix with rows and columns arranged to maximize nestedness, while in (b) they are sorted to maximize similarity in patterns of interaction and module affiliation. The network is organized in seven distinct modules, indicated with different shades of gray. Small squares indicate observed pairwise interactions; four species (three animals, in rows, and one plant, in columns) are highlighted as bullets to illustrate species that have interactions with species in different modules. These species are hubs, and in the nested representation of the network are part of the core. Based on Olesen et al. (2007).

Given the above description, it would be interesting to know how different modules are combined as networks assemble over evolutionary time. Unfortunately, there is a scarcity of information on the build-up of mutualistic networks. One indirect evidence of the evolutionary trace left by this assembly can be obtained by looking at the phylogenetic signal of species within and between modules. The comparative method, or phylogenetic independent contrasts, allows us to study the evolutionary history of mutualistic networks in general, and of modularity in particular. For example, in modules dominated by competition and limited resources, species may become incorporated into modules if they have traits that cause increased phylogenetic overdispersion, such that new species have traits that make them able to harvest resources in novel ways. Alternatively, in modules dominated by convergent traits we may see phylogenetic aggregation, i.e., species becoming attached to the modules if they have traits already present in the modules. This would result in increased phylogenetic aggregation. We turn to this topic next.

11.5 Phylogenetic signal of modularity

Phylogenetic approaches to mutualistic network structure were introduced by Rezende et al. (2007). These authors found that past evolutionary history embedded in the phylogenies are important in understanding the architecture of mutualistic networks. Specifically, closely related species tend to have a similar number of interactions and tend to interact with the same subset of species (see Figure 11.7). Finding such a phylogenetic signal is not only important in terms of assessing the suite of mechanisms generating network structure, but also illuminates the potential community-wide consequences of species extinctions. Coextinction cascades following simulated species extinctions tend to involve phylogenetically close species. This results in a higher loss of taxonomic diversity and a nonrandom pruning of the evolutionary tree (Rezende et al. 2007).

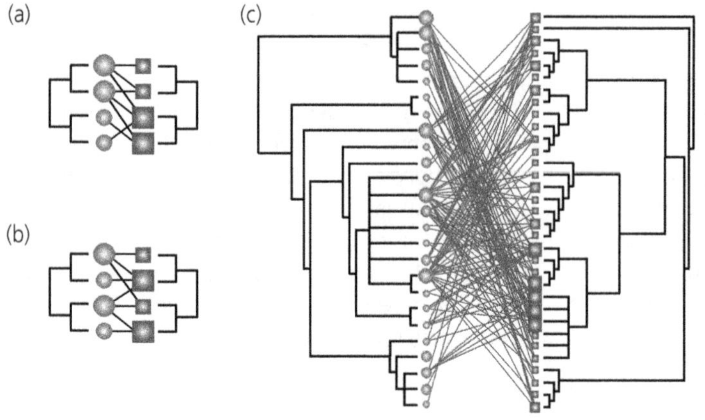

Figure 11.7 Phylogenetic patterns in the architecture of mutualistic networks. A significant phylogenetic signal indicates that past evolutionary history is important for understanding network structure. This is illustrated in panel (a), where phylogeny predicts the number of interactions a species has. In contrast, panel (b) illustrates a case without such a phylogenetic signal. Panel (c) represents a significant phylogenetic signal in the number of interactions a species has for a real plant–frugivore network. A similar significant phylogenetic signal was found in more than one-third of the mutualistic networks studied by Rezende et al. (2007).

A similar phylogenetic approach to mutualistic networks was undertaken by Donatti et al. (2011). These authors studied a seed-dispersal network in the Brazilian Pantanal that includes several taxonomic groups. They first characterized the modular structure of this network, organized around five modules. Of these modules, two are represented by bird species and their host plants, two others by mammal species and their host plants, and the last by fish species and their host plants (see Figure 11.3). Donatti et al. then explored a suite of mechanisms correlated with this modular organization. Overall, these modules were significantly associated with a combination of phylogenetic constraints and trait convergences among unrelated species. Phylogeny was found to explain the association of species within modules dominated by birds, but not by mammals. This points toward the interesting result that finding a phylogenetic signal may be a scale-dependent problem. Different modules correspond to different large taxonomic groups at a coarse scale. At a lower scale, some modules have phylogenetically close species, while this is not the case in other modules. A combination of network and evolutionary approaches is a profitable way to shed light on the nature of syndromes and coevolutionary units.

Donatti et al. (2011) also found that plants within all modules differed from each other significantly in fruit and seed traits. The trait that varied the most among modules was fruit mass, followed by fruit length. Regarding dispersers, their body mass was significantly correlated with module assignment. This result is consistent with studies of Rezende et al. (2009) on a marine food web. In the Rezende et al. (2009) food web, body size, phylogeny, and spatial structure were jointly associated with the compartmentalized structure. Interestingly, the phylogenetic signal was positive for bony fishes, meaning that related bony fishes tended to be in the same module. However, the phylogenetic signal was negative in the case of sharks; that is, closely related shark species tend to be segregated in different modules (Rezende et al. 2009).

Analyses of phylogenetic signals in network architecture are still quite preliminary in the sense that they have only looked at one component of network structure, such as modularity. Thus, they have left unanswered the question of whether other components of network structure (e.g., nestedness) are equally phylogenetically constrained. Rohr and Bascompte (2014) attempt to fill this gap by building a unified framework in which phylogenetic signal can be quantified in a standardized, comparative way across different network types, species roles, and architectural components. Interestingly, both in mutualistic and antagonistic interactions, phylogenetic signal in the modular component of network structure was higher than the equivalent for nestedness (Rohr and Bascompte 2014).

So far we have revealed a phylogenetic signal of modularity, which means that past evolutionary history embedded in the phylogenies may structure networks into a modular structure. Modules are a mix of species varying in their ecological and evolutionary history. Modules may certainly be the product of coevolution, but they may also result from ecological fitting (*sensu* Janzen 1985) and coincidental sampling of species. At present, it is not clear how one could tease apart these potential mechanisms. An overview of the few studies that have searched for ecological traits that correlate with modularity helps us to construct a list of the potential causes of modularity. Besides evolutionary history encapsulated in the phylogenetic signal, one can list the following potential mechanisms:

1. Local abiotic heterogeneity, i.e., divergent selection regimes (Pimm and Lawton 1980, Rezende et al. 2009, Hagen et al. 2012). Local sites may impose unique selection regimes, leading to unique subgroups of species that interact more strongly among themselves than with species from other sites.
2. Regional/geographical heterogeneity (Dupont and Olesen 2009, Dupont et al. 2009, Dalsgaard et al. 2013, Trøjelsgaard et al. 2013, Trøjelsgaard and Olesen 2013). This would act similarly to the above, but at a larger spatial scale.
3. Variation in abundance (Montesinos-Navarro et al. 2012). Different subgroups of species that happen to be more abundant may interact more strongly among themselves than with other, less abundant species, thus forming a module.
4. Variation in life-history traits, e.g., body size (Rezende et al. 2009, Bellisario et al. 2012, Martín et al. 2012). Species with life-history traits that facilitate their interaction (e.g., through coevolutionary convergence) may interact more strongly than with other species, and thus will form a module.
5. Sociality (Fortuna et al. 2009). Intraspecific social bonds create subgroups of strongly interacting individuals. Contrary to the other mechanisms from this list, this applies to social networks where nodes are individuals instead of species.
6. Impact of alien species (Valdovinos et al. 2009, Traveset et al. 2013). If an invasive species interacts strongly with a small group of local species, this may create a module within the overall network.
7. Human management (Macfadyen et al. 2011). Human intervention in an ecosystem may favor a small group of species, increasing the intensity of their interactions.
8. Influence from community assembly history. History may have led an initial subgroup of species to have interacted strongly, leaving a signal in the final network as this initial subgroup forms a module.
9. Seasonality. Only species that overlap in time can interact. Strong seasonality may create subgroups of interacting species that cannot interact with species present at another time.

Needless to say, the above list of causes is neither exhaustive, nor mutually exclusive, since several mechanisms may be correlated. Further research will undoubtedly shed light on the most common mechanisms leading to modularity. We now review evidence for two of the above mechanisms, variation in abundance and strong seasonality.

Regarding the former, abundance of both animals and plants is an important driver of modularity in some but not all mutualistic networks. For example, in the pollination network studied by Martín et al. (2012), abundance was not a significant driver of modularity. In other types of mutualisms, interaction frequency is best explained by abundance, but in combination with phylogenetic distance (Montesinos-Navarro et al. 2012). Thus, generalized, abundant species interact both with other abundant species, but also with distantly related specialist species, resulting in a nested pattern. For example, in a host plant–arbuscular mycorrhizal fungus (AMF) network, modules either contained one or a few common, generalized plants interacting with many rare specialized AMFs, or vice versa for fungi (Montesinos-Navarro et al. 2012), revealing a nested structure.

Seasonality can be a very important driver of modularity (Bosch et al. 2009). Species with relatively long phenophases, such as bumblebees, act as the backbone of pollination modules in which they interact with a series of shorter-phenophase flowering plant species (Olesen et al. 2008, Petanidou et al. 2008, Martín et al. 2012). Both plant and pollinator phenologies have been found to be drivers of modularity in pollination networks (Olesen et al. 2008, Petanidou et al. 2008, Martín et al. 2012). Thus, a seasonal series of modules was distinguished with a spring module of plants and their dipteran and bee pollinators, and mid- and late-season modules dominated by peaking beetle populations. A few bee species act as the glue joining different modules (see section on species roles below). These bees interact with seasonal series of flowering plants, each with a short flowering season. The same temporal structure of modularity was observed by Dupont and Olesen (2012) in a temperate heathland.

Having reviewed a suite of causes for the overall level of modularity of a network, we now focus on the shape of individual modules.

11.6 Types of modules

In addition to the level of modularity, other factors shape the topological structure of the individual modules. One common type of plant–pollinator module is the "star" (Olesen et al. 2007). It most often consists of 1–2 hub plants linked to many specialized pollinator species. However, the star "flickers in time," i.e., its links come and go (Olesen et al. 2008, Dupont and Olesen 2012). At any given point in time, only between one-fourth and three-fourths of all links may be present. Stars are assembled through very strong preferential attachment, i.e., all new species are only linked to the hub(s). That is, there is complete hub attraction. These star modules may consist of an abundant and/or widespread plant species with a large and long-lasting floral display visited by many rare and specialized animals. Similarly, in parasitic networks, such stars may be constituted by an abundant host with a stable population infected by a range of specialized parasites (Anderson and Sukhdeo 2011). A stable host makes it possible for parasites to complete their life cycles and even to evolve more complex life cycles.

Another type of plant–pollinator module is composed by compact sets of partners that are organized with high levels of connectance (20%–50%) and nestedness.

Pollination modules are linked together by connectors (see Section 11.7) such as large bees (Olesen et al. 2007) or plants varying in floral morphology (Kratochwil et al. 2009). Some modules are modular at different scales of resolution, revealing a hierarchical structure of nature (Bezerra et al. 2009, Olesen et al. 2010). A few modules, however, are satellites, unlinked to other modules (Fonseca and Ganade 1996, Cagnolo et al. 2011). These may in some cases be formed by tightly coevolved species groups displaying a high interaction intimacy.

Some modules in a network may deviate considerably in their taxonomic or functional content from the network they belong to, and are only weakly—or not at all—linked to other modules. This could be the result of a particularly strong specificity among the members of a module or the coevolution of specific trait combinations. Around 360 species of Apidae and Melittidae bees are floral oil collectors (Mello et al. 2013). They obtain the oil from specialized flowers and use it for larval provision. Thus, oil bees and oil flowers have both a high specificity and very complicated trait combinations, such as floral oil glands and elaborated bee leg combs involved in the harvesting of oil (Bezerra et al. 2009). These bees and their flowers make up distinct modules in the Brazilian rainforest.

So far we have reviewed the modular structure of an entire network and the shape of individual modules. Next, we zoom in to a single module to characterize the roles of specific species within it.

11.7 Species roles within modular networks

Once we have detected the partitioning of a network into well-defined modules, modularity analysis allows the ranking of nodes as a function of their topological importance, that is, their relative importance from the point of view of network architecture. Here, one must consider two dimensions, the importance of a node within its module and its importance at the level of the entire network (see Box 11.2 by Roger Guimerà for quantitiative details, and Figures 11.4 and 11.8 for specific examples). The importance of a node within its module is quantified as the within-module degree, i.e., the normalized number of interactions that any particular node has with the other nodes of its module. Biologically, this would be a surrogate for the generalization level of the species. The importance of a node in the network as a whole is quantified through its participation coefficient, which describes how spread out the interactions of that

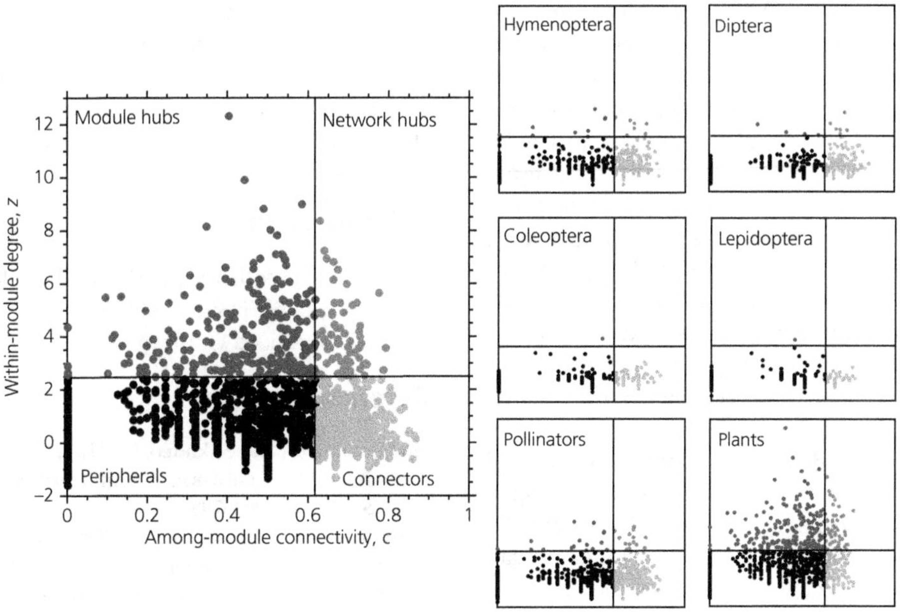

Figure 11.8 Species roles in a dataset of plant–pollinator networks. In all panels, each dot indicates a species. The large panel contains all cummulative data for the 29 modular networks. It comprises 8233 species. Small panels show the same information for given taxonomic groups. From Olesen et al. (2007).

node are across different modules. For a specific number of interactions, the node interacting with nodes from a larger number of different modules is the one with the highest participation coefficient. Nodes with high participation coefficients can be thought of as the glue that keeps the different modules well connected (Figure 11.2). In a strongly modular network, the effect of losing a species varies according to their participation coefficients. For example, the extinction of a module connector should be of huge consequence to the entire network, since it can result in the fragmentation of the original network in several, disjoint subnetworks. This is another illustration of the potential of network studies to inform conservation biology (see Box 11.1 and Chapter 14).

Olesen et al. (2007) classified species in their pollination network dataset in terms of their topological importance. Only about 15% of the almost 10,000 species in these networks were important either at the module level, the network level, or both (Figure 11.8). Species that were important at both the module and entire network were mainly social bees, especially *Apis* spp. Module connectors included beetles and small flies. These are species whose extinction can have implications at the level of the entire network, not only for the species with which they directly interact. Other things being equal and assuming there is no reorganization of the community, their disappearance could induce the fragmentation of the mutualistic network in the different modules or clusters of modules.

What would be the consequences? One could argue that as long as all the species would be present, this would not make a difference, or could even be positive in terms of strengthening the tendency of a modular structure to buffer the effects of a perturbation now entirely confined within the subnetwork. However, Melián and Bascompte (2002) have discussed the dual nature of robustness; this fragmentation, indeed, could also have negative consequences. For example, the isolation of subnetworks would reduce the number of alternative paths for system response to perturbations and affect the coevolutionary process. This is particularly relevant if, as argued by John Thompson (2006), networks need to reach a critical size for some mutualistic life histories to appear. An example of this scenario suggested by Thompson would be that some bee species require an unbroken seasonal succession of flowering plants in order to persist. Thus, if network size is strongly reduced after human alteration, the resulting network may not be large enough to harbor all the previously existing life histories.

The modular description of a network and the consequent classification of species according to their roles, have been traditionally represented as snapshots fixed in time. More recently, we have started adding a temporal dimension to this portrait. Preliminary results in Brazilian pollination networks already suggest that the species' roles may be highly dynamic. The same species do not always play the same network topological role in the different habitats where they occur. For example, the oil flower *Byrsonima sericea* (Malpighiaceae) varied from being a peripheral species in some networks to being a highly connected hub in others (Mello et al. 2013). Among bees, species of the genus *Centris* are more often hubs than are members of other genera. They may, in general, play a central role in module build-up and evolution. Such modules have a long coevolutionary history and function relatively independent of the general pollination network in their habitat.

11.8 Conclusions

The study of mutualistic networks has provided a community-wide framework to understand mutualistic interactions in species-rich communities. It has helped to dispel the assumption that mutualism between free-living species has to lead to either highly specific one-on-one interactions, or to highly diffuse assemblages intractable to study. Studies of mutualistic networks have illustrated that despite the apparent complexity of these interactions, there are repeated patterns in network structure. In particular, the study of modularity has identified units midway between pairwise interactions and full networks. These structures are candidates to be understood as coevolutionary units in cases where coevolution is likely to be an important force shaping these networks. Within a modular network, species play distinct roles related to their topological positions both within their module and within the entire network. These contrasting positions help us to identify subsets of species that are likely to play keystone roles in communities, and hence that should be the focus for conservation efforts.

Acknowledgments

We thank our colleague Pedro Jordano for multiple discussions and insight. This work was funded by an ERC Advanced Grant to J.B.

References

Anderson, T.K., and Sukhdeo, M.V.K. (2011). Host centrality in food web networks determines parasite diversity. *PLoS ONE*, **6**, e0026798.

Arroyo, M.T.K., Primack, R., and Armesto J. (1982). Community studies in pollination ecology in the high temperate Andes of central Chile. I. Pollination mechanisms and altitudinal variation. *American Journal of Botany*, **69**, 82–97.

Ashworth, L., Aguilar, R., Galetto, L., and Aizen, M.A. (2004). Why do pollination generalist and specialist plant species show similar reproductive susceptibility to habitat fragmentation? *Journal of Ecology*, **92**, 717–719.

Baker, H.G. (1983). An outline of the history of anthecology, or pollination biology. In: Real, L. (ed.) *Pollination Biology*. Orlando: Academic Press, pp. 7–28.

Bascompte, J., Jordano, P., Melián, C.J., and Olesen, J.M. (2003). The nested assembly in plant-animal mutualistic networks. *Proceedings of the National Academy of Sciences of the USA*, **100**, 9383–9387.

Bascompte, J., and Jordano, P. (2007). Plant-animal mutualistic networks: the architecture of biodiversity. *Annual Review of Ecology, Evolution, and Systematics*, **38**, 567–593.

Bascompte, J., and Jordano, P. (2014). *Mutualistic Networks*. Princeton, NJ: Princeton University Press.

Bastolla, U., Fortuna, M.A., Pascual-García, A., Ferrera, A., Luque, B., and Bascompte, J. (2009). The architecture of mutualistic networks minimizes competition and increases biodiversity. *Nature*, **458**, 1018–1020.

Bellisario, B., Cerfolli, F., and Nascetti, G. (2012). The interplay between network structure and functioning of detritus-based communities in patchy aquatic environment. *Aquatic Ecology*, **46**, 431–441.

Bezerra, E.L.S., Machado, I.C., and Mello, M.A.R. (2009). Pollination networks of oil-flowers: a tiny world within the smallest of all worlds. *Journal of Animal Ecology*, **78**, 1096–1101.

Bosch, J., Martín, A.M.G., Rodrigo, A., and Navarro, D. (2009). Plant-pollinator networks: adding the pollinator's perspective. *Ecology Letters*, **12**, 409–419.

Cagnolo, L., Salvo, A., and Valladares, G. (2011). Network topology: patterns and mechanisms in plant-herbivore and host-parasitoid food webs. *Journal of Animal Ecology*, **80**, 342–351.

Chave, J. (2013). The problem of pattern and scale in ecology: what have we learned in 20 years? *Ecology Letters*, **16**, 4–16.

Cook, J.M., and Rasplus, J.-Y. (2003). Mutualists with attitude: coevolving fig wasps and figs. *Trends in Ecology and Evolution*, **18**, 241–248.

Darwin, C. (1862). *On the Various Contrivances by Which British and Foreign Orchids are Fertilised by Insects, and on the Good Effects of Intercrossing*. London: John Murray.

Dalsgaard, B., Trøjelsgaard, K., Martín, A.M.G., et al. (2013). Historical climate-change influences modularity of pollination networks. *Ecography*, **36**, 001–010.

Donatti, C.I., Guimarães, P.R., Galetti, M., Pizo, M.A., Marquitti, F.M.D., and Dirzo, R. (2011). Analysis of a hyper-diverse seed dispersal network: modularity and underlying mechanisms. *Ecology Letters*, **14**, 773–781.

Dupont, Y.L., and Olesen, J.M. (2009). Ecological modules and roles of species in heathland plant-insect-flower visitor networks. *Journal of Animal Ecology*, **78**, 346–353.

Dupont, Y.L., Padrón, B., Olesen, J.M., and Petanidou, T. (2009). Spatio-temporal variation in the structure of pollination networks. *Oikos*, **118**, 1261–1269.

Dupont, Y.L., and Olesen, J.M. (2012). Stability of modularity and structural keystone species in temporal cumulative plant-flower visitor networks. *Ecological Complexity*, **11**, 84–90.

Elberling, H., and Olesen, J.M. (1999). The structure of a high latitude plant-flower visitor system: the dominance of flies. *Ecography*, **22**, 314–323.

Faegri, K., and van der Pijl, L. (1971). *The Principles of Pollination Ecology*. Oxford: Pergamon Press.

Fonseca, C.R., and Ganade, G. (1996). Asymmetries, compartments and null interactions in an Amazonian ant-myrmecophyte community. *Journal of Animal Ecology*, **65**, 339–347.

Fortuna, M.A., Popa-Lisseanu, G., Ibañez, C., and Bascompte, J. (2009). The roosting spatial network of a bird-predator bat. *Ecology*, **90**, 934–944.

Fortuna, M.A., Stouffer, D.B., Olesen, J.M., Jordano, P., Mouillot, D., Krasnov, B.R., Poulin, R., and Bascompte, J. (2010). Nestedness versus modularity in ecological networks: two sides of the same coin?*Journal of Animal Ecology*, **79**, 811–817.

Guimerà, R., Stouffer, D.B., Sales-Pardo, M., Leicht, E., Newman, M.E.J., and Amaral, L.A.N. (2010). Origin of compartmentalization in food webs. *Ecology*, **91**, 2941–2951.

Hagen, M., Kissling, W.D., Rasmussen, C., et al. (2012). Biodiversity, species interactions and ecological networks in a fragmented world. *Advances in Ecological Research*, **46**, 89–210.

Herrera, C.M. (1982). Seasonal variation in the quality of fruits and diffuse coevolution between plants and avian dispersers. *Ecology*, **63**, 773–785.

Janzen, D.H. (1980). When is it coevolution? *Evolution*, **34**, 611–612.

Janzen, D.H. (1985). On ecological fitting. *Oikos*, **45**, 308–310.

Joppa, L.N., Bascompte, J., Montoya, J.M., Solé, R.V., Sanderson, J., and Pimm, S.L. (2009). Reciprocal specialization in ecological networks. *Ecology Letters*, **12**, 961–969.

Jordano, P. (1987). Patterns of mutualistic interactions in pollination and seed dispersal: connectance, dependence asymmetries, and coevolution. *American Naturalist*, **129**, 657–677.

Jordano, P. (1995). Angiosperm fleshy fruits and seed dispersers: a comparative analysis of adaptation and constraints in plant-animal interactions. *American Naturalist*, **145**, 163–191.

Jordano, P., Bascompte, J., and Olesen, J.M. (2003). Invariant properties in coevolutionary networks of plant–animal interactions. *Ecology Letters*, **6**, 69–81.

Kratochwil, A., Beil, M., and Schwabe, A. (2009). Complex structure of pollinator-plant interaction-webs: random, nested, with gradients or modules? *Apidologie*, **40**, 634–650.

Levin, S.A. (1992). The problem of pattern and scale in ecology. *Ecology*, **73**, 1943–1967.

Lomáscolo, S.B., Levey, D.J., Kimball, R.T., Bolker, B.M., and Alborn, H.T. (2010). Dispersers shape fruit diversity in *Ficus* (Moraceae). *Proceedings of The National Academy of Sciences of the USA*, **107**, 14668–14672.

Macfadyen, S., Gibson, R.H., Symondson, W.O.C., and Memmott, J. (2011). Landscape structure influences modularity patterns in farm food webs: consequences for pest control. *Ecological Applications*, **21**, 516–524.

Martín, A.M.G., Allesina, S., Rodrigo, A., and Bosch, J. (2012). Drivers of compartmentalization in a Mediterranean pollination network. *Oikos*, **121**, 2001–2013.

May, R.M. (1972). Will a large complex system be stable? *Nature*, **238**, 413–414.

Melián, C.J., and Bascompte, J. (2002). Complex networks: two ways to be robust? *Ecology Letters*, **5**, 705–708.

Melián, C.J., and Bascompte, J. (2004). Food web cohesion. *Ecology*, **85**, 352–358.

Mello, M.A.R., Bezerra, E.L.S., and Machado, I.C. (2013). Functional roles of Centridini oil bees and Malpighiaceae oil flowers in biome-wide pollination networks. *Biotropica*, **45**, 45–53.

Memmott, J. (1999). The structure of a plant-pollinator food web. *Ecology Letters*, **2**, 276–280.

Montesinos-Navarro, A., Segarra-Moragues, J.G., Valiente-Banuet, A., and Verdú, M. (2012). Plant facilitation occurs between species differing in their associated arbuscular mycorrhizal fungi. *New Phytologist*, **196**, 835–844.

Olesen, J.M., Bascompte, J., Dupont, Y.L., and Jordano, P. (2007). The modularity of pollination networks. *Proceedings of the National Academy of Sciences of the USA* **104**, 19891–19896.

Olesen, J.M., Bascompte, J., Elberling, H., and Jordano, P. (2008). Temporal dynamics of a pollination network. *Ecology*, **89**, 1573–1582.

Olesen, J.M., Dupont, Y.L., O'Gorman, E.J., et al. (2010). From Broadstone to Zackenberg: space, time and hierarchies in ecological networks. *Advances in Ecological Research*, **42**, 1–69.

Ollerton, J., Alarcón, R., Waser, N.M., Price, M.V., Watts, S., Cranmer, L., Hingston, A.B., Peter, C.I., and Rotenberry, J. (2009). A global test of the pollination syndrome hypothesis. *Annals of Botany* **103**, 1471–1480.

Pellmyr, O. (2003). Yuccas, yucca moths and coevolution: a review. *Annals of the Missouri Botanical Garden*, **90**, 35–55.

Petanidou, T., Kallimanis, A.S., Tzanopoulos, J., Sgardelis, S.P., and Pantis, J.D. (2008). Long-term observation of a pollination network: fluctuation in species and interactions, relative invariance of network structure, and implications for estimates of specialization. *Ecology Letters*, **11**, 564–575.

Pimm, S.L. (1979). The structure of food webs. *Theoretical Population Biology*, **16**, 144–158.

Pimm, S.L., and Lawton, J.H. (1980). Are food webs divided into compartments? *Journal of Animal Ecology*, **49**, 879–898.

Raffaelli, D., and Hall, S. (1992). Compartments and predation in an estuarine food web. *Journal of Animal Ecology*, **61**, 551–560.

Rezende, E.L., Lavabre, J.E., Guimarães Jr., P.R., Jordano, P., and Bascompte, J. (2007). Non-random coextinctions in phylogenetically structured mutualistic networks. *Nature*, **448**, 925–928.

Rezende, E.L., Albert, E.M., Fortuna, M.A., and Bascompte, J. (2009). Compartments in a marine food web associated with phylogeny, body mass, and habitat structure. *Ecology Letters*, **12**, 779–788.

Rohr, R.P., and Bascompte, J. (2014). Components of phylogenetic signal in antagonistic and mutualistic networks. *The American Naturalist*, **184**, 556–564.

Rohr, R.P., Saavedra, S., and Bascompte, J. (2014). On the structural stability of mutualistic systems. *Science*, 345.

Sprengel, C.K. (1793). *Das entdeckte Geheimniss der Natur im Bau und der Befruchtung der Blumen*. Berlin: F. Vieweg.

Stouffer, D.B., and Bascompte, J. (2011). Compartmentalization increases food-web persistence. *Proceedings of the National Academy of Sciences of the USA*, **108**, 3648–3652.

Thompson, J.N. (2005). *The Geographic Mosaic of Coevolution*. Chicago: University of Chicago Press.

Thompson, J.N. (2006). Mutualistic webs of species. *Science*, **312**, 372–373.

Traveset, A., Heleno, R., Chamorro, S., Vargas, P., McMullen, C.K., Castro-Urgal, R., Nogales, M., Herrera, H.W., and Olesen, J.M. (2013). Invaders of pollination networks in the Galápagos Islands: emergence of novel communities. *Proceedings of the Royal Society B*, **280**, 1471–2954.

Trøjelsgaard, K., and Olesen, J.M. (2013). Macroecology of pollination networks. *Global Ecology & Biogeography*, **22**, 149–162.

Trøjelsgaard, K., Báez, M., Espadaler, X., Nogales, M., Oromí, P., La Roche, F., and Olesen, J.M. (2013). Island biogeography of mutualistic interaction networks. *Journal of Biogeography*, **40**, 2020–2031.

Valdovinos, F.S., Ramos-Jiliberto, R., Flores, J.D., Espinoza, C., and López, G. (2009). Structure and dynamics of pollination networks: the role of alien plants. *Oikos*, **118**, 1190–1200.

Van der Pijl, L. (1969). *Principles of Dispersal in Higher Plants VII*. New York: Springer-Verlag.

Vázquez, D.P., and Aizen, M. (2004). Asymmetric specialization: a pervasive feature of plant-pollinator interactions. *Ecology*, **85**, 1251–1257.

Vogel, S. (1996). Christian Konrad Sprengel's theory of the flower: the cradle of floral ecology. In: D.G. Lloyd and S.C.H. Barrett (eds.) *Floral Biology. Studies on Floral Evolution in Animal-Pollinated Plants*. New York: Chapman & Hall.

Waser, N.M., Chittka, L., Price, M.V., Williams, N., and Ollerton, J. (1996). Generalization in pollination systems, and why it matters. *Ecology*, **77**, 1043–1060.

Waser, N.M. (2006). Specialization and generalization in plant-pollinator interactions: a historical perspective. In: N.M. Waser and J. Ollerton (eds.) *Plant–Pollinator Interactions: From Specialization to Generalization*. Chicago: University of Chicago Press, pp. 3–17.

CHAPTER 12

Mutualisms and ecosystem-level processes

Nancy Collins Johnson

12.1 Introduction

This chapter seeks consilience of population, community, and ecosystem ecology. Mutualisms are commonly studied at the scale of populations or communities with little consideration of the flow of matter and energy through the study system. However, an ecosystem perspective may improve the predictive abilities of population and community models because it accounts for variance generated by abiotic context. Similarly, ecosystem studies rarely consider mutualisms even though understanding the nature of biotic interactions can provide a more mechanistic understanding of ecosystem dynamics. Consilience of mutualism science and ecosystem science is valuable because designing human-dominated ecosystems to emulate mutualisms will help generate a more sustainable world order and a brighter future for all species on earth.

Ecosystem science examines the flow of matter and energy through living and non-living entities. Organisms move organic and inorganic molecules through land, air, and water through the processes of production, consumption, respiration, and decomposition. Throughout earth's history, mutualisms have increased ecosystem production and helped shape the global environment. Mutualisms help organisms overcome limitations caused by unfavorable conditions, insufficient resources, or biotic antagonism. The types and complexity of interactions among organisms can influence the productivity of ecosystems and whether matter and energy will be stored or exported from the system (Fath 2007).

Scale is unrestricted in ecosystem studies, but it is important to define. Ecosystems may be as small as a single cell or as large as the entire earth. Observers delineate system boundaries in space and time, and these boundaries generally reflect areas with natural habitat edges, such as cell walls, coral reefs, and forests. Ecosystems are further characterized as open or closed; matter and energy freely enter and leave an open system like a gastrointestinal tract, but not a closed system such as a bottle of beer undergoing fermentation. The law of conservation of matter controls the movement of matter, and the laws of thermodynamics rule the flux of energy (Lindeman 1942). Transfer of matter and energy in ecosystems creates food webs that can be empirically studied using a method called ecological network analysis (Patten 1991). This approach accounts for all observed and indirect interactions by classifying relationships between pairwise objects as positive, negative, or neutral depending on the net gain or loss experienced by each object. A fascinating outcome of network analysis is that including both trophic and non-trophic interactions in the network often generates a greater occurrence of mutualism than competition, making the whole system more positive than the direct relations that they are composed of (Fath and Patten 1998). This synergistic property of ecosystems is called network mutualism (Fath 2007).

Ecosystem properties emerge from complex interactions and feedbacks among non-living and living components of systems. At a global scale, these feedbacks drive the evolution of the abiotic and biotic characteristics of the earth, including the creation of an oxygen-rich atmosphere,

Mutualism. Edited by Judith L. Bronstein.

the development of multicellular organisms, and the evolution of humans as an ultrasocial species (Gowdy and Krall 2013). Individual organisms can be considered ecosystems because they host complex communities of smaller organisms, and interactions among these organisms and their abiotic environment determine the host organism's fitness as well as the fitness of all of its symbionts. There is increasing recognition that the phenotypes of organisms emerge from the net effects of interacting communities of associated organisms. For example, humans and other mammals can be considered ecosystems that host distinctive microbial communities in and on their bodies (Costello et al. 2009). Microbial inhabitants of the gastrointestinal tract are important mutualists that help digest food and synthesize essential vitamins and other cofactors (Bäckhed et al. 2005; Box 2.2). Other microbes inhabit the skin and generate characteristic body odors that may potentially influence their host's fitness (Archie and Theis 2011). Many factors, including social context, ultimately determine whether particular microbial symbionts are beneficial or detrimental to the reproductive success of their host. As shown in Chapter 10, context dependency is a critical determinant of whether or not a particular association will be mutualistic (Hussa and Goodrich-Blair 2013).

An ecosystem perspective helps provide context for studies of mutualisms. Studies that focus solely on two interacting species are unlikely to accurately capture the dynamics of species interactions in nature because they miss critical feedbacks that emerge from constraints imposed by the abiotic environment. This limitation of reductionist science has delayed recognition of the pivotal role of mutualisms in structuring ecosystems of all sizes. Also, bias toward competitive and antagonistic interactions has delayed recognition of the importance of mutualisms in the evolution of species' niches. Hutchinson (1957) defined the fundamental niche as the range of environmental conditions within which a species can live indefinitely in the absence of interspecific interactions, and the realized niche as the range of conditions that is restricted by antagonistic interactions experienced by organisms (Figure 12.1a). By definition, Hutchinson's realized niche was smaller than the fundamental niche; however, more recently, ecologists have noted that biotic interactions often *increase* niche breadth because

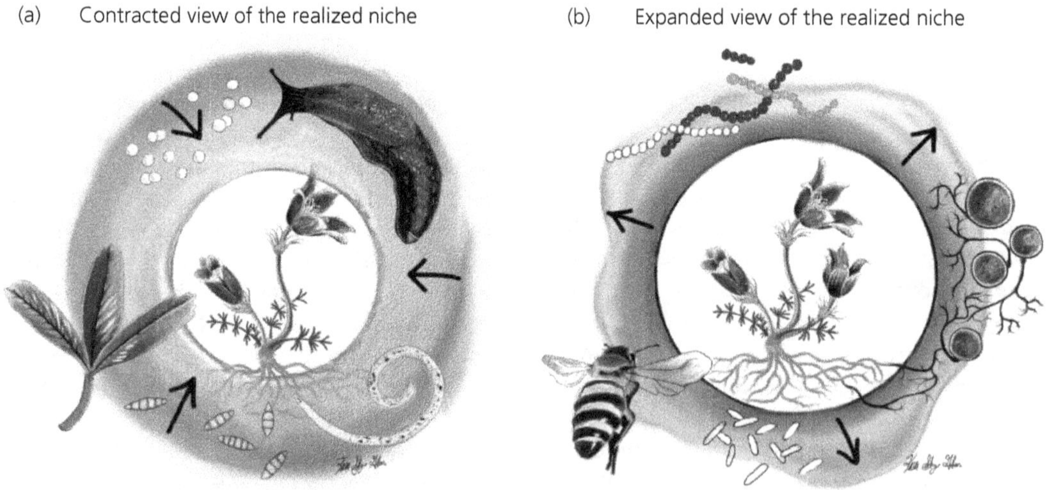

Figure 12.1 Interspecific interactions can contract or expand a species' realized niche relative to its fundamental niche. For example, the conventional view of the realized niche of Pasque flowers (*Pulsatilla patens*) emphasizes niche constriction by herbivores, competitors, pathogens, and parasites (a). However, recent work shows that mutualist species such as pollinators and mycorrhizal fungi can expand the niche breadth of this species (b). Illustrations by Kara Skye Gibson.

beneficial relationships can enhance the availability of essential resources and expand the range of conditions under which organisms can survive and reproduce (Figure 12.1b; Bruno et al. 2003, Poisot et al. 2011). Examples described in this chapter illustrate how mutualistic partnerships generate opportunities for species to radiate into environments that would be uninhabitable without a mutualist. Indeed, in the most extreme examples, the fundamental niches of some species would not exist in the absence of their obligate mutualists (Margulis and Fester 1991, Diamond 1998, Gowdy and Krall 2013, Hussa and Goodrich-Blair 2013).

This chapter explores how mutualisms structure ecosystems and analyzes some mechanisms that are believed to encourage the formation of mutualisms. The chapter is organized in three sections. The first section examines evidence that throughout time, mutualisms have helped to shape the physical, chemical, and biotic properties of the earth by increasing primary production and stimulating nutrient cycling. This section also examines examples of how highly successful mutualisms in competitively dominant species can reduce the diversity of communities. The second section further explores these ideas in the context of agricultural mutualisms. The final section reviews theories that help determine the relative costs and benefits of mutualisms. These theories may contribute to the design of human institutions that preserve ecosystem services. Insights gained from the study of mutualisms may help inform the creation of sustainable partnerships within human societies and among humans and their biotic and abiotic environment (Li et al. 2012).

12.2 Mutualisms structure ecosystems

Since the dawn of life, mutualistic partnerships have influenced the cycling of matter and energy through ecosystems of all sizes ranging from individual cells to the entire globe. The origin of mitochondria from endosymbiotic partnerships approximately two billion years ago was a major innovation in the evolution of life. This advance occurred when a heterotrophic α-proteobacterium symbiont joined with an autotrophic archean host in a metabolically driven mutualism (Martin and Russell 2003). One hypothesis suggests that a methanogen hosted a methanotroph and the reciprocal exchange of methane and hydrogen ions created a win–win symbiosis (Dyall et al. 2004). Another scenario posits that oxygen tension may have driven the partnership between an anaerobic host and an aerobic symbiont (Dyall et al. 2004). Either way, *syntrophy*, defined as obligately mutualistic metabolism, based upon utilization of metabolic wastes, permitted these primitive organisms to radiate into previously uninhabitable niches (Morris et al. 2013). This is an obvious example of how mutualisms expand the realized niche of species. But how do they influence ecosystems?

Lotka (1922) hypothesized that natural selection should maximize energy flux through ecosystems because groups of organisms that exploit the residue of untapped sources of matter and energy will be favored. Mutualisms may play a key role in the efficient use of matter and energy. Fukui et al. (2007) used an ecosystem perspective and simultaneous ordinary differential equations to compare model cells with and without endosymbionts and concluded that endosymbionts accelerate the flow of biomaterials. Furthermore, their "cytological ecosystem" showed that the bifurcation between parasitism and mutualism is controlled by the efficiency of metabolic assistance provided by the endosymbiont. Pairs of organisms can initiate mutualistic relationships as long as the value of the cooperative trait exceeds the reproductive costs to the partners; once the threshold is exceeded, evolution of cooperation proceeds rapidly (Frank 1995; Chapter 4). In the case of the mitochondrial endosymbiosis, the benefits were so great that, over time, the genomes of the host and symbiont blended and generated eukaryotic cells. Natural selection favored mitochondriate eukaryotes because they produced ATP more efficiently in the increasingly oxygenated environment that developed during the Proterozoic (Andersson et al. 2003). This innovation paved the way for the evolution of aerobic multicellular organisms that radiated across the globe and exploited untapped sources of matter and energy during the Cambrian period (Zhuravlev and Riding 2001).

Both living and non-living characteristics of the earth evolve over time, and mutualisms help drive these changes. The earth's atmosphere contained no free oxygen until about 3.5 billion years ago when cyanobacteria began transforming the anaerobic earth to the oxygen-rich world that we know today (Kasting and Siefert, 2002, Dyall et al. 2004). Oxygenic photosynthesis appears to have evolved only once in cyanobacteria, but endosymbiosis spread it throughout a variety of eukaryotic hosts (Falkowski et al. 2004). The origin of chloroplasts marks another major innovation in the evolution of life. Phylogenetic, structural, and biochemical evidence indicate that chloroplasts originated between 1.2 and 1.5 billion years ago when a cyanobacterium associated with a mitochondriate eukaryote (Dyall et al. 2004). Green algae arose from this initial primary endosymbiosis and became the progenitors of all land plants.

Over a century ago Mereschkowsky noted that chloroplasts function like "little green slaves" that nourish their host plant. As with mitochondria, this symbiosis vastly expanded the niches of both host and symbiont, but there is no going back; over time, neither host nor cyanobacteria could survive independently. Many genes were lost when the cyanobacterium took up residence in a host cell. Free-living cyanobacteria possess about 3000 genes but chloroplasts have only 100 to 200 genes (McFadden 1999). Many, but not all, of the indispensable genes were relocated into the plant nucleus. Thus, remnants of a distant endosymbiotic past are conserved in the greatly reduced chloroplast genome. The innovation of chloroplasts initiated the radiation of multicellular land plants and transformation of the air, water, and soil of the global ecosystem.

Oxygen released by photosynthetic cyanobacteria and early photosynthetic eukaryotes would have been a deadly toxin for the anaerobic organisms that dominated the anoxic Archean world three billion years ago. But organisms influence the physical and chemical conditions in their immediate environment; through the process of syntrophy, the wastes of one type of organism nourish other organisms with different metabolic and resource requirements. Fossil stromatolites and present-day microbial mats in marine habitats provide salient examples of syntrophic ecosystems that maximize energy flux through efficient utilization of metabolic wastes and modification of environmental conditions. Interactions between biotic and abiotic processes generate microbial mats in which distinct microgradients of light, oxygen, sulfide, pH, redox potential, nutrients, and other factors correspond with stratified communities of aerobic and anaerobic bacteria and other microorganisms (Figure 12.2; Stal 2000).

It has been argued that market theory (see Chapter 4) applies very well to microbial interactions (Wagner et al. 2014). Exchange of specialized commodities (resources and services) among physiologically distinctive groups of microbial trading partners is an important driver of the formation

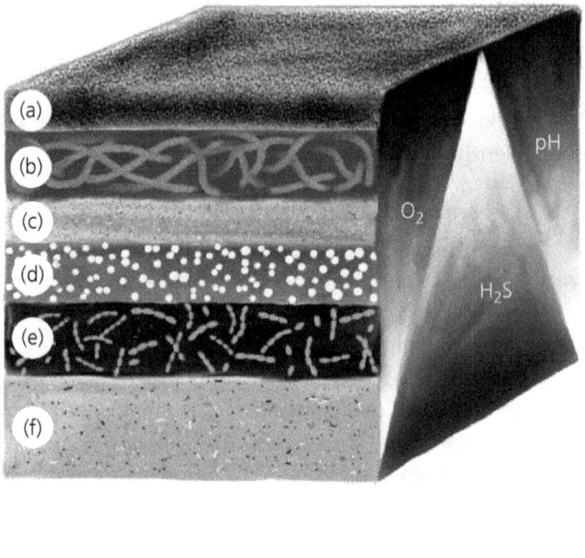

Figure 12.2 Microbial mats commonly form in illuminated marine habitats. Stratified communities of obligate and facultative aerobic and anaerobic organisms assemble in response to gradients of light, oxygen, redox potential, pH, and other physical and chemical factors. Illustration modified from Stal (2000) shows characteristic biotic and abiotic layers. The composition and properties of the layers labeled (a) through (f) is discussed in the text.

of microbial mats, as illustrated in Figure 12.2. The surface layer, represented as (a) in the figure, is comprised of a mucilaginous layer of sand, aerobic bacteria, and pigments, followed by filamentous photosynthetic cyanobacteria, layer (b), that benefit from the protective sunscreen provided by the surface organisms. Oxidized iron formed as a by-product of cyanobacterial photosynthesis accumulates in layer (c) and creates a barrier between the aerobic and anaerobic zones. Purple sulfur bacteria in layer (d) are metabolically versatile and can use either chemical energy or light energy. They utilize the far-red portion of the electromagnetic spectrum that is not used by cyanobacteria and passes though the upper layers. Green sulfur bacteria in layer (e) and sulfate-reducing bacteria in layer (f) thrive in the anaerobic, sulfate-rich environment created by organisms in the upper layers. Inquiry-based science teachers often use old plastic bottles filled with a mixture of soil, shredded paper, and chalk to create Winogradsky columns as classroom demonstrations of the ability of microorganisms to spontaneously create stratified communities based upon syntrophic conversion of wastes into essential resources (Anderson and Hairston 1999). These colorful, and sometimes odoriferous, columns clearly illustrate the concepts of succession and nutrient cycling; they may also be useful in demonstrating that stable and repeatable communities arise when organisms mutually benefit from each others' metabolic wastes.

Throughout earth's history, mutualisms have helped shape the atmosphere, hydrosphere, and lithosphere as a consequence of their influence on primary production and the flow of nutrients through ecosystems. Mycorrhizal symbioses between fungi and plant roots are examples of mutualisms that have played key roles in the movement of matter and energy through terrestrial ecosystems. Evidence suggests that mycorrhizas facilitated the colonization of land by the earliest terrestrial plants during the mid-Ordovician, 475 million years ago (Malloch et al. 1980, Bidartondo et al. 2011). Over time, the fungal symbionts in mycorrhizas progressed from the most primitive, Endogone-like Mucoromycotina and Glomeromycota to Ascomycota, Basidomycota, and Zygomycota as more recent clades of fungi and host plants evolved (Bidartondo et al. 2011, Brundrett 2002). With a few exceptions, mycorrhizal fungi trade minerals acquired from the soil for plant-derived photosynthate (Brundrett 2002). Mycorrhizas increase plant productivity in nutrient-poor environments because compared to plant roots, fine fungal hyphae more efficiently acquire essential nutrients such as nitrogen, phosphorus, copper, and zinc. In the process of mobilizing these essential elements, the mycorrhizal fungi enhance mineralization (Schimel and Bennett 2004, Talbot et al. 2008) and weathering of rock and soil parent material (Hoffland et al. 2004). As Jonathan Leake discusses in Box 12.1, over time, these side effects of mycorrhizal symbioses have been important drivers of soil formation and the composition of the earth's atmosphere.

Emergent properties of mutualisms may have variable effects on biodiversity and community structure. For example, mycorrhizas can either increase (van der Heijden et al. 1998) or decrease (Hartnett and Wilson 2002) plant diversity depending on the type of feedback between plant and fungal communities (Bever et al. 2010) and on the mycorrhizal dependency of the competitively dominant plant species (Urcelay and Diaz 2003). Highly successful mutualisms may exacerbate the success of invasive species and reduce the diversity of communities (Richardson et al. 2000: Chapter 14). This type of positive (self-reinforcing) feedback can have lasting effects on net ecosystem production and nutrient cycles. As Box 12.2 discusses, there is increasing evidence that plants support biotic communities that can help them survive and reproduce in their local environment. This finding has immediate implications for land managers involved with ecosystem restoration.

Associations between plants and nitrogen-fixing bacteria demonstrate how nutritional mutualisms can facilitate the success of invasive species and ultimately reduce plant diversity. Plants are nitrogen limited in most terrestrial ecosystems; consequently, partnerships with nitrogen-fixing bacteria can substantially improve plant fitness. For example, vegetation in ash-derived soils in Hawaii is generally nitrogen limited yet none of the native plants host nitrogen-fixing symbioses (Vitousek et al. 1987). The exotic tree *Myrica faya*

Box 12.1 Biogeochemical effects of coevolution of plants and fungi

Jonathan R. Leake

The coevolution of plants in association with symbiotic soil fungi led to a greening of the earth's continents by terrestrial plants and ecosystems that increased in biomass and complexity as they evolved. This transformed global biogeochemical cycles and biosphere–geosphere–ocean–atmosphere interactions over the past 400 million years (Box 12.1 Figure 1). The basis of the mycorrhizal mutualism is a stable cooperation between partners in which the plants preferentially reward, with photosynthate, fungal partners that mobilize and translocate to them growth-limiting nutrients such as phosphorus. Selection pressure for establishing these symbioses was extremely strong when plants first colonized the land, as the mutual benefits to the partners are amplified under the CO_2-rich atmosphere which was up to fifteen times higher than preindustrial concentrations at that time (Box 12.1 Figure 1; Humphreys et al. 2010, Field et al. 2012).

Among the earliest colonizers of the land ~475 million years ago were the liverworts, a group of small, rootless, non-vascular plants that form mutualistic symbioses with soil fungi (Field et al. 2012). The most basal extant liverwort clades form

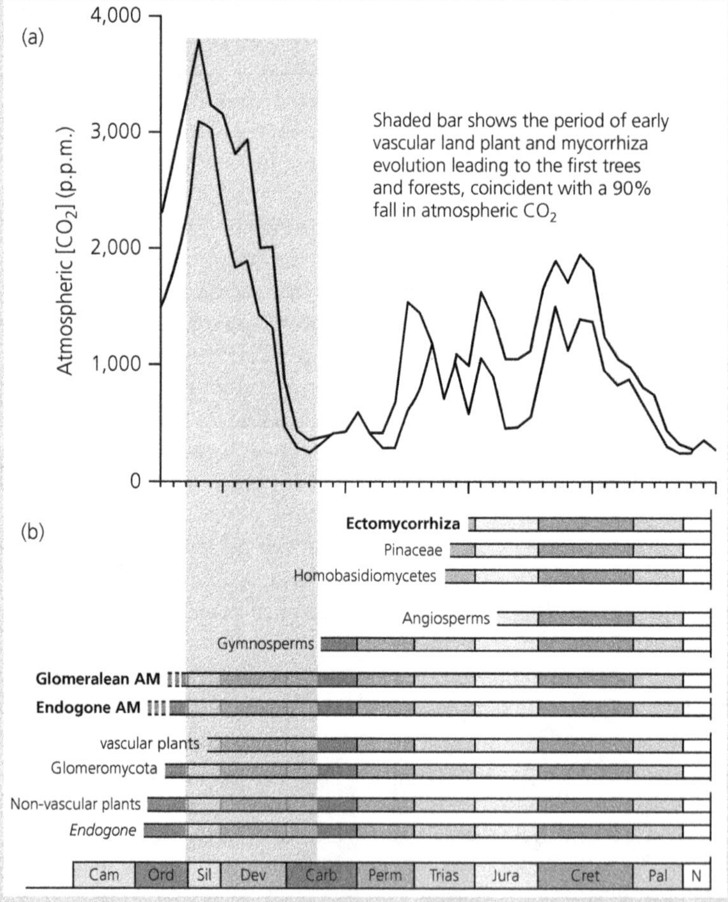

Figure 1 Coevolution of land plants and soil fungal symbioses in relation to dynamic changes in atmospheric CO_2 concentrations. (a) Depicts the modeled global atmospheric CO_2 concentration for the past 500 Ma (from Field et al. 2012), and (b) shows the approximate stratigraphic ranges of the major types of mutualistic soil fungal partners and land plants (adapted from Taylor et al. 2009, Bidartondo et al. 2011).

Box 12.1 *Continued*

arbuscular mycorrhiza-like associations exclusively with *Endogone*, a very ancient group of fungi in the Mucoromycota predating the Glomeromycotean fungi that are partners of slightly more advanced thalloid liverworts, and until recently thought to be the earliest plant–fungal symbionts (Wang et al. 2010, Bidartondo et al. 2011). Several of the major clades of obligately biotrophic Glomeralean arbuscular mycorrhizal fungi had diverged by 411 million years ago, based on fossil evidence from the exceptionally well-preserved Rhynie chert early vascular plant flora, together with evidence of formation of arbuscules and hyphae in plant tissues even before most plants had evolved roots (reviewed by Field et al. 2012). Phylogenetic analysis indicates vertical inheritance of mycorrhization genes from thalloid liverworts through to modern vascular land plants, supporting the view that symbiosis with fungal partners is a basal trait possibly stretching back to a single common ancestral land plant (Wang et al. 2010).

The coevolution of mycorrhizal fungal symbionts with vascular land plants and their rooting systems has played a pivotal role in the expansion of terrestrial ecosystems, during which plants progressively grew in stature, biomass, nutrient demand, and rooting depth as they colonized the continental land masses (Field et al. 2012). These plant–fungal partnerships therefore became increasingly potent partners in biogeochemical cycling of carbon and essential elements, as plant-driven fungal weathering plays a crucial role in the release of P from its primary source, the calcium phosphate mineral apatite (Quirk et al. 2012), and the release of other essential nutrients from minerals. Indeed, a 90% reduction in the atmospheric CO_2 concentration caused by enhanced release of Ca and Mg from silicate minerals, and their export to the oceans where they were precipitated into carbonate sediments (Berner 2006, Taylor et al. 2009), were coincident with the rise of mycorrhizal vascular land plants and forested ecosystems from 410–340 million years ago (Taylor et al. 2009, Field et al. 2012; Box 12.1 Figure 1). Falling atmospheric CO_2 concentrations have now been shown to drive biosphere–geosphere–ocean–atmosphere-stabilizing feedbacks through restricting the capacity of roots and mycorrhizal fungi to weather minerals by a reduced supply of photosynthate (Quirk et al. 2014). As global temperatures fall chemical weathering rates also decrease, thus limiting further CO_2 drawdown. The atmosphere was later replenished, for example by major episodes of volcanic degassing of CO_2. Intensification of weathering by the rise of mycorrhizal vascular plants and the stabilization of land surfaces by vegetation-reducing erosion causes the primary mineral apatite P source to become depleted (Turner et al. 2013). A gradual loss of the weathered P to the oceans causes large areas of terrestrial ecosystems on stable land surfaces to become extremely P limited (Turner et al. 2013). Phosphorus limitation is implicated in the evolution and selection of plant–mycorrhiza partnerships such as the rise of angiosperms at the expense of gymnosperms, and ectomycorrhizal fungi at the expense of arbuscular mycorrhizal fungi (Box 12.1 Figure 1), that have increased the intensity of mineral weathering (Quirk et al. 2012). These findings reveal plant–mycorrhizal coevolutionary advances as potent cogs in the transformations of the biosphere, geosphere, oceans, and atmosphere by the evolution and diversification of terrestrial ecosystems primarily through impacts on the flows of nutrients, carbon, and energy over hundreds of millions of years.

References

Berner, R.A. (2006). GEOCARBSULF: a combined model for Phanerozoic atmospheric O_2 and CO_2. *Geochimemica et Cosmochimica Acta*, **70**, 5653–5664.

Bidartondo, M.I., Read, D.J., Trappe, J.M., Merckx, V., Ligrone, R., and Duckett, J.G. (2011). The dawn of symbiosis between plants and fungi. *Biology Letters*, **7**, 574–577.

Field, K.J., Cameron, D.D., Leake, J.R., Tille, S., Bibartondo, M.I., and Beerling, D.J. (2012). Contrasting arbuscular mycorrhizal responses of vascular and non-vascular plants to a simulated Palaeozoic CO_2 decline. *Nature Communications*, **3**, 1–8.

Humphreys, C.P., Franks, P.J., Rees, M., Bidartondo, M.I., Leake, J.R., and Beerling, D.J. (2010). Mutualistic mycorrhiza-like symbiosis in the most ancient group of land plants. *Nature Communications*, **1**, 103.

Quirk J., Beerling, D.J., Banwart, S.A., Kakonyi, G., Romero-Gonzalez, M.E., and Leake, J.R. (2012). Evolution of trees and mycorrhizal fungi intensifies silicate mineral weathering. *Biology Letters*, **8**, 1006–1011.

Quirk, J., Leake, J.R., Banwart, S.A., Taylor, L.L., and Beerling, D.J. (2014) Weathering by tree root-associating fungi diminishes under simulated Cenozoic atmospheric CO_2 decline. *Biogeosciences*, **11**, 1–11.

Taylor, L.L., Leake, J.R., Quirk, J., Hardy, K., Banwart, S.A., and Beerling, D.J. (2009). Biological weathering and the long-term carbon cycle: integrating mycorrhizal evolution and function into the current paradigm. *Geobiology*, **7**, 171–191.

Turner, B.L., Lambers, H., Condron, L.M., Cramer, M.D., Leake, J.R., Richardson, A.E., and Smith, S.E. (2013). Soil microbial biomass and the fate of phosphorus during long-term ecosystem development. *Plant and Soil*, **367**, 225–234.

Wang, B., Yeun, L.H., Xue, J.-Y., Liu, Y., Ane', J.-M., and Qiu, Y.-L. (2010). Presence of three mycorrhizal genes in the common ancestor of land plants suggests a key role of mycorrhizas in the colonization of land by plants. *New Phytologist*, **186**, 514–525.

Box 12.2 Community and ecosystem phenotypes of plants and their symbionts

Thomas G. Whitham and Louis J. Lamit

Although mutualistic relationships are often viewed and studied at the scale of pairwise interactions, it is important to emphasize that these fundamental relationships can have far-reaching implications for the whole community and ecosystem. These interactions ultimately affect biodiversity, community stability, and the ability of communities to survive global challenges. Quantifying intraspecific genetic variation and its community consequences are essential to understanding these issues. The genetic variation that exists within a plant population is often extensive, affecting phytochemistry, ontogeny, phenology, induced plant defenses, morphology, sink–source relationships, and other traits. These traits combine to produce a multivariate phenotype (Holeski et al. 2012) that plays a major role in affecting mutualists, other community members such as insects, and their interactions. For example, Gehring and Whitham (1991) found that insect-resistant pinyon pine (*Pinus edulis*) supported higher colonization of ectomycorrhizae (EM) than susceptible trees. However, when insects were experimentally removed from insect-susceptible trees, EM colonization rebounded to resemble resistant trees. Similar experiments also showed that, independent of the impacts of herbivory on EM colonization, the genetics of the tree determined the composition of the EM community (Gehring et al. 2014). Furthermore, the abiotic environment of climate and the biotic environment of adjacent shrub competitors also interacted with plant genotype to affect EM composition, which in turn affected tree performance (Gehring et al. 2014). Thus, in the same system, plant genetics, insect herbivores, climate change, and competition with shrubs all interact to affect a crucial mutualism.

Although the role of plant genetics in defining mutualisms is poorly known, in their review, Whitham et al. (2012) found that in a wide array of plants ranging from grasses to trees growing from the tropics to the tundra, and from deserts to mountain tops, 85% showed a significant genotypic signal in which different genotypes supported different communities of microbes to insects and vertebrates. For example, Lamit et al. (2014) found that different genotypes of narrowleaf cottonwood (*Populus angustifolia*) supported different communities of twig endophytes, such that the endophyte community represented a heritable plant trait (Box 12.2 Figure 1). These and other studies with the soil microbial community (Schweitzer et al. 2008), arthropods (Keith et al. 2010), understory plants of trees (Lamit et al. 2011), epiphytic plants in the canopies of tropical trees (Zytynska et al. 2011), leaf pathogens (Busby et al. 2013), spiders living under the exfoliating bark of eucalypts (Barbour et al. 2009), and others are, in part, genotype dependent. Furthermore, Crawford and Rudgers (2013) found that genetic diversity within the plant *Ammophila breviligulata* more strongly influenced arthropod communities than plant species diversity. These and other studies argue that individual plant genotypes vary greatly in their interactions and the communities they support, which in turn can significantly alter ecosystem processes (Schweitzer et al. 2012, Fischer et al. 2014).

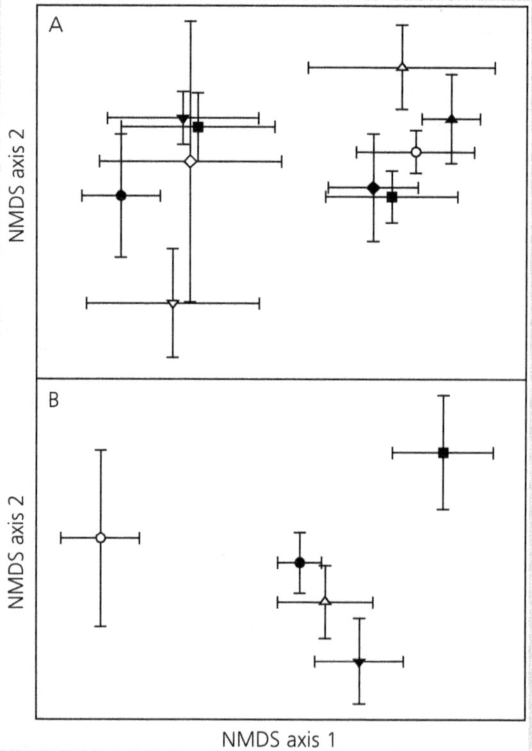

Figure 1 In each of two years of study (ten tree genotypes in 2006, panel A; five tree genotypes in 2008, panel B), different tree genotypes of *Populus angustifolia* supported different endophyte communities in their twigs. Symbols with bars represent means ±1SE of ordination axes scores for endophyte communities in clonally replicated tree genotypes growing in a common garden. Broad-sense heritability estimates indicate that approximately 10% to 25% of the variance in endophyte communities is attributable to differences among tree genotypes (adapted from Lamit et al. 2014).

Box 12.2 *Continued*

Since different microbial communities can result in different rates of decomposition and nitrogen mineralization, the potential exists for microbial communities to differentially feed back to affect the performance of individual plant genotypes. Three lines of evidence support this hypothesis with poplars and pines. First, Schweitzer et al. (2012) found that different microbial communities significantly differ in the nitrogen mineralization rates beneath individual trees, which can then affect nutrient uptake. Second, when seedlings were reciprocally grown in the soils of maternal and non-maternal seed donors, survival and subsequent growth was 250% and 17.5% greater, respectively, in maternal than non-maternal soils (Smith et al. 2012). Third, Gehring et al. (2014) found that climate change resulted in high pinyon mortality in which some tree genotypes were far more likely to survive than others. Importantly, those with specific mycorrhizal communities enjoyed both higher survival and growth rates than those with different mycorrhizal communities. Although much more research needs to be conducted on the mechanisms that drive such feedbacks, it appears that plant genotype can result in selection that creates locally adapted soil communities, which in turn leads to a "home-field advantage" at fine genetic scales (Smith et al. 2012, Schweitzer et al. 2012). These findings demonstrate the genetic linkages among diverse taxa that provide the context for understanding the evolution of complex communities (Shuster et al. 2006) in which mutualisms represent key candidates for future investigations.

References

Barbour, R.C., Forster, L.G., Baker, S.C., Steane, D.A., and Potts, B.M. (2009). Biodiversity consequences of genetic variation in bark characteristics within a foundation tree species. *Conservation Biology*, **23**, 1146–1155.

Busby, P.E., Newcombe, G., Dirzo, R., and Whitham, T.G. (2013). Genetic basis of pathogen community structure for foundation species in a common garden and in the wild. *Journal of Ecology*, **101**, 867–877.

Crawford, K.M., and Rudgers, J.A. (2013). Genetic diversity within a dominant plant outweighs plant species diversity in structuring an arthropod community. *Ecology*, **94**, 1025–1035.

Fischer, D.G., Chapman, S.K., Classen, A.T., Gehring, C.A., Grady, K.C., Schweitzer, J.A., and Whitham, T.G. (2014). Plant genetic effects on soils under climate change. *Plant and Soil*, **379**, 1–19.

Gehring, C.A., and Whitham, T.G. (1991). Herbivore-driven mycorrhizal mutualism in insect-susceptible pinyon pine. *Nature*, **353**, 556–557.

Gehring, C.A., Flores-Rentería, D., Sthultz, C.M., Leonard, T.M., Flores-Rentería, L., Whipple, A.V., and Whitham, T.G. (2014). Plant genetics and interspecific competitive interactions determine ectomycorrhizal fungal community responses to climate change. *Molecular Ecology*, **23**, 1379–1391.

Holeski, L.M., Hillstrom, M.L., Whitham, T.G., and Lindroth, R.L. (2012). Relative importance of genetic, ontogenetic, induction and seasonal variation in producing a multivariate defense phenotype in a foundation tree species. *Oecologia*, **170**, 695–707.

Keith, A.R., Bailey, J.K., and Whitham, T.G. (2010). A genetic basis to community repeatability and stability. *Ecology*, **11**, 3398–3406.

Lamit, L.J, Wojtowicz, T., Kovacs, Z., Wooley, S.C., Zinkgraf, M., Whitham, T.G., Lindroth, R.L., and Gehring, C.A. (2011). Hybridization among foundation tree species influences the structure of associated understory plant communities. *Botany*, **89**, 165–174.

Lamit, L.J., Lau, M.K., Sthultz, C.M., Wooley, S.C., Whitham, T.G., and Gehring, C.A. (2014). Tree genotype and genetically-based growth traits structure twig endophyte communities. *American Journal of Botany*, **101**, 467–478.

Schweitzer, J.A., Bailey, J.K., Fischer, D.G., LeRoy, C.J., Lonsdorf, E.V., Whitham, T.G., and Hart, S.C. (2008). Plant-soil microorganism interactions: a heritable relationship between plant genotype and associated soil microorganisms. *Ecology*, **89**, 773–781.

Schweitzer, J.A., Bailey, J.K., Fischer, D.G., LeRoy, C.J., Whitham, T.G., and Hart S.C. (2012). Functional and heritable consequences of plant genotype on community composition and ecosystem processes. In T. Ohgushi, O. Schmitz, and R.D. Holt (eds.), *Trait-Mediated Indirect Interactions: Ecological and Evolutionary Perspectives*, pp. 371–390. Cambridge: Cambridge University Press.

Shuster, S.M., Lonsdorf, E.V., Wimp, G.M., Bailey, J.K., and Whitham, T.G. (2006). Community heritability measures the evolutionary consequences of indirect genetic effects on community structure. *Evolution*, **60**, 991–1003.

Smith, D.S., Schweitzer, J.A., Turk, P., Bailey, J.K., Hart, S.C., Shuster, S.M., and Whitham, T.G. (2012). Soil-mediated local adaptation alters seedling survival and performance. *Plant and Soil*, **352**, 243–251.

Whitham, T.G., Gehring, C.A., Lamit, L.J., Wojtowicz, T., Evans, L.M., Keith, A.R., and Smith, D.S. (2012). Community specificity: life and afterlife effects of genes. *Trends in Plant Science*, **17**, 271–281.

Zytynska, S.E., Fay, M.F., Penney, D., and Preziosi, R.F. (2011). Genetic variation in a tropical tree species influences the associated epiphytic plant and invertebrate communities in a complex forest ecosystem. *Philos. Trans. R. Soc. Lond. B Biol. Sci.*, **366**, 1329–1336.

has been wildly successful in replacing native Hawaiian vegetation largely because of its mutualism with *Frankia*, a nitrogen-fixing gram-positive bacteria inhabiting nodules on *Myrica* roots (Vitousek et al. 1987). Symbiotic nitrogen-fixing bacteria may be cointroduced with their invasive host plant, or they may already reside in the invaded environment, or both. For example, invasive *Acacia* in South Africa has been linked to cointroduced rhizobial symbionts as well as recruitment of novel symbionts in their new environment (Ndlovu et al. 2013).

In addition to the loss of native plant species, mutualism-enhanced invasions by exotic plants are likely to generate long-term effects on the flow of matter and energy through local ecosystems. The *Frankia–Myrica* mutualism quadruples the amount of nitrogen entering the soil and this has lasting effects on biotic communities and ecosystem processes in Hawaii (Vitousek et al. 1987). Even without nodule formation, plant partnerships with nitrogen-fixing bacteria can facilitate successful plant invasion and generate nutrient-cycle feedbacks. The highly invasive Johnsongrass (*Sorghum halepense*) has many characteristics that contribute to its successful invasion of native prairies in North America. One of these characteristics is its mutualism with endophytic nitrogen-fixing bacteria in the rhizomes (belowground stems) and roots. This extra source of nitrogen supports a much higher growth rate of Johnsongrass compared to the native prairie species that it displaces. Positive plant–soil feedbacks further reinforce the invasion because, within just a few years, soil chemistry is changed to the advantage of Johnsongrass and disadvantage of the prairie species (Rout and Chrzanowski 2009). Thus, nutritional mutualisms between plants and bacteria can vastly increase the niche breadth and abundance of both partners. Agriculture can be considered a type of nutritional mutualism between a farmer species and a food-producing species.

12.3 Agricultural mutualisms

The previous section shows that mutualisms may increase ecosystem productivity by cooperatively exploiting untapped sources of matter and energy, and that highly successful mutualisms have the potential to either increase or decrease community diversity. This section further explores these ideas with an analysis of farming mutualisms involving ants and humans. Associations between humans and their crops share uncanny similarities with associations between leafcutter ants and fungi (Diamond 1998, Gowdy and Krall 2013). In the same way that human agriculture co-opts an ever-growing proportion of arable lands, huge colonies of fungus-farming leafcutter ants dominate forests, grasslands, and pastures throughout tropical and subtropical regions of the New World (Hölldobler and Wilson 2011). Other groups of ants tend aphids and other Hemiptera in much the same way that humans tend livestock for milk and meat (Buckley 1987).

Between 50 and 60 million years ago, a single ant lineage gave rise to approximately 200 species of leafcutter (attine) ants that cultivate fungi in complex subterranean gardens (Mueller and Gerardo 2002; see also Box 5.1). Most of the cultivated fungi belong to the basidiomycete family Lepiotaceae, and all higher attine ants cultivate a derived monophyletic group of fungi in the tribe Leucocoprineae (Hölldobler and Wilson 2011). Leafcutter ants in the genus *Atta* are highly social with an extreme level of cooperation, coordination, and division of labor. Colonies of *Atta* construct elaborate networks of trails and superhighways so that workers can travel far and efficiently deliver copious amounts of fresh vegetation to enormous nests extending deep underground. Inside the nest, an assembly line of progressively smaller subcastes of workers cut plant fragments into pieces, crush them into pellets, and add fecal liquid to create an optimal substrate for fungal growth (Hölldobler and Wilson 2011). The smallest and most abundant subcastes of workers delicately arrange the developing fungal hyphae and continuously manage and clean the gardens by removing alien "weed" fungi along with nematodes and other fungus-feeding pests (Hölldobler and Wilson 2011). Conversion of complex plant parts into nourishing ant food requires a suite of digestive enzymes. Both fungi and ants produce carbohydrases, and there is a high degree of complementarity in the enzymes produced by each partner. Fungal enzymes digest molecules like xylan and cellulose while ant enzymes digest sucrose and maltose (D'Ettorre et al. 2002). In much the same way that the ancestors of mitochondria and chloroplasts became locked into obligate symbiosis, millions of years of coevolution between leafcutter ants and their cultivated fungi have rendered both ant and fungus incapable of living independently of each other (Mueller and Gerardo 2002).

Despite rather dramatic differences between leafcutter ants and humans, certain aspects of our societies share interesting similarities. Like *Atta*, *Homo sapiens* societies function in some ways like superorganisms. Similar forces of multilevel selection have driven the development of ant and human agriculture (Gowdy and Krall 2013). Division of labor and mutually beneficial exchange of goods and services allow mutualist organisms to dramatically increase production through enhanced acquisition of limiting resources and amelioration of inclement conditions. In return for a constant supply of high-quality food, leafcutter ants build underground gardens where they accumulate plant substrates and manage the environment for fungal production. In exchange for high-yielding crops, human farmers fertilize their fields, irrigate dry soil, drain wet soil, and actively remove competitors (weeds) and pests from their crops.

Agricultural mutualisms have vastly increased the niche breadth and abundance of *Homo sapiens*. The human population has increased more than a thousandfold since the development of agriculture 10,000 years ago (Cohen 1997). Gowdy and Krall (2013) consider the development of agriculture as the major driver of our transition into an ultrasocial species with effective sharing of information and full-time division of labor where non-producers and producers form an articulated whole. The human race has blown its realized niche wide open through agricultural, industrial, and medical revolutions, and our exponentially growing population is radically transforming the earth. Although the human race thrives from agriculture, the impacts on natural ecosystems can be devastating. Worldwide, the total area of cultivated land increased 466% from 1700 to 1980 (Meyer and Turner 1992). It is estimated that humans have transformed 16% to 23% of the earth's land surface into row-crops, urban areas, or pastureland (Vitousek et al. 1997). Habitat destruction, pollution, and over-hunting associated with a burgeoning human population have caused the extinction of one-quarter of the earth's bird species during the past two thousand years (Vitousek et al. 1997). Humans have so significantly altered earth's systems that a new geological epoch, the Anthropocene, has been proposed (Crutzen 2002, Ellis 2011).

Agricultural mutualisms not only increase the fitness of the farmer species, they also dramatically increase the abundance of domesticated partner species. In terms of biomass and abundance, rice, wheat, and maize are among the most abundant plants on earth. Similarly, the biomass of large domestic animals is estimated to exceed that of the humans who maintain them (Cohen 1997). Although the fitness gained by domestication can be substantial, it comes with a price. As with all partnerships there is a cost to interdependence. Darwin noted that the modifications selected for through domestication might be so extreme as to preclude survival in the wild. It is likely that domesticated organisms such as maize and fryer chickens are incapable of independent survival (Gepts 2004). In other words, dependency upon the agricultural mutualism can be irreversible. At what point do crops and livestock lose so much independence that they become like chloroplasts, "little green slaves" inside plant cells? In turn, how well could the human race survive without our domesticated plants and animals?

12.4 Managing mutualisms for sustainable systems and ecosystem services

A better understanding of the role of agriculture in humanity's domination of the global ecosystem may help societies find ways to generate better cooperation among humans and other species. With sufficient understanding, mutualisms can be purposefully managed to conserve biodiversity and natural ecosystem services rather than to destroy them. As described previously, mutualisms generally increase the rate at which matter and energy move through ecosystems. In nutrient-limited ecosystems like tropical rain forests and coral reefs. Mutualisms help increase net ecosystem production and biodiversity through syntrophy and complex networks of species interactions that recycle wastes and prevent dominance by super-competitor species. In contrast, agricultural mutualisms reinforce the competitive dominance of the partner species and their burgeoning populations reduce the biodiversity of the ecosystem. This type of positive feedback occurs in highly successful farming mutualisms where farmer and cultivar populations increase at the expense of rare and vulnerable native species. In addition to biodiversity loss, industrialized farming operations become increasingly uncoupled from the natural ecosystem processes that recycle wastes and stabilize soil (Matson et al. 1997). Although the Green Revolution doubled global grain production, agricultural intensification is degrading the air, land, and water. These negative side effects are causing humans to reevaluate conventional agricultural practices and seek more sustainable approaches to food production (Tilman et al. 2002).

Successful development of sustainable food production requires an ecosystem perspective that accounts for the storage and flows of materials and energy through all components of the system. In other words, all externalities related to energy inputs and waste outputs must be accounted for in sustainable food production. The natural functioning of ecosystems provides many services that are not accounted for in traditional production-based agriculture. Classic examples include the role of native insects in pollinating flowering crops, or the role of naturally occurring mycorrhizal fungi and other soil organisms in providing plants essential nutrients and water. As Buck Sanford discusses in Box 12.3, in arid environments, biological soil crust communities and vascular plants derive mutual benefits from the exchange of resources and services, providing many important ecosystem services in the process.

Sustainable agricultural systems can benefit from harnessing multiple syntrophic associations in much the same way that stratified microbial mat communities benefit from reciprocal waste consumption through byproduct mutualisms (Figure 12.2). In this regard, mixed crop–livestock systems show great promise because they efficiently utilize animal wastes for fertilizing crops (Herrero et al. 2010). In Asia, the ancient technique of growing paddy rice in association with ducks is gaining popularity as markets for organic foods are creating economic incentives to reduce the use of pesticides and chemical fertilizers (Xi and Qin 2009). The "organic rice–duck mutualism system" involves maintaining flightless ducks in the rice paddies (Figure 12.3). The ducks eat weeds, insects, and other pests, their droppings provide plant nutrients, and their swimming activity increases rice growth (Tojo et al. 2007). In this example of a sustainable agroecosystem, the farmer actively manages rice plants and ducks, and indirectly manages a vast network of interactions among associated organisms in the soil and water; the end result is a stable system that maximizes food production and minimizes the use of fossil fuels, pesticides, and inorganic fertilizers. Broad consideration of all components of the system is a critical component for the success of this, and other, sustainable food production systems.

The quality of the environment structures biotic interactions and can favor or disfavor mutualisms. The evolution of mutually beneficial associations, and their persistence over time, is more likely to occur in some environments than others. In general, stress caused by harsh physical conditions and insufficient availability of resources stimulates mutualistic innovation and persistence. Mutualisms are most stable when the benefits exceed the costs of the cooperative partnership (Chapter 7; Sachs and Simms 2006). When the ambient environment makes life difficult for one or both partners, mutualisms are likely to have a high benefit to cost ratio and enlarge the realized niches of the partner organisms. In extremely stressful and unpredictable environments, such as alpine tundra or hydrothermal environments, mutualisms may be a prerequisite for survival (Callaway et al. 2002, Marquez et al. 2007).

Biological market theory provides a mechanistic understanding of exchange of goods and services between interacting organisms (see Chapter 4; Schwartz and Hoeksema 1998). Trade dynamics among mutualists is structured by the supply and demand of essential resources or services

Figure 12.3 The rice–duck mutualism has been practiced in Asia for centuries. This method has been shown to improve rice growth using no chemical pesticides or fertilizers (Tojo et al. 2007). Illustration by Kara Skye Gibson.

Box 12.3 Ecosystem services of soil crust mutualisms

Robert L. Sanford, Jr.

The Millennium Ecosystem Assessment (2005) report defines *ecosystem services* as benefits people obtain from ecosystems. Essentially, ecosystem services are products and processes of ecosystems, many of which may be assigned a monetary value, but which are usually not accounted for by society. Here, the main ecosystem services provided by biological soil crusts are described.

Biological soil crusts (BSC), also known as microphytic, cryptogamic, microbiotic, or cryptobiotic soil crusts, occur globally in all hot, cool, and cold regions with an arid or semi-arid precipitation regime (West 1990). They are composed of liverworts, lichens, mosses, green algae, microfungi, cyanobacteria, and other bacteria growing together, and occur in or near the soil surface. Most often they grow best in open spaces between vascular plants. They strongly influence key ecosystem processes including water infiltration, carbon and nitrogen fixation, and transformations (Eldridge et al 2010).

Essentially, BSC can be thought of as a mosaic of mutualisms that cohabit the surface centimeter of soil over vast expanses of arid lands. Given their abundance and importance in many arid systems, BSC qualify as ecosystem engineers and/or foundation communities. The ecosystem services that they provide are a result of their controlling influences on arid-land ecosystem processes. It is important that in contrast to many ecosystem services, BSC form *discrete* communities on the soil surface. As such, it is the interaction of the biotic community with the abiotic elements of the environment that provides these ecosystem services.

Many studies have shown how variable BSC are in their hydrological and erosional responses, which depend on crust composition, roughness, underlying material properties, and stage of development (Chamizo et al., 2010). In essence, surface roughness is the most important variable for understanding the influence of BSC on runoff and erosion. Rodriguez-Caballero et al. (2011) show conclusively how BSC modify soil surface microtopography, generating rough surfaces as a crust develops successionally, and ultimately demonstrating that surface micro-depressions which are induced by surface roughness act as temporal water sinks. In turn, these reduce source area connectivity and reduce runoff rates. Similarly, BSC have a robust capacity for reducing

continued

Box 12.3 *Continued*

wind erosion in dryland areas (Belnap and Gillette 1997). The mechanisms for this are: 1) soil aggregate accretion via BSC polysaccharide exudations, and 2) the still-air boundary layer that is created by rough BSC microtopography protecting sandy surface soils from wind erosion (Bowker et al. 2008).

BSC communities are important in reducing soil erosion, increasing soil fertility via dust capture, as well as fixing atmospheric nitrogen and carbon fixation into ions that become available to other life forms (Elbert et al. 2012). Vascular plants are benefited by BSC in several important ways. First, it has long been known that seedling establishment is increased where intact BSC is present. Secondly, vascular plants that grow as islands surrounded by BSC have higher nutrient concentrations compared to plants growing in uncrusted areas (Belnap and Harper 1995). This is because BSC produce amino acids, carbohydrates, and extracellular polymeric substances, which are carbon- and nitrogen-containing organic compounds (Delgado-Baquerizo et al. 2010). Not surprisingly, nitrogen in leaf tissues ranges from 9%–31% higher in the presence of BSC, and biomass of plants growing in BSC-dominated soils has been shown to increase two to four times more than individuals growing in soils without BSC (Harper and Pendelton 1993). Hence, BSC produce bioavailable, organic forms of C and N that are absorbed by plants and other organisms across most arid ecosystems. In the grand scheme of things, BSC provision food webs via nitrogen fixation and photosynthesis. This is particularly important in drylands where organic nutrients are in low supply. Taken together, the positive influence of BSC on vascular plant seedling establishment, nutrient supply, and biomass show a large net positive effect of BSC on many arid-land plants.

Almost all ecosystem services have been considered to be free services until recently. With increasing human population, the need for such services is rising while ecosystem service outputs decline. BSC are part of a collection of ecosystem services that are contributed by soil biota. A decade ago, van der Putten et al. (2004) estimated that soil biota provided ecosystem services that were valued between 1.5 and 13 trillion US dollars annually, on a global basis.

Management of BSC has become an important focus of many research groups over the last ten years. These (mostly experimental) management methods will hopefully be developed in time to conserve the hugely important and valuable ecosystem services of biological soil crusts.

References

Belnap, J., and Gillette, D.A. (1997). Disturbance of biological soil crusts: impacts on potential wind erodibility of sandy desert soils in southeastern Utah, USA. *Land Degradation and Development*, **8**, 355–362.

Belnap, J., and Harper, K.T. (1995). Influence of cryptobiotic soil crusts on elemental content of tissue in two desert seed plants. *Arid Soil Research and Rehabilitation*, **9**, 107–115.

Bowker, M.A., Belnap, J., Bala Chaudhary, V., and Johnson N.C. (2008). Revisiting classic water erosion models in drylands: the strong impact of biological soil crusts. *Soil Biology and Biochemistry*, **40**, 2309–2316.

Chamizo, S., Rodríguez-Caballero, E., Miralles-Mellado, I., Afana, A., Lázaro, R., Domingo, F., Calvo-Cases, A., Solé-Benet, A., and Cantón, Y. (2010). Características de las costras físicas y biológicas del suelo con mayor influencia sobre la infiltración y la erosión en ecosistemas semiáridos. *Pirineos*, **165**, 69–96.

Delgado-Baquerizo, M., Castillo-Monroy, A.P., Maestre, F.T., and Gallardo, A. (2010). Plants and biological soil crusts modulate the dominance of N forms in a semi-arid grassland. *Soil Biol. Biochem*, **42**, 376–378.

Elbert, W., Weber, B., Burrows, S., Steinkamp, J., Büdel, B., Andreae, M.O., and Pöschl, U. (2012). Impact of cryptogamic covers on the global cycles of carbon and nitrogen. *Nature Geoscience*, **5**, 459–462.

Elderidge, D.J., Bowker, M.A., Maestre, F.T., Alonso, P., Mau, R.L., Papadopoulos, J., and Escudero, A. (2010). Ineteractive effects of three ecosystem engineers on infiltration in a semi-arid Mediterranean grassland. *Ecosystems*, **13**, 499–510.

Harper, K.T., and Pendleton, R.L. (1993). Cyanobacteria and cyanolichens: can they enhance availability of essential minerals for higher plants? *Great Basin Naturalist*, **53**, 59–72.

Millennium Ecosystem Assessment. (2005). <http://www.millenniumassessment.org/en/index.html>.

Rodriguez-Caballero, E., Cantón, Y., Chamizo, S., Afana, A., and Sole-Benet, A. (2011). Effects of biological soil crusts on surface roughness and implications for runoff and erosion. *Geomorphology*, **145–146**, 81–89.

Van der Putten, W.H., Anderson, J.M., Bardgett, R.D., et al. (2004). In D.H. Wall (ed.), *Sustaining Biodiversity and Ecosystem Services in Soils and Sediments*, pp. 15–43. Washington, D.C.: Island Press.

West, N.E. (1990). Structure and function of soil microphytic crusts in wildland ecoystems of arid and semi-arid regions. *Advances in Ecological Research*, **20**, 179–223.

(commodities). The Sprengel–Lebig Law of the Minimum states that production is limited by the resource that is in most limited supply (van der Ploeg et al. 1999). Ecological stoichiometry, the quantitative relationship of the chemical constituents of organisms and their environment (Sterner and Elser 2002), can help quantify the relative value of the chemical resources traded in nutritional mutualisms. The elemental stoichiometry of living organisms and their environment helps predict the economics of resource exchange in mutualisms. A large disparity between the elemental ratios of organisms and their food supply will limit production because of a shortage of an essential resource; thus, there is a great benefit from nutritional mutualisms that access resources that are in limiting supply. In contrast, when no resources limit production there is little benefit from partnerships and nutritional mutualisms such as mycorrhizas will be unstable (Johnson 2010).

According to the stress-gradient hypothesis (Bertness and Callaway 1994), facilitation among species increases and competition decreases with increasing biotic or abiotic stress. A meta-analysis of 727 plant community studies involving many different stresses in a wide range of communities and ecosystems provides strong empirical support for this hypothesis (He et al. 2013). Abiotic stresses such as inclement temperature, extreme pH, salinity, and resource limitation select for mutualisms that enhance fitness in habitats with adverse conditions. In contrast, mutualisms are often less essential for survival in benign environments with ample resources. For example, the benefits of mycorrhizas for nutrient uptake and drought tolerance greatly exceed their costs in many natural ecosystems, but agricultural irrigation and fertilization diminish the mutualistic qualities of mycorrhizas (Figure 12.4).

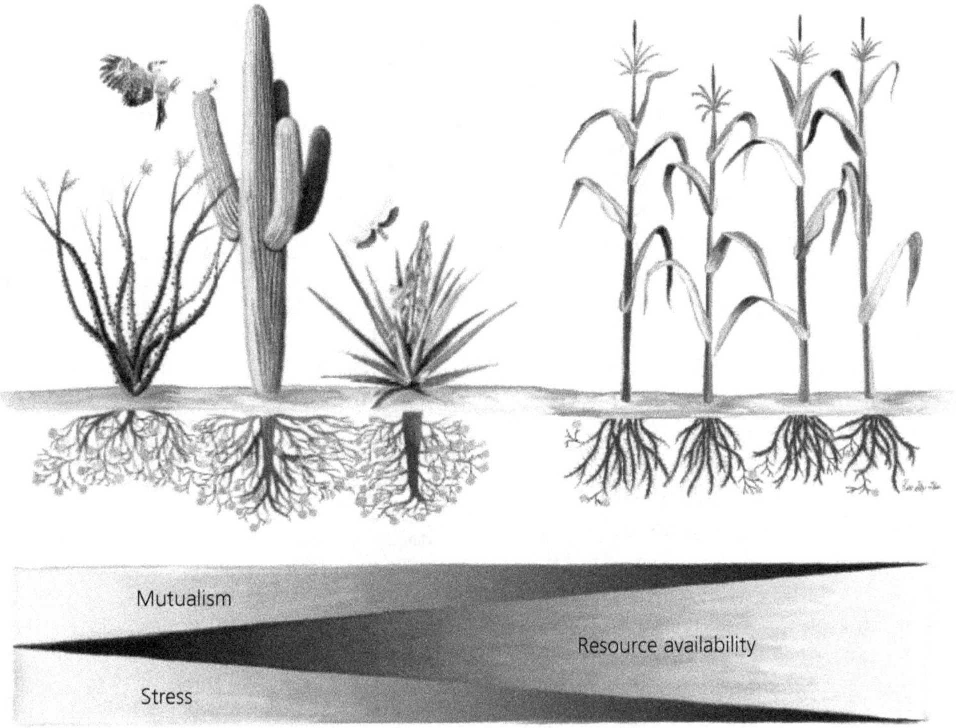

Figure 12.4 Mutualisms are most likely in stressful environments and when resources are in limited supply, for example in desert environments. Fertilized and irrigated agricultural systems are predicted to have fewer mutualisms because stresses are ameliorated and essential resources are enriched.

A better understanding of the factors that control mutualistic function may help ameliorate unintended consequences of anthropogenic environmental change (Johnson et al. 2013). If a mutualistic relationship consistently ameliorates inclement conditions or alleviates resource limitation, then the removal of adversity may have the unintended side effect of generating less-cooperative interactions (Andras et al 2007). For example, environmental changes such as anthropogenic nitrogen deposition and eutrophication can remove resource limitation, destabilize mutualisms, and generate undesirable changes to ecosystems (Chapter 13; Aerts 2002).

12.5 Concluding thoughts

Careful analysis of mutualisms holds great promise for the advancement of the emerging field of sustainability science. This chapter shows that throughout our planet's history, mutualisms have influenced the flow of energy and materials and played critical roles in the evolution of the living and non-living properties of earth. We have also seen how the nature of species interactions within systems can influence community dynamics. Highly successful mutualisms can displace whole communities and drive vulnerable species to extinction. On the other hand, informed management of the mutualisms that naturally provide ecosystem services could help design more sustainable food production systems.

Studies of mutualisms provide a vehicle for consilience among population, community, and ecosystem ecology. Furthermore, understanding the ecological mechanisms that generate and stabilize mutualisms can facilitate the design of long-term cooperative systems among humans, and between humans and the millions of other species that coinhabit the earth. But this is only one step toward the creation of sustainable societies. Consilience among all human institutions—science, philosophy, politics, and religion—is necessary to advance this lofty goal. Ecologists can take a leadership role in this endeavor by discovering and revealing how mutually beneficial interactions among organisms transform the flux of energy and cycling of matter in the global ecosystem.

Acknowledgments

This work was supported by the Ecosystems Program of the National Science Foundation, DEB-0842327.

References

Aerts, R. (2002). The role of various types of mycorrhizal fungi in nutrient cycling and plant competition. In M.G.A. van der Heijden and I. Sanders (eds.), *Mycorrhizal Ecology* (Ecological Studies, **volume 127**, pp. 117–133). Berlin Heidelberg: Springer-Verlag.

Anderson, D.C., and Hairston, R. (1999). The Winogradsky column and biofilms. *American Biology Teacher*, **61**, 453–459.

Andersson, S.G.E., Karlberg, O., Canback, B., and Kurland, C.G. (2003). On the origin of mitochondria: a genomics perspective. *Philosophical Transactions of the Royal Society of London Series B*, **358**, 165–179.

Andras, P., Lazarus, J., and Roberts, G. (2007). Environmental adversity and uncertainty favor cooperation. *BMC Evolutionary Biology*, **7**, 240–248.

Archie, E.A., and Theis, K.R. (2011). Animal behaviour meets microbial ecology. *Animal Behaviour*, **82**, 425–436.

Bäckhed, F., Ley, R.E., Sonnenburg, J.L., Peterson, D.A., and Gordon, J.I. (2005). Host-bacterial mutualism in the human intestine. *Science*, **307**, 1915–1920.

Bertness, M.D., and Callaway, R. (1994). Positive interactions in communities. *Trends in Ecology & Evolution*, **9**, 191–193.

Bever, J.D., Dickie, I.A., Facelli, E., et al. (2010). Rooting theories of plant community ecology in microbial interactions. *Trends in Ecology & Evolution*, **25**, 468–478.

Bidartondo, M.I., Read, D.J., Trappe, J.M., Merckx, V., Ligrone, R., and Duckett, G. (2011). The dawn of symbiosis between plants and fungi. *Biology Letters*, **7**, 574–577.

Brundrett, M.C. (2002). Coevolution of roots and mycorrhizas of land plants. Tansley review no. 134. *New Phytologist*, **154**, 275–304.

Bruno, J.F., Stachowicz, J.J., and Bertness, M.D. (2003). Inclusion of facilitation into ecological theory. *Trends in Ecology & Evolution*, **18**, 119–125.

Buckley, R.C. (1987). Interactions involving plants, Homoptera, and ants. *Annual Review of Ecology and Systematics*, **18**, 111–135.

Callaway, R.M., Brooker, R.W., Choler, P., et al. (2002). Positive interactions among alpine plants increase with stress. *Nature*, **417**, 844–848.

Cohen, J.E. (1997). Population, economics, environment and culture: an introduction to human carrying capacity. *Journal of Applied Ecology*, **34**, 1325–1333.

Costello, E.K., Lauber, C.L., Hamady, M., Fierer, N., Gordon, J.I., and Knight, R. (2009). Bacterial community variation in human body habitats across space and time. *Science*, **326**, 1694–1697.

Crutzen, P.J. (2002). Geology of mankind. *Nature*, **415**, 23.

D'Ettorre, P., Mora, P., Dibangou, V., and Rouland, C. (2002). The role of the symbiotic fungus in the digestive metabolism of two species of fungus-growing ants. *Journal of Comparative Physiology B*, **172**, 169–176.

Diamond, J. (1998). Ants, crops, and history. *Science*, **281**, 1974–1975.

Dyall, S.D., Brown, M.T., and Johnson, P.J. (2004). Ancient invasions: from endosymbionts to organelles. *Science*, **304**, 253–257.

Ellis, E.C. (2011). Anthropogenic transformation of the terrestrial biosphere. *Philosophical Transactions of the Royal Society of London Series A*, **369**, 1010–1035.

Falkowski, P.G., Katz, M.E., Knoll, A.H., Quigg, A., Raven, J.A., Schofield, O., and Taylor, F.J.R. (2004). The evolution of modern eukaryotic phytoplankton. *Science*, **305**, 354–360.

Fath, B.D. (2007). Network mutualism: positive community-level relations in ecosystems. *Ecological Modelling*, **208**, 56–67.

Fath, B.D., and Patten, B.C. (1998). Network synergism: emergence of positive relations in ecological system. *Ecological Modelling*, **107**, 127–143.

Frank, S.A. (1995). The origin of synergistic symbiosis. *Journal of Theoretical Biology*, **176**, 403–410.

Fukui, S., Fukatsu, T., Ikegami, T., and Shimada, M. (2007). Endosymbiosis as a compact ecosystem with material cycling: Parasitism or mutualism? *Journal of Theoretical Biology*, **246**, 746–754.

Gepts, P. (2004). Crop domestication as a long-term selection experiment. In J. Janick (ed.), *Plant Breeding Reviews* (**Vol. 24**, part 2, pp. 1–37). Hoboken, NJ: John Wiley and Sons.

Gowdy, J., and Krall, L. (2013). Agriculture as a major evolutionary transition to human ultrasociality. *Journal of Bioeconomics*, **16**, 179–202.

Hartnett, D.C., and Wilson, G.W.T. (2002). The role of mycorrhizas in plant community structure and dynamics: lessons from grasslands. *Plant and Soil*, **244**, 319–331.

He, Q., Bertness, M.D., and Altieri, A.H. (2013). Global shifts towards positive species interactions with increasing environmental stress. *Ecology Letters*, **16**, 695–706.

Herrero, M., Thornton, P.K., Notenbaert, A.M., et al. (2010). Smart investments in sustainable food production: revisiting in mixed crop-livestock systems. *Science*, **327**, 822–825.

Hoffland, E., Kyper, T.W., Wallander, H., et al. (2004). The role of fungi in weathering. *Frontiers in Ecology and the Environment*, **2**, 258–264.

Hölldobler, B., and Wilson, E.O. (2011). *The Leafcutter Ants: Civilization By Instinct*. New York: WW Norton & Company.

Hussa, E.A., and Goodrich-Blair, H. (2013). It takes a village: ecological and fitness impacts of multipartite mutualism. *Annual Review of Microbiology*, **67**, 161–178.

Hutchinson, G.E. (1957). Concluding remarks. *Cold Spring Harbour Symposium on Quantitative Biology*, **22**, 415–427.

Johnson, N.C. (2010). Tansley review: resource stoichiometry elucidates the structure and function of arbuscular mycorrhizas across scales. *New Phytologist*, **185**, 631–647.

Johnson, N.C., Angelard, C., Sanders, I.R., and Kiers, E.T. (2013). Predicting community and ecosystem outcomes of mycorrhizal responses to global change. *Ecology Letters*, **16**, 140–153.

Kasting, J.F., and Siefert, J.L. (2002). Life and the evolution of earth's atmosphere. *Science*, **296**, 1066–1068.

Li, S., Zhang, Y., Yang, Z., Liu, H., and Zhang, J. (2012). Ecological relationship analysis of the urban metabolic system of Beijing, China. *Environmental Pollution*, **170**, 169–176.

Lindeman, R.L. (1942). The trophic-dynamic aspect of ecology. *Ecology*, **23**, 399–417.

Lotka, A. (1922). Contribution to the energetics of evolution. *Proceedings of the National Academy of Sciences USA*, **8**, 147–151.

Malloch, D.W., Pirozynski, K.A., and Raven, P.H. (1980). Ecological and evolutionary significance of mycorrhizal symbioses in vascular plants (a review). *Proceedings National Academy of Science USA*, **77**, 2113–2118.

Margulis, L., and Fester, R. (eds.). (1991). *Symbiosis as a Source of Evolutionary Innovation*. Cambridge, MA: The MIT Press.

Marquez, L.M., Redman, R.S., Rodriguez, R.J., and Roossinck, M.J. (2007). A virus in a fungus in a plant: three-way symbiosis required for thermal tolerance. *Science*, **315**, 513–515.

Martin, W.E., and Russell, M.J. (2003). On the origins of cells: a hypothesis for the evolutionary transitions from abiotic geochemistry to chemoautotrophic prokaryotes, and from prokaryotes to nucleated cells. *Philosophical Transactions of the Royal Society of London Series B*, **358**, 59–85.

Matson, P.A., Parton, W.J., Power, A.G., and Swift, M.J. (1997). Agricultural intensification and ecosystem properties. *Science*, **277**, 504–509.

McFadden, G.I. (1999). Endosymbiosis and evolution of the plant cell. *Current Opinion in Plant Biology*, **2**, 513–519.

Meyer, W.B., and Turner, B.L.I. (1992). Human population growth and global land-use/cover change. *Annual Review of Ecology and Systematics*, **23**, 39–61.

Morris, B.E.L., Henneberger, R., Huber, H., and Moissl-Eichinger, C. (2013). Microbial syntrophy: interaction for the common good. *FEMS Microbiological Review*, **37**, 384–406.

Mueller, U.G., and Gerardo, N. (2002). Fungus-farming insects: multiple origins and diverse evolutionary histories. *Proceedings of the National Academy of Sciences USA*, **99**, 15247–15249.

Ndlovu, J., Richardson, D.R., Wilson, J.R.U., and Le Roux, J.J. (2013). Co-invasion of South African ecosystems by an Australian legume and its rhizobial symbionts. *Journal of Biogeography*, **40**, 1240–1251.

Patten, B.C. (1991). Network ecology: indirect determination of the life-environment relationship in ecosystems. In M. Higashi and T.P. Burns (eds.), *Theoretical Ecosystem Ecology: The Network Perspective*, pp. 288–351. London: Cambridge University Press.

Poisot, T., Bever, J.D., Nemri, A., Thrall, P.H., and Hochberg, M.E. (2011). A conceptual framework for the evolution of ecological specialization. *Ecology Letters*, **14**, 841–451.

Richardson, D.M., Allsopp, N., D'Antonio, C.M., Milton, S.J., and Rejmanek, M. (2000). Plant invasions—the role of mutualisms. *Biological Review*, **75**, 65–93.

Rout, M.E., and Chrzanowski, T.H. (2009). The invasive *Sorghum halepense* harbors endophytic N_2-fixing bacteria and alters soil biogeochemistry. *Plant and Soil*, **315**, 163–172.

Sachs, J.L., and Simms, E.L. (2006). Pathways to mutualism breakdown. *Trends in Ecology & Evolution*, **21**, 585–592.

Schimel, J.P., and Bennett, J. (2004). Nitrogen mineralization: challenges of a changing paradigm. *Ecology*, **85**, 591–602.

Schwartz, M., and Hoeksema, J. (1998). Specialization and resource trade: biological markets as a model of mutualisms. *Ecology*, **79**, 1029–1038.

Stal, L.J. (2000). Cyanobacterial mats and stromatolites. In B.A. Whitton and M. Potts (eds.), *The Ecology of Cyanobacteria*, pp. 61–120. The Netherlands: Kluwer Academic Publishers.

Sterner, R.W., and Elser, J.J. (2002). *Ecological Stoichiometry: The Biology of Elements from Molecules to the Biosphere*. Princeton, NJ: Princeton University Press.

Talbot, J.M., Allison, S.D., and Treseder, K.K. (2008). Decomposers in disguise: mycorrhizal fungi as regulators of soil C dynamics in ecosystems under global change. *Functional Ecology*, **22**, 955–963.

Tilman, D., Cassman, K.G., Matson, P.A., Naylor, R., and Polask, S. (2002). Agricultural sustainability and intensive production practices. *Nature*, **418**, 671–677.

Tojo, S., Yoshizawa, M., Motobayashi, T., and Watanabe, K. (2007). Effects of loosing Aigamo ducks on the growth of rice plants, weeks, and the number of arthropods in paddy fields. *Weed Biology and Management*, **7**, 38–43.

Urcelay, C., and Diaz, S. (2003). The mycorrhizal dependence of subordinates determines the effect of arbuscular mycorrhizal fungi on plant diversity. *Ecology Letters*, **6**, 388–391.

van der Heijden, M., Klironomos, J.N., Ursic, M., et al. (1998). Mycorrhizal fungal diversity determines plant biodiversity, ecosystem variability and productivity. *Nature*, **396**, 69–72.

van der Ploeg R.R., Bohm, W., and Kirkham, M.B. (1999). On the origin of the theory of mineral nutrition of plants and the law of the minimum. *Soil Science Society of America Journal*, **63**, 1055–1062.

Vitousek, P.M., Mooney, H.A., Lubchenco, J., and Melillo, J.M. (1997). Human domination of Earth's ecosystems. *Science*, **277**, 494–499.

Vitousek, P.M., Walker, L.R., Whiteaker, L.D., Mueller-Dombois, D., and Matson, P.A. (1987). Biological invasion by *Myrica faya* alters ecosystem development in Hawaii. *Science*, **238**, 802–804.

Warner, G.A., Strassmann, J.E., Ivens, A.B., Engelmoer, D.J.P., Verbruggen, E., Queller, D.C. Noë R., Johnson, N.C., Hammerstein, P., and Kiers, E.T. (2014). Evolution of microbial markets. *Proceedings of the National Academy of Sciences*, **16**, 1237–1244.

Xi, Y.-G., and Qin, P. (2009). Emergency evaluation of organic rice-duck mutualism system. *Ecological Engineering*, **35**, 1677–1683.

Zhuravlev, A.Y., and Riding, R. (2001). *The Ecology of the Cambrian Radiation. Perspectives in Paleobiology and Earth History*. New York: Columbia University Press.

SECTION 4

Introduction to Section 4

Judith L. Bronstein

It has become increasingly well recognized that not only species themselves, but species interactions as well, are under threat in a changing world. Extinction of any species should have immediate direct effects on its mutualistic partners. But anthropogenic change can disrupt mutualisms even when it doesn't lead to species loss. For example, mutualisms are fundamentally altered when native species shift to interacting with invasives, or when climate change shifts species' spatial and temporal distributions in ways that bring them into contact with new partner taxa. The potential clearly exists for effects on individual mutualisms to percolate through interaction networks, and through ecosystems as a whole.

It is easy to paint a bleak picture of the future of mutualisms in a rapidly changing world. However, these phenomena are remarkably complex, and there is a great deal that we do not yet know. What are the major drivers of change? Are the effects of change necessarily negative? It matters whose perspective one takes: for example, some invasive species are unquestionably superior partners to the mutualists with which certain taxa previously interacted. Will negative impacts, when they do exist, lead inexorably to species loss? There is surprisingly little evidence at present for coextinction in mutualisms, though it is predicted in some cases. In contrast, extensive evidence for mutualism resilience has started to accumulate. What will mutualisms look like in the future? When and how can we restore mutualisms when they are disrupted?

Anthropogenic threats to mutualism, and the possible responses to these threats, have been discussed throughout this volume. This is a striking example of how different the literature is at present from that covered in Boucher's 1985 edited volume on mutualism; it included a chapter on mutualism and agriculture, but nothing on anthropogenic threats or mutualism conservation. The final two chapters in this book address these challenging and urgent issues head on. In Chapter 13, Toby Kiers, Anthony Ives, and Atsushi Kawakita categorize the challenges that mutualisms face. Adding to their broad coverage, in Box 13.1 Diana Six delves into the effects of climate change in one iconic mutualism, the interaction between bark beetles and their symbiotic fungi. Kiers and her coauthors adopt an evolutionary perspective unusual in conservation biology. They stress that mutualisms that have persisted to the present day are not fundamentally fragile by nature; anthropogenic change is not the only type of variation that they have encountered in their evolutionary history. However, they make clear that some mutualisms are more easily disrupted than others, and will be harder to restore once they are; Box 13.2 by Eugenia Sampayo and John Paldofi highlights both the resilience and fragility of one particularly critical mutualism, the symbiosis between corals and *Symbiodinium* dinoflagellates that form the backbone of coral reef ecosystems. (See also Boxes 9.1 and 10.3, which explore other aspects of this interaction.) Rachael Winfree, Molly McLeod, Tina Harrison, and Daniel Cariveau pick up the theme of mutualism conservation and restoration in Chapter 14. They focus not on the preservation of individual species or even individual mutualisms, but rather on the twin goals of biodiversity conservation and (echoing a theme developed in Chapter 11) of ecosystem function. They conclude with specific suggestions for practical plans to manage communities structured by key mutualisms, including those that constitute our own microbiome (discussed by Brendan Bohannon in Box 14.2).

Both Chapters 13 and 14 stress the need for new tools and approaches for measuring both the consequences of and potential responses to mutualism disruption. Theoretical approaches include adoption of tools from network analysis, explored in different ways by Tony Ives in Box 13.3 and Jason Tylianakis in Box 11.1. In Box 14.1, David Kleijn presents a practical example of how intelligent agricultural management can contribute to pollinator restoration.

It may be too much to expect that a volume on mutualism twenty years hence won't need chapters on how mutualisms are faring in a rapidly changing world. However, we can at least hope that it will focus more on demonstrated successes than on looming threats.

CHAPTER 13

Global change and mutualisms

E. Toby Kiers, Anthony R. Ives, and Atsushi Kawakita

13.1 Introduction

The current biodiversity crisis is prompting calls for research enabling more accurate predictions of how organisms will respond to a rapidly changing world (Bellard et al. 2012). Human drivers such as habitat loss, exotic species introductions, and climate change are increasingly raising concerns about mutualism breakdown and the collapse of mutualistic networks worldwide. Despite the importance of mutualisms for global biodiversity, ecosystem services, and agriculture, our ability to predict their response to global change is limited; understanding these interactions remains a serious challenge and demands more study.

Here, we provide an overview of the major anthropogenic threats to mutualisms at local and global scales. We discuss the current state of the field, highlighting both the importance and difficulties in studying mutualisms. First, we define the types of breakdowns that mutualisms can experience under global change. We then identify the various drivers of these breakdowns. Although we find examples of mutualisms exhibiting resistance to global change, empirical research at the local scale is increasingly revealing their vulnerability to rapid environmental degradation. Whether these findings portend a pervasive global phenomenon is still unknown. We identify six factors that help increase the resilience of mutualisms to global change and conclude by discussing future approaches for studying the breakdown of mutualisms.

Because mutualisms can form and dissolve at both ecological and evolutionary scales, we will focus on both. On an ecological scale, partnerships undergo dramatic shifts in outcome (from mutualism to antagonism), and can even switch partners within their life times. On an evolutionary scale, selection pressures at the population level will shape partners through genetic change and adaption across generations. While much research has focused on the ecological responses of mutualisms to global change (Tylianakis et al. 2008), recent years have seen a growing interest in the long-term evolutionary responses of mutualisms (Kiers et al. 2010, Guimarães et al. 2011, Wernegreen 2012). We argue that a synergy of both vantage points is critical in (i) determining whether mutualisms are resilient enough to withstand the disturbances they face today, and (ii) being able to make reasonable predictions regarding their future responses. We suggest that a broad conservation approach is needed that looks beyond the set of species directly affected by environmental change. Identifying the common features that underlie mutualism vulnerability will be important in predicting the risks facing individual species and interactions, as well as the functioning of entire ecosystems.

13.2 Types of breakdowns

We delineate three distinct processes by which mutualisms can potentially break down under global change, from most to least severe. These are coextinction, mutualism loss, and mutualism reorganization. Throughout the chapter, we refer back to these processes.

Mutualism. Edited by Judith L. Bronstein.

13.2.1 Coextinction

Mutualism coextinction occurs when a decrease in the population of one functionally linked mutualist directly lowers the population size of its dependent partner(s), resulting in a positive feedback loop of negative consequences. This feedback makes mutualists vulnerable to disturbances and can lead to coextinction, for example as seen in specialized microbial mutualists associated with endangered animal species (Colwell et al. 2012). Theoretical analyses demonstrate the potential for cascading extinction chains of interacting mutualists (Rezende et al. 2007) and the existence of threshold levels of environmental degradation beyond which entire mutualist networks can collapse (Fortuna and Bascompte 2006, Colwell et al. 2012). However, in contrast to these theoretical predictions, direct empirical evidence for coextinction of mutualists, as well as parasite–host coextinction (e.g., Strona et al. 2013) is often lacking (see Section 13.3). One possible explanation is the difficulty of observing and identifying local coextinctions in the field. Many mutualisms, especially those involving microbes, are only just being described, or have yet to be discovered (Morse et al. 2012). Furthermore, the breakdown of mutualisms might occur slowly, so while we have not seen coextinctions, it is eminently possible that we will.

13.2.2 Mutualism loss

Coextinction scenarios assume that as one mutualist declines, its partner or partners must decline as well. However, evolutionary studies indicate that mutualists are not necessarily locked into a dance from which they cannot escape. Rather, species can drop partners and escape mutualism entirely (Sachs and Simms 2006). *Loss of mutualism* is the process by which one partner shifts to obtaining goods or services from a source other than from a mutualist, leaving its original partner(s) without a benefactor. Many mutualisms form and dissolve on a regular basis according to the short-term needs of one partner. One well-studied example from a nutritional mutualism involves access nutrients: when soil fertility rises, plants can sever their connections to mycorrhizal fungal partners because the nutrient benefits that the fungal mutualists provide become redundant with the abiotic source that does not require a costly "payment" (Johnson 1993). Mutualism loss can likewise happen on an evolutionary scale. As populations of pollinators decline, plants can evolve new modes of pollination, such as abiotic pollen vectors or self-pollination. While these patterns can be difficult to link, positive correlations have been found in some pollination mutualisms between population selfing rates and estimates of field-scale pollinator failure (Ruan et al. 2009).

13.2.3 Mutualism reorganization

Finally, in *mutualism reorganization* both partners retain biotic associations (i.e., mutualistic associations are not lost), however these associations are significantly altered: partner species may switch, or a mutualistic relationship may become less beneficial or even detrimental. Although subtle, these changes could decrease the efficacy of partnerships and/or lead to long-term, system-wide changes for both partners. For example, invasive species can replace native mutualists or degrade the species richness of a mutualistic guild, leaving partners with fewer options for future interactions. This is best documented in pollination mutualisms (e.g., Aizen et al. 2008). In other cases, species switch to entirely novel but potentially less beneficial partners, as seen in some marine sponge mutualisms (e.g., Webster et al. 2008). These reorganization processes are likely ubiquitous, if rarely studied, responses to global change, and an important question is what communities will look like after mutualism reorganization. Partner switches can, for example, spark a complex cascade of effects nearly impossible to anticipate (Colwell et al. 2012). A second unknown is whether anthropogenic factors are accelerating the process or increasing incidences of reorganization. This is where research is severely lacking.

13.3 Drivers of mutualism breakdown

The main current and anticipated anthropogenic drivers of mutualism breakdown are the same as those that drive much of the world's ecological degradation: increases in global temperatures and CO_2 concentrations, eutrophication, habitat fragmentation, overhunting, and species invasions. Here, we discuss the taxonomy of breakdown processes in relation to these specific anthropogenic selection pressures.

13.3.1 Rising temperatures

Anthropogenic climate change is argued to be among the most severe threats to the stability of complex ecological interactions. The increase in mean global surface temperature is estimated at ~0.2 °C per decade, and precipitation is expected to increase in some regions and decrease in others. Possibly even more ecologically worrying are predicted increases in both seasonal and inter-annual variation in temperature and precipitation. Changes in climate have the potential to disrupt mutualisms both by having direct negative impacts on one or both partners, or via mutualism reorganization, for example by decoupling phenologies of mutualist partners (see Chapter 14).

Mutualisms that depend on synchrony of partners' life-history events were among the first interactions to be closely studied in a climate-change framework (Colwell et al. 2012). The majority of research on decoupling of interactions has focused on pollination mutualism (see Rafferty et al. 2014). Research suggests that plant phenologies are shifting earlier as global temperatures increase (e.g., Burkle et al. 2013). It is not yet clear whether the timing of activities of pollinators and seed dispersers are shifting as well, or to the same degree. In a fifty-year data set for flowering phenology of four *Prunus* species and the cabbage white butterfly, flowers showed a trend towards appearing earlier, but the butterfly did not (Doi et al. 2008). In contrast, Bartomeus et al. (2011) found that generalist plant species and wild bee pollinators seem to be keeping pace with each other.

Within diverse communities, we would expect heterogeneity among species, with mismatch potential for some species but not for others. For example, Rafferty and Ives (2011) found that some plant species within a prairie grassland community experienced reduced pollinator visits when experimentally forced to flower earlier, while other species did not. They noted that the plant species that experienced reduced pollinator visits were also those showing no natural historic changes in phenology over the last seventy years.

One of the most iconic illustrations of a temperature-driven mutualism breakdown is the expulsion of photosynthesizing zooxanthellae partners from their coral reef hosts (coral bleaching). This can be triggered by temperatures only 1–2 °C warmer than normal. Mass bleaching in corals has been documented with increasing frequency and severity around the world (Burrows et al. 2011), with accelerated losses and forecasted extinctions of fish dependent on biogenic reef structure (Pratchett et al. 2008). The degree to which the coral mutualism can survive global climate change is under intense debate. The adaptive bleaching hypothesis proposes that coral bleaching reflects active mutualism management, with coral hosts adaptively switching to zooxanthellae partners that exhibit increased thermal stress tolerance. Reorganization in the form of partner switching, as a response to stress, is a common feature of mutualisms. Although potentially adaptive, partner switching nonetheless exposes the corals to potentially poorer partners. For example, expulsion of mutualist zooxanthellae during thermal stress events can increase susceptibility of corals to fast-growing, opportunistic symbionts (Stat et al. 2008). In marine sponges, replacement of symbiotic bacterial communities by pathogens and alien bacteria is predicted under rising ocean temperatures (Webster et al. 2008). A second problem is that loss of symbionts is often associated with reductions in calcification and lower growth in coral. This can increase susceptibility to pathogens, such as endolithic fungi that can perforate the skeleton, creating skeletal growth anomalies (McClanahan et al. 2009).

Although failing to garner the same media attention as coral bleaching, an analogous loss of mutualism may occur at high temperatures between insects and their microbial symbionts. Symbionts can help insects tolerate and adapt to changes in temperature via a combination of behavioral, physiological, and cellular responses (Wernegreen 2012). Facultative mutualists, for example the endosymbiont *Serratia symbiotica*, can facilitate population spikes in their aphid hosts by protecting them from heat stress (Russell and Moran 2006). Obligate mutualists can likewise provide temperature tolerance benefits, but present a potential danger: as insects become more dependent on their microbial mutualists, there is an increase in the likelihood of coextinction processes

(Wernegreen 2012). Obligate interactions are potentially dangerous because survival is dependent on successful adaptation of both organisms, not just one. This may be especially dangerous in cases where high dependence on symbionts drives trait loss and genome degradation of their host (Ellers et al. 2012). In such cases, hosts can suffer from associating with less adapted (but necessary) symbionts, while also having degraded genomes that provide less genetic flexibility of their own to cope with change.

Even in cases where the symbiotic partners are more evolutionarily flexible, small changes in temperature can have large effects. In some stinkbug species, vertically transmitted bacterial symbionts live in crypts of insect hindguts. In one study, a 5 °C increase in temperature completely eliminated symbionts from two stinkbug species, causing a significant reduction in host fitness (Prado et al. 2010). These experimental results complement ecological studies documenting the range changes of two Japanese stinkbug species that are shifting to more northern and to higher-elevation ecosystems (Tougou et al. 2009). While the exact mechanisms of temperature degradation on the mutualism are unknown, it is thought that gut symbionts malfunction under extremely high temperatures. In this case, mutualism loss is ultimately driving changes in current host distributions.

Some insects rely on multiple symbionts that can tolerate different ranges of temperature. In Box 13.1, Diana Six describes the association between bark beetles and their temperature-sensitive symbiotic fungi. In this system, fungal mutualists with different temperature tolerances also differ in the degree of benefits they provide (Six 2012). Such complex interactions are crucial to dissect because the identity of the dominant symbiont partner in these insect hosts can have huge ecosystem level effects.

Box 13.1 Consequences of climate change on bark beetle–fungus mutualisms in coniferous forests

Diana L. Six

Climate change is currently driving massive bark beetle (Coleoptera: Curculionidae, Scolytinae, Box 13.1 Figure 1) outbreaks in North America. The unprecedented severity of these outbreaks has led to interest in the effects of warming on beetle population dynamics. However, little attention has been paid to effects on the mutualistic fungal symbionts upon which many bark beetles depend for survival.

Conifer-infesting bark beetles are associated with an array of fungal partners (Six 2012). Some are mutualists and form consistent species-specific partnerships with beetle hosts (Six 2012). Adult beetles possess structures (mycangia) that transport mutualistic fungi from tree to tree ensuring continuity of the symbiosis across generations. The fungi provide nutrition, allowing beetles to use nutrient-poor woody tissues as food (Ayres et al. 2000, Bleiker and Six 2007).

These mutualisms often involve two fungi that differ in the degree of benefit they provide. Superior fungal partners support greater beetle survival, adult size, and reproduction than do inferior partners, and changes in the relative prevalence of each partner in a population influences beetle population dynamics. The fungi often have strikingly different upper and lower temperature limits as well as different optimal temperatures for growth, which affect their prevalence with the host over time and in a given location (Six 2012). These differences determine rates of resource capture by the fungi. The amount of area within the tree that each fungus captures, in turn determines the proportion of beetles that feed on each fungus and, thus, their subsequent fitness (Bleiker and Six 2007). Temperature also influences sporulation by the fungi and thus which fungal partner's spores are acquired in mycangia and dispersed. Cool-loving fungal partners are dispersed when it is cool; warm-loving partners are dispersed when it is warm; and both are carried by the beetle when temperatures are intermediate (Hofstetter et al. 2006, Six and Bentz 2007). When temperatures exceed tolerances for both fungi, beetles disperse without fungi with negative effects on the subsequent brood.

Temperature effects are not limited to the fungi. The length of a beetle life cycle is dependent upon thermal input. Life cycles must be timed so beetles enter winter as larvae to survive subfreezing temperatures. When temperatures are too cool or too warm, beetle life cycles become seasonally asynchronous, and the habitat becomes unsuitable (Bentz et al. 1991). This results not only in a range contraction for the beetle, but also for its dependent symbionts (Addison et al. 2013).

Box 13.1 *Continued*

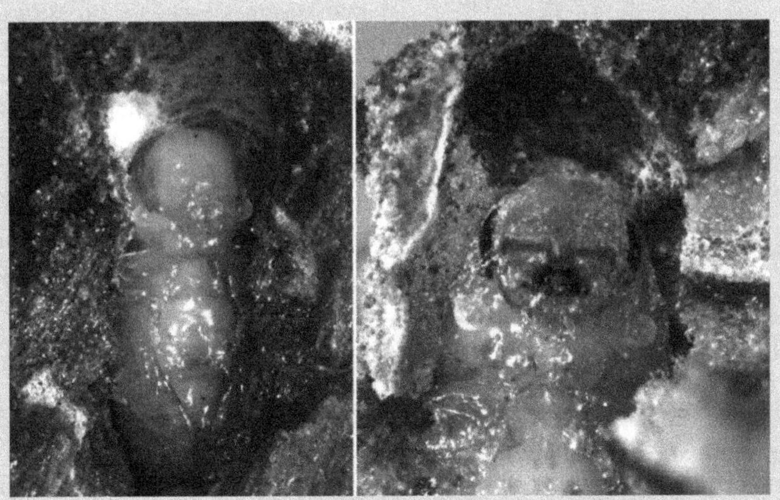

Figure 1 Western pine beetle (*Dendroctonus brevicomis*) pupae in pupal chambers lined with mutualistic fungi. Photo credit Ryan Bracewell.

How will climate change affect these temperature-sensitive mutualisms? Evidence shows that as sites get hotter, cool-loving symbionts become rare (Six and Bentz 2007). We also know that the prevalence of the fungi varies across latitudinal gradients in patterns that match their temperature tolerances (Roe et al. 2011). As temperatures increase, we can expect the extinction of cool-loving partners in many populations. The impact of the loss of a symbiont on the host will depend, in part, on whether the symbiont was the superior or the inferior partner. However, the loss of any symbiont may have serious consequences. For bark beetles, many of which live in highly variable environments, host dependence on one symbiont may not be ideal (Six and Bentz 2007). Possessing multiple partners may ensure at least one symbiont is present and functional at any given time. The loss of one may decrease ecological flexibility, restricting a beetle's range to only those conditions under which the remaining symbiont can exist. Warming may also lead to some symbioses decoupling entirely, which may lead to local extinctions of both beetles and fungi alike and, consequently, range contractions. Whether these associations possess sufficient flexibility and genetic variability to allow host switching and rapid adaptation is unknown. What can be expected is that forest ecosystems will be altered in very dramatic ways as climate change increasingly alters the dynamics and stability of these important symbioses.

References

Addison, A., Powell, J.A., Six, D.L., Moore, M., and Bentz, B. (2013). The role of temperature variability in stabilizing the mountain pine beetle-fungus mutualism. *Journal of Theoretical Biology*, **335**, 40–50.

Ayres, M.P., Wilkens, R.T., and Ruel, J.J. (2000). Nitrogen budgets of phloem-feeding bark beetles with and without symbiotic fungi. *Ecology*, **81**, 2198–2210.

Bentz, B.J., Logan, J.A., and Amman, G.D. (1991). Temperature dependent development of the mountain pine beetle (Coleoptera: Scolytidae), and simulation of its phenology. *Canadian Entomologist*, **123**, 1083–1094.

Bleiker, K., and Six, D.L. (2007). Dietary benefits of fungal associates to an eruptive herbivore: potential implications of multiple associates on host population dynamics. *Environmental Entomology*, **36**, 1384–1396.

Hofstetter, R.W., Klepzig, K.D., Moser, J.C., and Ayres, M.P. (2006). Seasonal dynamics of mites and fungi and their interactions with southern pine beetle. *Environmental Entomology*, **35**, 22–30.

Roe, A.D., James, P.M.A., Rice, A.V., Cooke, J.E.K., and Sperlin, F.A.H. (2011). Spatial community structure of mountain pine beetle fungal symbionts across a latitudinal gradient. *Microbial Ecology*, **62**, 347–360.

Six, D.L. (2012). Ecological and evolutionary determinants of bark beetle-fungus symbioses. *Insects*, **3**, 339–366.

Six, D.L, and Bentz, B.J. (2007). Temperature determines the relative abundance of symbionts in a multipartite bark beetle-fungus symbiosis. *Microbial Ecology*, **54**, 112–118.

13.3.2 Rising atmospheric CO_2 levels

Human activities have increased the concentration of CO_2 in the atmosphere from ~250 ppm to 395 ppm in the past century. Changes in atmospheric CO_2 levels are likely to drive a complex cascade of interactive effects, involving increasing temperature regimes, nutrient stoichiometry imbalances, and direct carbon fertilization effects. Linking shifts in mutualisms to specific interacting drivers will be a major challenge under rising CO_2. This is best illustrated by marine ecosystems where ocean acidification, warming, increasing disease pressure, and changes in water chemistry are interactively driving mutualism breakdown (Holcomb et al. 2010). Recent work on cnidarian symbioses has attempted to dissect these processes, testing the hypothesis that the effectiveness of host controls on symbiotic zooxanthalle proliferation rate becomes restricted in warm, nutrient-replete seawater conditions, especially under increased dissolved CO_2 concentrations in seawater (Wooldridge 2012). While the mechanisms are still unclear, there is evidence that hosts lose control over the zooxanthellae, resulting in their uncontrolled division and the disruption of photosynthate transfer to the host. This process, described by Eugenia Sampayo and John Pandolfi in Box 13.2, explores the potential of corals to acclimate and/or adapt to climate change, a topic of intense debate over the past decade.

Box 13.2 Adaptation of coral symbiosis to climate change

Eugenia M. Sampayo and John M. Pandolfi

Reef-building corals (Scleractinia) produce the hard calcium carbonate framework of tropical coral reef ecosystems. These invertebrate animals are one of the main contributors to primary production in tropical, nutrient-poor oceans. Corals are able to trap light energy through photosynthesis by living in an endosymbiotic relationship with single-celled dinoflagellates (primarily of the genus *Symbiodinium*). The symbionts, sometimes colloquially referred to as zooxanthellae, live inside

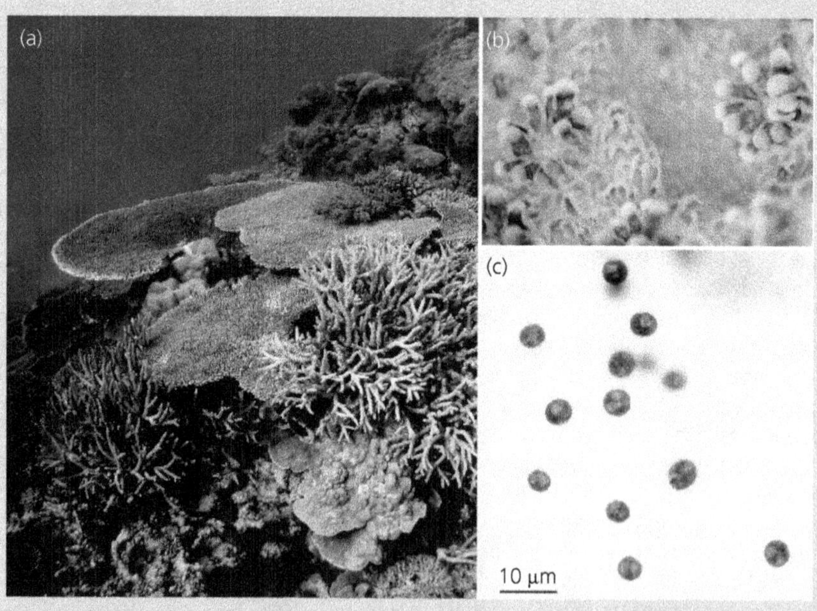

Figure 1 The brown color visible across various coral taxa (a) comes from the intracellular symbionts that are present in large densities throughout the coral polyp and the connective tissues [(b), *Stylophora pistillata*]. These single-celled endosymbionts are dinoflagellates of the genus *Symbiodinium* (c). [(a) ©George Roff; (b) ©Eugenia Sampayo; (c) ©Natalia Winkler.] See Plate 7 for color version.

Box 13.2 *Continued*

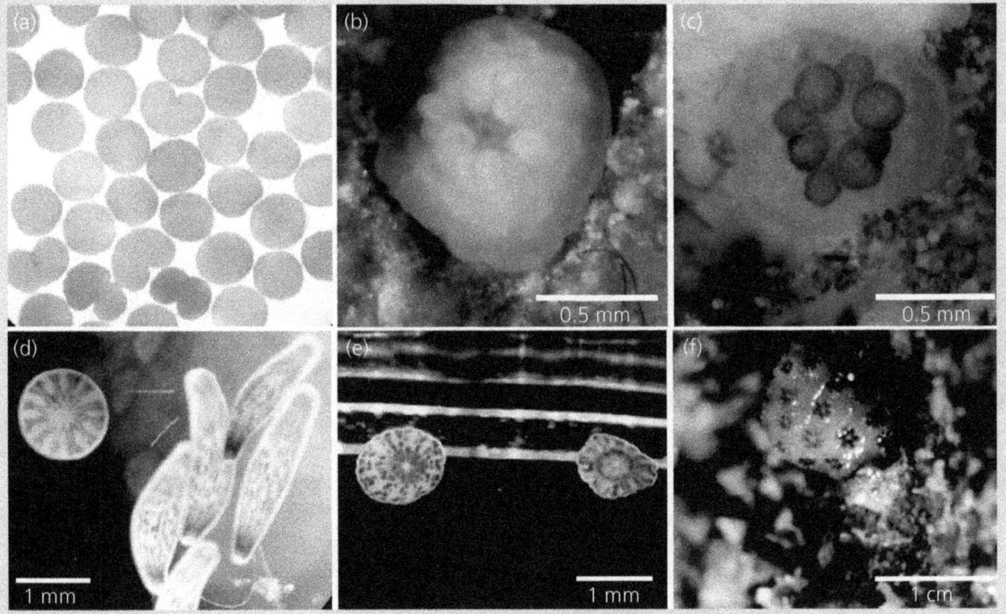

Figure 2 Symbiont acquisition occurs during the larval stages either from (a)–(c) the environment or (d)–(f) the maternal colony. The aposymbiotic egg/sperm bundles of the broadcast spawning coral *Acropora millepora* go through (a) various developmental stages before becoming swimming larvae that take up symbionts from the environment. Newly settled recruits (b) quickly develop a calcium carbonate skeleton (four-day old recruit shown) and (c) within ten days significant densities of symbionts are visible with the naked eye (starting in the tentacles). In contrast, (d) the swimming larvae released by the brooding coral *Stylophora pistillata* already contain a full complement of the maternal symbionts. Maternal transfer is linked to high levels of specificity, and likely improves settlement success to the local environment from (e) settlement into (f) adulthood (six-month-old juvenile colony). [(a), (d), (e) ©Eugenia Sampayo; (b), (c), (f) ©Natalia Winkler.] See Plate 8 for color version.

the endodermal layers of their host (Box 13.2 Figures 1 and 2). *Symbiodinium* may be the most abundant organism on reefs with cell densities of 0.5–2 million per cm^2 host tissue. The symbiosis is the fundamental basis of coral reef productivity.

Climate change has been highlighted as one of the most significant threats to the persistence of coral reefs (Hoegh-Guldberg et al. 2007, Pandolfi et al. 2011). Environmental stress leads to the breakdown of the coral symbiosis, whereby the brown *Symbiodinium* cells are lost from the host tissues, in a process called "coral bleaching." Bleaching events have been recorded throughout the world's oceans since the late 1970s, and are directly linked to climate-driven increases in sea surface temperatures. These acute events, predicted to increase in frequency, have the potential to decimate coral reefs by causing widespread direct mortality, and reduced growth rates and reproductive output in surviving individuals. This, coupled with other threats such as land-based nutrient runoff, pollution, and over-fishing, has raised widespread concern over the future of coral reefs.

Like other mutualisms, coral partnerships will survive if they can migrate, acclimatize, or adapt. However, the scientific consensus is that coral abundance and diversity will face dramatic declines in tropical regions despite these potential responses. Temperature increases are already pushing distribution ranges in pole-ward (Burrows et al. 2011). As benthic adults, movement of corals is restricted to the larval life stages. Realized migration distances are limited by local physical and biological barriers to gene flow (reducing connectivity and isolating populations along their range, Hellberg 2007). While the magnitude of climate change is not unprecedented over geological history,

continued

Box 13.2 *Continued*

current rates of change are (Hoegh-Guldberg et al. 2007, Pandolfi et al. 2011). Although recent observations that tropical coral species are appearing at higher latitudes seem promising (Yamano et al. 2011), it remains uncertain whether migrations can keep up with the exceedingly rapid pace of shifting temperature zones (Burrows et al. 2011).

The ability of corals to acclimate and/or adapt to climate change has been a topic of much debate over the past decade (Pandolfi et al. 2011). Due to the longevity and slow generation times of reef-building corals, the adaptive potential of corals will likely rely heavily on symbiont dynamics, primarily *Symbiodinium*. The genus *Symbiodinium* is highly diverse, occurs in high densities, and has short generation times. Stress-tolerant clades of *Symbiodinium* symbionts are found in coral reef systems (LaJeunesse et al. 2010). For example, specific types of "clade D" *Symbiodinium* can confer heat tolerance to their host and are frequently recorded during/after thermal stress events (Berkelmans and van Oppen 2006). These attributes make *Symbiodinium* a likely candidate to facilitate rapid acclimation toward more thermally tolerant coral populations.

Symbiodinium and other dinoflagellates may help corals acclimate, but only if they can successfully establish the symbiosis. In adult corals, maintaining a stable changeover in symbionts appears limited due to (i) high specificity and stability of established symbioses (LaJeunesse et al. 2010), (ii) reversion to original resident symbionts after stress-induced changes (Sampayo et al. 2008, Coffroth et al. 2010), and (iii) trade-offs associated with thermally tolerant symbionts (Ortiz et al. 2013). During the early life stages at the onset of symbiosis, an important window of opportunity exists for changes in established symbioses as coral recruits of many species acquire symbionts de novo from the environment (Box 13.2 Figure 2). Host–symbiont specificity is relaxed and multiple symbiont types can be take up simultaneously. Despite a wealth of evidence that specific *Symbiodinium* influence the ability of corals to thrive under certain conditions (e.g. LaJeunesse et al. 2010), we lack a detailed understanding of how distinct symbionts shape their hosts' acclimation during the early life stages, and whether generational symbiont turnover will suffice to align tolerance ranges with climatic changes (Berkelmans and van Oppen 2006, Coffroth et al. 2010). Nonetheless, the symbionts represent the most likely source of rapid adaptation, and getting a firm grasp on the evolutionary ecology of this unique symbiosis holds promise for predicting the corals' response to future climate change.

References

Berkelmans, R., and van Oppen, M.J.H. (2006). The role of zooxanthellae in the thermal tolerance of corals: a nugget of hope for coral reefs in an era of climate change. *Proceedings of the Royal Society London: Biological Sciences*, **273**, 2305–2312.

Burrows, M.T., Schoeman, D.S., Buckley, L., et al. (2011). The pace of shifting climate in marine and terrestrial ecosystems. *Science*, **334**, 652–655.

Coffroth, M.A., Poland, D.M., Petrou, E.L., Brazeau, D.A., and Holmberg, J.C. (2010). Environmental symbiont acquisition may not be the solution to warming seas for reef-building corals. *PLOS One*, **5**, e13258.

Hellberg, M.E. (2007). Footprints on water: the genetic wake of dispersal among reefs. *Coral Reefs*, **26**, 463–473.

Hoegh-Guldberg, O., Mumby, P.J., Hooten, A.J., et al. (2007). Coral reefs under rapid climate change and ocean acidification. *Science*, **318**, 1737–1742.

LaJeunesse, T.C., Pettay, D.T., Sampayo, E.M., Phongsuwan, N., Brown, B., Obura, D., Hoegh-Guldberg, O., and Fitt, W.K. (2010). Long-standing environmental conditions, geographical isolation and host-symbiont specificity influence the relative ecological dominance and genetic diversification of coral endosymbionts in the genus *Symbiodinium. Proceedings of the Royal Society London: Biological Sciences*, **277**: 2925–2934.

Ortiz, J.C., Gonzalez-Rivero, M., and Mumby, P.J. (2013). Can a thermally tolerant symbiont change the future of Caribbean coral reefs. *Global Change Biology*, **19**, 273–281.

Pandolfi, J.M., Connoly, S.R., Marshall, D.J., and Cohen, A.H. (2011). Projecting coral reef futures under global warming and ocean acidification. *Science*, **333**, 418–422.

Sampayo, E.M., Ridgway, T., Bongaerts, P., and Hoegh-Guldberg, O. (2008). Bleaching susceptibility and mortality of corals is determined by fine-scale differences in symbiont type. *Proceedings of the National Academy of Science USA*, **105**, 10444–10449.

Yamano H., Sugihara, K., and Nomura, K. (2011). Rapid poleward range expansion of tropical reef corals in response to rising sea surface temperatures. *Geophysical Research Letters*, **38**, L04601.

Decreases in the effectiveness with which hosts can "control" their symbionts may be a common theme in mutualism breakdown under anthropogenic change, especially in plant–microbial mutualisms. These symbioses are ubiquitous worldwide: 20,000 plant species are completely dependent on microbial mutualists for growth and survival, and 80% of all plant species are involved in mutualisms with mycorrhizal fungi (van der Heijden et al. 2008). In plant–microbial mutualisms, CO_2 fertilization may reduce carbon limitation in plant hosts. This means that the value of carbon for hosts will decrease, while the value of symbiont-provided nitrogen and phosphorus will increase. Theoretically, over evolutionary time scales this has the potential to result in relaxed selection against low-quality symbiont strains under elevated CO_2. There are data suggesting that strains of N_2-fixing *Rhizobium* bacteria evolving under elevated CO_2 become more competitive in initiating nitrogen-fixing relationships with their legume plants (Montealegre et al. 2000), but whether they allocate fewer nutrient resources (e.g., become low quality) for hosts is unknown. In contrast, in plant–mycorrhizal fungi mutualisms, a comparison of fungal strains isolated from eight years of ambient or elevated field CO_2 treatments found that symbiotic fungi from elevated plots improved the N nutrition of their host plants compared to those in plots treated with ambient CO_2 (Gamper et al. 2005). However, the presence of these higher-quality strains that improve N nutrition does not guarantee that they will be the dominant strains colonizing the host under field conditions. On an ecological time scale, CO_2 enrichment has been shown to generate a rapid shift in the community composition of symbiotic fungal partners inside plant roots, favoring those which extract more carbon from hosts (Drigo et al. 2010). These studies reiterate the need to understand how hosts control and discriminate among competing symbiont strains (Kiers and Denison 2008), in particular asking how changing environmental conditions affect a host's ability to interact with the highest-quality partner.

13.3.3 Habitat fragmentation

Humans have significantly altered the global landscape (Ellis 2011), removing or fragmenting over half of the world's temperate broadleaf and mixed forests, and roughly one-quarter of the tropical rainforest biome (Wade et al. 2003). This can have direct effects on mutualisms by decreasing the area and quality of habitat for forest-dwelling mutualistic partners. For example, in the last two centuries fragmentation of native oak forest by agricultural landscapes and invasive forest species in Northern Italy have led to massive loss of lichens, a mutualism between a fungus and a photosynthetic partner; the vast majority of lichens sampled during the nineteenth century are now extinct (Marini et al. 2011).

Indirect effects of fragmentation can be just as severe. Although habitat fragmentation will affect a broad array of mutualisms, pollination has been given most attention to date. Biotic pollination is critical for the majority of flowering plants worldwide (Ollerton et al. 2011). Mutualistic assemblages in fragmented habitats tend to exhibit even greater variations of non-random, nested patterns of species composition (see Chapter 11 and Valladares et al. 2012); this nestedness means that a small number of pollinators have a disproportionate role in pollination, making the entire pollination network more vulnerable if the disproportionately important pollinators' experience declines (Gates and Nason 2012). Disruption of pollination mutualisms will potentially affect plant recruitment and patterns of gene flow, and diminish the degree to which plants can respond evolutionarily to environmental change. Obligate pollination mutualisms may be especially sensitive to extinction risk. Because pollination services are also potentially important for agriculture, breakdown of this mutualism poses acute global ecological and economic threats (see Chapter 12).

Habitat fragmentation can also have direct and severe effects on seed dispersal mutualisms. These processes are particularly important in tropical forest ecosystems, where up to 90% of the plant species produce fruits that are eaten by animals (Howe and Smallwood 1982). Mutualistic seed dispersal can be crucial for plant populations because it allows for colonization of new sites and reduces the density-dependent mortality that recruits might suffer nearby their mother trees. A major question is how tropical plant–frugivore networks differ between forest interior and disturbed forest edge. Recent work in the Kakamega Forest of Kenya found that plant–frugivore networks at forest edges were more connected, more nested, and more robust against species extinctions than those in the forest interior (Menke et al. 2012). An increase in forest-visiting birds from nearby fields drove changes in network structure, leading to the prediction that forest edges are better buffered against random species extinctions compared to forest interiors (Menke et al. 2012). However, it is not yet clear if forest visitors provide seed dispersal services comparable to those of forest specialists; it is likely that forest specialists are more efficient seed dispersers (McConkey and Brockelman 2011). A second problem is if visitors disperse seeds into "sinks" of surrounding farmlands as opposed to forests, ultimately reducing the quality of dispersal. Finally, dispersal services of small-fruited species were favored in this study, raising the concern that changes in dispersal could favor small-seeded and non-native tree species (Menke et al. 2012). Thus, while seed-dispersal networks may not experience drastic cascading extinctions from fragmentation, clear shifts in composition of both plant and animal dispersers are anticipated.

Will habitat fragmentation ultimately drive coextinction of mutualistic partners? While no clear coextinction events have been documented, parallel declines in abundance of mutualist species are clearly occurring. In the Usambara Mountains of Tanzania, forest fragmentation-driven declines of frugivorous birds appear to have curtailed recruitment in an endemic tree species (Cordeiro and Howe 2003), but coextinction is not anticipated. An important line of research is to determine the relationship (if any) between the size and age of habitats and mutualism resilience.

For example, in the mutualism between plant hosts and mycorrhizal fungi, fragment size has been shown to be positively correlated with fungal abundance and diversity (Grilli et al. 2012). Similarly, in lichens, habitat area is one of the strongest predictors of lichen richness (Marini et al. 2011). However, contrary to expectations, networks of ant–plant mutualists in 25-year-old fragments were found to be similar to those in continuous forest, suggesting stronger mutualism resilience to the biotic and abiotic changes that accompany fragmentation (Passmore et al. 2012). In Box 13.3, Anthony Ives uses a population-level model of mutualist networks to explore how network structure affects mutualist community resilience. One important point from the model is that animals that experience greater benefits from mutualistic interactions with plants are protected against environmental degradation, but more generally the results highlight the difficulty of generalizing conclusions about the effects of mutualist network structure on network resilience.

13.3.4 Overhunting

Escalating bushmeat and game hunting throughout the world's tropical forests is reducing or eliminating entire guilds of larger-bodied seed-dispersing animals (Wright et al. 2007). The loss of animals can alter seed dispersal patterns and reduce seedling recruitment of tree species, resulting in community-wide shifts in tree recruitment that favor smaller-seeded species dispersed by the remaining small mammal and bird community (Terborgh et al. 2008). How such changes in recruitment will influence long-term forest community structure remains unknown. However, it is likely that the extirpation of entire guilds of large-bodied dispersers will influence the densities of plant species whose fruits and seeds are principally consumed and dispersed by overhunted animals.

A critical issue is the extent to which market trade in wildlife—estimated at $8–10 billion per year, second only to the illegal traffic of drugs and arms—has cascading indirect effects on non-hunted organisms. Hunting can reduce both the quantity of seeds removed and the distances over which they are transported. As a result, huge caches of undispersed seeds at parental trees can become vulnerable to attack by various pests (Terborgh et al. 2008). Manipulative studies are becoming an important tool to assess how spatial distributions of seeds drive seedling recruitment. A recent study mimicked dispersal patterns based on the predicted extirpation of monkeys in an Afrotropical forest and found that the "no dispersal treatments" decreased seedling survival by 78% relative to natural dispersal treatments (Poulsen et al. 2012).

Box 13.3 A population-level model of mutualist network resilience

Anthony R. Ives

To illustrate how network structure affects mutualist community resilience, a simple population-level model of mutualist networks is instructive. Many models have addressed the resilience of mutualism networks and the possibility of cascading secondary extinctions, as mutualists lose their partners and consequently go extinct themselves (see Section 11.1). In contrast to the majority of such models, our model explicitly incorporates the abundances of all mutualist partners, rather than simply the presence or absence of interactions between mutualist guilds. Furthermore, even though some previous models have included the possible non-random loss of species from the mutualist network, such as the loss of rare species, they have not addressed how the mutualistic interactions themselves may make species more or less prone to extinction. Mutualisms provide numerous benefits to interacting partners in the face of environmental changes (see Section 13.4). How do the benefits accrued from mutualisms weigh against the risks of cascading secondary extinctions in determining the overall resilience of mutualist networks to environmental changes?

Model

Although the model is not specific for a particular type of mutualism, we can call one set of organisms animals and their mutualistic partners plants, as would be the case for pollination or seed-dispersal mutualisms. Suppose there are $n = 15$ plant and $n = 15$ animal species engaged in mutualistic interactions. The dynamics of both plants and animals are governed by Lotka–Volterra-like discrete-time equations,

$$x_i(t+1) = \begin{cases} x_i(t)\exp[r - x_i(t)] & \text{if } \sum_{k=1}^{n} p_{ik} y_k(t) > m \\ 0 & \text{otherwise} \end{cases} \quad (1)$$

$$y_i(t+1) = y_i(t)\exp\left[c_i(t) - y_i(t) + \sum_{k=1}^{n} p_{ki} x_k(t) - \sum_{j=1, j\neq i}^{n} \beta_{ij} y_j(t)\right] \quad (2)$$

where $x_i(t)$ and $y_i(t)$ are the densities of plant i and animal i at time t. The per capita population growth rate of plant i depends on r, the intrinsic rate of increase (assumed to be the same for all plant species), and decreases with increasing density of conspecifics, $x_i(t)$. Plant populations can only survive if they receive a minimum m of mutualistic benefit (e.g., pollination services) from the animals. The strength of interaction between plant i and animal j is given by p_{ij}, which might represent, for example, the per capita visitation rate of pollinators of species j on plants of species i. The sum $\sum_{k=1}^{n} p_{ik} y_k(t)$ is the combined benefit to plant i from all animals; if this sum is not greater than m, the population of plant i goes extinct.

The per capita population growth rate of animal i depends on $c_i(t)$, its intrinsic rate of increase, which may differ among animal species and may depend on time t. The per capita population growth rate declines with increasing intraspecific density, $y_i(t)$, and increases with the net benefit provided by mutualist plants, $\sum_{k=1}^{n} p_{ki} x_k(t)$; here, the benefit provided to animal i from plant k is proportional to the benefit provided to plant k from animal i, so the benefits of mutualism are symmetrical. Competition among animals for plant mutualists is given by $\sum_{j=1, j\neq i}^{n} \beta_{ij} y_j(t)$. Here, the per capita competitive effect of animal j on animal i, β_{ij}, is proportional to their shared use of plants and the abundance of plants,

$$\beta_{ij} = \sum_{k=1}^{n} p_{ki} p_{kj} x_k(t). \quad (3)$$

This formulation of competition coefficients β is equivalent to classical models of resource overlap in competitive models.

To generate a network of animal and plant mutualists with a nested structure, we started with strictly nested interactions and then added variability to interaction strengths by multiplying by a random variable (Box 13.3 Figure 1a). For a network with a non-nested structure, the interaction strengths from the nested model were randomized among animals and plants (Box 13.3 Figure 1b). To explore the consequences of the mutualistic benefits accrued by partners on network resiliency, animals were removed from the network under two different scenarios. For the baseline case in which animals are lost regardless of their possible mutualistic benefits, every fifty time steps one animal species was removed independently of its abundance or connectedness. For the case in which

continued

Box 13.3 *Continued*

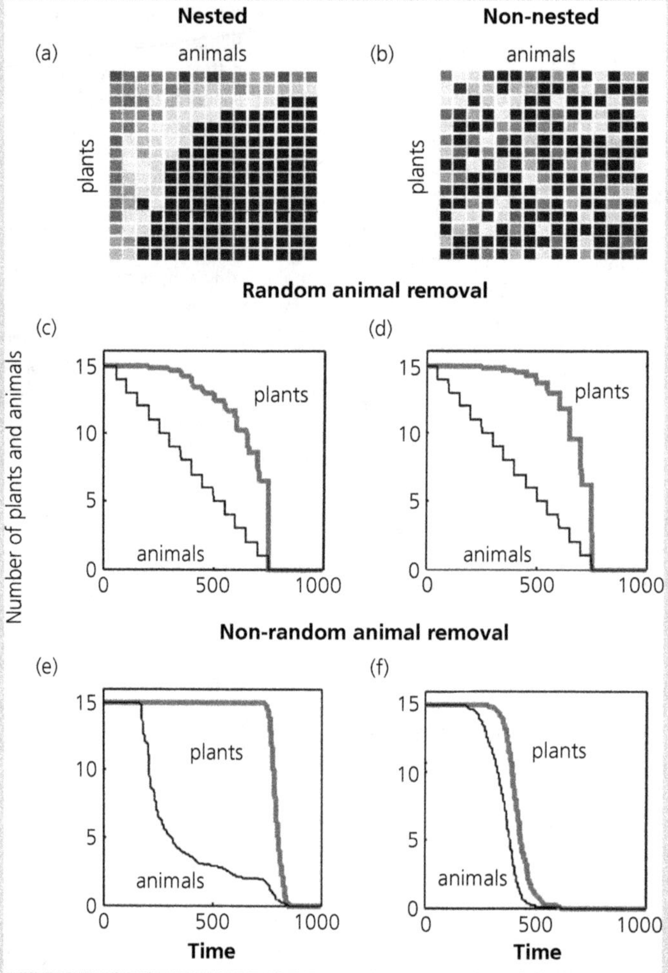

Figure 1 Collapse of nested and non-nested mutualist networks, illustrated by single model realizations (a) and (b), in which stronger interactions are denoted by lighter shades of gray. (c) and (d) summarize the collapse of 100 simulated pairs of nested and non-nested networks when animals were removed randomly at intervals of 50 time steps. (e) and (f) give the average across 100 simulations of the numbers of animal and plant species when animals were removed due to decreases in their intrinsic rates of increase when their densities drop to low values during environmental degradation. The parameter values used for simulating networks were $r = 0.5$ and $m = 0.01$; for random animal extinctions, $c_i(t) = 0.1$, and for non-random animal extinctions, $c_i(t) = 0.1 - 0.002t$.

mutualistic benefits decrease the chances of species loss by increasing their abundance, it is necessary to model the environmental change in terms of how it affects abundances. We assumed that the intrinsic rates of increase of all animal species, $c_i(t)$, decreased at the same linear rate through time, leading to reduced populations and eventual extinction. This scenario corresponds to the case in which all animals are equally sensitive in terms of their intrinsic rates of increase to the same environmental degradation. This does not exclude the possibility that animals may, through different strengths of interactions with plants, differ in the ultimate effect of environmental degradation

Box 13.3 *Continued*

on their extinction proneness. Although previous models have considered the primary extinction of rare species (e.g., Memmott et al. 2007), in our case, species that have strong mutualistic interactions consequently have higher abundances and are less prone to extinction. Thus, the model includes not only the risks of cascading secondary extinctions, but also the benefits of mutualism in reducing the risks of primary extinctions.

In the model, the effect of the nestedness structure of the network depended on the manner in which animal species experienced primary extinction. For the baseline case when animals were removed randomly (Box 13.3 Figure 1c, d), the average pattern of secondary plant extinctions was similar for nested and non-nested networks. In contrast, when animal species were lost through environmental degradation, the patterns of both primary extinctions of animals and secondary extinctions of plants depended strongly on network nestedness (Box 13.3 Figure 1e, f). For the nested network, the initial loss of animals occurred early and suddenly, as animals with few mutualistic interactions went extinct first. Nonetheless, the animals that had many mutualistic plant species were relatively protected from environmental degradation and persisted for a long time. In contrast, the onset of animal extinctions for the non-nested networks was more gradual, but in the absence of highly connected animal species that obtain high benefits from mutualists, the entire animal community went extinct much sooner than in the nested network. The extinctions of plants reflected these patterns; in particular, the plant species in the nested community all persisted until almost all animal species had been lost.

These results highlight the difficulty of generalizing conclusions about the effects of mutualist network structure on network resilience (Winfree et al. 2014 and references therein). Nestedness may (Box 13.3 Figure e, f) or may not (Box 13.3 Figure 1c, d) have a large influence depending on the other basic assumptions about mutualisms and the drivers of primary extinctions. Much of the literature on mutualistic networks (Chapter 11) focuses on how mutualisms affect secondary extinctions; nonetheless, in our simple model animals that experienced greater benefits from mutualistic interactions with plants were protected against environmental degradation. This shows that mutualistic outcomes may also play a role in setting the vulnerability of mutualists to primary extinctions.

References

Memmott, J., Craze, P.G., Waser, N.M., and Price, M.V. (2007). Global warming and the disruption of plant-pollinator interactions. *Ecology Letters*, **10**, 710–717.

Winfree, R., Dushoff, J., Williams, N., and Kremen, C. (2014). Species abundance, not diet breadth, drives the persistence of the most linked pollinators as plant-pollinator networks disassemble. *American Naturalist*, **183**, 600–611.

One of the greatest difficulties in studying mutualism disruption in long-lived species is the potential lag time in effects when one partner is lost. The vulnerability of long-lived trees is often concealed by the current abundance of established individuals. Brodie et al. (2009) show that in the national parks of Thailand experiencing severe hunting, long-term population growth and persistence of the study species *Choerospondias axillaris* has been reduced, but only slightly. The authors argue that the overhunting of tropical frugivores is creating an "extinction debt" in which long-lived trees remain extant (and even common). As they fail to reproduce, however, they risk extinction in the long term. This finding has been confirmed on an island ecosystem in which, even after the loss of its only seed disperser, populations of the shrub *Daphne rodriguezii* are remaining stable (Rodriguez-Perez and Traveset 2012). However, the authors argue that where dispersal is limiting, environmental changes (e.g., weather patterns, habitat quality, etc.) will impose even greater potential disruptions.

Rodent communities can sometimes provide effective long-distance seed dispersal services (Jansen et al. 2012). In a Neotropical forest, seeds cached by agoutis were initially found dispersed only short distances from the trees. However, further analysis found that agoutis would recover the seeds, move, and recache them up to 36 times. Individuals were found to repeatedly steal and recache each other's buried seeds, which ultimately improved seed dispersal services (Jansen et al. 2012). This gives hope to the idea that some taxa (e.g., rodents) may be better dispersers than previously assumed. Using a meta-analysis approach, Aslan et al. (2012) studied the effectiveness of 34 non-native vertebrate pollination and seed-dispersal mutualists in 20 geographic locations. They found that non-native vertebrate pollinators were generally less effective than native mutualists. However, when native mutualists had been extirpated, non-native mutualists helped increase seed set and seedling performance. While there may be different levels of effectiveness between native and non-native mutualists, reorganizational processes facilitate reliance on non-native vertebrates as natives become extinct, and these can help enhance some plant fitness measures.

13.3.5 Eutrophication

One of the most drastic changes to the world's ecosystems is the alteration of nutrient cycles. Acceleration of nutrient cycles is remarkably variable across the globe, with some regions experiencing little, and other regions such as western Europe, the northeastern US and east Asia witnessing ten- to fifteenfold increases in nutrient flows (Howarth 2008). Nutrient enrichment is transforming mutualistic interactions via three major routes: (i) by causing direct, pollution-type effects on mutualistic partners, (ii) by driving shifts in competitive dynamics so species outside the mutualism become dominant, and (iii) by shifting dynamics within the mutualism itself, particularly in cases where nutrient transfer is the main commodity of trade.

Nitrogen pollution has direct negative effects on a wide range of species, and consistent relationships between increases in nitrogen and reduction in species diversity have been found (e.g., Stevens et al. 2004). Surveys of lichen vegetation in highly nitrogen-polluted areas suggest that many lichens decline at high nitrogen levels because of direct physiological damage from nitrogen pollution (e.g., Hauck 2011). Some organisms, such as zooxanthellae in corals, initially respond well to nitrogen enrichment, and this will actually stimulate their growth. However, stimulated growth causes them to quickly deplete their other resources, resulting in phosphorus starvation that can ultimately be fatal for the mutualism (Wiedenmann et al. 2013).

Second, eutrophication can drive shifts in competitive dynamics, with nutrients favoring the growth of species outside the mutualism to the detriment of the mutualistic partners. Studies of periodic eruptions of volcanic ash into coral systems illustrate how whole communities can be reorganized from nutrient input of reduced compounds. Nutrients such as iron have been shown to cause localized outbreaks of the coral-killing cyanobacteriosponge, resulting in aggressive overgrowth of coral colonies (Schils 2012). Host resistance to pathogens may decrease under nutrient enrichment, causing shifts in competing populations of symbionts and pathogens. Field experiments have demonstrated how nutrient enrichment can increase the severity of pathogen virulence; experimentally increasing nutrient concentrations by two- to fivefold resulted in a near doubling of host tissue loss to yellow band disease in some reef-building corals (Bruno et al. 2003).

Lastly, nutrient enrichment can eliminate the nutrient limitation that makes most nutritional mutualists beneficial. By changing nutrient stoichiometry, a once-beneficial relationship for both partners may become less beneficial or even detrimental. This depends on shifting cost:benefit ratios (Chapter 1). Ultimately, this may drive the loss of mutualisms (Kiers and Denison 2008). In marine systems, planktonic diatoms and dinoflagellates adopt nitrogen fixers as partners to obtain organic nitrogen used for vitamins and nitrogen-rich defensive chemistry. However, when oceanic nitrogen is abundant, phytoplankton abandon bacterial partners on an ecological time scale, suggesting bacterial partners require high maintenance costs (Hay et al. 2004). In plant–mycorrhizal mutualisms, nutrient enrichment can cause the loss of mutualism on both ecological and evolutionary time scales. Hosts may avoid partnering with mycorrhizal fungi if soil nutrient conditions are already high. This is a flexible response that can be altered within the lifetime of the host. Evolutionary loss of the mycorrhizal condition has also been documented, occurring repeatedly in independent lineages, most notably in species colonizing in nutrient-rich environments (Wang and Qiu 2006). We expect more plant lineages to adopt this non-mycorrhizal strategy under a future of anticipated nutrient enrichment (Ellers et al. 2012).

Subtle reorganization of mycorrhizal communities has also been documented under rising nutrient regimes. Long-term monitoring of nutrient-enriched agricultural sites has documented changes in mycorrhizal populations towards more parasitic strains (Egerton-Warburton et al. 2001). Concurrent decreases in mycorrhizal species richness and diversity have also been noted, perhaps signifying a loss of mutualist richness due to abandonment processes. Studies of long-term fertilization of ectomycorrhizal forest communities have found a similar decrease in fungal taxa (Deslippe et al. 2011). This is worrying because forest fertilization is increasingly being used in North America and Europe as a means to increase timber harvest. However, researchers have argued that ectomycorrhizal communities are hyperdiverse, and that there is appreciable overlap in physiological processes. Therefore, shifts in species composition with fertilization are not expected to dramatically reduce the functional ability of the ectomycorrhizal fungal community to degrade soil organic matter (Jones et al. 2012). This raises the issue of the role of functional redundancy in mutualisms, and demands that conservationists examine the critical question of whether it is the mutualistic services or the mutualistic species we aim to preserve (see Chapter 14).

13.3.6 Invasive species

Exotic species invasion is now considered to be a leading factor in mutualism disruptions (Bascompte and Jordano 2007). Introduced species have been shown to alter closely interlinked ecological relationships (Inderjit and van der Putten 2010). Mutualisms and introduced species can interact in at least three ways: (i) invasive species can degrade native mutualisms, (ii) native mutualisms can facilitate invasion of exotics, and (iii) introduced species may fail to establish and invade if they lack mutualistic partners. We will discuss these in turn.

The introduction of a non-native species often reduces the densities of native mutualist communities. Entitled the "degraded mutualism hypothesis," this can cause a disruption of a native partnership and a reorganization of the mutualist guild (Cipollini et al. 2012). One of the best-studied invasives, the Argentine ant *Linepithema humile*, alters mutualisms worldwide. In the diverse shrublands in South Africa, these ants depress the abundance of native ants that effectively disperse large-seeded plants, inducing a community-wide shift in plant recruitment favoring smaller-seeded species and potentially triggering the onset of linked community extinctions (Christian 2001). In this case, the native mutualistic partner is out-competed and its services are replaced by an invasive species. The ability of invasive *L. humile* to respond to alarm signals of native ant species has facilitated its invasion (Mondor and Addicott 2007). For critically endangered plants, such as the endemic *Roussea simplex* of Mauritius, disruption of pollination and seed-dispersal services by invasive *Technomyrmex* ants raises the specter of host extinction (Hansen and Muller 2009). Less studied, but potentially as severe, are the impacts of the Asian needle ant *Pachycondyla chinensis* which disrupts ant–seed dispersal mutualisms by displacing native ant species; native ants suffered a 96% reduction in invaded areas (Rodriguez-Cabal et al. 2012).

Disruption can also be driven by an invasive species outside the mutualism. In North American forests, the loss of massive stands of hemlocks to the invasive woolly adelgid (*Adelges tsugae*) has led to a reduction in species richness of mutualistic ectomycorrhizal fungal symbiont communities, potentially stunting growth of other species, such as oaks that also depend on this mutualistic guild (Lewis et al. 2008). In these cases, loss of mutualistic diversity from an invasive species is decreasing partner performance. In the Ogasawara archipelago in Japan, the alien predatory lizard *Anolis carolinensis* has disrupted the endemic pollination system by depleting island insect fauna, including native pollinators (Abe et al. 2011). Introduced honeybees now dominate flower visitation, and whereas endemic pollinators preferred native flowers, honeybees favor alien plants, driving an increase in alien fruit set (Abe et al. 2011). Such positive feedbacks among invasive species in a single community can drive massive cascade effects (Grosholz 2005). In some cases these systems appear to operate normally (i.e., there are still plants and pollinators on the island). However, the functioning of these invasive–invasive mutualisms remains an open line of research. An outstanding question is how vulnerable mutualisms are to breakdown when the number of inhabitants lacking an evolutionary history together increases.

The second outcome is known as an "enhanced mutualism response" in which native mutualisms facilitate the success of invading species (Reinhart and Callaway 2006). Although this may not be a general rule, some invasives may offer potentially better services as they sometimes provide high nectar rewards whose nutritional value attract native pollinators (Lopezaraiza-Mikel et al. 2007). Similarly, fleshy-fruited invasive plant species are often supported by populations of local birds (Merow et al. 2011). Aslan (2011) documented the emergence of novel mutualisms between three potentially invasive trees in California and local bird populations—a prime example of a bird-mediated invasion. Likewise, but belowground, native mycorrhizal fungi have been shown to facilitate the spread of non-native plants from low to super-high abundance because an invading plant species benefits more from the symbioses than the indigenous plants. *Centaurea maculosa* is a noxious perennial plant that appears to prosper from native mycorrhizal fungi, facilitating its spread throughout much of the native prairie in the northwestern USA (Harner et al. 2010). Similarly, the competitive ability of invasive *C. stoebe* has been shown to increase when in association with particular native fungal endophytes (Aschehoug et al. 2012).

Plants in the family Leguminosae are often cited as the classic example of profiting from an "enhanced mutualism response"; partnering with nitrogen-fixing bacterial mutualists adopted from a new range allows them to rank among the most notorious global invaders (Parker et al. 2007). It is thought that a low degree of symbiotic specificity allows legumes to establish novel symbiotic relationships when introduced into new geographical ranges. However, research is challenging the assumption that legumes are invasive because of their ability to partner with native mutualists. Molecular studies have revealed that the rhizobial communities of invasive legumes are low in diversity and composed primarily of exotic symbionts, likely introduced with the plants themselves (Wei et al. 2009, Rodriguez-Echeverria 2010). Research on these processes in a Mediterranean coastal dune ecosystem found that the invasive legumes did not rely on novel mutualisms with native mutualists, but rather their success hinged on the cointroduction of compatible symbionts during the invasion process. However, these exotic rhizobia are able to colonize native legumes in invaded areas even though they are not effective partners for the natives. Ultimately, these introduced symbionts can reduce native legume fitness by disrupting the native mutualism (Rodriguez-Echeverria et al. 2012). Future research should aim to better elucidate the frequency with which invasive plants carry their own mutualists to new habitats (Uchitel et al. 2011) and investigate the level of host specificity of key mutualists so that mutualistic interactions between introduced species and native mutualists can be better predicted (Birnbaum et al. 2012).

A third issue involving invasions is that an invasive species is unable to establish because the new habitat lacks the appropriate partner. In these cases, the *lack* of a specific mutualist can be a critical factor in slowing invasion of fragile habitats. For aboveground mutualisms, dependence on pollinators may be an important constraint limiting range expansion of invasives. Invasions can fail until the moment non-native pollinators are introduced. Belowground mutualisms with microbial symbionts can show similar patterns. A highly cited example is that of exotic invasion of Pinaceae being limited by availability of ectomycorrhizal fungal inoculum in native soils. Near pine plantations where fungal inoculum has been introduced, invasives can establish, but as distance from the plantations increases, low availability of symbionts has slowed invasions (Nunez et al. 2009). This and other work (e.g., Inderjit and van der Putten 2010) highlight the critical but often overlooked role of microbes in mediating ecosystem invasibility.

13.4 Mutualism resiliency

The current speed and the magnitude of global change are unprecedented in the evolutionary history of many mutualisms. But many mutualisms are not collapsing. What are the characteristics that allow some partnerships to be resilient to this rapid anthropogenic change? Our ability to predict changes and conserve mutualisms will be significantly increased if we better understand the traits mutualisms exhibit under change and identify key attributes of resilient partnerships. Here, we discuss the major characteristics that lend resilience to mutualisms under anthropogenic change. We discuss each of these in turn, citing recent examples to illustrate how these traits help mutualisms cope with changing biotic and abiotic conditions.

13.4.1 Lack of strict dependence on specific partners

Facultative mutualisms potentially provide greater flexibility to rapidly changing environments compared to obligate ones. In obligate mutualisms, there is a decreased chance to escape and survive lapses in services. This can prevent organisms from acclimating to new environments. One of the most useful model systems for studying the costs and benefits of strict dependency is endosymbiotic mutualisms in insect hosts because the interactions can range from facultative to obligate. In these relationships, the endosymbiont is thought to benefit from reduced exposure to predators and competition, a consistent supply of nutrients, and greater environmental stability inside the host. The hosts can benefit nutritionally or physically, for example the endosymbiont may help the host withstand higher temperatures (Wernegreen 2012). It has been suggested that in the future we will see a shift towards facultative symbionts replacing obligate symbionts, as these potentially provide insect hosts with a larger repertoire of benefits to deal with extreme temperatures (Wernegreen 2012). Wholesale replacement of obligate endosymbionts in some weevils has been documented (Conord et al. 2008), and in Cedar aphids integration of the facultative *Serratia symbiotica* endosymbiont with residing obligate partners has been noted (Lamelas et al. 2011). In some cases, facultative symbionts even protect the obligate symbionts themselves from stress. The facultative *S. symbiotica* has been shown to protect both its aphid host and its obligate symbiont *Buchnera* from effects of heat (Burke et al. 2010).

The ability to pair with multiple partners also provides a sort of insurance against rapid change and cryptic symbiotic partners in low abundances can play an important role in host survival. While their role is not entirely clear, such partners may represent a class of symbionts that are utilized only under extreme conditions. In some reef-building corals, hosts can shift the composition of their algal symbionts in response to changing environmental conditions. Previously, it was thought that this adaptive capacity was rare, with only 25% of coral species associating with multiple algal clades. However, recent deeper genetic sequencing revealed the presence of additional symbionts at very low abundances (Silverstein et al. 2012). Such low-frequency partnerships provide a potential "reservoir" of benefits to hosts.

13.4.2 Rapid evolution and/or plasticity

If one or more partners can respond rapidly to environmental change, this capacity can potentially rescue the mutualistic interaction. Rapid evolution of one partner, or set of partners, can increase mutualism resilience by allowing species to adapt to changing environmental conditions (Irwin et al. 2014). This is best illustrated by a recent experimental evolution study that tested for adaptive plant responses to predicted anthropogenic drought stress (Lau and Lennon 2012). The authors demonstrated a low evolutionary plant response to experimentally imposed drought. However, they found that a rapidly adapting soil microbial community helped compensate for this lack of adaptation. Associating with a "drought-adapted" microbial community, which was generated after only three plant generations of selection, increased host fitness, even in naïve plant populations not previously exposed to the drought pressure (Lau and Lennon 2012).

While there have been some examples in the literature suggesting that rapid evolution is facilitating species persistence under anthropogenic changes (Bradshaw and Holzapfel 2006), others argue that genetic evidence for climate-change-driven evolution is still scarce (e.g., Merila 2012). New molecular approaches are aiming to identify specific, rapidly evolving gene networks and proteins that allow mutualistic partners to cope with changing environments (Voolstra et al. 2011). The ultimate goal of this gene-centric approach is to identify the rates, patterns, and mechanisms facilitating rapid partner adaptation.

One potential problem is that mismatches in evolutionary rates can limit the potential for both partners to survive environmental changes. Evolutionary traps can also occur in which formerly adaptive behaviors (or partnerships) become maladaptive when cues become decoupled from fitness consequences (Fletcher et al. 2012). Theoretical models (*sensu* Louca et al. 2012) are useful tools to explore the potential for the emergence of evolutionary traps between specific mutualistic systems.

13.4.3 Broad or novel niches

Under global change, any adaptation that increases a partner's ability to exploit new niches or broaden a partner's tolerance to environmental conditions is likely to be favored on an evolutionary time scale. Niches may broaden by developing new partnerships or even losing old ones.

Mutualism reorganization, for example developing new partnerships or moving to more specialized (or more generalized) relationships, has proved a successful strategy in the race to colonize new habitats. Among the most well-studied cases are host plants that associate with fungal endophytes that confer heat and drought tolerance in increasingly high-stress environments (e.g., Rodriguez et al. 2008). The co-opting of native mutualists by invasive species provides numerous examples of mechanisms by which species extend their host range (Inderjit and van der Putten 2010). Symbionts may likewise expand their ranges by colonizing novel hosts. The ectomycorrhizal death cap fungus (*Amanita phalloides*) expanded its range from its native Europe to North America by broadening its host range to include pines and oaks (Wolfe and Pringle 2012). Likewise, the acquisition of bacterial gut symbionts is thought to have driven, or at least facilitated, the evolution of herbivory in ants by opening up a novel feeding niche (Russell et al. 2009).

In contrast, partners may lose a mutualistic association and this can potentially allow for broader niche exploitation. For example, in ant–plant mutualisms, microenvironments can shift the outcomes of partnerships: populations experiencing reduced benefits may ultimately abandon the partnership to increase their success (Moraes and Vasconcelos 2009). Cold climates have been shown to filter out myrmecophily (mutualisms between ants and butterflies) because it is potentially more difficult to find an ant partner in a resource-limited environment (Pellissier et al. 2012).

13.4.4 Strict control over partners

Many mutualisms are characterized by asymmetrical dominance in which hosts maintain strict control over their symbionts. This asymmetry can help stabilize cooperation and prevent less-effective partners from co-opting the relationship. There is accumulating evidence that under anthropogenic change, hosts may have less control over the growth and fitness of their symbionts, potentially destabilizing the partnership leading to mutualism reorganization. In cnidarian symbioses, the effectiveness of host controls on symbiotic zooxanthalle proliferation rates decreases in warm, nutrient-replete seawater conditions (Wooldridge 2012). Humans may also inadvertently select for a decrease in host control in agricultural settings by breeding crops that fail to optimize resource exchange between partners. There is evidence that plant breeding practices have selected against beneficial mycorrhizal associations, sometimes leading to mutualism loss (Xing et al. 2012) and decades of breeding in fertilized soils have produced soybean varieties that fail to prevent parasitism by rhizobial symbionts (Kiers et al. 2007). Over the long term, these artificially selected crops may facilitate the spread of less-effective mutualists (i.e., root symbionts that provide fewer benefits) into agricultural and natural habitats. For mutualisms to be resilient, associating with high-quality partners, not so-called cheaters (see Chapter 6), is critical.

13.4.5 Protection from environmental variation

Mutualistic partners can be found in environments that are both well buffered against global changes and environments that are exposed to dramatic oscillations in climate; how does this level of environmental variation affect their survival? Recent work suggests that in aquatic systems, the magnitude and pace of temperature changes can be as great as, and even exceed, those seen in terrestrial habitats; land-locked seas are warming at rates that are two to four times the global mean rate (Merila 2012), but there is a lack of research exploring how this rapid increase affects mutualisms. There is also a question of how heterogeneity in warming rates results in a larger exposure to extreme selection pressures. In contrast, some mutualisms are buffered against global change by their habitat. Partners in some mutualisms engineer optimal conditions; for example, underground ant–aphid cultivation mutualisms (Ivens et al. 2012) will experience less variation in environmental extremes.

13.4.6 Flexibility in the face of environmental variation

How flexible are partners in mutualisms? Flexibility can arise, for example, when a single partner offers multiple services. Mycorrhizal fungal symbionts are primarily known for their ability to transfer nutrients, such as phosphorus and nitrogen, to their host plants. However, the last two decades have revealed the various other services mycorrhizal fungi can confer to their hosts, such as pathogen protection, protection from drought, and decreases in herbivore pressures (Sikes et al. 2010). This type of diversity in function suggests that—even if a single service such as nutrient provisioning becomes less important in a nutrient-enriched rhizosphere—the mutualism may still be maintained if other benefits are provided. Flexibility can also come in the form of tolerance to poor partnerships or environmental conditions. Tolerance is particularly important when long-lived hosts, such as coral and trees, associate with short-lived partners like zooxanthellae and other microbes. If hosts are able to recover from poor partnerships (e.g., Edwards 2009) in the short term, this can translate into larger benefits over the host's lifetime as new ecological contexts arise (Palmer et al. 2010).

13.5 Mutualism breakdown in a multispecies context

Much of our current understanding of mutualistic interactions and their breakdowns focuses on pairs of interacting species. In ecological communities, however, most mutualisms involve large networks of species, posing challenges for disentangling direct and indirect pathways of interactions. Furthermore, a network perspective introduces the question of how to assess the resilience of mutualism: in terms of the persistence of individual species, the persistence of specific mutualistic relationships, persistence of the number of species, or persistence of the overall structure of the mutualist network.

In a network context, mutualistic interactions as a whole may be maintained even as certain species are extirpated. In other cases, the entire mutualist network may break down (see Chapter 11). Massive hemlock tree die-offs, triggered by an invasive insect, are being shown to degrade the abundance and richness of ectomycorrhizal fungal communities. Failure of replacement tree species, also dependent on this mutualism, to recolonize the forest has been hypothesized to significantly slow forest recovery (Lewis et al. 2008). Similarly, mass disturbances (bleaching, disease, overfishing) are causing reef systems to become overgrown by algae (Hoegh-Guldberg et al. 2007). Contrary to expectations, not all fish populations that feed upon and suppress algal growth are increasing in response to increased algal resources (Wilson et al. 2006), and this is further impeding coral recolonization. The generality of these community-breakdown phenomena, as well as their short- and long-term consequences, remain to be explored.

Can any predictions on community breakdown be made based on what we know about mutualism network structure (Chapters 11 and 14)? For example, plant–animal mutualist networks are often structured, with non-random patterns of interactions among individual members. These interactions are often nested, with specialists interacting with generalists and generalists interacting with each other (Bascompte and Jordano 2007). Given seemingly common structures among mutualist networks, we can ask how these structures may make these networks more or less vulnerable to total collapse. As explored in Box 13.3, we see that when animal species, such as pollinators, are lost through environmental degradation, the patterns of both primary extinctions of animals and secondary extinctions of plants depend strongly on network nestedness.

Numerous types of models have been derived to explore different features of mutualist networks, and different assumptions about how environmental changes may disrupt them. Two broad types of models can be distinguished: those that model networks in terms of the presence/absence of species and of mutualistic interactions between them, and those that explicitly model the densities of the interacting species. The former type of model generally includes the assumption that a species will persist as long as it has at least one mutualistic partner remaining. Therefore, if species are removed from the network at random, for example, it is possible to observe how network structure affects the temporal pattern of secondary extinctions (e.g., Sole and Montoya 2001). An advantage of this type of model is that it can be mapped onto existing datasets that give the species and interactions within a mutualist network, allowing the exploration of different scenarios of species removals and the patterns of secondary extinctions that these might cause (Dunne et al. 2002, Memmott et al. 2004). This gives a sense of realism to the exercise and makes it possible to add information about real species. For example, rather than remove species randomly, species with low abundances (or other attribute) could be removed first (Srinivasan et al. 2007). On the other hand, the apparent realism can be misleading, as there are still numerous assumptions made by models. These include the assumptions that (i) the mutualist network is static, with species unable to form new mutualisms, and (ii) the mutualistic benefits accrued by species do not themselves alter the chances that species resist primary extinctions. These assumptions can have major consequences for model prediction. For example if a species is misidentified as a specialist because it is not observed interacting with other partners, then models could erroneously predict that it would go extinct due to lack of partners, when in fact its partners are just rare (Blüthgen 2010).

A second, but less common, type of model explicitly accounts for species densities, taking a more population-ecological approach (Stouffer and Bascompte 2011). These models include not only who interacts with whom in the network, but also how their interactions affect population dynamics—how population densities change through time. Thus, assumptions are made about how strongly species interact, not just whether they interact or not. A disadvantage of these models is that they do not have a complementary collection of datasets that contain information on population dynamics within mutualist networks; therefore, they are constrained to give largely theoretical explorations of the resilience of mutualist networks. Nonetheless, many of the issues discussed in this chapter, such as mutualism loss and mutualism reorganization, will require more complicated, population-level models to investigate.

Box 13.3 gives a simple example of a population-level model of the breakdown of a mutualist network consisting of animals and plants. Modeling exercises such as this suggest two kinds of questions that should be asked of mutualistic networks if we are to conserve species: (i) Which species, if lost, are most likely to cause a cascade of secondary extinctions? (ii) Which mutualistic partners confer protection against primary extinctions? While many models have addressed the first of these questions, far fewer have addressed the second. It would be useful to be able to identify these two categories of mutualists, since this will aid in management for preventing mutualistic network collapse.

13.6 How can we anticipate change before it is too late?

While it is a useful exercise to identify characteristics that lend resilience to mutualisms under environmental change, it is difficult to unequivocally categorize partnerships, as well as mutualistic networks, as "resilient" or "vulnerable." Some aspects and traits of the mutualism may be vulnerable, while the partners themselves may be resilient, or vice versa. For example, the mutualism may only be found in one particular habitat, but it may flourish there. This is best illustrated by the mutualism between the dioecious herb *Borderea chouardii* (a relict of Tertiary tropics) and its ant mutualists. Worldwide, the mutualism is restricted to only a single population growing on two adjacent vertical cliffs of the Spanish Pyrenees (Garcia et al. 2012). The host plant depends on two ant species for pollination and one ant species for seed dispersal.

This type of dual mutualistic role for ants appears to be very rare in nature and typically only found in habitats of impoverished animal diversity, such as deserts and islands. By most measures, this would be considered an unequivocally vulnerable mutualism. However, unique aspects of the *B. chouardii*–ant mutualism keep it well protected. The cliff location protects it from antagonist species, such as large herbivores, and because the cliff side is a climatically very stable habitat, it displays long-term demographic stability, with some individuals exceeding three hundred years in lifespan.

The *B. chouardii*–ant mutualism illustrates the complexity of trying to conserve mutualisms based on only a few select traits. It is also unrealistic to expect detailed ecological and evolutionary studies on vast numbers of globally important mutualisms. So what criteria should we use to set mutualism conservation priorities? How can we anticipate change before it is too late? Below, we suggest a focus on four approaches.

13.6.1 Understudied habitats and model systems

First, we need a dramatic broadening of mutualism climate research to novel habitats and emerging model systems. For example, compared to terrestrial ecosystems, we know less about mutualisms in aquatic systems, particularly marine symbioses. This is unfortunate because climate-related temperature increases in aquatic systems can exceed those of terrestrial ecosystems (Merila 2012). Hotspots of high marine biodiversity often show greater shifts in response to climate changes (Burrows et al. 2011), resulting in potentially large changes in mutualism dynamics. We need to make better use of the huge heterogeneity in the warming trends of marine ecosystems to investigate heterogeneity in selection pressures, and thus in evolutionary responses of mutualistic partners.

We should also be willing to focus on emerging model systems, such as mutualisms involving microbes, rather than plant–pollinator/plant–seed-disperser mutualisms, which have been the historical focus of phenologies and phenological disruption research (Rafferty et al. 2014). Microbes can offer unparalleled opportunities for studying how mutualistic partners evolve in response to specific selection pressures (Wernegreen 2012), as they can be easily manipulated in both the lab and field, and tracked over multiple generations (e.g., Lau and Lennon 2012). Microbial systems can provide a framework for understanding mutualism evolution under global change (Six 2012). In some cases, microbes can allow us to test specific theoretical models in microcosm settings, such as the existence of tipping points after which mutualistic partnerships collapse. Given that we are discovering novel microbial mutualisms all the time (Morse et al. 2012), this is a potentially important, but understudied approach to global change research.

13.6.2 Trait-based focus

Recently there has been a call in global change research to focus on life-history traits (Berg and Ellers 2010), with the goal of identifying the key attributes that govern tolerances. The difficulty in predicting mutualism response to global change is that partnerships can be composed of species mixes with radically different phylogenies and physiologies. The persistence of a mutualism largely depends upon the degree of overlap between partners' ranges of tolerance.

The life-history traits that help buffer stochastic perturbation are important in determining tolerances and how prone an organism is to extinction. Those traits that help an organism interact with its mutualistic partner will be important in determining the stability of the interaction. Studies are needed to explicitly test how partner traits vary along gradients of global change (Moran et al. 2004). The goal would be to develop a trait-based framework for understanding the effects of disturbance on mutualism interactions, with the aim of identifying the traits most important in increasing the stability of partnerships across high environmental variability.

Van Allen et al. (2012) have used this approach to test the hypothesis that species with life histories in which the stage-specific vital rates are evenly important to population growth rate (in contrast to greater variation in the importance of each life-history stage) are less likely to decline when faced with increasing stochastic environmental events and major landscape changes. Similar hypotheses could be tested for traits of mutualistic partners, allowing scientists to hone in on the characteristics that increase and decrease vulnerability of mutualisms to global change. The ultimate goal is to use traits as a basis to aggregate and analyze mutualistic interactions to improve our predictions.

13.6.3 Focus on interacting drivers

Understanding how specific drivers of global change (fragmentation, eutrophication, increases in CO_2, etc.) affect the persistence of mutualisms is a daunting task; the sheer number of potential interactions makes a comprehensive approach unfeasible. Even more difficult—but arguably more important—is to understand how these global change drivers interact. Synergisms among simultaneous environmental changes are posing unprecedented threats to ecosystems around the world. The problem is that synergism between drivers, such as rapid temperature rise and habitat fragmentation, potentially disrupts mutualisms more rapidly than one driver alone. For example, the effects of systemic pesticides on bee behavior and reproductive output (Henry et al. 2012) may also accelerate their decline caused by alien pathogens (Cameron et al. 2011).

Approaches that study multiple drivers simultaneously are likely to be highly valued in the sustained effort to gather predictive information. Recent work in the coral mutualism has sought to understand why increases in nitrogen concentrations are linked to a significant loss of coral cover and diversity, despite the fact that corals can survive in high-nutrient water with no obvious negative impacts on their physiology (Wiedenmann et al. 2013). The authors found that changes in the stoichiometry of nutrients (e.g., increases in nitrogen under limited phosphate conditions) increased the susceptibility of corals to temperature- and light-induced bleaching; the imbalanced nutrient supply led to phosphate starvation of the coral's symbiotic algae (Wiedenmann et al. 2013). These findings are of high applied value because adjustments can be made to local agricultural and wastewater outlets to achieve less harmful nutrient ratio profiles.

Unfortunately, in other cases, drivers will be harder to modify. This is especially true in ecosystems where deforestation is increasing species vulnerability to climate change (Feeley and Rehm 2012), or where temperature increases are affecting species susceptibility to disease (Sekercioglu et al. 2012). Understanding the interactive effects of multiple global change drivers needs to be a high priority for future research.

13.6.4 Historical data

Lastly, we advocate learning from historical data. Paleontological and phylogenetic data demonstrate that over geologic time scales, mutualisms have undergone dramatic changes in their outcome; some organisms have abandoned mutualism, while others have shifted to novel partners. A little-explored question is how past climatic shifts can be linked to these evolutionary events. The Cenozoic era has seen previous dramatic changes in global temperature and CO_2 of magnitudes comparable to current anthropogenic changes, and these provide ideal references from which to learn how mutualisms respond to global change. With rapid growth in the availability of dated molecular phylogenies, combined with improved paleoclimatic reconstructions (Zachos et al. 2008), historical data can provide an alternative approach to refine our understanding on future mutualism responses.

For example, climatic changes may favor or restrict partner survival, causing global spread or contraction of mutualist lineages. Because many mutualisms are able to survive major geologic events, we can study these historic processes. The dated phylogenetic tree for coral symbiont genus *Symbiodinium* reveals that most modern lineages had appeared already by the late Eocene (~35 Ma). However, major radiation of extant *Symbiodinium* lineages occurred only after the mid-Miocene (~15 Ma) when global temperature and CO_2 levels remained consistently low (Pochon et al. 2006). This period coincides with rapid increase in the fossil record of coral types known to have low CO_2 tolerance (Wooldridge 2012). Notably, none of the extant symbionts, except those harbored by soritid foraminifera, increased their diversity during the period between late

Oligocene warming and mid-Miocene climatic optimum (25–15 Ma), consistent with present observation that long-term increase in water temperature and CO_2 levels negatively affect coral symbioses (Pochon et al. 2006).

Historical data can also be informative about the causes and processes underlying mutualism change. For example, many plants are adapted to different pollinators in different parts of their ranges (Anderson et al. 2010) or have selfing morphs in some populations (Barrett et al. 2009). Identifying the processes of such evolutionary transitions can help us predict how and under what circumstances partners reorganize or abandon mutualisms under anthropogenic change.

Together, these four approaches will strengthen our ability to identify the warning signs that indicate a future mutualism collapse, as well as strengthen our ability to mediate the factors promoting mutualism persistence.

Acknowledgments

Research was supported by the Japanese Society for the Promotion of Science Visiting Fellowship, Netherlands Organisation for Scientific Research grants 836.10.001 and 864.10.005, and an ERC Starting grant 335542 to ETK.

References

Abe, T., Wada, K., Kato, Y., Makino, S., and Okochi, I. (2011). Alien pollinator promotes invasive mutualism in an insular pollination system. *Biological Invasions*, **13**, 957–967.

Aizen, M.A., Morales, C.L., and Morales, J.M. (2008). Invasive mutualists erode native pollination webs. *PLoS Biology*, **6**, 396–403.

Anderson, B., Alexandersson, R., and Johnson, S.D. (2010). Evolution and coexistence of pollination ecotypes in an African Gladiolus (Iridaceae). *Evolution*, **64**, 960–972.

Aschehoug, E.T., Metlen, K.L., Callaway, R.M., and Newcombe, G. (2012). Fungal endophytes directly increase the competitive effects of an invasive forb. *Ecology*, **93**, 3–8.

Aslan, C.E. (2011). Implications of newly-formed seed-dispersal mutualisms between birds and introduced plants in northern California, USA. *Biological Invasions*, **13**, 2829–2845.

Aslan, C.E., Zavaleta, E.S., Croll, D., and Tershy, B. (2012). Effects of native and non-native vertebrate mutualists on plants. *Conservation Biology*, **26**, 778–789.

Barrett, S.C.H., Ness, R.W., and Vallejo-Marin, M. (2009). Evolutionary pathways to self-fertilization in a tristylous plant species. *New Phytologist*, **183**, 546–556.

Bartomeus, I., Ascher, J.S., Wagner, D., et al. (2011). Climate-associated phenological advances in bee pollinators and bee-pollinated plants. *Proceedings of the National Academy of Sciences of the United States of America*, **108**, 20645–20649.

Bascompte, J., and Jordano, P. (2007). Plant-animal mutualistic networks: the architecture of biodiversity. *Annual Review of Ecology Evolution and Systematics*, **38**, 567–593.

Bellard, C., Bertelsmeier, C., Leadley, P., Thuiller, W., and Courchamp, F. (2012). Impacts of climate change on the future of biodiversity. *Ecology Letters*, **15**, 365–377.

Berg, M.P., and Ellers, J. (2010). Trait plasticity in species interactions: a driving force of community dynamics. *Evolutionary Ecology*, **24**, 617–629.

Birnbaum, C., Barrett, L.G., Thrall, P.H., and Leishman, M.R. (2012). Mutualisms are not constraining cross-continental invasion success of *Acacia* species within Australia. *Diversity and Distributions*, **18**, 962–976.

Blüthgen, N. (2010) Why network analysis is often disconnected from community ecology: a critique and an ecologist's guide. *Basic and Applied Ecology*, **11**, 185–195

Bradshaw, W.E., and Holzapfel, C.M. (2006). Climate change—evolutionary response to rapid climate change. *Science*, **312**, 1477–1478.

Brodie, J.F., Helmy, O.E., Brockelman, W.Y., and Maron, J.L. (2009). Bushmeat poaching reduces the seed dispersal and population growth rate of a mammal-dispersed tree. *Ecological Applications*, **19**, 854–863.

Bruno, J.F., Petes, L.E., Harvell, C.D., and Hettinger, A. (2003). Nutrient enrichment can increase the severity of coral diseases. *Ecology Letters*, **6**, 1056–1061.

Burke, G., Fiehn, O., and Moran, N. (2010). Effects of facultative symbionts and heat stress on the metabolome of pea aphids. *Isme Journal*, **4**, 242–252.

Burkle, L.A., Marlin, J.C., and Knight, T.M. (2013). Plant-pollinator interactions over 120 years: loss of species, co-occurrence, and function. *Science*, **339**, 1611–1615.

Burrows, M.T., Schoeman, D.S., Buckley, L.B., et al. (2011). The pace of shifting climate in marine and terrestrial ecosystems. *Science*, **334**, 652–655.

Cameron, S.A., Lozier, J.D., Strange, J.P., et al. (2011). Patterns of widespread decline in North American bumble bees. *Proceedings of the National Academy of Sciences of the United States of America*, **108**, 662–667.

Christian, C.E. (2001). Consequences of a biological invasion reveal the importance of mutualism for plant communities. *Nature*, **413**, 635–639.

Cipollini, D., Rigsby, C.M., and Barto, E.K. (2012). Microbes as targets and mediators of allelopathy in plants. *Journal of Chemical Ecology*, **38**, 714–727.

Colwell, R.K., Dunn, R.R., and Harris, N.C. (2012). Coextinction and persistence of dependent species in a changing world. *Annual Review of Ecology, Evolution, and Systematics*, **43**, 183–203.

Conord, C., Despres, L., Vallier, A., et al. (2008). Long-term evolutionary stability of bacterial endosymbiosis in curculionoidea: additional evidence of symbiont replacement in the dryophthoridae family. *Molecular Biology and Evolution*, **25**, 859–868.

Cordeiro, N.J., and Howe, H.F. (2003). Forest fragmentation severs mutualism between seed dispersers and an endemic African tree. *Proceedings of the National Academy of Sciences of the United States of America*, **100**, 14052–14056.

Deslippe, J.R., Hartmann, M., Mohn, W.W., and Simard, S.W. (2011). Long-term experimental manipulation of climate alters the ectomycorrhizal community of *Betula nana* in Arctic tundra. *Global Change Biology*, **17**, 1625–1636.

Doi, H., Gordo, O., and Katano, I. (2008). Heterogeneous intra-annual climatic changes drive different phenological responses at two trophic levels. *Climate Research*, **36**, 181–190.

Drigo, B., Pijl, A.S., Duyts, H., et al. (2010). Shifting carbon flow from roots into associated microbial communities in response to elevated atmospheric CO_2. *Proceedings of the National Academy of Sciences of the United States of America*, **107**, 10938–10942.

Dunne, J.A., Williams, R.J., and Martinez, N.D. (2002). Network structure and biodiversity loss in food webs: robustness increases with connectance. *Ecology Letters*, **5**, 558–567.

Edwards, D.P. (2009). The roles of tolerance in the evolution, maintenance and breakdown of mutualism. *Naturwissenschaften*, **96**, 1137–1145.

Egerton-Warburton, L.M., Graham, R.C., Allen, E.B., and Allen, M.F. (2001). Reconstruction of the historical changes in mycorrhizal fungal communities under anthropogenic nitrogen deposition. *Proceedings of the Royal Society B-Biological Sciences*, **268**, 2479–2484.

Ellers, J., Kiers, E.T., Currie, C.R., McDonald, B.R., and Visser, B. (2012). Ecological interactions drive evolutionary loss of traits. *Ecology Letters*, **15**, 1071–1082.

Ellis, E. (2011). Anthropogenic transformation of the terrestrial biosphere. *Philosophical Transactions Royal Society A*, **369**, 1010–1035.

Feeley, K.J., and Rehm, E.M. (2012). Amazon's vulnerability to climate change heightened by deforestation and man-made dispersal barriers. *Global Change Biology*, **18**, 3606–3614.

Fletcher, R.J., Orrock, J.L., and Robertson, B.A. (2012). How the type of anthropogenic change alters the consequences of ecological traps. *Proceedings of the Royal Society B-Biological Sciences*, **279**, 2546–2552.

Fortuna, M.A., and Bascompte, J. (2006). Habitat loss and the structure of plant-animal mutualistic networks. *Ecology Letters*, **9**, 278–283.

Gamper, H., Hartwig, U.A., and Leuchtmann, A. (2005). Mycorrhizas improve nitrogen nutrition of *Trifolium repens* after 8 yr of selection under elevated atmospheric CO_2 partial pressure. *New Phytologist*, **167**, 531–542.

Garcia, M.B., Espadaler, X., and Olesen, J.M. (2012). Extreme reproduction and survival of a true cliffhanger: the endangered plant *Borderea chouardii* (Dioscoreaceae). *PLoS One*, **7**, e44657.

Gates, D.J., and Nason, J.D. (2012). Flowering asynchrony and mating system effects on reproductive assurance and mutualism persistence in fragmented fig-fig wasp populations. *American Journal of Botany*, **99**, 757–768.

Grilli, G., Urcelay, C., and Galetto, L. (2012). Forest fragment size and nutrient availability: complex responses of mycorrhizal fungi in native-exotic hosts. *Plant Ecology*, **213**, 155–165.

Grosholz, E.D. (2005). Recent biological invasion may hasten invasional meltdown by accelerating historical introductions. *Proceedings of the National Academy of Sciences of the United States of America*, **102**, 1088–1091.

Guimarães, P.R., Jr., Jordano, P., and Thompson, J.N. (2011). Evolution and coevolution in mutualistic networks. *Ecology Letters*, **14**, 877–885.

Hansen, D.M., and Muller, C.B. (2009). Invasive ants disrupt gecko pollination and seed dispersal of the endangered plant *Roussea simplex* in Mauritius. *Biotropica*, **41**, 202–208.

Harner, M.J., Mummey, D.L., Stanford, J.A., and Rillig, M.C. (2010). Arbuscular mycorrhizal fungi enhance spotted knapweed growth across a riparian chronosequence. *Biological Invasions*, **12**, 1481–1490.

Hauck, M. (2011). Eutrophication threatens the biochemical diversity in lichens. *Lichenologist*, **43**, 147–154.

Hay, M.E., Parker, J.D., Burkepile, D.E., et al. (2004). Mutualisms and aquatic community structure: The enemy of my enemy is my friend. *Annual Review of Ecology Evolution and Systematics*, **35**, 175–197.

Henry, M., Beguin, M., Requier, F., et al. (2012). A common pesticide decreases foraging success and survival in honey bees. *Science*, **336**, 348–350.

Hoegh-Guldberg, O., Mumby, P.J., Hooten, A.J., et al. (2007). Coral reefs under rapid climate change and ocean acidification. *Science*, **318**, 1737–1742.

Holcomb, M., McCorkle, D.C., and Cohen, A.L. (2010). Long-term effects of nutrient and CO_2 enrichment on the temperate coral *Astrangia poculata* (Ellis and Solander,

1786). *Journal of Experimental Marine Biology and Ecology*, **386**, 27–33.

Howarth, R.W. (2008). Coastal nitrogen pollution: a review of sources and trends globally and regionally. *Harmful Algae*, **8**, 14–20.

Howe, H.F., and Smallwood, J. (1982). Ecology of seed dispersal. *Annual Review of Ecology and Systematics*, **13**, 201–228.

Inderjit, and van der Putten, W.H. (2010). Impacts of soil microbial communities on exotic plant invasions. *Trends in Ecology & Evolution*, **25**, 512–519.

Irwin, R.E., Warren, P.S., Carper, A.L., and Adler, L.S. (2014). Plant-animal interactions in suburban environments: implications for floral evolution. Oecologia, **174**, 803–815.

Ivens, A.B.F., Kronauer, D.J.C., Pen, I., Weissing, F.J., and Boomsma, J.J. (2012). Reproduction and dispersal in an ant-associated root aphid community. *Molecular Ecology*, **21**, 4257–4269.

Jansen, P.A., Hirsch, B.T., Emsens, W.J., Zamora-Gutierrez, V., Wikelski, M., and Kays, R. (2012). Thieving rodents as substitute dispersers of megafaunal seeds. *Proceedings of the National Academy of Sciences of the United States of America*, **109**, 12610–12615.

Johnson, N.C. (1993). Can fertilization of soil select less mutualistic mycorrhizae? *Ecological Applications*, **3**, 749–757.

Jones, M.D., Phillips, L.A., Treu, R., Ward, V., and Berch, S.M. (2012). Functional responses of ectomycorrhizal fungal communities to long-term fertilization of lodgepole pine (*Pinus contorta* Dougl. ex Loud. var. latifolia Engelm.) stands in central British Columbia. *Applied Soil Ecology*, **60**, 29–40.

Kiers, E.T., and Denison, R.F. (2008). Sanctions, cooperation, and the stability of plant-rhizosphere mutualisms. *Annual Review of Ecology Evolution and Systematics*, **39**: 215–236.

Kiers, E.T., Hutton, M.G., and Denison, R.F. (2007). Human selection and the relaxation of legume defences against ineffective rhizobia. *Proceedings of the Royal Society B-Biological Sciences*, **274,** 3119–3126.

Kiers, E.T., Palmer, T.M., Ives, A.R., Bruno, J.F., and Bronstein, J.L. (2010). Mutualisms in a changing world: an evolutionary perspective. *Ecology Letters*, **13,** 1459–1474.

Lamelas, A., Gosalbes, M.J., Manzano-Marin, A., Pereto, J., Moya, A., and Latorre, A. (2011). *Serratia symbiotica* from the aphid Cinara cedri: a missing link from facultative to obligate insect endosymbiont. *PLoS Genetics*, **7**, e1002357–e1002357.

Lau, J.A., and Lennon, J.T. (2012). Rapid responses of soil microorganisms improve plant fitness in novel environments. *Proceedings of the National Academy of Sciences of the United States of America*, **109**, 14058–14062.

Lewis, J.D., Licitra, J., Tuininga, A.R., Sirulnik, A., Turner, G.D., and Johnson, J. (2008). Oak seedling growth and ectomycorrhizal colonization are less in eastern hemlock stands infested with hemlock woolly adelgid than in adjacent oak stands. *Tree Physiology*, **28**, 629–636.

Lopezaraiza-Mikel, M.E., Hayes, R.B., Whalley, M.R., and Memmott, J. (2007). The impact of an alien plant on a native plant-pollinator network: an experimental approach. *Ecology Letters*, **10**, 539–550.

Louca, S., Ibanez, S., Piau, D., and Despres, L. (2012). Specialized nursery pollination mutualisms as evolutionary traps stabilized by antagonistic traits. *Journal of Theoretical Biology*, **296**, 65–83.

Marini, L., Nascimbene, J., and Nimis, P.L. (2011). Large-scale patterns of epiphytic lichen species richness: Photobiont-dependent response to climate and forest structure. *Science of the Total Environment*, **409**, 4381–4386.

McClanahan, T.R., Weil, E., and Maina, J. (2009). Strong relationship between coral bleaching and growth anomalies in massive Porites. *Global Change Biology*, **15**, 1804–1816.

McConkey, K.R., and Brockelman, W.Y. (2011). Nonredundancy in the dispersal network of a generalist tropical forest tree. *Ecology*, **92**, 1492–1502.

Memmott, J., Waser, N.M., and Price, M.V. (2004). Tolerance of pollination networks to species extinctions. *Proceedings of the Royal Society B-Biological Sciences*, **271**, 2605–2611.

Menke, S., Bohning-Gaese, K., and Schleuning, M. (2012). Plant-frugivore networks are less specialized and more robust at forest-farmland edges than in the interior of a tropical forest. *Oikos*, **121**, 1553–1566.

Merila, J. (2012). Evolution in response to climate change: in pursuit of the missing evidence. *Bioessays*, **34**, 811–818.

Merow, C., LaFleur, N., Silander, J.A., Wilson, A.M., and Rubega, M. (2011). Developing dynamic mechanistic species distribution models: predicting bird-mediated spread of invasive plants across northeastern North America. *American Naturalist*, **178**, 30–43.

Mondor, E.B., and Addicott, J.F. (2007). Do exaptations facilitate mutualistic associations between invasive and native species? *Biological Invasions*, **9**, 623–628.

Montealegre, C.M., van Kessel, C., Blumenthal, J.M., Hur, H.G., Hartwig, U.A., and Sadowsky, M.J. (2000). Elevated atmospheric CO_2 alters microbial population structure in a pasture ecosystem. *Global Change Biology*, **6**, 475–482.

Moraes, S.C., and Vasconcelos, H.L. (2009). Long-term persistence of a Neotropical ant-plant population in the absence of obligate plant-ants. *Ecology*, **90**, 2375–2383.

Moran, C., Catterall, C. P., Green, R. J., et al. (2004). Functional variation among frugivorous birds: implications for rain- forest seed dispersal in a fragmented subtropical landscape. *Oecologia*, **141**, 584–595.

Morse, S.F., Dick, C.W., Patterson, B.D., and Dittmar, K. (2012). Some like it hot: evolution and ecology of novel endosymbionts in bat flies of cave-roosting bats (Hippoboscoidea, Nycterophiliinae). *Applied and Environmental Microbiology*, **78**, 8639–8649.

Nunez, M.A., Horton, T.R., and Simberloff, D. (2009). Lack of belowground mutualisms hinders Pinaceae invasions. *Ecology*, **90**, 2352–2359.

Ollerton, J., Winfree, R., and Tarrant, S. (2011). How many flowering plants are pollinated by animals? *Oikos*, **120**, 321–326.

Palmer, T.M., Doak, D.F., Stanton, M.L., et al. (2010). Synergy of multiple partners, including freeloaders, increases host fitness in a multispecies mutualism. *Proceedings of the National Academy of Sciences of the United States of America*, **107**, 17234–17239.

Parker, M.A., Wurtz, A.K., and Paynter, Q. (2007). Nodule symbiosis of invasive *Mimosa pigra* in Australia and in ancestral habitats: a comparative analysis. *Biological Invasions*, **9**, 127–138.

Passmore, H.A., Bruna, E.M., Heredia, S.M., and Vasconcelos, H.L. (2012). Resilient networks of ant-plant mutualists in Amazonian forest fragments. *PLoS One*, **7**, e40803.

Pellissier, L., Litsios, G., Fiedler, K., et al. (2012). Loss of interactions with ants under cold climate in a regional myrmecophilous butterfly fauna. *Journal of Biogeography*, **39**, 1782–1790.

Pochon, X., Montoya-Burgos, J.I., Stadelmann, B., and Pawlowski, J. (2006). Molecular phylogeny, evolutionary rates, and divergence timing of the symbiotic dinoflagellate genus *Symbiodinium*. *Molecular Phylogenetics and Evolution*, **38**, 20–30.

Poulsen, J.R., Clark, C.J., and Bolker, B.M. (2012). Experimental manipulation of seed shadows of an Afrotropical tree determines drivers of recruitment. *Ecology*, **93**, 500–510.

Prado, S.S., Hung, K.Y., Daugherty, M.P., and Almeida, R.P.P. (2010). Indirect effects of temperature on stink bug fitness, via maintenance of gut-associated symbionts. *Applied and Environmental Microbiology*, **76**, 1261–1266.

Pratchett, M.S., Munday, P.L., Wilson, S.K., et al. (2008). Effects of climate-induced coral bleaching on coral-reef fishes—ecological and economic consequences. *Oceanography and Marine Biology: An Annual Review*, **46**, 251–296.

Rafferty, N.E., CaraDonna, P.J., and Bronstein, J.L (2014). Phenological shifts and the fate of mutualisms. *Oikos*, **124**, 14–21.

Rafferty, N.E., and Ives, A.R. (2011). Effects of experimental shifts in flowering phenology on plant-pollinator interactions. *Ecology Letters*, **14**, 69–74.

Reinhart, K.O., and Callaway, R.M. (2006). Soil biota and invasive plants. *New Phytologist*, **170**, 445–457.

Rezende, E.L., Lavabre, J.E., Guimaraes, P.R., Jr., Jordano, P., and Bascompte, J. (2007). Non-random coextinctions in phylogenetically structured mutualistic networks. *Nature*, **448**, 925–926.

Rodriguez, R.J., Henson, J., Van Volkenburgh, E., et al. (2008). Stress tolerance in plants via habitat-adapted symbiosis. *ISME Journal*, **2**, 404–416.

Rodriguez-Cabal, M.A., Stuble, K.L., Guenard, B., Dunn, R.R., and Sanders, N.J. (2012). Disruption of ant-seed dispersal mutualisms by the invasive Asian needle ant (*Pachycondyla chinensis*). *Biological Invasions*, **14**, 557–565.

Rodriguez-Echeverria, S. (2010). Rhizobial hitchhikers from Down Under: invasional meltdown in a plant-bacteria mutualism? *Journal of Biogeography*, **37**, 1611–1622.

Rodriguez-Echeverria, S., Fajardo, S., Ruiz-Diez, B., and Fernandez-Pascual, M. (2012). Differential effectiveness of novel and old legume-rhizobia mutualisms: implications for invasion by exotic legumes. *Oecologia*, **170**, 253–261.

Rodriguez-Perez, J., and Traveset, A. (2012). Demographic consequences for a threatened plant after the loss of its only disperser. Habitat suitability buffers limited seed dispersal. *Oikos*, **121**, 835–847.

Ruan, C.J., Mopper, S., da Silva, J.A.T., Qin, P., Zhang, Q.X., and Shan, Y. (2009). Context-dependent style curvature in *Kosteletzkya virginica* (Malvaceae) offers reproductive assurance under unpredictable pollinator environments. *Plant Systematics and Evolution*, **277**, 207–215.

Russell, J.A., and Moran, N.A. (2006). Costs and benefits of symbiont infection in aphids: variation among symbionts and across temperatures. *Proceedings of the Royal Society B-Biological Sciences*, **273**, 603–610.

Russell, J.A., Moreau, C.S., Goldman-Huertas, B., Fujiwara, M., Lohman, D.J., and Pierce, N.E. (2009). Bacterial gut symbionts are tightly linked with the evolution of herbivory in ants. *Proceedings of the National Academy of Sciences of the United States of America*, **106**, 21236–21241.

Sachs, J.L., and Simms, E.L. (2006). Pathways to mutualism breakdown. *Trends in Ecology & Evolution*, **21**, 585–592.

Schils, T. (2012). Episodic eruptions of volcanic ash trigger a reversible cascade of nuisance species outbreaks in pristine coral habitats. *PLoS One*, **7**, e46639.

Sekercioglu, C.H., Primack, R.B., and Wormworth, J. (2012). The effects of climate change on tropical birds. *Biological Conservation*, **148**, 1–18.

Sikes, B.A., Powell, J.R., and Rillig, M.C. (2010). Deciphering the relative contributions of multiple functions within plant-microbe symbioses. *Ecology*, **91**, 1591–1597.

Silverstein, R.N., Correa, A.M.S., and Baker, A.C. (2012). Specificity is rarely absolute in coral-algal symbiosis: implications for coral response to climate change. *Proceedings of the Royal Society B-Biological Sciences*, **279**, 2609–2618.

Six, D.L. (2012). Ecological and evolutionary determinants of bark beetle-fungus symbioses. *Insects*, **3**, 339–366.

Sole, R.V., and Montoya, J.M. (2001). Complexity and fragility in ecological networks. *Proceedings of the Royal Society B-Biological Sciences*, **268**, 2039–2045.

Srinivasan, U.T., Dunne, J.A., Harte, J., and Martinez, N.D. (2007). Response of complex food webs to realistic extinction sequences. *Ecology*, **88**, 671–682.

Stat, M., Morris, E., and Gates, R.D. (2008). Functional diversity in coral-dinoflagellate symbiosis. *Proceedings of the National Academy of Sciences of the United States of America*, **105**, 9256–9261.

Stevens, C.J., Dise, N.B., Mountford, J.O., and Gowing, D.J. (2004). Impact of nitrogen deposition on the species richness of grasslands. *Science*, **303**, 1876–1879.

Stouffer, D.B., and Bascompte, J. (2011). Compartmentalization increases food-web persistence. *Proceedings of the National Academy of Sciences of the United States of America*, **108**, 3648–3652.

Strona, G., Galli, P., and Fattorini, S. (2013). Fish parasites resolve the paradox of missing coextinctions. *Nature Communications*, **4**, 1718.

Terborgh, J., Nunez-Iturri, G., Pitman, N.C.A., et al. (2008). Tree recruitment in an empty forest. *Ecology*, **89**, 1757–1768.

Tougou, D., Musolin, D.L., and Fujisaki, K. (2009). Some like it hot! Rapid climate change promotes changes in distribution ranges of *Nezara viridula* and *Nezara antennata* in Japan. *Entomologia Experimentalis et Applicata*, **130**, 249–258.

Tylianakis, J.M., Didham, R.K., Bascompte, J., and Wardle, D.A. (2008). Global change and species interactions in terrestrial ecosystems. *Ecology Letters*, **11**, 1351–1363.

Uchitel, A., Omacini, M., and Chaneton, E.J. (2011). Inherited fungal symbionts enhance establishment of an invasive annual grass across successional habitats. *Oecologia*, **165**, 465–475.

Valladares, G., Cagnolo, L., and Salvo, A. (2012). Forest fragmentation leads to food web contraction. *Oikos*, **121**, 299–305.

Van Allen, B.G., Dunham, A.E., Asquith, C.M., and Rudolf, V.H.W. (2012). Life history predicts risk of species decline in a stochastic world. *Proceedings of the Royal Society B-Biological Sciences*, **279**, 2691–2697.

van der Heijden, M.G.A., Bardgett, R.D., and van Straalen, N.M. (2008). The unseen majority: soil microbes as drivers of plant diversity and productivity in terrestrial ecosystems. *Ecology Letters*, **11**, 296–310.

Voolstra, C.R., Sunagawa, S., Matz, M.V., et al. (2011). Rapid evolution of coral proteins responsible for interaction with the environment. *PLoS One*, **6**, e20392.

Wade, T.G., Riitters, K.H., Wickham, J.D., and Jones, K.B. (2003). Distribution and causes of global forest fragmentation. *Conservation Ecology*, **7**, 7.

Wang, B., and Qiu, Y.L. (2006). Phylogenetic distribution and evolution of mycorrhizas in land plants. *Mycorrhiza*, **16**, 299–363.

Webster, N.S., Cobb, R.E., and Negri, A.P. (2008). Temperature thresholds for bacterial symbiosis with a sponge. *Isme Journal*, **2**, 830–842.

Wei, G., Chen, W., Zhu, W., Chen, C., Young, J.P.W., and Bontemps, C. (2009). Invasive *Robinia pseudoacacia* in China is nodulated by *Mesorhizobium* and *Sinorhizobium* species that share similar nodulation genes with native American symbionts. *Fems Microbiology Ecology*, **68**, 320–328.

Wernegreen, J.J. (2012). Mutualism meltdown in insects: bacteria constrain thermal adaptation. *Current Opinion in Microbiology*, **15**, 255–262.

Wiedenmann, J., D'Angelo, C., Smith, E.G., Hunt, A.N., Legiret, F., Postle, A.D., and Achterberg, E.P. (2013). Nutrient enrichment can increase the susceptibility of reef corals to bleaching. *Nature Climate Change*, **3**, 160–164.

Wilson, S.K., Graham, N.A.J., Pratchett, M.S., Jones, G.P., and Polunin, N.V.C. (2006). Multiple disturbances and the global degradation of coral reefs: are reef fishes at risk or resilient? *Global Change Biology*, **12**, 2220–2234.

Wolfe, B.E., and Pringle, A. (2012). Geographically structured host specificity is caused by the range expansions and host shifts of a symbiotic fungus. *ISME Journal*, **6**, 745–755.

Wooldridge, S.A. (2012). A hypothesis linking sub-optimal seawater $pCO_{(2)}$ conditions for cnidarian-*Symbiodinium* symbioses with the exceedence of the interglacial threshold (> 260 ppmv). *Biogeosciences*, **9**, 1709–1723.

Wright, S.J., Hernandez, A., and Condit, R. (2007). The bushmeat harvest alters seedling banks by favoring lianas, large seeds, and seeds dispersed by bats, birds, and wind. *Biotropica*, **39**, 363–371.

Xing, X.K., Koch, A.M., Jones, A.M.P., Ragone, D., Murch, S., and Hart, M.M. (2012). Mutualism breakdown in breadfruit domestication. *Proceedings of the Royal Society B-Biological Sciences*, **279**, 1122–1130.

Zachos, J.C., Dickens, G.R., and Zeebe, R.E. (2008). An early Cenozoic perspective on greenhouse warming and carbon-cycle dynamics. *Nature*, **451**, 279–283.

CHAPTER 14

Conserving and restoring mutualisms

Rachael Winfree, Molly MacLeod, Tina Harrison, and Daniel P. Cariveau

14.1 Introduction

Conservation and restoration goals range from preserving endangered species and biodiversity to maintaining and enhancing ecosystem function and services (Kremen 2005). Mutualistic interactions such as pollination, seed dispersal, and plant–mycorrhizal interactions are increasingly recognized as essential to maintaining both biodiversity and ecosystem function. Indeed, because mutualisms are so essential to maintaining ecosystems and also so ubiquitous, it is hard to imagine how conservation and restoration could be successful without the inclusion of mutualisms. A few summary statistics readily make this point. Almost 90% of the world's angiosperm species rely on pollination mutualisms for their reproduction (Ollerton et al. 2011), and over 90% of vascular plant species participate in mycorrhizal mutualisms (Wang and Qiu 2006). Mutualistic associations between plants and bacteria account for an estimated 60% of annual terrestrial nitrogen inputs (Zahran 1999). Coral reefs, which are built via a mutualism with photosynthetic microalgae, or zooxanthellae, support such a diversity of marine species that they are referred to as "the rainforests of the sea." Thus, mutualists are becoming targets of conservation and restoration in their own right, as well as tools for conserving and restoring other species.

In this chapter we review the emerging field focused on conserving and restoring mutualisms. We first summarize the conservation and restoration challenges that are specific to mutualisms. We next present a framework for understanding approaches to the restoration (and conservation) of mutualisms in terms of contrasting goals for the restoration outcome and contrasting scales at which restorations are manipulated. The contrasting goals may be either increasing biodiversity per se, which generally involves targeting rare or endangered species, or increasing the ecosystem functions that mutualists provide, irrespective of which species provide them. (Consistent with the literature, we use the term "ecosystem function" without any teleological implication.) For either type of goal, restoration actions may occur at smaller scales, by manipulating particular species and their interactions, or at larger scales by manipulating entire communities, networks, or ecosystems. The factorial combinations of goals and scales provide a framework for understanding the trade-offs and consequences of management choices. Lastly, we use insights from network analysis, which is emerging as an important methodology for studying mutualism, to explore the implications of common network features such as nestedness and asymmetry for the conservation and restoration of mutualisms.

Throughout the chapter we make three simplifying assumptions, which are necessary due to the very limited literature that exists thus far on the conservation and restoration of mutualisms. First, we use the word "mutualism" to refer broadly to classes of interactions that can be mutualistic, for example those between plants and their pollinators, or between plants and protective ants. Thus we do not distinguish between true mutualisms and cryptic commensalism or parasitism, which no doubt occur in some cases but are beyond the focus of this chapter. Second, we note that different types of mutualism are represented unevenly

Mutualism. Edited by Judith L. Bronstein.

in the conservation and restoration literature and consequently in this chapter, which may or may not reflect which mutualisms are of the greatest ecological and economic importance. Mutualisms studied in a conservation context thus far may have been chosen for many other reasons, including tractability as a study system, assumptions about their importance, or the backgrounds of conservation researchers. Finally, while we made a substantial effort to locate studies concerned with the conservation and restoration of the mutualistic interaction per se, the scarcity of such studies led us to rely largely on projects focused on conserving one or both partner species. Indeed, this latter approach may be the practical one in a management situation.

We discuss both restoration and conservation studies throughout the chapter. Conservation and restoration have similar goals and objectives including increasing biodiversity and enhancing or maintaining ecosystem functioning and services. However, these disciplines differ in methodology and in the context in which they occur (Young 2000). Conservation has traditionally focused on reducing threats to populations of rare species (Young 2000) or specific parcels of land that are imperiled and have high biodiversity or endemism (Noss et al. 2005). Restoration ecology emphasizes interventions with the goal of the recovery of biodiversity, rare species, or ecosystem functioning in human-impacted ecosystems (Brudvig 2011). Our review of the literature comprised both types of studies, although we chose to focus on restoration studies as the most informative for two reasons. First, restoration projects more often focus on sets of interacting species or even the interactions (mutualisms) per se, whereas conservation tends to focus on particular rare species or on land area of high biodiversity value. Second, restoration projects by definition involve a treatment (the human intervention) and perhaps for this reason more frequently included controls, thus aiding ecological interpretation.

14.2 Conservation/restoration challenges specific to mutualisms

Mutualisms face double jeopardy: they require the conservation of both partners in order to persist. This double jeopardy gives rise to several potential challenges inherent to conserving and restoring mutualisms, as compared with other types of biotic interactions. First, direct threats to one partner species become indirect threats to the other partner. For example, large-bodied seed dispersers have declined due to habitat loss and hunting, which has in turn caused dispersal limitation in large-seeded tropical trees (Markl et al. 2012). Declines in the indirectly affected species can add further pressure to the directly affected species. For example, pollinating insects and their obligate plant partners have declined in parallel in recent decades in northwestern Europe, likely through a combination of drivers internal and external to the mutualism (Biesmeijer et al. 2006). Accounting for such positive feedbacks between declining partners may be critical for mutualism restoration efforts.

A second challenge to conserving mutualisms is that both partners need to maintain sufficient population sizes over time if the interaction is to persist. When the abundance of one partner species falls below a critical threshold, it can fail to attract its mutualist and therefore experience positive density dependence (an Allee effect), resulting in further fitness declines. Allee effects have been shown for stands of pathogen-endangered whitebark pine (*Pinus albicaulis*) that are too small or sick to produce an abundance of seeds, and thus have low visitation rates from their obligate disperser, the generalist Clark's Nutcracker (*Nucifraga columbiana*; Barringer et al. 2012). Similarly, pollinators may avoid small and isolated plant populations, causing pollination to fail, and resulting in population growth rate declines for the plant (Groom 1998). Models suggest that large populations of fig trees are required to allow pollinating fig wasp populations to survive stochastic events, especially if seasonality in fig flowering creates temporal gaps in flower availability (Anstett et al. 1997). The existence of mutualism-mediated Allee effects means that researchers must consider a critical spatial scale or population size with regard to both partners, below which conservation or restoration efforts would not be effective.

Third, even if both partners maintain adequate population sizes, if individuals do not overlap spatially or temporally, they will no longer be able to interact. This is a particular concern in the context of climate change. For example, coral species and their endosymbiotic algae can respond differently to ocean warming (Hughes et al. 2003, Baker et al. 2008, West and Salm 2003), which could lead to mutualism breakdown if partners can no longer coexist under warmed conditions (Baker et al. 2008, Hughes et al. 2003). Likewise, if plants and pollinators use different environmental cues to time their spring emergence, early-blooming plants could be left without pollinators or vice versa as the climate warms. However, the few empirical studies to use large data sets on both plants and pollinators to explore relative rates of change have found evidence for the robustness of the mutualism: plants and pollinators seem to be advancing their seasonal phenologies at similar rates (Bartomeus et al. 2011), appear to use similar environmental cues to time their emergence (Forrest and Thomson 2011), or in cases where partners use different environmental cues, this does not lessen the period of overlap (Iler et al. 2013). Lastly, partners may be separated spatially through differential response to human disturbance or through stochastic patterns of occupancy in remnant habitat fragments. This apparently occurred in New Zealand, where the endangered kakapo parrot (*Strigops habroptilus*) and the threatened root parasite *Dactylanthus taylorii* do not co-occur due to extensive range contractions since the arrival of humans, but interacted in the past as evidenced by the presence of *D. taylorii* pollen in fossilized kakapo parrot feces (Wood et al. 2012). In the Cape Floristic Region of South Africa, an oil-collecting bee (*Rediviva peringueyi*) is absent from smaller, urban conservation areas, but the orchid plants that depend on *R. peringueyi* for pollination persist, although with very low seed set rates possibly indicating extinction debt (Pauw 2007).

Fourth, global change can affect the nature of the interaction between partners, such that it either grades from mutualism to parasitism or results in one partner ending the interaction (Kiers et al. 2010). In this case both partners might persist in the same place and time, but the benefits provided by one of the partners would be greatly reduced or even lost and may result in the dissolution of the mutualism itself. This is already occurring for corals due to rising sea surface temperatures. Temperature-induced coral bleaching causes photosynthetic endosymbiotic algae to be expelled from corals (Baker et al. 2008, Hughes et al. 2003; see also Boxes 9.1 and 13.2 in this volume), and the loss of the mutualism in turn leads to the decline of the corals themselves (Anthony et al. 2008). In another example, N and P fertilization can lead to the reduction of root colonization by mycorrhizal fungi, presumably because plants reduce expensive sugar rewards in response to freely available soil nutrients (Hoeksema et al. 2010). Fertilization can also change the relative abundance of mycorrhizal fungi species and lead to an increase in species that provide less benefit to their plant hosts (Johnson 1993). Lastly, when the extirpation of elephants was simulated with large mammal exclosures, *Acacia drepanolobium* trees reduced their investment in rewards for protecting ants, leading to the loss of the most effectively protective ant species (Palmer et al. 2008; see Chapter 9). In one sense, these examples demonstrate the resiliency of mutualistic species in adapting to a shifted cost–benefit balance. However, the breakdown of mutualistic interactions could eventually result in the local extinction of a partner species, or at a minimum might be difficult to re-establish, which could create conservation challenges in fluctuating environmental conditions.

14.3 Approaches to conserving and restoring mutualisms

Although sometimes not explicitly stated, published approaches to the conservation and restoration of mutualisms must generally make a choice of goals (biodiversity versus ecosystem function) and a choice of scale at which treatments are applied (individual species versus entire system). The traditional goal of conservation efforts has been biodiversity preservation, which in turn requires focusing on rare species, as these are the most vulnerable. However, restoration of function is an increasingly common goal. Whether biodiversity and rare species will be protected when conservation goals focus on ecosystem function remains an open question.

Studies on the conservation and restoration of mutualisms vary in the scale at which they apply conservation or restoration treatments. Conservation and restoration treatments, such as habitat conservation, habitat amelioration, or even direct species additions, may be designed to manipulate either one to a few mutualist species (hereafter, "species"), or entire guilds and networks of mutualists ("systems"). Species-scale manipulations may be efficient when using species known to host many mutualistic partners or to provide high ecosystem function. In other cases, lack of ecological knowledge about mutualists may make this approach less tractable. For example, documenting rare species' life histories, or determining which specific mutualisms provide most function, is data-intensive and expensive. Thus, managers may adopt broad-scale "system" manipulations of mutualist guilds, without attention to species identity, in the hope that specific rare species or ecosystem service providers will benefit. System-scale manipulations are also appropriate when species identity is not important to the conservation goals.

Here we consider the two goals (biodiversity and ecosystem function) and two scales (species and systems) as a two-factorial conceptual framework in which to understand the consequences and trade-offs of different approaches to mutualism conservation. Key published studies representing each of these four categories are presented in Table 14.1. We classify a study's goal according to the units of the outcome variables. Studies with a goal of preserving biodiversity measure species richness, or population abundance of focal species of conservation concern. Studies with a goal of preserving ecosystem function measure in units specific to their focal mutualism, for example "rate

Table 14.1 Key published examples of conservation and restoration studies taking the four types of approaches defined by the goal (maintenance of biodiversity versus ecosystem function) and the scale (species-level or system-level) of the project.

Citation	Mutualism	Goal	Scale	Description
Gibbs et al. 2008	Plant & tortoise	Biodiversity	Species	Juvenile cactus population restored by reintroduction of captive-bred giant tortoises
La Mantia et al. 2012	Microbes & plants	Biodiversity	Species	Rare plants restored using inoculation of specific microbes
Swarts and Dixon 2009	Plants & mycorrhizal fungi	Biodiversity	Species	Rare orchids propagated using highly specific mycorrhizal fungi
Richter and Stutz 2002	Plants & mycorrhizal fungi	Biodiversity	Species	Dominant plant of an endangered ecosystem restored by adding a few mycorrhizal fungi species
Forup et al. 2008	Plants & pollinators	Biodiversity & function	System	Heath land vegetation restored and it had similar function to remnants but different biodiversity
Robinson and Handel 1993	Birds & plants	Biodiversity	System	Tree seedlings restored by addition of bird perches that increased bird use and seed dispersal
Berkelmans and van Oppen 2006	Corals & zooxanthellae	Biodiversity	System	Bleaching-tolerant zooxanthellae increased in abundance by transplanting corals from cool to warm sites
Korb et al. 2003	Plants & mycorrhizal fungi	Biodiversity	System	AMF communities restored through fire and thinning and was associated with more diverse plant communities
Baskett et al. 2011	Thistle & pollinators	Function	Species	Target plant pollination increased by removing exotic plant species that competed for pollinators
Baker et al. 2008	Coral	Function	Species	Particular coral species restored to speed reef recovery
Iqbal et al. 2012	Plant & fungal endophyte	Function	Species	Tall fescue infection by specific fungal endophyte associated with increased C and N sequestration
Aschero and Garcia 2012	Cattle & forests	Function	Species	Forest regeneration increased using cattle as seed dispersers
Gaines et al. 2010	Coral	Function	System	Functionally important coral species conserved using Marine Protected Areas
Rickson and Rickson 1998	Ant & plant	Function	System	Ant protection of cashew crop plants positively associated with higher habitat heterogeneity and lower pesticide use
Wilson et al. 2009	Plant & mycorrhizal fungi	Function	System	Fire increases hyphae density and C and N sequestration

of pollen grain deposition on stigmas" or "rate of limestone deposition in a coral reef." We classify a study's scale by determining whether the manipulated mutualist partner is represented by a single species or aggregates of many species. We hope this framework can inform future studies that still have a choice of approach, and define the decision space that land managers work within when planning conservation and restoration actions. In Sections 14.3.1.1 through 14.3.2.2, we draw primarily on restoration studies to fill out this framework. In Section 14.4, we expand this framework further using recent developments in network theory, before summarizing our conclusions in Section 14.5.

14.3.1 Conserving/restoring mutualist biodiversity

14.3.1.1 Species-scale manipulations

Conserving biodiversity requires species-scale manipulations when rare, target mutualist species are dependent upon a specific partner. For example, plant–mycorrhizal fungi interactions can require species-specific manipulations to meet conservation goals. Orchid seeds, because they are so small, are highly limited by nutritional resources and require immediate association with mycorrhizal fungi in order to germinate successfully. Furthermore, many rare orchid species associate with particular species of mycorrhiza. Conservation managers thus identify and propagate the specific species of mycorrhizal fungus each orchid requires, as part of each restoration effort (Swarts and Dixon 2009). A challenge of this approach is that it requires detailed biological understanding of the pairwise interaction, and furthermore, benefits of such conservation efforts may be limited to the two partner species.

When exclusive mutualistic relationships are central to entire ecosystems, species-scale manipulations represent a crucial tool for restoring biodiversity beyond the partner species. A successful example comes from Española Island in the Galapagos, where the endangered cactus *Opuntia megasperma*, a keystone resource for most vertebrates on the island (Gibbs et al. 2008), has declined due in part to lack of seed dispersal. The cactus's seed disperser, an endangered giant tortoise (*Geochelone nigra hoodensis*), was reintroduced to the island from a captive breeding program in 1971, and the interaction between the cactus and tortoise appears to have been restored, as measured by improved juvenile cactus recruitment in tortoise-occupied areas (Gibbs et al. 2008, Figure 14.1). Thus a species-scale restoration has been successful in restoring not only particular species but also ecosystem function. Species-scale manipulations can also be an effective way to restore populations of more generalist species. Richter and Stutz (2002) inoculated greenhouse-grown *Sporobolus wrightii* (Poaceae) with a mycorrhizal inoculum dominated by only two fungal species, leading to increased restoration success as compared with non-inoculated plants. The restoration of *S. wrightii* is particularly important as it is a dominant grass of the floodplain grasslands in southwestern North America, an ecosystem that has undergone a 95% reduction in its historic range (Richter and Stutz 2002).

In some cases, the conservation of a single target species will require the conservation of multiple partner species. This can occur within a single mutualism, as has been shown for the African tree *Acacia drepanolobium*, for which an individual tree's fitness is maximized when it interacts with a sequence of four different protective ant species over the course of its lifetime (Palmer et al. 2010; see Chapter 9). Alternatively, a single species can rely on multiple categories of specialized mutualisms, as for orchids that require specific mycorrhizal partners and specific pollinators (Swarts and Dixon 2009). Some mutualists require so many partners that their conservation is best approached at a system scale, as discussed below.

14.3.1.2 System-scale manipulations

System-scale manipulations are appropriate when the goal is to increase mutualist biodiversity generally, or to restore mutualist species that have diffuse interactions with an entire community of mutualist partners. A common restoration goal is simply to increase the number of species in severely simplified anthropogenic habitats such as former agricultural fields, clear-cut or burned forest, or over-fished reefs. In these types of restorations, managers seek to restore a mutualist guild, without attention to species identity, with the hope that a minimal initial investment will initiate continuous positive feedback between the restored

Figure 14.1 Images of restorations involving mutualisms. (1) Removal of black locust trees and their nitrogen-fixing symbionts restores natural nutrient cycling (Malcolm et al. 2008). (2) Identifying effective mycorrhizal partner species can be critical for successfully culturing rare plants such as these orchid seedlings (Swartz and Dixon 2009). (3) Corals are cultivated for future transplanting in reef restorations (Shaish et al. 2010). (4) Bird perches installed in isolated restoration patches can encourage seed dispersal from source fragments (Bustamante-Sanchez and Armesto 2012). (5) Importing non-native giant tortoises to islands where endemic tortoise species are extinct can restore important dispersal function to endemic large-seeded plants (Griffiths et al. 2011). (6) An experiment to assess the effectiveness of pollinator restoration plantings (MacLeod et al. *In preparation*). Photo credits: (1) Spotlight News. (2) Nontachaiyapoom et al. (2011). Symbiotic seed germination of *Grammatophyllum speciosum* Blume and *Dendrobium draconis* Rchb. f., native orchids of Thailand. *Scientia Horticulturae*, 130, 303–308. Elsevier. (3) Shaish et al. 2010. Elsevier. (4) Orangutan Information Centre (OIC) <http://orangutancentre.org/>. (5) Hansen et al. (2010). Ecological history and latent conservation potential: large and giant tortoises as a model for taxon substitutions. *Ecography*, 33, 272–284. John Wiley & Sons (permission via RightsLink®). (6) Rachael Winfree. <http://winfreelab.com>. See Plate 9 for color version.

mutualists and their partners. For example, to restore mutualist biodiversity in cleared landscapes, managers use small, scattered fruit tree plantings or perches to attract seed dispersers to isolated restoration sites and thereby encourage natural colonization of bird-dispersed plant species from source patches (Bustamante-Sanchez and Armesto 2012, Figure 14.1). This "nucleation" approach is attractive because it is cost effective relative to widespread restoration planting, and draws from pools of naturally occurring plant species rather than artificial seed mixes (Robinson and Handel 1993). Another widely used approach is to restore pollinators on agricultural lands through pollinator-attractive plantings, which increase pollinator abundance and species richness. This should in turn provide pollination partners for many native plants. Although significant government funding goes into pollinator restorations in the USA, it is an even more widespread practice in the EU, where detailed research has identified the floral partners required for restoration of both common and rare bee species. In Box 14.1, David Kleijn discusses lessons that restoration ecologists have learned from the experience of restoring plants and pollinators in agricultural landscapes across Europe.

Box 14.1 Lessons from pollinator restoration in the EU

David Kleijn

A large-scale experiment in the restoration of biodiversity on agricultural lands is taking place in Europe, where 25% of farmland is enrolled in agri-environment schemes, many of which are explicitly aimed at ecological restoration on agricultural lands. These programs, which are similar to the Conservation Reserve Program administered by the United States Department of Agriculture (USDA), offer government-funded financial incentives to farmers for enhancing biodiversity through restoring habitat and/or practicing low-intensity agriculture (European Commission 2005). Although agri-environment schemes mostly target biodiversity in general, the restoration of pollinators and the flowering plants they use is an explicit goal of some of them. Thus, we can use studies of agri-environment scheme effectiveness to explore the factors that are most important to consider in the restoration of the plant–pollinator mutualism. A recent meta-analysis by Scheper et al. (2013) based on 71 studies suggests, first, that, overall, species richness and abundance of pollinators was enhanced by these schemes, and second, that the most important factor in restoring pollinators is the restoration of flowering plants. In managed grasslands, establishing wildflower strips increased pollinator abundance and species richness more strongly than did organic farming, temporary abandonment of farming (set-aside), or restrictions on agrochemical inputs. Furthermore, effect size increased with the number of flowering plant species that were restored. The relative benefit of restoring flowers, as compared with other restoration measures, is greater in grasslands than in croplands, possibly because croplands are so depauperate in pollinator habitat that any type of restoration measure taken will enhance pollinators (Scheper et al. 2013).

An important lesson emerging from agri-environment schemes is that the composition of the landscape determines the potential for success for a given local-scale restoration. This is because most pollinators do not nest or overwinter on agricultural lands, at least when these are intensively managed (Tscharntke et al. 2005). Rather, pollinators migrate into crop fields from nearby semi-natural habitats on a seasonal or daily basis. A minimum number of semi-natural habitats is thus required for successful restoration because restored sites need to be colonized from these habitats (Kleijn et al. 2011).

An unresolved issue is whether nest site availability limits pollinator populations. In intensively managed agricultural landscapes, patches of bare ground preferred for nesting by most ground-nesting bees, or dead wood and plant stems used by most cavity-nesting bees, are increasingly hard to find. We currently have little idea how important this factor is relative to floral resource availability (but see Steffan-Dewenter and Schiele 2008) and how it can be manipulated to mitigate pollinator loss in agricultural landscapes.

Plant species that successfully establish on agricultural lands are mostly common species. Enhancing floral resources by introducing flowering plants on farmland may therefore only effectively restore the pollinators that use these common plants. This is illustrated by the host plant preferences of ten species of European bumblebees (Box 14.1 Figure 1), which suggests that most could be enhanced by agri-environment schemes because they use plant species that can easily be established on former arable land. This includes threatened species such as *Bombus ruderatus*, which specializes on red clover (*Trifolium pratense*) and which indeed has been found to benefit from the creation of flower strips targeted at bumblebees (Carvell et al. 2007). Two other threatened species, however, *Bombus soroensis* and *B. jonellus*, are unlikely to benefit from agri-environment schemes because their host plants are adapted to nutrient-poor conditions and do poorly on agricultural land (*Campanula* species and *Vaccinium/Calluna* species, respectively; Box 14.1 Figure 1). Most threatened bee species similarly specialize on plants from non-agricultural habitats, and/or are absent from agricultural landscapes, and thus restoration on farmland offers little benefit to them (Kleijn et al. 2006).

Even if agri-environment schemes do not always succeed in enhancing the abundance of species of conservation concern, they can be expected to succeed in restoring ecosystem services such as pollination. The common pollinator species that do generally benefit from agri-environment schemes may build up larger populations or concentrate in larger numbers. If the implementation of these measures is spatially targeted toward fields with insect-pollinated crops this may result in higher pollination rates (Garibaldi et al. 2013) and ultimately benefit yield and income. So far, little research has been done, however, to test whether measures which mitigate pollinator loss, such as agri-environment schemes, actually result in an increase of pollination services.

Box 14.1 *Continued*

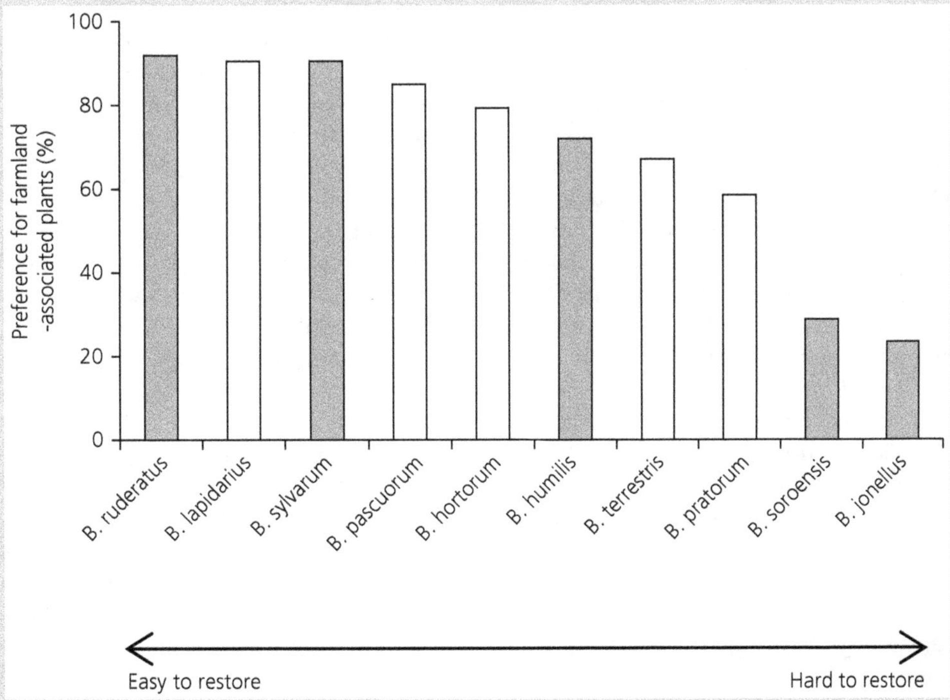

Figure 1 European bumblebee species ordered by their restoration potential, which is measured as their preference for plant species that can easily be re-established on (former) agricultural land. Bumblebee preference for plants associated with farmland was established by examining pollen loads from specimens in entomological collections collected before 1950 (see Kleijn and Raemakers, 2008). Bee species that have become rare since 1950 are shown with hatched bars. This study shows that some rare bee species are good candidates for restoration on agricultural lands whereas others are not, and that particular mutualistic interactions are differentially threatened by agricultural expansion.

References

Carvell, C., Meek, W.R., Pywell, R.F., Goulson, D., and Nowakowski, M. (2007). Comparing the efficacy of agri-environment schemes to enhance bumble bee abundance and diversity on arable field margins. *Journal of Applied Ecology*, **44**, 29–40.

European Commission 2005. Agri-environment measures—Overview on general principles, types of measures, and application. European Commission—Directorate General for Agriculture and Rural Development.

Garibaldi, L.A., Steffan-Dewenter, I., Winfree, R., et al. (2013). Wild pollinators enhance fruit set of crops regardless of honey bee abundance. *Science*, **339**, 1608–1611.

Kleijn, D., Baquero, R.A., Clough, Y., et al. (2006). Mixed biodiversity benefits of agri-environment schemes in five European countries. *Ecology Letters*, **9**, 243–254.

Kleijn, D., and Raemakers, I. (2008). A retrospective analysis of pollen host plant use by stable and declining bumble bee species. *Ecology*, **89**, 1811–1823.

Kleijn, D., Rundlof, M., Scheper, J., Smith, H.G., and Tscharntke, T. (2011). Does conservation on farmland contribute to halting the biodiversity decline? *Trends in Ecology & Evolution*, **26**, 474–481.

Scheper, J., Holzschuh, A., Kuussaari, M., et al. (2013). Environmental factors driving the effectiveness of European agri-environmental measures in mitigating pollinator loss—a meta-analysis. *Ecology Letters*, **16**, 912–920.

Steffan-Dewenter, I., and Schiele, S. (2008). Do resources or natural enemies drive bee population dynamics in fragmented habitats? *Ecology*, **89**, 1375–1387.

Tscharntke, T., Klein, A.M., Kruess, A., Steffan-Dewenter, I., and Thies, C. (2005). Landscape perspectives on agricultural intensification and biodiversity—ecosystem service management. *Ecology Letters*, **8**, 857–874.

In some situations, the systems approach will focus on environmental factors rather than biotic manipulations as a means of conserving many mutualist species. For example, because coral reefs contain so many interacting species and are threatened by ocean-wide global changes, an effective strategy might be to choose the location of protected areas based on system-wide physical properties that maximize bleaching resilience (Berkelmans and van Oppen 2006). Such locations include those already acclimated to thermal stress (Oliver and Palumbi 2011) or protected from it due to shading or currents (McClanahan et al. 2012, West and Salm 2003). Similarly, in forest ecosystems, manipulating (increasing) fire frequency can lead to rapid colonization of mycorrhizal fungi, which is in turn positively associated with plant species richness (Korb et al. 2003). Through such abiotic manipulations, networks of mutualist species can be conserved even when the identities of many species are unknown.

14.3.2 Conserving/restoring the ecosystem functions mutualisms provide

In some situations the ecosystem function provided by the mutualism is the target for conservation and/or restoration, irrespective of which species provide it. There are fewer ecological studies directed toward conserving and restoring mutualism function as compared with biodiversity, most likely due to the additional data requirements it imposes of measuring the function provided by many species in addition to assessing community composition.

14.3.2.1 Species-scale manipulations

Given the cost and effort involved in ecological restoration, there is great interest in whether ecosystem function can be restored by restoring a small number of species that provide much of the total function, because many ecological communities are strongly dominated by a small number of species, not only in terms of abundance but also potentially in terms of function (Balvanera et al. 2005). For example, a global synthesis of studies measuring wild bee visitation rate to crop flowers found that although on average each study recorded 24 bee species, the single most abundant species accounted for 40% of all flower visits (Kleijn et al. *In review*). Thus, restoring the functionally dominant species could be a practical approach to restoring function. In an example from natural habitats, a study of restored versus remnant heathlands found that pollination of native plants was similar between the two habitats, i.e., function was successfully restored. This occurred because 90% of the interactions were provided by only three species of abundant, generalist bees, which readily colonized the restorations (Forup et al. 2008).

In contrast, efforts to conserve functionally dominant species in coral reefs have been less successful. The coral–zooxanthellae mutualism provides the ecosystem function of coral reef building; specifically, photosynthesis by zooxanthellae facilitates coral calcification and resilience (Baker et al. 2008). *Stylophora pistillata* is a fast-growing coral species in the Red Sea that fills a key functional role as a reef framework builder, thus providing habitat for a large number of fish and invertebrate species (Epstein et al. 2005). However, thus far efforts to conserve this species by protecting it within small reef areas have been insufficient due to a lack of a buffer from human activities. Successful conservation of this species might require protection within larger marine protected areas (Epstein et al. 2005) or those arranged as networks (Gaines et al. 2010) in combination with active reef restoration using *S. pistillata* transplants (Epstein et al. 2001).

There is an interesting, but not yet realized, opportunity to conserve coral reef-building function despite ocean warming by facilitating coral's use of particular, warming-tolerant zooxanthellae "species." Among zooxanthellae, members of the lineage *Symbiodinium* clade D have been recognized for their thermal tolerance and thus their role in bleaching resilience, although there is also an expected trade-off in calcification rate for corals that associate with clade D (Ortiz et al. 2013; see also Box 13.2). When exposed to warmer temperatures, some corals can adaptively shift their zooxanthellae communities to favor clade D symbionts (Hughes et al. 2003, Baker et al. 2008), but so far there has been little empirical work on manipulating the corals' "choice" of clade D abundance in the context of restoration efforts (Baker et al. 2008). Furthermore, recent experimental work suggests that thermal tolerance of corals may be driven by local acclimatization and long-term adaptation to thermal stress within the corals themselves, independent of the identity of their zooxanthellae partners (Palumbi et al. 2014).

Another type of species-scale restoration of ecosystem function involves increasing mutualistic function between partners that are both already common. For example, the common tall fescue, *Schedonorus arundinaceous* (Poaceae), regularly associates with the fungal endophyte *Neotyphodium coenphialum*. Fungal endophytes often provide alkaloids to their plant hosts that can influence important ecosystem processes such as litter decomposition rates and in turn affect carbon and nitrogen cycling (Lemons et al. 2005). Iqbal et al. (2012) found that tall fescue pastures infected by the fungal endophyte had higher soil organic carbon and soil nitrogen, due to increased sequestration caused by the mutualism. This study demonstrates how function can be increased by making a mutualistic interaction more common, even in the absence of furthering biodiversity goals.

Finally, if restoring ecosystem function is the goal, and species identity is not a concern, then missing taxa could be substituted by fast-growing, habitat generalist taxa that fill the same functional role. For example, the transplantation of fast-growing, habitat generalist coral species has the potential to speed up reef restoration while also attracting herbivorous fish (Shaish et al. 2010). Most examples of fast-growing, habitat-generalist taxa that replace declining or extirpated mutualists involve invasive, non-native, or otherwise human-associated taxa. For example, livestock can be important seed dispersers in many semi-wild ecosystems, sometimes offsetting their lethal destruction to plants in their seedling and sapling life stages (Rohner and Ward 1999, Stroh et al. 2012, Aschero and Garcia 2012). Even human hikers are important dispersers of many epizoochorous plants (Auffret 2011). Examples from pollination include African honey bees (*Apis mellifera scutellata*), which have become the primary pollinators of tropical rainforest trees that remain standing in Amazonian cattle pastures, and Japanese silvereyes (*Zosterops lateralis*), which are replacing locally extinct, endemic vertebrates as pollinators of New Zealand trees (Pattemore and Wilcove 2012, Dick 2001). Of course, introduced species also destructively displace native mutualists, as discussed in Chapter 13. Exploiting introduced species in restoration efforts should be limited to preexisting and practically irreversible introductions, or in systems that have been greatly altered from their preinvaded state. In these contexts, managing the benefits conferred by introduced species is like squeezing a little lemonade from shiploads of lemons.

14.3.2.2 System-scale manipulations

In cases where the entire network of mutualist species is important to ecosystem function, or where the relative importance of different mutualist species is unknown, conserving or restoring the entire system would seem the most effective way to restore mutualist function. However, there is a surprising paucity of examples in the conservation and restoration literature of this approach being taken. Thus we use this section to review the ecological evidence that suggests such a system-scale approach is necessary in some contexts, and may also be practical.

One reason to target mutualist communities, rather than species, is that the identity of the mutualists most important to function can vary over time, and even over the lifetime of an individual organism. For example, the species pollinating a single native plant may vary widely across the plant's range (Moeller 2005), and the pollinator species most important to pollinating a single crop plant differ between years (Kremen et al. 2002). In addition, individual plants can require a series of different mutualist species over their lifetimes, for both mycorrhizal associations (Swarts and Dixon 2009) and ant protection (Palmer et al. 2010). A related issue, which is particularly important in the context of conservation and restoration, is that different species involved in the same mutualism may respond differently to global change factors such as climate warming and human land use (Cariveau et al. 2013, Rader et al. 2013; see Chapter 13 of this volume). Thus, conserving the mutualism into the future may require conserving multiple species now, such that those species that may be robust to global change—the identities of which are largely unknown now—will still be present in the future.

A second reason to target entire assemblages of mutualists is that there can be complementarity and/or positive interactions among the mutualists providing a function, such that function increases with the number of species providing it. A global review of crop pollination services found that fruit set in a farm field increases with the number of pollinator species found there (Garibaldi et al. 2013), and experimental studies with caged pollinators and plants suggests complementarity among pollinator species as the mechanism behind this effect (Albrecht et al. 2012). In restorations of both former mines and highly disturbed beaches, mycorrhizal inoculation that includes a combination of *Rhizobia* bacteria and arbuscular mycorrhizal fungi (AMF) greatly improves the resulting ecosystem function provided by the restored plants, suggesting that the bacterial and fungal mutualists play complementary roles (Matias et al. 2009, La Mantia et al. 2012). In fact, such restorations, which are inherently experimental, are contributing new knowledge about the basic natural history of mutualisms. Different seed-disperser species can also provide complementary function in terms of the distance over which seeds are dispersed, as exemplified by the contrast between bulbuls (*Pycnonotus xanthopygos*), which disperse most seeds of the desert shrub *Ochradenus baccatus* on a local scale, and grackles (*Onychognathus tristramii*), which provide long-distance dispersal thus linking spatially separated plant populations (Spiegel and Nathan 2007).

Even greater complementarity among mutualist species will likely be found when the scope of research expands to define "ecosystem function" at the community or network scale. At present, most studies are restricted to a narrow definition of function as delivered to a single partner species. This is merely a logistical impediment and is due to the challenges of measuring function across multiple partner species. In reality, many mutualisms exist as networks, with one-to-one exclusivity being rare (Chapter 11). For example, most pollinator species visit many different plant species. If "ecosystem function" were defined to be the pollination of an entire native plant community, as opposed to the pollination of a single plant as it is defined in field studies, we would find that more pollinator species are important to function, because pollinator species differ in their preferences for plants. A similar argument can be made for the coral–zooxanthellae mutualism, for which the primary function is reef building in a structural sense. Conserving the core dominant coral species could provide much of this function, because natural regeneration of coral cover after a bleaching event is typically driven by a few dominant reef-building coral taxa (Baker et al. 2008). However, the maintenance of biodiversity of reef-dwelling organisms, which is arguably one of the key ecosystem functions of coral reefs (though not a direct result of the mutualism per se), might be more dependent on a diversity of coral species. For instance, a study of coral restoration in the Seychelles found that even in areas where coral cover was restored, shifts in dominance of coral taxa affected the presence of reef-dwelling fish and invertebrates that depend on certain coral morphologies (Wilson et al. 2012).

System-scale conservation measures might be more practical to implement than species-scale conservation in many cases, especially when the identities of many mutualists are unknown, because they can be accomplished through habitat manipulation and without a thorough ecological inventory of the species involved. An example of how system-scale habitat manipulation can conserve mutualist function comes from the literature on ecosystem services delivered to agriculture. A global synthesis of crop pollination services provided by wild pollinators shows that simply providing diverse cropping systems as well as high-quality semi-natural habitat surrounding a farm controls the abundance of wild pollinators of many species on crop flowers (Kennedy et al. 2013). Similarly, cashew (*Anacardium occidentale*) cropping systems in southeast Asia have a higher diversity of protective ants in plantations with heterogeneous habitats and low to moderate pesticide application (Rickson and Rickson 1998). Finally, a system-scale manipulation of fire frequency increases the abundance of the hyphae of arbuscular mycorrhizae, which in turn increases soil carbon and nitrogen sequestration (Wilson et al. 2009). Thus, even though the identity of the important function-providers was unknown in many of these cases, ecosystem function can be increased through system-scale habitat manipulation. A final example of restoring ecosystem function through a whole-system manipulation comes from human medicine, in which physicians seek to restore the functioning of the human gut microbiota, often without knowing the ecological details of species composition and pairwise interactions with the host. This procedure is known as "rewilding" of the human gut and has high success rates with medical patients. In Box 14.2, Brendan Bohannan expands upon the interpretation of this medical procedure as an ecological process.

Box 14.2 Restoration of the human microbiome

Brendan Bohannan

Organisms as ecosystems

All plants and animals are intimately associated with a complex array of microorganisms, collectively known as an organism's *microbiome* (Costello et al. 2012). Humans are no exception. We are covered with microorganisms, inside and out. Our microbial companions can play an important role in our health, influencing our development (including maturation of our immune system), providing resistance to invasion by pathogens, contributing significantly to our nutrition, and even potentially altering our behavior. These benefits are not thought to be conferred by any one specific organism, but rather arise as an aggregate property of numerous members of the microbiome. The human microbiome system is thus analogous to an ecosystem, with health emerging as an ecosystem-level function. The human body is thought to encourage the growth of certain members of our microbiome, which in turn provide benefits to their host. Thus, in a general sense, the relationship between humans and their microbiome can be thought of as mutualistic, although at a level of complexity higher than the pairwise interspecific mutualisms classically studied by ecologists.

Collapse and restoration of host–microbe ecosystems

Although the human microbiome is dynamic, certain shifts in composition are associated with disease. These shifts are referred to as *dysbiosis* (Kaur et al. 2011). It is not clear what causes dysbiosis but it has been linked to antibiotic use, infection, and changes in diet, among other factors. Dysbiosis is associated with a number of disorders, including inflammatory bowel disease, obesity, and atopic diseases (such as eczema, asthma, and food allergies), and it is often accompanied by a decrease in microbiome diversity (Kaur et al. 2011). There is growing interest in treating dysbiosis (and its associated diseases) through restoration of the microbiome, but this is extremely challenging because it involves restoring a complex ecosystem.

The restoration of non-host ecosystems has relied on three primary strategies: (1) altering the environment to promote the growth of desired species, (2) removal of undesirable species that may be interfering with the establishment or growth of desired species, and (3) introduction of the desired species into the target habitat. Our understanding of microbiome restoration is still in its infancy; however, versions of these three restoration strategies have been attempted in host–microbe ecosystems. Altering the host environment has been used to restore microbiomes in a number of ways; for example, alterations in gut microbiomes have been attempted through changing host diet (Maslowski and Mackay 2011) or the administration of *prebiotics* (non-digestible dietary supplements that may selectively stimulate particular members of the gut microbiome; Preidis and Versalovic 2009) and attempts to restore vaginal microbiomes through altering the vaginal environment (for example, through application of acidic compounds) have a long history (Ma et al. 2012). Antibiotics have been used in host–microbe systems to remove existing (undesirable) members of an altered microbiota (Preidis and Versalovic 2009), and to allow the successful establishment of desired members, although with mixed results (Manichanh et al. 2010). There is an entire industry devoted to introducing desirable members into the human microbiome, especially that of the gut, mouth, and vagina. Commonly referred to as *probiotics*, these mixtures of presumably beneficial bacteria are ingested or applied topically, and include pharmaceutical preparations as well as "live" foods (such as yogurt; Preidis and Versalovic 2009). There is clinical evidence that probiotics can effectively treat diseases associated with dysbiosis, and it has been assumed that this occurs through restoration of microbiome membership (i.e., the successful invasion of the microbiome by the probiotic strains). However, recent research suggests that this need not happen, and that probiotics can restore microbiome functioning even without successful invasion (McNulty et al. 2011).

Rewilding the human gut

One of the most ambitious (and fascinating) current frontiers in the restoration of host–microbe ecosystems is the "rewilding" (Donlan et al. 2005) of the human gut (van Nood et al. 2013). In an ecological context, rewilding projects have as a goal the complete restoration of ecosystems on very large scales (e.g., prairie ecosystems across North America). In the context of the human gut, rewilding has as a goal the complete restoration of an entire gut ecosystem. Typically this involves removing as much as possible of the existing gut microbiome of an individual by administering antibiotics and colonic irrigation, and then replacing this microbiome with that of a donor. The donor's microbiome is essentially "transplanted" into the recipient by inoculating the recipient with fecal material from the donor. First attempted in the 1950s, this approach has been increasingly used to treat diseases associated with dysbiosis, and has recently been shown to

continued

Box 14.2 *Continued*

be effective in clinical trials (van Nood et al. 2013). Recent research suggests that a "fecal substitute" of as few as 33 different bacterial species from a healthy donor can also be effective, while reducing the risks associated with full fecal transplants (the accidental introduction of pathogens, for example; Petrof et al., 2013).

References

Costello, E.K., Stagaman, K., Dethlefsen, L., Bohannan, B.J., and Relman, D.A. (2012). The application of ecological theory toward an understanding of the human microbiome. *Science*, **336**, 1255–1262.

Donlan, J., Berger, J., Bock, C.E., et al. (2005). Re-wilding North America. *Nature*, **436**, 913–914.

Kaur, N., Chen, C.C., Luther, J., and Kao, J.Y. (2011). Intestinal dysbiosis in inflammatory bowel disease. *Gut Microbes*, **2**, 211–216.

Ma, B., Forney, L.J., and Ravel, J. (2012). Vaginal microbiome: rethinking health and disease. *Annual Review of Microbiology*, **66**, 371–389.

Manichanh, C., Reeder, J., Gibert, P., et al. (2010). Reshaping the gut microbiome with bacterial transplantation and antibiotic intake. *Genome Research*, **20**, 1411–1419.

Maslowski, K.M., and Mackay, C.R. (2011). Diet, gut microbiota and immune responses. *Nature Immunology*, **12**, 5–9.

McNulty, N.P., Yatsunenko, T., Hsiao, A., et al. (2011). The impact of a consortium of fermented milk strains on the gut microbiome of gnotobiotic mice and monozygotic twins. *Science Translational Medicine*, **3**, 106ra106.

Petrof, E.O., Gloor, G.B., Vanner, S.J., et al. (2013). Stool substitute transplant therapy for the eradication of *Clostridium difficile* infection: "RePOOPulating" the gut. *Microbiome*, **1**, 1–12.

Preidis, G.A., and Versalovic, J. (2009). Targeting the human microbiome with antibiotics, probiotics, and prebiotics: gastroenterology enters the metagenomics era. *Gastroenterology*, **136**, 2015–2031.

van Nood, E., Vrieze, A., Nieuwdorp, M., et al. (2013). Duodenal infusion of donor feces for recurrent *Clostridium difficile*. *New England Journal of Medicine*, **368**, 407–415.

A concluding point that is essential to keep in mind while discussing the biodiversity versus function debate is that the goals of these two approaches will converge when rare species are important to function (Lyons et al. 2005). An important example is large-bodied animals, which tend to have low natural abundances, high vulnerability to anthropogenic change, but also provide high per capita function in mutualistic interactions. For example, large-bodied frugivores are disproportionately important for seed dispersal in many systems, but also face disproportionate pressure from human hunting and habitat loss (Galetti et al. 2008). Thus in this case the same species are targeted through both the biodiversity and function approaches.

A general argument, however, leads to the opposite expectation that often the goals of the two approaches could diverge, when a few common species not in need of biodiversity protection provide most of the ecosystem function. This expectation follows from two observations. First, species abundance distributions are universally skewed in natural communities, such that a few species are numerically dominant and a large number of species are rare (McGill et al. 2007). Second, interaction frequency is the main determinant of the total function provided by a mutualist species; function provided per interaction is less important (Vázquez et al. 2005). Essentially this is because the variance in the species abundance distribution exceeds the variance of the per-interaction function distribution (Vázquez et al. 2005). Considered in combination, these two observations predict that distribution of species function will also be highly skewed, with a few functionally dominant species and many rare species that contribute little to function. This appears to be the case for the small number of mutualist-based functions for which species-function distributions have been explored (Balvanera et al. 2005, Winfree et al. *in press*, Kleijn et al. *In review*). In particular, a study of wild bee pollination of watermelon crops found that biodiversity and ecosystem function were decoupled over space and time due to the underlying species abundance distribution. Biodiversity was driven by rare species that contributed little to function, whereas function was driven by a few common species (Winfree et al. *in press*). Thus in this mutualist system the biodiversity and ecosystem function approaches would target different species. If this finding proves a general one, a positive outcome of having common species be functional dominants is that function might be relatively easy to conserve or restore, insofar as common species have lower risk of extinction. However, in cases where a common species has declined, aggressive conservation or restoration actions may be required to prevent or reverse a dramatic loss of function.

Lastly, even if rare species are not essential to an ecosystem function, conserving function could still protect biodiversity if the same conservation or restoration actions benefit both rare and service-providing mutualists. This possibility has been little explored as yet.

14.4 The role of network analysis in conservation and restoration

The use of network approaches to study mutualist interactions is increasing rapidly (see Chapter 11). In this section, we discuss how network approaches might be used in the conservation and restoration of mutualisms (see also Box 11.1). The main conservation predictions to emerge from bipartite network analysis concern the pivotal role of "core" species, which are the small subset of generalist species that interact with the greatest number of partners, including both other generalists and specialists. Because the core species interact with many partners, they are predicted to be an effective tool for the conservation or restoration of other species. The persistence of core species within networks undergoing species loss is also predicted to stabilize networks against secondary extinctions. Core species are often the most abundant species in the network, and recent work has explored the role of abundance per se in determining the properties attributed to core species.

14.4.1 Structural features of networks: nestedness, asymmetry, and modularity

As discussed in Chapter 11, plant–pollinator, plant–mycorrhizal and plant–seed disperser networks are often highly nested (Vázquez et al. 2009, Chagnon et al. 2012, Montesinos-Navarro et al. 2012). One aspect of nestedness is that most species in the network interact with a small core of often numerically abundant generalists. Simulation models of species loss from mutualist networks predict that the extinction rate of the more rare and specialized species in the network should depend largely on whether the core species are retained (Kaiser-Bunbury et al. 2010, Memmott et al. 2004). These and similar studies suggest that conservation and restoration could focus on these core species, as an efficient step toward assuring that all species in the network have available partners.

A second structural feature shown by nested networks is that interactions are asymmetrical, such that specialists tend to interact with partner species that are themselves generalists (Vázquez et al. 2009). Asymmetry reinforces the hypothesis that restoring the core generalists could help to restore their more rare and specialist partners. It is important to remember, however, that in most cases a conservation focus on core generalist species will only indirectly help species that are conservation targets. The core species themselves tend to be abundant and not in need of conservation action.

A third feature shown by many mutualistic networks is modularity: sets of interacting taxa form separate "compartments" that are linked together as a network by one or few taxa (Donatti et al. 2011, Vázquez et al. 2009, Chagnon et al. 2012); see Box 11.1 in this volume. The most effective restoration plan for modular networks, then, might be to focus on the core species within each compartment, and the species that link across compartments (Montoya et al. 2012). Modular networks might also be more resilient than non-modular networks, because the loss of one or a few species might only affect one compartment (Guimarães et al. 2007), as has been argued for some ant–plant systems undergoing forest fragmentation (Passmore et al. 2012).

14.4.2 Unanswered questions about the use of network approaches

Several questions remain to be answered before network approaches can be used effectively in conservation/restoration practice. First, the predictions of network analysis focus on conserving/restoring core species, but it is not known to what extent the network properties of core species simply follow from their abundance. "Core" status could be due to trophic generalism, abundance, or both. Furthermore, generalism and abundance are easily confounded due to sampling effects, insofar as abundant species will be observed interacting with more partners by chance alone (Blüthgen 2010, Winfree et al. 2014). It is also important to recognize that the core generalists themselves are often common, even weedy species, which may not be in need of conservation action (Vázquez et al. 2009).

Second, network analyses may overestimate the reliance of rare species on particular partners due to sampling bias. In many sampling designs, rarity is confounded with specialization because a species for which few individuals are observed will necessarily be observed interacting with few partners (Blüthgen 2010). Similarly, many network studies are based on short-term data, whereas longer-term studies may reveal a change in partner identity and therefore a greater total number of partners over time (Petanidou et al. 2008). What this means in a conservation/restoration context is that species that appear to be specialists in a network analysis might appear that way due to sampling error, and may in reality interact with more partners (Colwell et al. 2012).

14.5 Conclusions

This chapter has by necessity focused on the types of mutualisms that have been studied in the context of conservation and restoration. These better-studied mutualisms include the transportation and nutritional interactions that support plants, such as pollination, seed dispersal, and mycorrhizal associations. This emphasis in the literature is not surprising insofar as these ubiquitous mutualisms provide a crucial underpinning to Earth's primary production. In marine systems, there is a large literature on coral restoration, which is likewise not surprising given the high biodiversity value of coral reefs and the threat of ocean warming and acidification. In contrast, we did not find a significant conservation or restoration literature on interactions between heterotrophs, such as insect–fungus interactions or interactions between microbes and animals. It is possible that these mutualisms, not directly supporting primary productivity, are indeed less critical to ecosystem biodiversity and functioning. We think it more likely that these interactions are simply not well enough understood to even imagine how we might manipulate them. As microbial mutualisms are identified and their functions discovered, we look forward to novel conservation and restoration approaches that focus on these interactions.

The mutualistic interactions we cover have enormous effects beyond the mutualistic partner species or guilds, since they support primary producers that interact, directly or indirectly, with almost every living organism. Despite this, mutualisms have received little explicit consideration in conservation and restoration projects that are aimed at wider communities and ecosystems. The importance of mutualisms in structuring ecological communities has been well demonstrated by cases in which land managers purposefully disrupt mutualisms that facilitate the spread of exotic, pest, or dominant species. For example, suppression of a mycorrhizal mutualism has been used to reduce the dominance of grass species and restore plant diversity (McCain et al. 2011). Likewise, disruption of a mutualism between hemipterans and the invasive ants has enabled native trees to recover from scale damage (Gaigher and Samways 2012), and the removal of nitrogen-fixing black locust (*Robinia pseudoacacia*) has been shown to restore naturally low soil nitrogen levels (Malcolm et al. 2008, Figure 14.1). These examples, as well as those presented elsewhere in this book, demonstrate the ecological and evolutionary importance of mutualistic interactions, as well as their particular fragilities and strengths in the face of environmental change. As our understanding of these interactions continues to grow, explicit integration of mutualisms will become an essential tool to ensure the success of conservation and restoration programs.

Acknowledgments

We would like to thank Natalie Lemanski, Sean Griffin, Bethanne Bruninga-Socolar, Natalie Howe, George Condon, and Joanna Jorge for their useful comments during the development of this book chapter.

References

Albrecht, M., Schmid, B., Hautier, Y., and Müller C.B. (2012). Diverse pollinator communities enhance plant reproductive success. *Proceedings of the Royal Society, Biological Sciences*, **279**, 4845–4852.

Anstett, M.-C., Hossaert-McKey, M., and McKey, D. (1997). Modeling the persistence of small populations of strongly interdependent species: figs and fig wasps. *Conservation Biology*, **11**, 204–213.

Anthony, K.R.N., Kline, D.I., Diaz-Pulido, G., Dove, S., and Hoegh-Guldberg, O. (2008). Ocean acidification causes bleaching and productivity loss in coral reef builders. *Proceedings of the National Academy of Sciences of the United States of America*, **105**, 17442–17446.

Aschero, V., and Garcia, D. (2012). The fencing paradigm in woodland conservation: consequences for recruitment of a semi-arid tree. *Applied Vegetation Science*, **15**, 307–317.

Auffret, A.G. (2011). Can seed dispersal by human activity play a useful role for the conservation of European grasslands? *Applied Vegetation Science*, **14**, 291–303.

Baker, A.C., Glynn, P.W., and Riegl, B. (2008). Climate change and coral reef bleaching: an ecological assessment of long-term impacts, recovery trends and future outlook. *Estuarine Coastal and Shelf Science*, **80**, 435–471.

Balvanera, P., Kremen, C., and Martinez-Ramos, M. (2005). Applying community structure analysis to ecosystem function: examples from pollination and carbon storage. *Ecological Applications*, **15**, 360–375.

Barringer, L.E., Tomback, D.F., Wunder, M.B., and McKinney, S.T. (2012). Whitebark pine stand condition, tree abundance, and cone production as predictors of visitation by clark's nutcracker. *PLoS One*, **7**, e37663.

Bartomeus, I., Ascher, J.S., Wagner, D., et al. (2011). Climate-associated phenological advances in bee pollinators and bee-pollinated plants. *Proceedings of the National Academy of Sciences of the United States of America*, **108**, 20645–20649.

Baskett, C.A., Emery, S.M., and Rudgers, J.A. (2011). Pollinator visits to threatened species are restored following invasive plant removal. *International Journal of Plant Sciences*, **172**, 411–422.

Berkelmans, R., and van Oppen, M.J.H. (2006). The role of zooxanthellae in the thermal tolerance of corals: a "nugget of hope" for coral reefs in an era of climate change. *Proceedings of the Royal Society B Biological Sciences*, **273**, 2305–2312.

Biesmeijer, J.C., Roberts, S.P.M., Reemer, M., et al. (2006). Parallel declines in pollinators and insect-pollinated plants in Britain and the Netherlands. *Science*, **313**, 351–354.

Blüthgen, N. (2010). Why network analysis is often disconnected from community ecology: a critique and an ecologist's guide. *Basic and Applied Ecology*, **11**, 185–195.

Brudvig, L.A. (2011). The restoration of biodiversity: where has research been and where does it need to go? *American Journal of Botany*, **98**, 549–558.

Bustamante-Sanchez, M.A., and Armesto, J.J. (2012). Seed limitation during early forest succession in a rural landscape on Chiloe Island, Chile: implications for temperate forest restoration. *Journal of Applied Ecology*, **49**, 1103–1112.

Cariveau, D.P., Williams, N.M., Benjamin, F.E., and Winfree, R. (2013). Response diversity to land use occurs but does not consistently stabilise ecosystem services provided by native pollinators. *Ecology Letters*, **16**, 903–911.

Chagnon, P.L., Bradley, R.L., and Klironomos, J.N. (2012). Using ecological network theory to evaluate the causes and consequences of arbuscular mycorrhizal community structure. *New Phytologist*, **194**, 307–312.

Colwell, R.K., Dunn, R.R., and Harris, N.C. (2012). Coextinction and persistence of dependent species in a changing world. *Annual Review of Ecology, Evolution, and Systematics*, **43**, 183–203.

Dick, C. (2001). Genetic rescue of remnant tropical trees by an alien pollinator. *Proceedings of the Royal Society of London Series B Biological Sciences*, **268**, 2391–2396.

Donatti, C.I., Guimaraes, P.R., Galetti, M., Pizo, M.A., Marquitti, F.M.D., and Dirzo, R. (2011). Analysis of a hyperdiverse seed dispersal network: modularity and underlying mechanisms. *Ecology Letters*, **14**, 773–781.

Epstein, N., Bak, R.P.M., and Rinkevich, B. (2001). Strategies for gardening denuded coral reef areas: the applicability of using different types of coral material for reef restoration. *Restoration Ecology*, **9**, 432–442.

Epstein, N., Vermeij, M.J.A., Bak, R.P.M., and Rinkevich, B. (2005). Alleviating impacts of anthropogenic activities by traditional conservation measures: can a small reef reserve be sustainedly managed? *Biological Conservation*, **121**, 243–255.

Forrest, J.R.K., and Thomson, J.D. (2011). An examination of synchrony between insect emergence and flowering in Rocky Mountain meadows. *Ecological Monographs*, **81**, 469–491.

Forup, M.L., Henson, K.S.E., Craze, P.G., and Memmott, J. (2008). The restoration of ecological interactions: plant-pollinator networks on ancient and restored heathlands. *Journal of Applied Ecology*, **45**, 742–752.

Gaigher, R., and Samways, M.J. (2012). Strategic management of an invasive ant-scale mutualism enables recovery of a threatened tropical tree species. *Biotropica*, **45**, 128–134.

Gaines, S.D., White, C., Carr, M.H., and Palumbi, S.R. (2010). Designing marine reserve networks for both conservation and fisheries management. *Proceedings of the National Academy of Sciences of the United States of America*, **107**, 18286–18293.

Galetti, M., Donatti, C.I., Pizo, M.A., and Giacomini, H.C. (2008). Big fish are the best: seed dispersal of *Bactris glaucescens* by the pacu fish (*Piaractus mesopotamicus*) in the Pantanal, Brazil. *Biotropica*, **40**, 386–389.

Garibaldi, L.A., Steffan-Dewenter, I., Winfree, R., et al. (2013). Wild pollinators enhance fruit set of crops regardless of honey bee abundance. *Science*, **339**, 1608–1611.

Gibbs, J.P., Marquez, C., and Sterling, E.J. (2008). The role of endangered species reintroduction in ecosystem restoration: tortoise-cactus interactions on Española Island, Galapagos. *Restoration Ecology*, **16**, 88–93.

Griffiths, C. J., Hansen, D.M., Jones, C.G., Zuel, N., and Harris, S. (2011). Resurrecting extinct interactions with extant substitutes. *Current Biology*, **21**, 762–765.

Groom, M.J. (1998). Allee effects limit population viability of an annual plant. *American Naturalist*, **151**, 487–496.

Guimarães, P.R., Rico-Gray, V., Oliveira, P.S., Izzo, T.J., dos Reis, S.F., and Thompson, J.N. (2007). Interaction intimacy affects structure and coevolutionary dynamics in mutualistic networks. *Current Biology*, **17**, 1797–1803.

Hoeksema, J.D., Chaudhary, V.B., Gehring, C.A., et al. (2010). A meta-analysis of context-dependency in plant response to inoculation with mycorrhizal fungi. *Ecology Letters*, **13**, 394–407.

Hughes, T.P., Baird, A.H., Bellwood, D.R., et al. (2003). Climate change, human impacts, and the resilience of coral reefs. *Science*, **301**, 929–933.

Iler, A.M., Inouye, D.W., Hoye, T.T., Miller-Rushing, A.J., Burkle, L.A., and Johnston, E.B. (2013). Maintenance of temporal synchrony between syrphid flies and floral resources despite differential phenological responses to climate. *Global Change Biology*, **19**, 2348–2359.

Iqbal, J., Siegrist, J.A., Nelson, J.A., and McCulley, R.L. (2012). Fungal endophyte infection increases carbon sequestration potential of southeastern USA tall fescue stands. *Soil Biology & Biochemistry*, **44**, 81–92.

Johnson, N.C. (1993). Can fertilization of soil select less mutualistic mycorrhizae? *Ecological Applications*, **3**, 749–757.

Kaiser-Bunbury, C.N., Muff, S., Memmott, J., Muller, C.B., and Caflisch, A. (2010). The robustness of pollination networks to the loss of species and interactions: a quantitative approach incorporating pollinator behaviour. *Ecology Letters*, **13**, 442–452.

Kennedy, C.M., Lonsdorf, E., Neel, M.C., et al. (2013). A global quantitative synthesis of local and landscape effects on wild bee pollinators in agroecosystems. *Ecology Letters*, **16**, 584–599.

Kiers, T.E., Palmer, T.M., Ives, A.R., Bruno, J.F., and Bronstein, J.L. (2010). Mutualisms in a changing world: an evolutionary perspective. *Ecology Letters*, **13**, 1459–1474.

Kleijn, D., Scheper, J., Winfree R, et al. (In press). Delivery of pollination services is an insufficient argument for pollinator conservation. *Nature Communications*.

Korb, J.E., Johnson, N.C., and Covington, W.W. (2003). Arbuscular mycorrhizal propagule densities respond rapidly to ponderosa pine restoration treatments. *Journal of Applied Ecology*, **40**, 101–110.

Kremen, C. (2005). Managing ecosystem services: what do we need to know about their ecology? *Ecology Letters*, **8**, 468–479.

Kremen, C., Williams, N.M., and Thorp, R.W. (2002). Crop pollination from native bees at risk from agricultural intensification. *Proceedings of the National Academy of Sciences of the United States of America*, **99**, 16812–16816.

La Mantia, T., Messana, G., Billeci, V., et al. (2012). Combining bioengineering and plant conservation on a Mediterranean islet. *Iforest - Biogeosciences and Forestry*, **5**, 296–305.

Lemons, A., Clay, K., and Rudgers, J.A. (2005). Connecting plant-microbial interactions above and belowground: a fungal endophyte affects decomposition. *Oecologia*, **145**, 595–604.

Lyons, K.G., Brigham, C.A., Traut, B.H., and Schwartz, M.W. (2005). Rare species and ecosystem functioning. *Conservation Biology*, **19**, 1019–1024.

Malcolm, G.M., Bush, D.S., and Rice, S.K. (2008). Soil nitrogen conditions approach preinvasion levels following restoration of nitrogen-fixing black locust (*Robinia pseudoacacia*) stands in a pine-oak ecosystem. *Restoration Ecology*, **16**, 70–78.

Markl, J.S., Schleuning, M., Forget, P.M., et al. (2012). Meta-analysis of the effects of human disturbance on seed dispersal by animals. *Conservation Biology*, **26**, 1072–1081.

Matias, S.R., Pagano, M.C., Muzzi, F.C., et al. (2009). Effect of rhizobia, mycorrhizal fungi and phosphate-solubilizing microorganisms in the rhizosphere of native plants used to recover an iron ore area in Brazil. *European Journal of Soil Biology*, **45**, 259–266.

McCain, K.N.S., Wilson, G.W.T., and Blair, J.M. (2011). Mycorrhizal suppression alters plant productivity and forb establishment in a grass-dominated prairie restoration. *Plant Ecology*, **212**, 1675–1685.

McClanahan, T.R., Donner, S.D., Maynard, J.A., et al. (2012). Prioritizing key resilience indicators to support coral reef management in a changing climate. *PLoS One*, **7**, e42884.

McGill, B.J., Etienne, R.S., Gray, J.S., et al. (2007). Species abundance distributions: moving beyond single prediction theories to integration within an ecological framework. *Ecology Letters*, **10**, 995–1015.

Memmott, J., Waser, N.M., and Price, M.V. (2004). Tolerance of pollination networks to species extinctions.

Proceedings of the Royal Society of London B, Biological Sciences, **271**, 2605–2611.

Moeller, D.A. (2005). Pollinator community structure and sources of spatial variation in plant-pollinator interactions in *Clarkia xantiana* ssp. *xantiana*. *Oecologia*, **142**, 28–37.

Montesinos-Navarro, A., Segarra-Moragues, J.G., Valiente-Banuet, A., and Verdu, M. (2012). The network structure of plant-arbuscular mycorrhizal fungi. *New Phytologist*, **194**, 536–547.

Montoya, D., Rogers, L., and Memmott, J. (2012). Emerging perspectives in the restoration of biodiversity-based ecosystem services. *Trends in Ecology & Evolution*, **27**, 666–672.

Noss, R.F., Beier, P., Covington, W.W., et al. (2005). Ecology and conservation biology in ponderosa pine forests of the Southwestern United States. *Restoration Ecology*, **14**, 4–10S.

Oliver, T.A., and Palumbi, S.R. (2011). Do fluctuating temperature environments elevate coral thermal tolerance? *Coral Reefs*, **30**, 429–440.

Ollerton, J., Winfree, R., and Tarrant, S. (2011). How many flowering plants are pollinated by animals? *Oikos*, **120**, 321–326.

Ortiz, J.C., Gonzalez-Rivero, M., and Mumby, P.J. (2013). Can a thermally tolerant symbiont improve the future of Caribbean coral reefs? *Global Change Biology*, **19**, 273–281.

Palmer, T.M., Doak, D.F., Stanton, M.L., et al. (2010). Synergy of multiple partners, including freeloaders, increases host fitness in a multispecies mutualism. *Proceedings of the National Academy of Sciences of the United States of America*, **107**, 17234–17239.

Palmer, T.M., Stanton, M.L., Young, T.P., Goheen, J.R., Pringle, R.M., and Karban, R. (2008). Breakdown of an ant-plant mutualism follows the loss of large herbivores from an African Savanna. *Science*, **319**, 192–195.

Palumbi, S., Barshis, D., Traylor-Knowles, N., and Bay, R. (2014). Mechanisms of reef coral resistance to future climate change. *Science*, **344**, 895–898.

Passmore, H.A., Bruna, E.M., Heredia, S.M., and Vasconcelos, H.L. (2012). Resilient networks of ant-plant mutualists in Amazonian forest fragments. *Plos One*, **7**, e40803.

Pattemore, D.E., and Wilcove, D.S. (2012). Invasive rates and recent colonist birds partially compensate for the loss of endemic New Zealand pollinators. *Proceedings of the Royal Society B Biological Sciences*, **279**, 1597–1605.

Pauw, A. (2007). Collapse of a pollination web in small conservation areas. *Ecology*, **88**, 1759–1769.

Petanidou, T., Kallimanis, A.S., Tzanopoulos, J., Sgardelis, S.P., and Pantis, J.D. (2008). Long-term observation of a pollination network: fluctuation in species and interactions, relative invariance of network structure and implications for estimates of specialization. *Ecology Letters*, **11**, 564–575.

Rader, R., Reilly, J., Bartomeus, I., and Winfree, R. (2013). Native bees buffer the negative impact of climate warming on honey bee pollination of watermelon crops. *Global Change Biology*, **19**, 3103–3110.

Richter, B.S., and Stutz, J.C. (2002). Mycorrhizal inoculation of big sacaton: implications for grassland restoration of abandoned agricultural fields. *Restoration Ecology*, **10**, 607–616.

Rickson, F.R., and Rickson, M.M. (1998). The cashew nut, *Anacardium occidentale* (Anacardiaceae), and its perennial association with ants: extrafloral nectary location and the potential for ant defense. *American Journal of Botany*, **85**, 835–849.

Robinson, G.R., and Handel, S.N. (1993). Forest restoration on a closed landfill - rapid addition of new species by bird dispersal. *Conservation Biology*, **7**, 271–278.

Rohner, C., and Ward, D. (1999). Large mammalian herbivores and the conservation of arid Acacia stands in the Middle East. *Conservation Biology*, **13**, 1162–1171.

Shaish, L., Levy, G., Katzir, G., and Rinkevich, B. (2010). Employing a highly fragmented, weedy coral species in reef restoration. *Ecological Engineering*, **36**, 1424–1432.

Spiegel, O., and Nathan, R. (2007). Incorporating dispersal distance into the disperser effectiveness framework: frugivorous birds provide complementary dispersal to plants in a patchy environment. *Ecology Letters*, **10**, 718–728.

Stroh, P.A., Mountford, J.O., and Hughes, F.M.R. (2012). The potential for endozoochorous dispersal of temperate fen plant species by free-roaming horses. *Applied Vegetation Science*, **15**, 359–368.

Swarts, N.D., and Dixon, K.W. (2009). Terrestrial orchid conservation in the age of extinction. *Annals of Botany*, **104**, 543–556.

Vázquez, D.P., Bluthgen, N., Cagnolo, L., and Chacoff, N.P. (2009). Uniting pattern and process in plant–animal mutualistic networks: a review. *Annals of Botany*, **103**, 1445–1457.

Vázquez, D.P., Morris, W.F., and Jordano, P. (2005). Interaction frequency as a surrogate for the total effect of animal mutualists on plants. *Ecology Letters*, **8**, 1088–1094.

Wang, B., and Qiu, Y.L. (2006). Phylogenetic distribution and evolution of mycorrhizas in land plants. *Mycorrhiza*, **16**, 299–363.

West, J.M., and Salm, R.V. (2003). Resistance and resilience to coral bleaching: implications for coral reef conservation and management. *Conservation Biology*, **17**, 956–967.

Wilson, S.K., Graham, N.A.J., Fisher, R., et al. (2012). Effect of macroalgal expansion and marine protected areas on coral recovery following a climatic disturbance. *Conservation Biology*, **26**, 995–1004.

Wilson, G.W.T., Rice, C.W., Rillig, M.C., Springer, A., and Hartnett, D.C. (2009). Soil aggregation and carbon sequestration are tightly correlated with the abundance of arbuscular mycorrhizal fungi: results from long-term field experiments. *Ecology Letters*, **12**, 452–461.

Winfree, R., Fox, J., Williams, N.M., Reilly, J., and Cariveau, D.P. (In press). Abundance of common species, not species richness, drives delivery of a real-world ecosystem service. *Ecology Letters*.

Winfree, R., Williams, N.M., Dushoff, J., and Kremen, C. (2014). Species abundance, not diet breadth, drives the persistence of the most linked pollinators as plant-pollinator networks disassemble. *American Naturalist*, **183**, 600–611.

Wood, J.R., Wilmshurst, J.M., Worthy, T.H., Holzapfel, A.S., and Cooper, A. (2012). A lost link between a flightless parrot and a parasitic plant and the potential role of coprolites in conservation paleobiology. *Conservation Biology*, **26**, 1091–1099.

Young, T.P. (2000). Restoration ecology and conservation biology. *Biological Conservation*, **92**, 73–83.

Zahran, H.H. (1999). Rhizobium-legume symbiosis and nitrogen fixation under severe conditions and in an arid climate. *Microbiology and Molecular Biology Reviews*, **63**, 968.

Index

The manufacturer's authorised representative in the EU for product safety is Oxford University Press España S.A. of el Parque Empresarial San Fernando de Henares, Avenida de Castilla, 2 – 28830 Madrid (www.oup.es/en or product.safety@oup.com). OUP España S.A. also acts as importer into Spain of products made by the manufacturer.

www.ingramcontent.com/pod-product-compliance
Lightning Source LLC
Chambersburg PA
CBHW082026210826
50213CB00005B/327
9780199675654